Computational Pharmaceutics: Application of Molecular Modeling in Drug Delivery

Computational Pharmaceutics: Application of Molecular Modeling in Drug Delivery

Editor: Gulliver Altman

www.statesacademicpress.com

www.statesacademicpress.com

Cataloging-in-Publication Data

Computational pharmaceutics : application of molecular modeling in drug delivery / edited by Gulliver Altman.
   p. cm.
Includes bibliographical references and index.
ISBN 978-1-63989-797-1
1. Drug delivery systems. 2. Drugs--Design. 3. Computational biology.
4. Molecules--Models. 5. Pharmaceutical technology. I. Altman, Gulliver.
RS199.5 .C66 2023
615.6--dc23

States Academic Press,
109 South 5th Street,
Brooklyn, NY 11249, USA

ISBN 978-1-63989-797-1 (Hardback)

Contents

Preface

Computational pharmaceutics is a new discipline that integrates big data, artificial intelligence, and multi-scale modeling techniques into pharmaceutics. It facilitates the understanding of drug delivery mechanisms, and the development of various novel systems in drug delivery. The conventional trial-and-error experiments, which are still used for formulation development in drug delivery systems, are time-consuming, costly and unpredictable. There are various molecular modeling techniques that are widely used to develop new drug delivery systems. These include solid dispersions, polymorphism prediction, dendrimer-based delivery systems, surfactant-based micelle, polymeric drug delivery systems, liposome, protein, and non-viral gene delivery systems. This book contains some path-breaking studies in the field of computational pharmaceutics. The various studies that are constantly contributing towards the application of molecular modeling in drug delivery are examined in detail. Scientists and students actively engaged in this field will find this book full of crucial and unexplored concepts.

This book has been the outcome of endless efforts put in by authors and researchers on various issues and topics within the field. The book is a comprehensive collection of significant researches that are addressed in a variety of chapters. It will surely enhance the knowledge of the field among readers across the globe.

It gives us an immense pleasure to thank our researchers and authors for their efforts to submit their piece of writing before the deadlines. Finally in the end, I would like to thank my family and colleagues who have been a great source of inspiration and support.

Editor

In Silico Strategies in Tuberculosis Drug Discovery

Stephani Joy Y. Macalino [1,2]⬦, **Junie B. Billones** [2,*]⬦, **Voltaire G. Organo** [2] and
Maria Constancia O. Carrillo [2]

[1] Chemistry Department, De La Salle University, 2401 Taft Avenue, Manila 0992, Philippines;
stephanimacalino@gmail.com

[2] OVPAA-EIDR Program, "Computer-Aided Discovery of Compounds for the Treatment of Tuberculosis in
the Philippines", Department of Physical Sciences and Mathematics, College of Arts and Sciences, University
of the Philippines Manila, Manila 1000, Philippines; vgorgano@up.edu.ph (V.G.O.);
mtobrerocarrillo@up.edu.ph (M.C.O.C.)

* Correspondence: jbbillones@up.edu.ph

Academic Editor: Rainer Riedl

Abstract: Tuberculosis (TB) remains a serious threat to global public health, responsible for an
estimated 1.5 million mortalities in 2018. While there are available therapeutics for this infection,
slow-acting drugs, poor patient compliance, drug toxicity, and drug resistance require the discovery
of novel TB drugs. Discovering new and more potent antibiotics that target novel TB protein targets
is an attractive strategy towards controlling the global TB epidemic. *In silico* strategies can be applied
at multiple stages of the drug discovery paradigm to expedite the identification of novel anti-TB
therapeutics. In this paper, we discuss the current TB treatment, emergence of drug resistance,
and the effective application of computational tools to the different stages of TB drug discovery when
combined with traditional biochemical methods. We will also highlight the strengths and points
of improvement in *in silico* TB drug discovery research, as well as possible future perspectives in
this field.

Keywords: tuberculosis; druggability; docking; pharmacophore; MD simulation; QSAR; DFT

1. Introduction

In 1882, *Mycobacterium tuberculosis* (*Mtb*) was identified by Robert Koch as the causative agent
of tuberculosis (TB), an infectious disease that continuous to be a relevant threat to global public
health, especially in low- to middle-income countries. The pathogenesis of TB has several risk factors,
including HIV infection, malnutrition, air pollution, type 2 diabetes, alcoholism, and smoking [1–5].

TB is encountered either as latent TB infection (LTBI), which is non-communicable and
asymptomatic [6], or active TB, which is communicable and has symptoms such as fever, weight loss,
productive cough, and hemoptysis [7]. Active infection is also classified depending on the strain:
(1) drug-sensitive, (2) multidrug-resistant TB (MDR-TB), which is resistant to isoniazid and rifampicin,
and (3) extensively drug-resistant TB (XDR-TB), which shows resistance to isoniazid, rifampicin, any
fluoroquinolone, and aminoglycoside. Around 1.7 billion people are projected to suffer from LTBI and
are at risk of progressing into active TB infection [8]. The World Health Organization (WHO) stated
that active TB disease can be found in approximately 10 million people and has caused approximately
1.5 million deaths in 2018. An estimated half million individuals have rifampicin-resistant TB (RR-TB),
of which 78% had MDR-TB. Furthermore, approximately 6.2% are suggested to have XDR-TB from
these MDR cases [8].

2. Current Tuberculosis Management

One of the major challenges in managing TB is the estimated three million 'missing' individuals who have developed active infections but remained undetected or undiagnosed. TB can be deadly if not treated. With the help of conventional regimen, an estimated 58 million infected individuals were saved from 2000 to 2018. Global treatment outcome in 2017 shows a success rate of 85% for new TB cases and 56% for those with drug-resistant TB [8].

2.1. Latent Tuberculosis Infection

Treatment for LTBI are only provided for select groups that have a high risk of transitioning to active TB infection, including HIV-positive patients, people who were exposed to those with active TB, patients undergoing dialysis for end-stage renal disease, taking anti-tumor necrosis factor (TNF) medications, preparing for transplant surgery, or those with silicosis. Depending on whether it is beneficial or not, especially for children below 5 years of age, exposure to patients with active MDR-TB would require personalized treatment regimens and close observation. WHO recommended several different treatment regimens for LTBI, including 3 months of rifapentine and isoniazid, 3–4 months of isoniazid and rifampicin, 3–4 months of rifampicin, and 6–9 months of isoniazid [9,10]. While all these have established efficacy, poor patient compliance continues to be an issue especially with long treatment periods [9–11].

2.2. Active Drug-Sensitive Tuberculosis

In the last several decades, the treatment strategy for active drug-sensitive TB has not changed from the standard regimen of first-line drugs rifampicin, isoniazid, pyrazinamide and ethambutol (Figure 1) for the first 2 months continued by isoniazid and rifampicin for the next 4 months [12,13]. While this treatment procedure is highly efficacious and successful, its long duration primarily leads to poor patient compliance. This has long been an issue in TB management, necessitating monitoring protocols like the directly observed therapy (DOT), wherein a health professional directly supervises each dose intake [14]. Another issue brought about by the prolonged treatment is drug toxicity resulting in numerous adverse effects such as skin rash, gastrointestinal intolerance, neuropathy, arthralgia, increase in liver enzymes, hepatitis, immune thrombocytopaenia, agranulocytosis, haemolysis, renal failure, optic neuritis, and ototoxicity [15,16].

First-line TB drugs

Isoniazid   Rifampicin   Ethambutol   Pyrazinamide

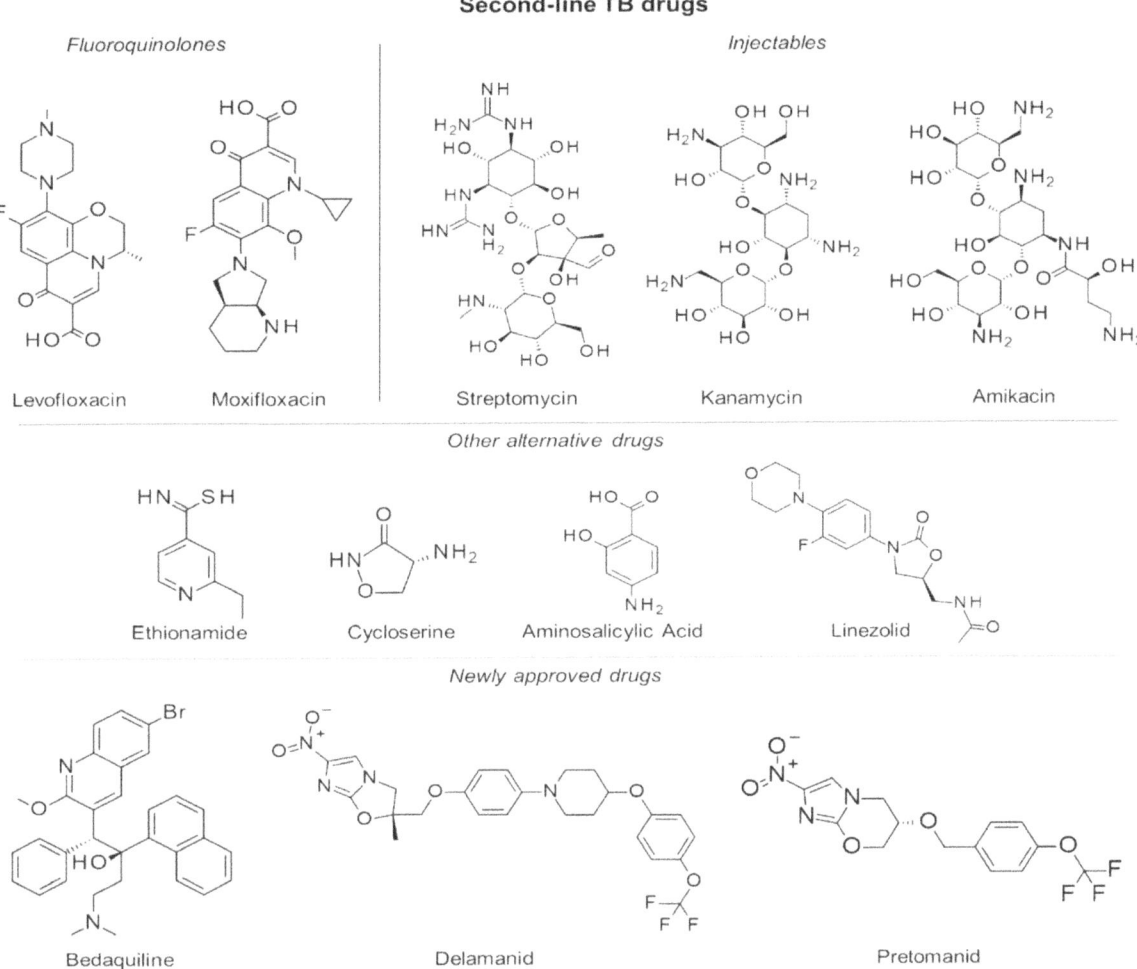

Figure 1. First- and second-line drugs approved for tuberculosis treatment.

2.3. Multiple and Extensively Drug-Resistant Tuberculosis

Failure to complete the full TB regimen leads to disease relapse and drug resistance, which is more challenging to treat. A specific regimen can be designed depending on the resistance profile of the TB strain in a patient [17,18]. These treatments are often of longer duration (18 months or more) and utilize the more expensive second-line drugs (Figure 1) which have uncertain efficacy and high toxicity, resulting in poorer compliance and undesirable outcomes. To mitigate these issues, an updated seven-drug regimen guideline for the treatment of drug-resistant TB lasting 9 to 12 months was released by the WHO last 2016 [19].

With the increasing threat of treatment-resistant TB infection, a number of drugs have been fast-tracked to aid with the efforts in controlling TB worldwide. At the end of 2012, the US Food and Drug Administration (FDA) conferred accelerated approval to the drug bedaquiline for the treatment of resistant TB [20]. Bedaquiline's anti-mycobacterial activity is due to its inhibition of the mycobacterial ATP synthase, a key enzyme in ATP synthesis, resulting in bacterial death. However, its use was shown to have an increased risk of death, thereby causing concerns about its approval. During clinical trials, roughly 11.4% of patients who took bedaquiline died as compared with 2.5% of those who took placebo treatments [21]. In 2014, the use of delamanid, a nitro-dihydro-imidazooxazole derivative, in the treatment of MDR-TB in adults was given conditional approval by the European Medicines Agency (EMA) [22]. Delamanid inhibits mycolic acid biosynthesis to block the formation of mycobacterial cell wall leading to improved drug permeation and more effective treatment [23]. Just recently, pretomanid in combination with bedaquiline and linezolid has also been approved by the FDA for treatment-resistant TB [24]. Pretomanid is a prodrug activated by nitroreductase, which reduces pretomanid's imidazole ring to generate active metabolites. Specifically, a des-nitro metabolite leads to elevated levels of nitric oxide, which displays antimycobacterial activities due to its work as a poison for bacterial respiration under anaerobic conditions [25]. In aerobic conditions, it works like delaminid by targeting cell wall mycolic acid biosynthesis [26], and while there were several potential targets for this drug, its exact protein target is not yet known [27].

An increasing number of XDR-TB cases, such as in India, China, South Africa, Russia, and in eastern Europe, have proved difficult to treat even with the more intensive drug-resistant TB treatment regimen [18]. Novel therapeutics such as bedaquiline, delamanid, and pretomanid might help in curing these patients, though a suitable treatment regimen still has to be carefully designed. However, there is an additional difficulty in acquiring these drugs, especially in developing countries, resulting in a pool of patients that may remain untreated. Essentially, TB can be cured completely with the use of currently available and newly approved anti-tubercular drugs. However, difficulties in diagnosing and reporting infection, long treatment durations leading to drug toxicity and poor patient compliance, emergence of drug resistant strains, and limited acquisition of required treatment urgently necessitates the discovery and development of newer and effective drugs for TB.

3. Rise of Computer-Aided Drug Design in TB Drug Discovery

The drug discovery paradigm covers a wide range of fields, including biochemistry, chemical and structural biology, chem- and bioinformatics, computational chemistry, physical chemistry, organic synthesis, and others. The whole process entails large investments of time, money, and effort in order to produce promising candidates for the pipeline. Over the years, the drug discovery process for new antitubercular therapeutics have changed due to the increase in biological and chemical data, number of identified and validated targets, and advances in high-throughput screening technologies and software development. Moreover, the progress in data storage capacities, supercomputing powers, and parallel processing in the last several years allowed computer-aided drug design (CADD) to become an integral part of TB pharmaceutical research. This continuing expansion in computing power can soon potentially allow the exploration of the vast chemical space, thought to comprise of approximately 10^{60} organic molecules below 500 Da, in order to identify therapeutically interesting scaffolds [28]. Moreover, the boom in protein structural data, including over 150,000 macromolecular structures found in the Protein Data Bank (PDB, www.rcsb.org) [29], proved beneficial in elucidating important molecular and computational concepts for drug design studies. As with any other disease, TB has been the subject of continuous and numerous drug discovery studies, including thousands of published CADD investigations. Despite this, a paper by Ekins et al. noted gaps in the application of these methods in TB research [30], resulting in the slow output of candidates into the TB drug pipeline despite the apparent need and urgency for this disease. This suggests that more rigorous efforts are needed in TB drug discovery to maximize the advantages provided by computational tools.

Computational or *in silico* methods are knowledge-driven, rationally exploring available data to investigate protein function and design new molecular entities (NMEs) that can effectively regulate its behavior. Computational drug discovery approaches are generally divided into structure—(SBDD) and ligand-based drug design (LBDD), depending on the availability of structural data (Figure 2). However, it has been a common practice to integrate these methods in a complementary manner in order to increase the success rate of current drug discovery projects (Figure 2). SBDD requires the target's three-dimensional (3D) structure to be able to examine and use the binding pocket for screening and design of suitable ligands, which can then be experimentally validated and optimized. In the absence of protein structural data, LBDD utilizes knowledge gained from a collection of diverse ligands with known activity to create predictive models for hit discovery and lead optimization [31]. Different types of SB and LB strategies, or a combination thereof, can be applied at different stages of TB drug discovery and development in order to alleviate the challenges involved with experimental methods. With the availability of TB genome and proteome, as well as abundant structural data, data mining and docking strategies can be employed for target identification. Virtual screening (VS) can then be applied to pick out the best potential candidates from a database containing millions of molecules for a chosen TB target. After validation of candidates, structure-activity or -property relationship (SAR/SPR) studies can be implemented to understand mechanism of action and ADMET (absorption, distribution, metabolism, excretion, and toxicity) properties in order to design compounds with better activity and pharmacokinetics. Data (both positive and negative results) taken from these investigations can be kept and used for further iteration and method optimization in the design of novel TB compounds. Both commercial and free software and webservers have been developed covering different SBDD and LBDD techniques, some of which are listed in Table 1.

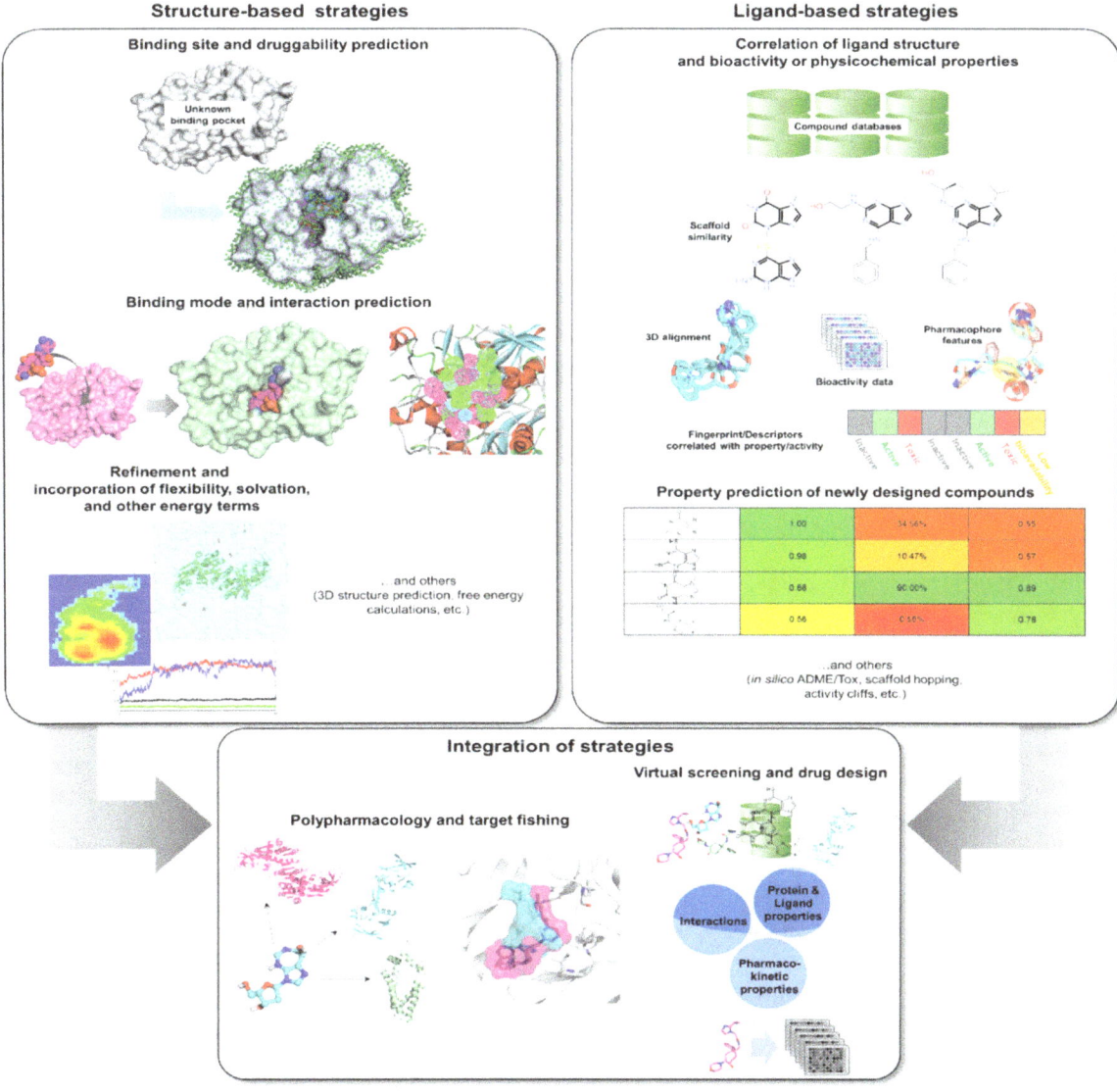

Figure 2. *In silico* tools that can be applied to TB drug design and development.

Table 1. Free and commercially available programs, webservers, and source codes for SBDD and LBDD.

Function	Software/ Webserver Name	Availability	Website
Comparative modeling	SWISS-MODEL [32]	Free webserver	https://swissmodel.expasy.org/
Structural geometry confirmation	MODELLER [33]	Free standalone program for academic license or commercially available through BIOVIA	https://salilab.org/modeller/ https://www.3dsbiovia.com/
	Robetta [34]	Free webserver	http://new.robetta.org/
	Prime [35]	Commercially available through Schrödinger	https://www.schrodinger.com/prime
	I-TASSER [36–41]	Free webserver or standalone program for academic license	https://zhanglab.ccmb.med.umich.edu/I-TASSER/
	HHPred [42–44]	Free webserver	https://toolkit.tuebingen.mpg.de/tools/hhpred
Structural geometry confirmation	PROCHECK [45]	Free webserver and source code	https://www.ebi.ac.uk/thornton-srv/software/ PROCHECK/
Druggability and binding site prediction Druggability and binding site prediction	ProSA [46]	Free webserver	https://prosa.services.came.sbg.ac.at/prosa.php
	VERIFY3D [47]	Free webserver	https://servicesn.mbi.ucla.edu/Verify3D/
	ERRAT [48]	Free webserver	https://servicesn.mbi.ucla.edu/ERRAT/
	PockDrug [49]	Free webserver	http://pockdrug.rpbs.univ-paris-diderot.fr/cgi-bin/index.py?page=home
	DoGSiteScorer [50]	Free webserver	https://proteins.plus/
	fpocket [51,52]	Free/open source platform	https://github.com/Discngine/fpocket
	CASTp [53–55]	Free webserver	http://sts.bioe.uic.edu/castp/calculation.html
	PocketQuery [56]	Free webserver	http://pocketquery.csb.pitt.edu/
	PASS [57]	Free/open source platform	http://www.ccl.net/cca/software/UNIX/pass/ overview.html
	SiteMap [58]	Commercially available through Schrödinger	https://www.schrodinger.com/sitemap
Docking, pharmacophore, and virtual screening Docking, pharmacophore, and virtual screening	ConCavity [59]	Free webserver	https://compbio.cs.princeton.edu/concavity/
	PrankWeb [60]	Free webserver	http://prankweb.cz/
	ProFunc [61]	Free webserver	http://www.ebi.ac.uk/thornton-srv/databases/ProFunc/
	AutoDock [62] and AutoDock Vina [63]	Free standalone program	http://autodock.scripps.edu/
	DOCK [64]	Free/open source platform	http://dock.compbio.ucsf.edu/
	GOLD [65]	Commercially available through CCDC	https://www.ccdc.cam.ac.uk/solutions/csd-discovery/components/gold/
	Glide [66]	Commercially available through Schrödinger	https://www.schrodinger.com/glide/
	Induced Fit [67]	Commercially available through Schrödinger	https://www.schrodinger.com/induced-fit
	FlexX [68]	Commercially available through BioSolveIT	https://www.biosolveit.de/flexx/index.html
	RosettaLigand [69]	Free/open source platform for academic license	https://www.rosettacommons.org/software
	CDOCKER [70]	Commercially available through BIOVIA	https://www.3dsbiovia.com/
	SwissDock [71,72]	Free webserver	http://www.swissdock.ch/docking
	Pharmer [73]	Free/open source platform	http://smoothdock.ccbb.pitt.edu/pharmer/
	CATALYST [74]	Commercially available through BIOVIA	https://www.3dsbiovia.com/products/ collaborative-science/biovia-discovery-studio/ pharmacophore-and-ligand-based-design.html
	PharmGist [75]	Free webserver	http://bioinfo3d.cs.tau.ac.il/pharma/php.php
	LigandScout [76]	Commercially available through Inte:Ligand	http://www.inteligand.com/ligandscout/
	SwissSimilarity [77]	Free webserver	http://www.swisssimilarity.ch/

Table 1. *Cont.*

Function	Software/ Webserver Name	Availability	Website
	LEA3D [78]	Free webserver	https://chemoinfo.ipmc.cnrs.fr/LEA3D/index.html
	PyRx [79]	Free (no support) or commercially available	https://pyrx.sourceforge.io/
	Phase [80]	Commercially available through Schrödinger	https://www.schrodinger.com/phase
Molecular Dynamics	AMBER [81,82]	Commercially available	https://ambermd.org/
	CHARMM [83]	Free or commercially available through CHARMM or BIOVIA	http://charmm.chemistry.harvard.edu/ https://www.3dsbiovia.com/products/collaborative-science/biovia-discovery-studio/simulations.html
	CHARMMing [84]	Free webserver	https://www.charmming.org/charmming
	GROMACS [85,86]	Free/open source platform	http://www.gromacs.org/
	NAMD [87]	Free/open source platform	https://www.ks.uiuc.edu/Research/namd/
	Desmond [88]	Commercially available through Schrödinger	https://www.schrodinger.com/desmond
	SwissParam [89]	Free webserver	http://www.swissparam.ch/
	CHARMM-GUI [90]	Free webserver	http://www.charmm-gui.org/
	ParamChem CGenFF [91–93]	Free webserver	https://cgenff.umaryland.edu/
	VMD [94]	Free/open source platform	https://www.ks.uiuc.edu/Research/vmd/
Molecular Descriptors, Fingerprints, and Quantitative Structure-Activity Relationship	Dragon [95]	Commercially available through Talete	http://www.talete.mi.it/products/dragon_description.htm
	E-Dragon [96]	Free webserver	http://146.107.217.178/lab/edragon/start.html
	Canvas [97]	Commercially available through Schrödinger	https://www.schrodinger.com/canvas
	RDKit [98]	Free/open source platform	https://www.rdkit.org/docs/source/rdkit.ML.Descriptors.MoleculeDescriptors.html
	PyDescriptor [99]	Free/open source platform	https://ochem.eu/home/show.do
	Mordred [100]	Free/open source platform	https://github.com/mordred-descriptor/mordred
	Open3DQSAR [101]	Free/open source platform	http://open3dqsar.sourceforge.net/?Home
	ChemSAR [102]	Free webserver	http://chemsar.scbdd.com/
	SeeSAR [103]	Commercially available through BioSolveIT	https://www.biosolveit.de/SeeSAR/
Pharmacokinetic properties	QikProp [104]	Commercially available through Schrödinger	https://www.schrodinger.com/qikprop
	ADMET Predictor [105]	Commercially available through SimulationsPlus, Inc.	https://www.simulations-plus.com/software/overview/
	ACD Percepta [106]	Commercially available through ACD/Labs	https://www.acdlabs.com/products/percepta/index.php
	FAF-Drugs4 [107]	Free webserver	http://fafdrugs4.mti.univ-paris-diderot.fr/
	PatchSearch [108]	Free webserver	http://mobyle.rpbs.univ-paris-diderot.fr/cgi-bin/portal.py#forms::PatchSearch
	TOPKAT [109] and ADMET [110]	Commercially available through BIOVIA	https://www.3dsbiovia.com/products/collaborative-science/biovia-discovery-studio/qsar-admet-and-predictive-toxicology.html
	PASS Online [111]	Free webserver or commercially available standalone program	http://pharmaexpert.ru/Passonline/index.php
	SwissADME [112]	Free webserver	http://www.swissadme.ch/
	MetaSite [113]	Commercially available through Molecular Discovery	https://www.moldiscovery.com/software/metasite/
	ToxPredict [114]	Free webserver	https://apps.ideaconsult.net/ToxPredict#
	VirtualToxLab [115–118]	Free standalone software	http://www.biograf.ch/index.php?id=home
	admetSAR [119–121]	Free webserver	http://lmmd.ecust.edu.cn/admetsar1/home/
	MetaTox [122,123]	Free webserver	http://way2drug.com/mg2/

More available tools and detailed descriptions for the programs and servers can be found at https://www.click2drug.org/.

3.1. Databases

The era of big data has greatly affected the current drug discovery paradigm through innovations in data storage, management, and mining. Moreover, drastic cost reductions in sequencing technologies allowed the study of multi-omics (e.g., genomics, transcriptomics, proteomics, and metabolomics) for several species including *M. tuberculosis* [124–129]. In order to take advantage of the benefits provided by SB and LB techniques, these biological and/or chemical data must be acquired for analysis via numerous publicly accessible databases on the internet [130–132]. Given that TB is an old disease, vast amounts of data points have already been gathered and are waiting to be used in the fight against this infection.

One of the most extensive and widely-used protein information resource is UniProt (https://www.uniprot.org/) [133], which consists of annotations from several other databases for protein function, omics, and structural data. More specific to TB, the TB Database (http://tbdb.bu.edu/tbdb_sysbio/MultiHome.html) [134,135] contains information on mycobacterium genomes, genes, gene expression correlation, gene epitopes, and experimental and computational models of TB molecular pathways. Alternatively, genomic and proteomic data for various pathogenic mycobacteria can also be found in the Mycobrowser (https://mycobrowser.epfl.ch/) [136], which is linked to UniProt for mycobacterium protein information. On the other hand, patient clinical data is provided by the TB Portals (https://tbportals.niaid.nih.gov/) [137], which is an open-access platform containing socioeconomic, geographic, clinical, laboratory, radiological, and genomic data from patients infected with drug-resistant TB, from the National Institute of Allergy and Infectious Diseases (NIAID) in collaboration with data scientists and clinicians and scientists from countries suffering from heavy TB burden.

Advances in structural and computational biology techniques led to the surge in structural data, resulting in thousands of three-dimensional protein structures generated from X-ray crystallography, nuclear magnetic resonance (NMR), cryo-electron microscopy (EM), homology modeling, and molecular dynamics (MD) simulations. Data from these experiments are customarily deposited to structure databases such as PDB [29], PDBsum [138], etc. Associated with this, the size of the virtual chemical space [28] and improvements in combinatorial chemistry [139] also permitted the availability of chemical libraries (Table 2).

Table 2. Publicly available compound libraries.

Database	Size (Approximate)	Website
GDB-17 [140]	166 billion	http://gdb.unibe.ch/
Enamine REAL [141]	700 million	https://enamine.net/
PubChem [131]	97 million	https://pubchem.ncbi.nlm.nih.gov/
ChemSpider [142]	77 million	http://www.chemspider.com/
ZINC [143]	230 million	http://zinc.docking.org/
ChEMBL [144]	1.9 million	https://www.ebi.ac.uk/chembl/
NCI [145]	460,000	https://cactus.nci.nih.gov/download/roadmap/

Protein subcellular localization databases are also available for study, such as eSLDB (eukaryotic Subcellular Localization database) for general eukaryotes [146], LOCATE for human and other mammals [147], and PSORTdb for bacteria and archaea [148]. Lead optimization and drug repurposing researches can also benefit from protein binding databases like ReLiBase, which consists of interaction information for receptor-ligand complexes from PDB [149], BindingDB, which describes interactions and affinity information between protein target and drug-like molecules [150,151], and Database of Interacting Proteins (DIP) [152], Biological General Repository for Interaction Datasets (BioGRID) [153] and Search Tool for the Retrieval of Interacting Genes/Proteins (STRING) [154], which contains data on protein-protein interactions.

3.2. Structure-Based Tools

3.2.1. Comparative Modeling, Binding Site Prediction, and Druggability

Employment of SBDD tools (Figure 2 and Table 1) require not only the availability of 3D structural data but also information on its druggability and potential binding sites. In the absence of structural data obtained from experiments, such as X-ray crystallography and NMR, a computational model can also be generated either through homology modeling or protein threading techniques (Table 1), which are well-established methods in protein comparative modeling. Homology modeling entails the use of a structural template with suitably similar sequence as the target protein [155]. The most critical stage of any homology modeling procedure is the initial sequence alignment. While there are numerous bioinformatics tools available for this, such as NCBI Blast [156–158], COBALT [159], Clustal Omega [160], KAlign [161], etc., manual inspection and modification of the alignment is crucial, especially if a researcher's knowledge about specific protein folds and domains need to be further incorporated. Then, the secondary structures (i.e., alpha helices, beta strands, loops, etc.) are copied from the template based on the final sequence alignment in order to approximate the target structure. The final model is then refined through minimization or MD and its stereochemical quality is checked using tools like those listed in Table 1 until the structure has improved and is suitable for further computational studies.

Druggability is the capacity of a protein target to be modulated by a ligand. It is important to characterize this property as it helps avoid intractable proteins and allows the identification and prioritization of significant targets. Some predictive approaches that include this property are listed in Table 1. A druggability database, the Druggable Cavity Directory (DCD), is also publicly available to allow researchers to submit protein pocket and druggability information, which is later verified and made available to other researchers [162]. After ascertaining that a given target is indeed tractable, binding pocket information should be acquired either from protein structures complexed with natural substrates or known inhibitors, or from mutational data distinguishing key interaction residues. Ideally, a binding pocket is a concave area in the receptor that is characterized by chemical features with which a ligand can desirably interact to attain the required receptor behavior (e.g., inhibition or activation) [163]. However, if binding information is unknown, several *in silico* methods and webservers (Table 1) are available to identify potential receptor binding sites from a given structure and have been described in detail elsewhere [31]. Otherwise, a number of studies have also used 'blind' docking [164], wherein the whole protein is set as the binding site, allowing ligands to freely bind anywhere in the structure in the hopes of finding a suitable pocket.

It is also prudent to remember that other potential binding sites may be present on the target surface, i.e., allosteric sites. Conventional drug discovery efforts often target the primary (orthosteric) binding site to block substrate binding. But, as in the case of uncompetitive and noncompetitive inhibitors, ligands can also allosterically modulate activity within the protein structure. Such is the case in the study done by Shi and colleagues, where they identified a second druggable binding site in *Mtb* UDP-galactopyranose mutase (UGM) [165]. NMR and kinetics studies classified MS-208, a known *Mtb*UGM inhibitor, as a noncompetitive/mixed inhibitor and therefore binds in another site in the enzyme to allosterically affect substrate binding. Blind docking in AutoDock Vina [63] was performed to identify possible allosteric sites for MS-208. Two regions, A- and S-site, were initially identified and docked complexes featuring binding to either sites were further subjected to simulation studies using Amber [81]. The A-site-bound structure exhibited the most stable complex formation with excellent interaction energy, as well as the most number of contacts, suggesting that this is the second druggable binding site in *Mtb*UGM [165].

3.2.2. Pharmacophore Modeling and Molecular Docking

Virtual screening is one of the most popular *in silico* drug discovery approaches as it allows researchers to quickly extract data from unexplored chemical space in a cost-effective manner. It has become customary to complement high-throughput screening with VS for the prioritization and identification of novel ligands with the most potential as starting points for drug discovery efforts [166]. Two of the most common VS methods are pharmacophore modeling [167] and docking [168].

Pharmacophore models can be generated from a receptor alone or a receptor-ligand complex. Previously, due to lack of protein structural data, ligand-based pharmacophore has been more customarily used (see Section 3.3.2). A pharmacophore is a group of geometrically-mapped chemical features, such as *H*-bond donors and acceptors, hydrophobicity, and ionizable groups, required for optimal interactions to elicit a response between a receptor target and its partner molecule [169,170]. In line with 3D-mapping in a pharmacophore, exclusion volumes can also be included in the model to incorporate binding site shape [171]. After generating a 3D model, large chemical databases can be efficiently searched for candidates that match pharmacophore elements. However, the pharmacophore database screen only provides a fit score that cannot be translated as affinity. The fit score weighs the alignment quality between the ligand substructures and center of model features. Weights and penalties can also be employed for features deemed significant to activity [171]. These models can also be used for scaffold hopping, allowing for the discovery of novel chemotypes based on fit of interaction and geometric characteristics [171,172].

Another well-established *in silico* method is docking, which can be used to facilitate the investigation of how ligands can fit and complement receptor binding pocket features in order to modulate its activity [168]. Numerous docking methods (Table 1) have been developed and comparative studies and detailed reviews about these have been published elsewhere [173–177]. In the early days of computational drug discovery, docking was developed to be able to predict the bioactive conformation within a set of docking results. However, protein-ligand interactions need to be evaluated using a scoring function to find the best pose using estimated affinity, distinguish actives from inactives, and prioritize candidates for further testing and optimization. This was soon discovered to be the most challenging part of docking due to approximations applied to other crucial factors, such as protein flexibility, solvent involvement, and system entropy [168]. Despite advances in scoring functions through increased understanding of protein-ligand interactions, it is difficult to handle all these aspects while still maintaining method efficiency. Moreover, scoring functions often depend on the protein families and ligand sets from which it was generated and validated [31,178]. And while there is currently no ideal scoring function that can be utilized across all druggable targets, implementation of method validation before starting any VS project establishes whether a chosen docking method and scoring function can be applicable or not [171].

Both pharmacophore modeling and docking have been applied in combination with other *in silico* tools for the identification of novel antimycobacterial agents. Pharmacophore screening followed by docking can be employed as complementary screening tools, resulting in faster processing and more optimized results. A recent study published by our group exemplifies both pharmacophore-based and docking VS by targeting *Mtb* 7,8-diaminopelargonic acid aminotransferase (BioA), an important enzyme in its lipid biosynthesis pathway with no corresponding human ortholog [179]. A receptor-based pharmacophore was generated in Discovery Studio [180] using the BioA structure, characterizing 25 functionalities (nine hydrophobic, nine *H*-bond donors, and seven *H*-bond acceptors), and was employed to screen 4.5 million compounds from the Enamine REAL database. Compounds with good pharmacophore fit, as well as the co-crystallized inhibitor, were subsequently docked to the BioA protein via CDOCKER [70] and ligands with better binding energy values than the known inhibitor were chosen for the TOPKAT protocol [109] to filter out potentially toxic compounds. This step-by-step screening led to the identification of 45 virtual hits, 17 of which were available for purchasing and validation. Whole-cell assay was performed to eliminate compounds that cannot penetrate the distinctive thick, waxy lipid layer of the mycobacterium, identifying compound **7** ((*Z*)-*N*-(2-isopropoxyphenyl)-2-oxo-2-((3-(trifluoromethyl)-cyclohexyl)amino)acetimidic acid) as a potential BioA inhibitor with a minimum inhibitory concentration of ~25 μM [179].

Throughout the years, various improvements have been harnessed to enhance ranking performance, such as rescoring or consensus scoring. Given the different strengths and limitations of each scoring function, rescoring with the help of a separate scoring function not used in a docking study provides users with a different perspective for selection of final hits. For instance, a faster scoring function can be employed for pose prediction while another one is used for affinity prediction and ranking [171]. Currently, consensus scoring is more commonly used for docking studies and has already been examined by several groups in the last couple of decades as an improved protocol for finding potential hits [181–185]. This strategy aims to characterize the intricacy of target molecular recognition based on various energy functions which can be covered by several scoring schemes, resulting in decrease in false positives [184]. However, there is also a risk of rejecting true positives, which have favorable scores in only 1 function used. Thus, it is also imperative to validate a consensus scoring workflow against specific targets [171]. An exemplary case features salicylate synthase MbtI, a critical enzyme in the biosynthesis of siderophore mycobactins, which is used by *Mtb* to chelate iron required for growth and survival in the host. Absence of siderophores prevents bacterial growth in the persistent state after engulfment by macrophages [186]. Previously identified MbtI inhibitors include those based on the MbtI reaction intermediate isochorismate [187], benzimidazole-2-thione [188], and chromane scaffolds [189]. Chiarelli et al. [190] discovered furan-based MbtI inhibitors through structure-based pharmacophore and consensus docking. The pharmacophore model was generated from important binding features in the MbtI-inhibitor complex, including interaction with the conserved Y385, lipophilic interactions, and ionizable interaction with the Mg^{2+} ion. Screening of 1.5 M compounds from Enamine [141] led to over 2,000 pharmacophore hits which were subjected to consensus docking. Docking methods including AutoDock [62], AutoDock Vina [63], DOCK [64], FRED [191], GOLD [65] (comprising four scoring functions), and PLANT [192] were first evaluated using several MbtI-inhibitor complexes to identify which methods are most reliable. GOLD and PLANT were employed for the consensus procedure of the pharmacophore hits, and those with similar binding modes and consistent scores across all scoring functions were further examined if the docked conformation still matched the 3D arrangement of the pharmacophore model. From these, five virtual hits progressed to bioassays, wherein two compounds showed potent MbtI inhibitory activity. MD simulation was additionally applied to study enzyme-ligand interactions and provide information for further optimization. The furan scaffold from the more potent hit was used as a starting point for lead optimization, resulting in a candidate with promising activity against MbtI and suitable antimycobacterial activity [190].

Inclusion of limited protein flexibility, such as in the binding pocket, while still maintaining efficiency has been considered in methods like induced fit docking (IFD) and ensemble docking. IFD incorporates the principle that ligand binding induces changes in residue side chain conformations within the specified pocket, thereby inciting tighter binding with the receptor [193,194]. However, backbone movement should also be considered as it increases the accuracy of side chain positioning and orientations [31]. An example of this is shown in Figure 3, in which *Mtb* InhA exhibits backbone and side chain conformational differences between its apo (PDB ID: 4DRE) [195], fatty acyl substrate-bound (PDB ID: 1BVR) [196] and isoniazid (INH)-bound (PDB ID: 4TRO) structures [197]. These movements can change the binding pocket shape and volume, potentially affecting VS results. The utilization of several experimentally—(X-ray crystallography, NMR, or cryo-electron microscopy) or computationally-derived (MD trajectory) protein structures to integrate both backbone and side chain movements has also been employed in ensemble docking [198]. Structural ensembles provide better reproducibility of experimental conditions as rigid protein structures, such as those obtained from X-ray crystallography experiments, only provide a snapshot of a dynamic ensemble of conformations.

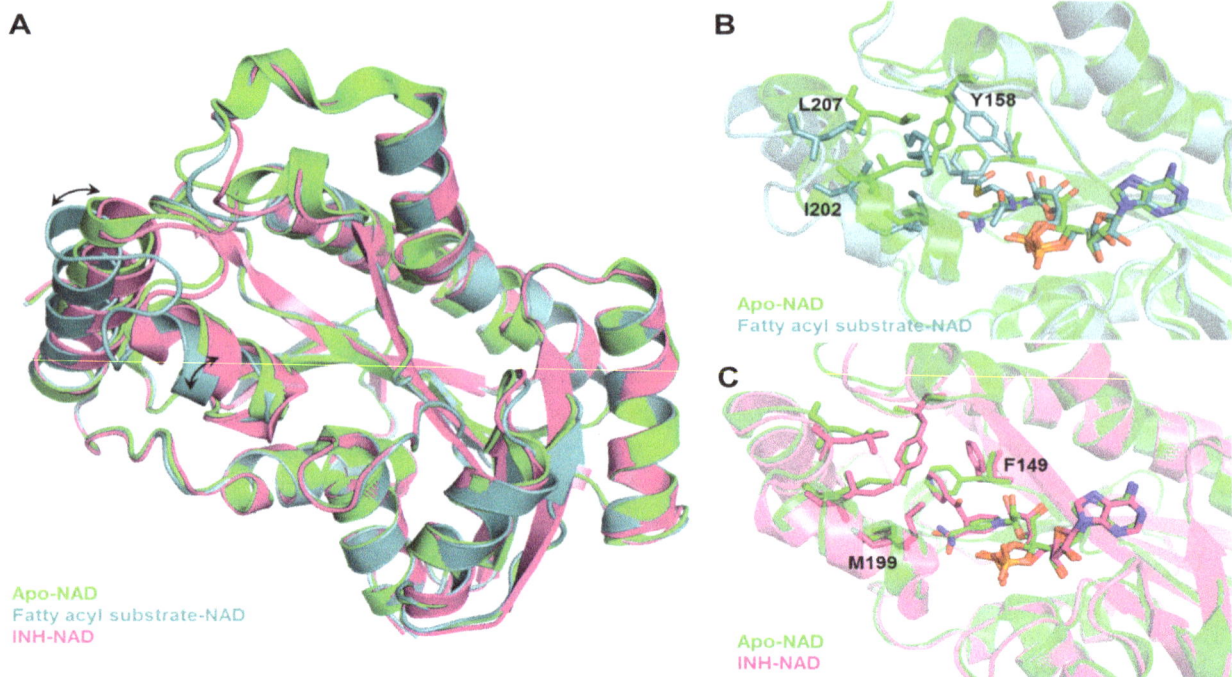

Figure 3. Backbone and sidechain flexibility shown by Mtb InhA, apo vs. fatty acyl-bound vs. INH-bound. (**A**) Structural overlay of apo, fatty acyl-bound, INH-bound Mtb InhA shows backbone movement upon substrate (fatty acyl) binding. Binding site comparison of (**B**) fatty acyl-bound and (**C**) INH-bound vs. apo Mtb InhA structure shows distinct changes in residue side chain positions and conformations. Black arrows indicate movement of alpha helices, side chains that showed large conformational change upon fatty acyl or INH binding are labelled in black.

Various docking approaches were applied by Brindha et al. [199] for drug repurposing against *Mtb* murE, which is an attractive target due to its significance in the peptidoglycan biosynthesis of tuberculosis bacteria and lack of eukaryotic homolog. VS of compounds from DrugBank [200] was first implemented through the parallel use of Glide Standard Precision (SP) [66] and AutoDock Vina [63]. To improve prioritization of compounds through the incorporation of binding site flexibility, common hits from both methods were further subjected to IFD [67]. Final rankings were done using Glide eXtra Precision (XP) scoring and AutoDock Vina binding energy prediction, resulting in 17 common top hits identified as repurposed antitubercular drug candidates [199]. In another example, ensemble docking using three enzyme structures was performed to better elucidate ligand binding interactions, especially due to the binding site flexibility of the *Mtb* Type II dehydroquinase (MtDHQase), an essential virulence factor in TB [201]. Conformation of key residues were determined by analyzing superimposed MtDHQase structures and rotamer distribution of each residue from the penultimate rotamer library [202]. The benzene sulfonamide containing compound with the best activity, a Schaeffer's acid amide, was docked using GOLD [65] and scored using ChemPLP [203]. Varying the side chain flexibility during the docking procedure led to the identification of residues that are required to transition the binding site into an open conformation, which is the preferred conformation of the inhibitor to display its activity. Along with the interaction and flexibility data, Schaeffer's acid amide can be optimized into a potent antitubercular therapeutic compound [201].

3.2.3. Molecular Dynamics

The availability of 3D structural information has greatly helped in structure-based drug design by presenting atomic-level insights into molecular interactions. Nonetheless, these provide only partial interpretations of biomolecular structures, as well as related aspects of molecular recognition and binding. In physiological conditions, proteins frequently undergo conformational changes upon binding with a partner, such as a small molecule, peptide, or another protein, to perform a specific function. At times, these transformations only involve side chain conformations and small to medium movements in the backbone. However, there are cases in which significant deviations are seen in the overall protein fold and/or subunit arrangement [204–207].

Molecular dynamics simulation, a method that was first developed in the 70s [208], can be employed to analyze these protein dynamics and study the binding energy landscape. The availability of MD platforms (Table 1) allowed for the routine assimilation of simulation studies for systems containing ~50,000–100,000 atoms. The investigation of even larger systems is made possible using graphics processing units (GPUs), which are high-performance processors that can support heavy computational load, and high-performance computing (HPC) technologies featuring messaging passing interface (MPI), a system which employs multiple cores in parallel to distribute computational load and reduce the time required for simulation [206,209]. Popular MD packages have been adapted for these tools, and while MD simulation projects commonly use a combination of both, the speedy development of more advanced GPUs increasingly allows for the use of personal workstations [206].

To start an MD simulation, a 3D protein structure of the required system (e.g., apo protein, protein-ligand or protein-protein complex) must be obtained experimentally or through homology modeling, and represented based on the duration and details of study [206,210]. Another critical aspect of system preparation in MD is the solvent model, which can be explicit or implicit. Explicit solvent is more frequently used due to its simplicity and its proficiency in recovering native solvent effects to protein structure [211]. However, large system size resulting from this model makes conformational sampling challenging. To speed up conformational sampling, an implicit solvent model can be generated by adding approximations to the system, but this may affect the free energy landscapes [212]. Once a solvent model has been chosen, the next step is to select an appropriate force field, which is used to define the forces acting on every atom in the system and to calculate the potential energy within the molecular structure. While there are numerous force fields that have been and are still being developed and improved, some of the most popular force fields applied in simulation systems are currently CHARMM [213], Amber [214], GROMOS [215], and OPLS-AA [216,217]. Different force fields use different parameterization to characterize atomistic molecular simulations, distinguishing their applicability in atomistic molecular simulations of diverse target structures and systems, but are often equivalent [206]. To ensure efficiency while keeping the calculations accurate, simple molecular representations in force fields include springs depicting bond length and angles, periodic functions depicting rotations and Lennard-Jones potentials, and Coulomb's law characterizing van der Waals and electrostatic interactions within the system [206]. Newton's law of motion is then employed for the computation of accelerations and velocities during atom movement. Minimization and equilibration are typically performed ahead of a production run to adjust the system to the applied force field, relax steric clashes, and to stabilize system temperature and pressure. Once the prepared system is correctly minimized and equilibrated, the production run can be performed for a suitable amount of time (ps, ns, μs, etc.) depending on research needs (Figure 4). A timestep of 1 or 2 fs is frequently used for atomistic MD simulations [206].

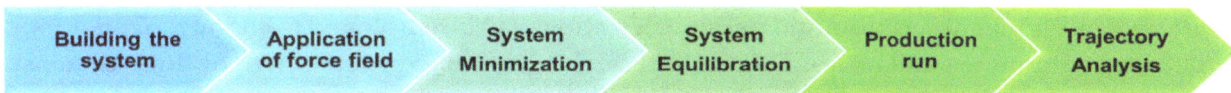

Figure 4. Typical molecular dynamics simulation workflow.

After obtaining simulation trajectories, this information can be used for additional analyses, including but not limited to: (1) verifying stability through root-mean-square deviation (RMSD), root-mean-square fluctuation (RMSF), or radius of gyration (RoG) [218,219], which can identify critical components for protein flexibility, (2) investigating protein structural or energy networks through network analysis [220], which can pinpoint residues that are pivotal in allosteric or long-range communications within the protein, (3) studying protein energy landscape by mapping trajectories [221], providing information about protein folding and function, as well as most stable populations within a given trajectory. Findings from each analysis provide crucial information that would otherwise be imperceptible with other techniques, thereby increasing our understanding of a given system.

With current computational capabilities, simulation times are nearly of biological relevance, allowing researchers to observe biological events, such as allosteric regulation, transient protein changes and binding, and enzyme catalysis. A number of MD studies have already been done for the structural and functional elucidation of validated TB targets and design new antitubercular agents [222–224]. One such study investigated the differences of inhibitor binding against wild-type and mutant structures of InhA, a very well-known TB target [223]. Mutations for this protein led to lower affinity for its co-factor, NADH, resulting in isoniazid resistance. MD simulations of the wild-type and mutant structures of InhA bound to NADH were performed using Amber [81] to understand the underlying aspects affected by structural mutations. Schroeder and colleagues found that mutations in the glycine-rich loop (I21V and I16T) disturbed the NADH binding conformation, specifically that of its pyrophosphate moiety, and decreased its direct and indirect *H*-bond contacts within the binding pocket. Isoniazid requires the formation of a covalent adduct with NADH within the InhA binding site. Changes in binding interactions and conformation of NADH can negatively affect this, hence, contributing to isoniazid resistance [223]. A very recent MD study for TB drug discovery using Amber [81] has been published by Cruz et al. [224], wherein the binding mechanisms of Tam1 and its analogs against polyketide synthase 13 (Pks13), an enzyme that carries out the final step in mycolic acid biosynthesis [225], were investigated to obtain insights that can aid in the design of new antitubercular agents [224]. Fluctuation analysis revealed distinct flexibility in the protein lid domain of *Mtb* Pks13, which was decreased upon ligand binding, suggesting that residues from this domain are critical for ligand interaction. Binding free energy calculations from trajectory data agreed with experimental data, identifying Tam16 as the most potent of the Tam analogs due to conformational stability offered by *H*-bond interactions at both ends of the ligand structure that was not observed for other compounds. Energy decomposition analysis further specifically identified residues that greatly contributed to inhibitor binding, which can then be targeted for optimization of Tam16 and the design of other analogs [224].

3.3. Ligand-Based Tools

3.3.1. Similarity-Based and Quantitative Structure-Activity/Property Relationship Methods

Even before the upsurge of available target structural data, rational inhibitor design has been employed with the help of substrate or product structures, and thus termed as ligand-based drug discovery and design (Figure 2). The simplest and most inexpensive LBDD approach is the similarity-based method, wherein compound candidates were designed and optimized through the principle of chemical similarity, i.e., similar (untested) ligand structures are posited to have similar activities as known inhibitors or modulators [226,227]. A typical workflow requires one or more reference structures with existing bioactivity information against a specific target. This is then used as a template to select new potential candidates from a chemical database to prioritize for assay testing.

Different molecular descriptors and parameterizations can be employed to characterize compounds and efficiently determine similarities. Descriptors can be generated as one-, two- or three-dimensional (1D, 2D, or 3D), wherein 1D descriptors comprise of global ligand properties (e.g., molecular weight, logP, number of *H*-bonds, etc.), 2D descriptors include topological and connectivity properties (e.g., aromaticity, degree of branching, etc.), and 3D descriptors involve geometrical properties (e.g., shape, volume, surface area, etc.) [228,229]. Additionally, fingerprints can also be used to depict template and database compounds by rendering structural features, such as those based on substructure (i.e., scaffold or functional groups), topology or path (i.e., fragmentation following a linear path of bonds), circular or radial (i.e., surrounding features of an atom up to a certain radius), and pharmacophoric (i.e., distance-based features, incorporating molecular shape, and interactions required for biological activity) elements [230]. Well-known platforms that generate molecular descriptors and fingerprints are listed in Table 1.

To compare structural similarity after obtaining simplified molecular features, similarity coefficient and weighing scheme are required to measure and highlight the importance of certain aspects of a compound's structure in relation with its activity. Given that there are multiple tools available for both components, careful selection of analysis tools is crucial to have successful VS campaign. Tanimoto coefficient, which uses the ratio of shared features in both fingerprints to the total number of features between each fingerprint sets, is used as a standard for similarity evaluation of any two vectors. This coefficient returns values between 1 and 0 to depict chemical similarity [231]. Other known similarity measures include Manhattan distance, Euclidean distance, Dice index, Cosine coefficient, and others [232]. In terms of weighing schemes, some features may be 'silenced' or set as optional depending on its importance for a specific activity. There is not one method that can be considered the best for the full range of known targets and chemicals, as each method have their own data set applicability [233,234]. In this case, data fusion can be employed to obtain a consensus of outputs from different methods [235,236]. Both 2D and 3D similarity methods have been successful in identifying hits for various targets and have been established to have comparable or even better enrichment than docking [237]. While similarity-based approaches are known for their efficiency, there is a risk of obtaining low diversity hits as most similarity methods are highly dependent on the input structures used to calculate descriptors [238]. Moreover, there is a potential occurrence of 'activity cliffs,' in which small modifications in a ligand structure lead to significant difference in activity [239–241].

The similarity concept is also applied in studies involving quantitative structure-activity/property relationship (QSAR/QSPR), a method which is used to investigate the correlation between structural and physicochemical properties of ligands with known biological activities. QSAR modeling depends on the premise that ligand 2D and 3D properties can provide information to establish a statistical model of the desired biological activity, which can then be employed for activity prediction of ligand candidates [242,243]. The statistical model is generated based on an appropriate data set, consisting of compound structures with known bioactivity against a specific target, which must be checked and pre-processed. This data set should contain an adequate number of samples (i.e., a minimum of 20 experimentally-validated compounds) and, if from separate studies, identified using same assay protocols such that equivalent activities are obtained. Included in the data set preparation, especially for higher dimensions of QSAR modeling, is the conformational selection and alignment which allows the identification of scaffolds and functional groups that are critical to activity and therefore has more weight in the statistical model [242]. In this case, it is important to remember that the lowest energy conformation is not always equivalent to the bioactive conformation [244,245] and that ligands chosen for the training set should interact with the same binding site [242]. Typical alignment methods include the analysis of molecular fields, structural shapes, or pharmacophores. Pharmacophore generation for ligand alignment is more favorable as it aligns compounds based on feature similarity rather than chemical substructure [246]. As with similarity-based methods, QSAR makes use of molecular descriptors with dimensionality depending on the information available. Descriptors applied to the model should be carefully chosen to avoid autocorrelation and over-fitting. Before proceeding with and *in silico* prediction study, the model must be validated with internal and/or external data sets to establish its predictivity and applicability against a desired target [31,247].

Different QSAR methods have been developed and incorporated in various open-source platforms and commercial software (Table 1). The earliest QSAR-based algorithms include Comparative Molecular Field Analyses (CoMFA) [248] and Comparative Molecular Similarity Indices Analysis (CoMSIA) [249], both of which are still widely used today for various drug discovery endeavors [250–253]. 3D-QSAR CoMFA was used by Singh and Supuran [252] for the discovery of novel *Mtb* carbonic anhydrase inhibitors. A number of sulfonamides that target *Mtb* carbonic anhydrase 2 to regulate bacterial growth were used as the data set for QSAR modeling. The best developed model had excellent predictivity and good fit with an r^2 value of 0.93 and cross-validated coefficient q^2 value of 0.88. From the CoMFA results, it was also determined that several steric and electrostatic features play critical roles in the inhibition of *Mtb* carbonic anhydrase 2. Using this information, nine compounds were designed and later observed to have better predicted inhibitory activities compared to the test set used. However, experimental validation is still required to determine the feasibility of these findings [252].

3.3.2. Ligand-Based Pharmacophore Modeling

Pharmacophore modelling have already demonstrated its value in ligand-based drug discovery studies. Ligand substructures required for optimal bioactivity can be aligned and characterized as a spatial arrangement of features in 3D space, which can be directly used for screening or applied to 3D-QSAR modeling [169,170,246]. Several software and webservers (Table 1) are available for the generation of ligand-based pharmacophores.

As with any other ligand-based methods, a data set of diverse ligands with known bioactivity against a specific target is required for pharmacophore generation. The training ligands used for ligand-based pharmacophore modeling must bind to the same pocket and have similar binding interactions, much like in QSAR studies. After obtaining a pharmacophore, its validity is assessed using a separate test set. It is then employed for virtual search of candidate compounds from libraries of untested molecules, wherein compounds are taken as potential hits if it 'fits' well with the pharmacophore (Figure 5). The main advantage of pharmacophore modeling is the use of molecular features rather than structural groups in depicting critical functionalities for activity, which allows for the identification of novel ligands with diverse structures (i.e., scaffold hopping) [172]. Moreover, pharmacophores can also be used for target profiling and polypharmacological studies to avoid adverse effects resulting from off-target binding [254]. This is especially useful when designing antitubercular and other antibiotic or antiviral agents as to avoid harmful interactions with human proteins.

Due to the numerous parallels between pharmacophore and 3D-QSAR modeling, these methods have been used in combination for a number of ligand-based drug discovery efforts. A study by Tawari et al. used PHASE [80] to target the *Mtb* aryl acid adenylating enzymes known as MbtA, which are involved in siderophore biosynthesis in tuberculosis [255]. A set of nucleoside bisubstrate analogs with known whole cell assay activity and bioactivity against siderophore biosynthesis in *Mtb* were used for pharmacophore and QSAR model development. *H*-bond donor, *H*-bond acceptor, and aromatic features were found to be critical for the inhibition of MbtA. The pharmacophore was also used to align molecules for the 3D-QSAR model, which exhibited suitable predictability and applicability with a Q^2 value of 0.71, RMSE of 0.65, and Pearson-R of 0.85 when assessed against a test set. The SAR studies additionally revealed the disadvantageous effects of bulky groups at the adenyl moiety *C*-6 position. Information taken from these models can be used for the rational design of new MbtA bisubstrate inhibitors as antitubercular agents [255].

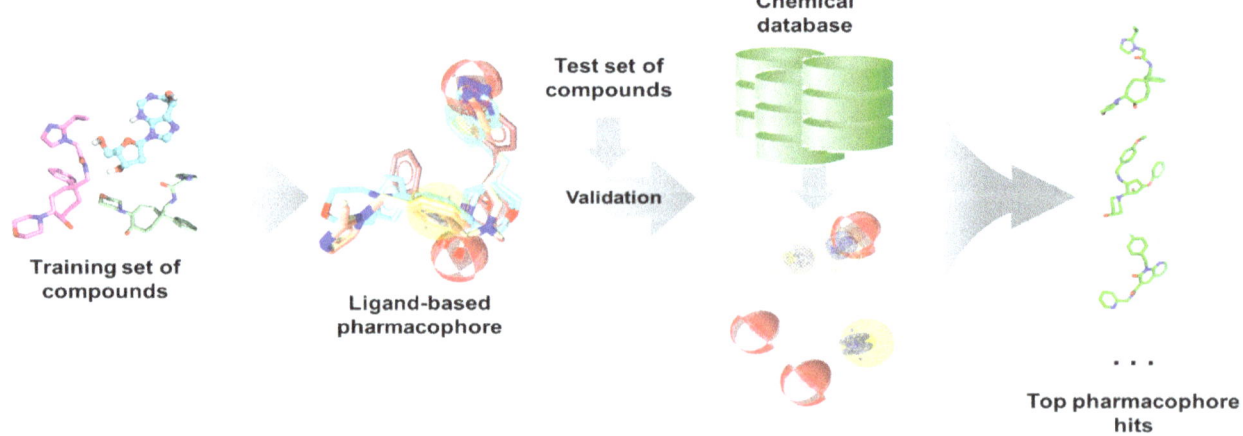

Figure 5. Typical ligand-based pharmacophore generation and screening workflow.

3.3.3. Density Functional Theory

Density functional theory (DFT) is quantum mechanical method established in the 1960s [256,257], which can be used in material science, computational chemistry, and computational physics to study the electronic properties of a many-particle (e.g., atom, molecule, condensed phase) system. DFT is based on two Hohenberg-Kohn (HK) theorems. First, the ground state properties of the many-particle system can be determined using only three spatially determined electron densities. Second, the HK theorem describes an energy functional, which can be minimized by the correct ground state electron density [258]. The use of DFT circumvents the computational expense of conventional methods like Hartree-Fock (HF) Theory since DFT relies on the premise that energies, intricate motions, and pair correlations can be derived directly from the electron probability density alone, instead of using wavefunctions. Theoretically, quantum mechanical wavefunctions consist of all the information required from a target system, and while the Schrödinger equation can be solved for a simple system, such as that of a hydrogen atom, it needs extensive computational efforts and it is impossible to solve this for a many-body system. In this case, DFT is used as an equivalent and efficient alternative to the Schrödinger equation [259], making DFT a popular tool in several computational fields [260].

In tuberculosis research, DFT has found uses in studies involving catalytic mechanisms [261,262], structure-activity relationship analysis [263], and inhibitor potency [264]. Chi and colleagues applied DFT to support their initial observations regarding a change in inhibitor binding mode in the MbtI enzyme after the addition of a substituted enolpyruvyl group to the parent compound structure previously designed from isochorismate [264]. X-ray crystal structures of MbtI complexed with its inhibitors depicted two different binding modes (Mode 1 and 2), suggesting binding site flexibility to accommodate ligand binding. The global minimum conformation of (E)-3-(1-carboxyprop-1-enyloxy)-2-hydroxybenzoic acid (AMT), Z-methyl-AMT, and E-methyl-AMT inhibitors in solution were calculated using Gaussian09 [265] with the B3LYP hybrid functional [266,267]. Global minimum conformation of free Z- and E-methyl-AMT were found to be similar to its bound conformation (Mode 2), indicating prearranged conformations to facilitate its binding to MbtI. Calculation of conformational entropy values for the three compounds revealed that Z-methyl-AMT is the least disordered, which may be due to the methyl conformational lock in its structure. Although a pure Z-isomer has not yet been obtained to experimentally differentiate it from the E-isomer, this finding rationalizes potent binding of methyl-AMT to MbtI and offers more information for the future design of novel and potent MbtI inhibitors [264].

Despite the success and popularity of DFT, it still has deficiencies due to approximations used in the development of functionals. Systems predominantly comprised of dispersion (van der Waals) forces, such as gaseous systems, or those wherein dispersion has a considerable contribution, such as biomolecular systems, are challenging to characterize using DFT [268]. However, several studies have already investigated the inclusion of van der Waals to improve this method [269–271]. Other major limitations of DFT application in computational chemistry include the characterization of charge transfer excitations, transition states, and global potential energy surfaces [272].

3.4. Integrated Tools

With the variety of available tools and structural data for drug discovery nowadays, it is more common to find studies that employ a combination of structure- and ligand-based approaches rather than exclusive application of each (Figure 2). Additionally, integration of these strategies often produces better results owing to more effective exploration of chemical and biological space. Moreover, the strengths of one method can overcome the limitations of the other, resulting in a highly complementary drug discovery process [171,238,273]. Integrated in silico workflows include sequential and parallel or data fusion methods [274–276], though hybrid methods have also already been developed [277,278]. Sequential methods involve the successive use of computational methods with the aim of increasing the selectivity of the VS workflow by continually reducing the number of potential hits before experimental evaluation [31]. However, it has been established that structure- and ligand-based methods have similar enrichment and frequently yield hits with different scaffolds [276], indicating that these methods are better applied in parallel rather than sequentially [279]. Parallel application, through simultaneous employment of various computational tools, often produce a more diverse hit profile [31,280]. Nonetheless, since results from these methods are often fused to produce a final ranking, a large number of virtual hits is obtained from this approach [31].

In silico methods have already been applied to tuberculosis studies in several different ways and combinations depending on the goal of the study implemented, such as for drug discovery [179,281], understanding protein structure and function [282,283], and others [284]. One example showing the integration of computational methods is a study implemented by Li et al. involving 3D-QSAR, binding pocket prediction, docking, and MD simulation studies for FtsZ, which is a validated *Mtb* target and plays a significant role in cell division [285]. Trisubstituted benzimidazoles were found to target this protein and used for 3D-QSAR CoMFA [248] analysis in order to elucidate important structural factors related to their inhibitory activities. Homology modeling using the SWISS-MODEL server [32] was required to obtain the GDP-bound structure of *Mtb* FtsZ using *S. aureus* FtsZ as template. Afterwards, binding site prediction using ProFunc [61] was performed to identify potential binding pockets (other than the GDP binding site) for the candidate compounds. Selected trisubstituted benzimidazole analogs were docked into the *Mtb* FtsZ model using AutoDock, after which the lowest binding energy docked complex was refined using MD simulation. Using the MD-refined *Mtb* FtsZ structure, all trisubstituted benzimidazoles were docked using Surflex-Dock [286]. In the lowest energy state of the compounds, the benzimidazole scaffold and cyclohexyl group were located in a highly hydrophobic pocket within FtsZ, while the carbamate groups were oriented towards the hydrophilic area. These interactions are posited to be crucial for ligand binding stabilization and inhibition of *Mtb* FtsZ. The results of this study display how the concerted application of different *in silico* methods can lead to better understanding of protein structure, ligand design, and inhibitory activities [285].

4. Edges and Pitfalls of *In Silico* Methods

There are roughly 2500 protein structures for tuberculosis in the PDB and perhaps thousands of ligand candidates published. All these pieces of information are available with a few keyboard strokes and a click of the mouse. Along with existing technologies, it is now possible to analyze TB enzymes and lead candidates at the atomic level in order to understand their function and how to regulate them. While computational methods have been widely used in drug discovery nowadays due to their successful applications [287–289], it is still important to remember that these tools are like any other experimental approaches—prone to limitations dependent on the system and other various parameters being studied [290–292].

VS has been known to successfully screen millions of compounds to identify potential inhibitors for a given target [287,289]. This lends efficiency to cost, time, and efforts used in drug discovery projects as only the most promising compounds are brought forward for more rigorous experimental testing and drug development. However, optimization and validation of these methods are far from perfect and are highly dependent on the protein system and compound classes used, leading to possible bias in the computational model. Thus, it is challenging to determine which method has the advantage over another; many benchmark studies have been published regarding this matter [293,294]. Other major limitations include difficulties in incorporating protein flexibility and solvent effects due to the computational burden attached to these factors [31]. Fortunately, available technologies seem to be catching up as enhance sampling methods, HPC, and MD platforms are now routinely applied in drug discovery projects and are known to calculate up to milliseconds of simulations for various protein targets [295–297]. In terms of ligand-based drug design, its main advantage is its simplicity and efficiency. Indeed, LBDD has a long history and numerous candidates have already been discovered even with the lack of protein structural information [298–300]. Nonetheless, several factors should be considered when applying ligand-based tools. Firstly, ligand alignments are based on the lowest conformation energy, which is often different from the bioactive conformation [244,245], as well as on the assumption that ligands bind in the same site and display the same conformation. Secondly, compounds should be evaluated by the same group (preferred) or tested using the same assay with the same parameters to be considered comparable [242]. Thirdly, the basic premise of 'similar structures display similar activities' are contradicted by the existence of activity cliffs [239–241], and so care should be taken when selecting potential candidates from a pool of virtual hits. Finally, it is also a challenge to incorporate the effects of solvation and protein flexibility due to the nature of the analysis.

As mentioned in the previous section, integration of several *in silico* methods have become common practice when designing and optimizing lead candidates to overcome the shortcomings of each individual tools. Despite requiring more computational resources, assimilation of computational methods result in better accuracy and enrichment of hits. In addition, the combination of a researcher's innate knowledge with the computational efficiency of these tools is perhaps the best integration of all, as a human's touch continues to be irreplaceable in the interpretation of all the data produced by *in silico* methods.

5. Conclusions and Future Perspectives

TB remains to be a relevant public health threat worldwide, necessitating accelerated discovery and design of novel antimycobacterial agents. Computer-aided drug design has become one of anchors of drug discovery research and continues to be a formidable tool in the hunt for promising drug leads, especially for tuberculosis. Continuous advancements in computing power and available software can enhance current computational tools and their application to different stages in the drug discovery pipeline. Nonetheless, these methods are not invincible as each tool have their own restrictions, and approximations are often used during the analysis. To overcome these, it is best to assimilate several *in silico* tools to complement the strength and limitations of each method used. The application of CADD in TB research has led to the identification of several antimycobacterial compounds that have already reached clinical evaluations, promoting its value in the drug discovery paradigm. Nonetheless, more work has to be done in order to expedite the discovery of anti-TB therapeutics.

Machine learning (ML) methods are making a comeback in drug discovery studies due to the upsurge in available data and enhanced computational powers. This has resulted in a wave drug discovery studies involving artificial intelligence (AI), wherein ML and deep learning (DL) techniques are applied to efficiently and 'intelligently' solve problems. This new shift in the drug discovery landscape is observed in personalized medicine and a number of relevant illnesses like cancer. While there are already several FDA-approved uses of AI in healthcare and diagnostics [301], it has yet to produce a successful drug candidate but it might not be far off. Currently, AI studies involving TB frequently covers diagnostics and treatment outcomes. This is perhaps one of the gaps that needs to be filled to be able to fast-track the discovery of novel and efficacious anti-TB drugs and finally alleviate the heavy burden of this infection around the globe.

Author Contributions: S.J.Y.M. and J.B.B. conceptualized and prepared the manuscript draft; J.B.B., V.G.O., and M.C.O.C. provided critical comments; all authors reviewed, edited, and approved the manuscript. All authors have read and agreed to the published version of the manuscript.

References

1. Pai, M.; Behr, M.A.; Dowdy, D.; Dheda, K.; Divangahi, M.; Boehme, C.C.; Ginsberg, A.; Swaminathan, S.; Spigelman, M.; Getahun, H.; et al. Tuberculosis. *Nat. Rev. Dis. Primers* **2016**, *2*, 16076. [CrossRef] [PubMed]
2. Havlir, D.V.; Getahun, H.; Sanne, I.; Nunn, P. Opportunities and challenges for HIV care in overlapping HIV and TB epidemics. *JAMA* **2008**, *300*, 423–430. [PubMed]
3. Leung, C.C.; Yew, W.W.; Chan, C.K.; Chang, K.C.; Law, W.S.; Lee, S.N.; Tai, L.B.; Leung, E.C.; Au, R.K.; Huang, S.S.; et al. Smoking adversely affects treatment response, outcome and relapse in tuberculosis. *Eur. Respir. J.* **2015**, *45*, 738–745. [CrossRef] [PubMed]

4.   Imtiaz, S.; Shield, K.D.; Roerecke, M.; Samokhvalov, A.V.; Lonnroth, K.; Rehm, J. Alcohol consumption as a risk factor for tuberculosis: Meta-analyses and burden of disease. *Eur. Respir. J.* **2017**, *50*, 1700216. [CrossRef] [PubMed]

5.   Restrepo, B.I. Diabetes and Tuberculosis. In *Understanding the Host Immune Response against Mycobacterium tuberculosis Infection*; Springer International Publishing: Cham, Switzerland, 2018; pp. 1–21.

6.   Getahun, H.; Matteelli, A.; Chaisson, R.E.; Raviglione, M. Latent *Mycobacterium tuberculosis* infection. *N. Engl. J. Med.* **2015**, *372*, 2127–2135. [CrossRef]

7.   Miller, L.G.; Asch, S.M.; Yu, E.I.; Knowles, L.; Gelberg, L.; Davidson, P. A population-based survey of tuberculosis symptoms: How atypical are atypical presentations? *Clin. Infect. Dis.* **2000**, *30*, 293–299. [CrossRef]

8.   World Health Organization (WHO). *Global Tuberculosis Report 2019*; World Health Organization (WHO): Geneva, Switzerland, 2019.

9.   World Health Organization (WHO). *Guidelines on the Management of Latent Tuberculosis Infection*; World Health Organization (WHO): Geneva, Switzerland, 2015.

10.   World Health Organization (WHO). *Latent Tuberculosis infection: Updated and Consolidated Guidelines for Programmatic Management*; World Health Organization (WHO): Geneva, Switzerland, 2018.

11.   Getahun, H.; Matteelli, A.; Abubakar, I.; Aziz, M.A.; Baddeley, A.; Barreira, D.; Den Boon, S.; Borroto Gutierrez, S.M.; Bruchfeld, J.; Burhan, E.; et al. Management of latent *Mycobacterium tuberculosis* infection: WHO guidelines for low tuberculosis burden countries. *Eur. Respir. J.* **2015**, *46*, 1563–1576. [CrossRef]

12.   World Health Organization (WHO). *Treatment of Tuberculosis: Guidelines*; World Health Organization (WHO): Geneva, Switzerland, 2010.

13.   Nahid, P.; Dorman, S.E.; Alipanah, N.; Barry, P.M.; Brozek, J.L.; Cattamanchi, A.; Chaisson, L.H.; Chaisson, R.E.; Daley, C.L.; Grzemska, M.; et al. Official American Thoracic Society/Centers for Disease Control and Prevention/Infectious Diseases Society of America Clinical Practice Guidelines: Treatment of Drug-Susceptible Tuberculosis. *Clin. Infect. Dis.* **2016**, *63*, e147–e195. [CrossRef]

14.   Volmink, J.; Garner, P. Directly observed therapy for treating tuberculosis. *Cochrane Database Syst. Rev.* **2007**, CD003343. [CrossRef]

15.   Horsburgh, C.R., Jr.; Barry, C.E., 3rd; Lange, C. Treatment of Tuberculosis. *N. Engl. J. Med.* **2015**, *373*, 2149–2160. [CrossRef]

16.   Saukkonen, J.J.; Cohn, D.L.; Jasmer, R.M.; Schenker, S.; Jereb, J.A.; Nolan, C.M.; Peloquin, C.A.; Gordin, F.M.; Nunes, D.; Strader, D.B.; et al. An official ATS statement: Hepatotoxicity of antituberculosis therapy. *Am. J. Respir. Crit. Care Med.* **2006**, *174*, 935–952. [CrossRef]

17.   Dheda, K.; Barry, C.E., 3rd; Maartens, G. Tuberculosis. *Lancet* **2016**, *387*, 1211–1226. [CrossRef]

18.   Dheda, K.; Gumbo, T.; Gandhi, N.R.; Murray, M.; Theron, G.; Udwadia, Z.; Migliori, G.B.; Warren, R. Global control of tuberculosis: From extensively drug-resistant to untreatable tuberculosis. *Lancet. Respir. Med.* **2014**, *2*, 321–338. [CrossRef]

19.   World Health Organization (WHO). *WHO Treatment Guidelines for Drug-Resistant Tuberculosis 2016 Update*; World Health Organization (WHO): Geneva, Switzerland, 2016.

20.   Walker, J.; Tadena, N. J&J Tuberculosis Drug Gets Fast-Track Clearance. *Wall St. J.* **2013**. Available online: https://www.wsj.com/articles/SB10001424127887323320404578213421059138236 (accessed on 27 November 2019).

21.   Mahajan, R. Bedaquiline: First FDA-approved tuberculosis drug in 40 years. *Int. J. Appl. Basic Med. Res.* **2013**, *3*, 1–2. [CrossRef]

22.   European Medicines Agency (EMA). *Deltyba Delamanid Summary of the European Public Assessment Report (EPAR) for Deltyba*; EMA: Amsterdam, The Netherlands, 2014; pp. 1–3. Available online: https://www.ema.europa.eu/en/medicines/human/EPAR/deltyba (accessed on 15 November 2019).

23.   Ryan, N.J.; Lo, J.H. Delamanid: First global approval. *Drugs* **2014**, *74*, 1041–1045. [CrossRef]

24.   US FDA. *FDA Approves New Drug for Treatment-Resistant Forms of Tuberculosis That Affects the Lungs*; US FDA: Silver Spring, MD, USA, 2019.

25. Baptista, R.; Fazakerley, D.M.; Beckmann, M.; Baillie, L.; Mur, L.A.J. Untargeted metabolomics reveals a new mode of action of pretomanid (PA-824). *Sci. Rep.* **2018**, *8*, 5084. [CrossRef]

26. Thompson, A.M.; Bonnet, M.; Lee, H.H.; Franzblau, S.G.; Wan, B.; Wong, G.S.; Cooper, C.B.; Denny, W.A. Antitubercular Nitroimidazoles Revisited: Synthesis and Activity of the Authentic 3-Nitro Isomer of Pretomanid. *ACS Med. Chem. Lett.* **2017**, *8*, 1275–1280. [CrossRef]

27. Manjunatha, U.; Boshoff, H.I.; Barry, C.E. The mechanism of action of PA-824: Novel insights from transcriptional profiling. *Commun. Integr. Biol.* **2009**, *2*, 215–218. [CrossRef]

28. Reymond, J.-L.; van Deursen, R.; Blum, L.C.; Ruddigkeit, L. Chemical space as a source for new drugs. *MedChemComm* **2010**, *1*, 30–38. [CrossRef]

29. Berman, H.M.; Westbrook, J.; Feng, Z.; Gilliland, G.; Bhat, T.N.; Weissig, H.; Shindyalov, I.N.; Bourne, P.E. The Protein Data Bank. *Nucleic Acids Res.* **2000**, *28*, 235–242. [CrossRef]

30. Ekins, S.; Freundlich, J.S.; Choi, I.; Sarker, M.; Talcott, C. Computational databases, pathway and cheminformatics tools for tuberculosis drug discovery. *Trends Microbiol.* **2011**, *19*, 65–74. [CrossRef]

31. Macalino, S.J.; Gosu, V.; Hong, S.; Choi, S. Role of computer-aided drug design in modern drug discovery. *Arch. Pharm. Res.* **2015**, *38*, 1686–1701. [CrossRef]

32. Schwede, T.; Kopp, J.; Guex, N.; Peitsch, M.C. SWISS-MODEL: An automated protein homology-modeling server. *Nucleic Acids Res.* **2003**, *31*, 3381–3385. [CrossRef]

33. Webb, B.; Sali, A. Comparative Protein Structure Modeling Using MODELLER. *Curr. Protoc. Protein Sci.* **2016**, *86*, 1–37. [CrossRef]

34. Kim, D.E.; Chivian, D.; Baker, D. Protein structure prediction and analysis using the Robetta server. *Nucleic Acids Res.* **2004**, *32*, W526–W531. [CrossRef]

35. Schrödinger. Prime. Available online: https://www.schrodinger.com/prime (accessed on 26 October 2019).

36. Zheng, W.; Zhang, C.; Bell, E.W.; Zhang, Y. I-TASSER gateway: A protein structure and function prediction server powered by XSEDE. *Future Gener. Comput. Syst.* **2019**, *99*, 73–85. [CrossRef]

37. Zhang, Y. I-TASSER server for protein 3D structure prediction. *BMC Bioinf.* **2008**, *9*, 40. [CrossRef]

38. Yang, J.; Zhang, Y. I-TASSER server: New development for protein structure and function predictions. *Nucleic Acids Res.* **2015**, *43*, W174–W181. [CrossRef]

39. Yang, J.; Yan, R.; Roy, A.; Xu, D.; Poisson, J.; Zhang, Y. The I-TASSER Suite: Protein structure and function prediction. *Nat. Methods* **2015**, *12*, 7–8. [CrossRef]

40. Roy, A.; Kucukural, A.; Zhang, Y. I-TASSER: A unified platform for automated protein structure and function prediction. *Nat. Protoc.* **2010**, *5*, 725–738. [CrossRef] [PubMed]

41. Zhang, Y. I-TASSER: Fully automated protein structure prediction in CASP8. *Proteins* **2009**, *77*, 100–113. [CrossRef] [PubMed]

42. Zimmermann, L.; Stephens, A.; Nam, S.Z.; Rau, D.; Kubler, J.; Lozajic, M.; Gabler, F.; Soding, J.; Lupas, A.N.; Alva, V. A Completely Reimplemented MPI Bioinformatics Toolkit with a New HHpred Server at its Core. *J. Mol. Biol.* **2018**, *430*, 2237–2243. [CrossRef] [PubMed]

43. Hildebrand, A.; Remmert, M.; Biegert, A.; Soding, J. Fast and accurate automatic structure prediction with HHpred. *Proteins* **2009**, *77*, 128–132. [CrossRef] [PubMed]

44. Soding, J.; Biegert, A.; Lupas, A.N. The HHpred interactive server for protein homology detection and structure prediction. *Nucleic Acids Res.* **2005**, *33*, W244–W248. [CrossRef] [PubMed]

45. Laskowski, R.A.; MacArthur, M.W.; Moss, D.S.; Thornton, J.M. PROCHECK: A program to check the stereochemical quality of protein structures. *J. Appl. Cryst.* **1993**, *26*, 283–291. [CrossRef]

46. Wiederstein, M.; Sippl, M.J. ProSA-web: Interactive web service for the recognition of errors in three-dimensional structures of proteins. *Nucleic Acids Res.* **2007**, *35*, W407–W410. [CrossRef]

47. Eisenberg, D.; Luthy, R.; Bowie, J.U. VERIFY3D: Assessment of protein models with three-dimensional profiles. *Methods Enzymol.* **1997**, *277*, 396–404.

48. Colovos, C.; Yeates, T.O. Verification of protein structures: Patterns of nonbonded atomic interactions. *Protein Sci.* **1993**, *2*, 1511–1519. [CrossRef]

49. Hussein, H.A.; Borrel, A.; Geneix, C.; Petitjean, M.; Regad, L.; Camproux, A.C. PockDrug-Server: A new web server for predicting pocket druggability on holo and apo proteins. *Nucleic Acids Res.* **2015**, *43*, W436–W442. [CrossRef]

50. Volkamer, A.; Kuhn, D.; Rippmann, F.; Rarey, M. DoGSiteScorer: A web server for automatic binding site prediction, analysis and druggability assessment. *Bioinformatics* **2012**, *28*, 2074–2075. [CrossRef] [PubMed]

51. Schmidtke, P.; Le Guilloux, V.; Maupetit, J.; Tuffery, P. fpocket: Online tools for protein ensemble pocket detection and tracking. *Nucleic Acids Res.* **2010**, *38*, W582–W589. [CrossRef] [PubMed]

52. Le Guilloux, V.; Schmidtke, P.; Tuffery, P. Fpocket: An open source platform for ligand pocket detection. *BMC Bioinf.* **2009**, *10*, 168. [CrossRef] [PubMed]

53. Tian, W.; Chen, C.; Lei, X.; Zhao, J.; Liang, J. CASTp 3.0: Computed atlas of surface topography of proteins. *Nucleic Acids Res.* **2018**, *46*, W363–W367. [CrossRef] [PubMed]

54. Binkowski, T.A.; Naghibzadeh, S.; Liang, J. CASTp: Computed Atlas of Surface Topography of proteins. *Nucleic Acids Res.* **2003**, *31*, 3352–3355. [CrossRef]

55. Dundas, J.; Ouyang, Z.; Tseng, J.; Binkowski, A.; Turpaz, Y.; Liang, J. CASTp: Computed atlas of surface topography of proteins with structural and topographical mapping of functionally annotated residues. *Nucleic Acids Res.* **2006**, *34*, W116–W118. [CrossRef]

56. Koes, D.R.; Camacho, C.J. PocketQuery: Protein-protein interaction inhibitor starting points from protein-protein interaction structure. *Nucleic Acids Res.* **2012**, *40*, W387–W392. [CrossRef]

57. Brady, G.P., Jr.; Stouten, P.F. Fast prediction and visualization of protein binding pockets with PASS. *J. Comput. Aided Mol. Des.* **2000**, *14*, 383–401. [CrossRef]

58. Halgren, T.A. Identifying and characterizing binding sites and assessing druggability. *J. Chem. Inf. Model.* **2009**, *49*, 377–389. [CrossRef]

59. Capra, J.A.; Laskowski, R.A.; Thornton, J.M.; Singh, M.; Funkhouser, T.A. Predicting protein ligand binding sites by combining evolutionary sequence conservation and 3D structure. *PLoS Comput. Biol.* **2009**, *5*, e1000585. [CrossRef]

60. Jendele, L.; Krivak, R.; Skoda, P.; Novotny, M.; Hoksza, D. PrankWeb: A web server for ligand binding site prediction and visualization. *Nucleic Acids Res.* **2019**, *47*, W345–W349. [CrossRef] [PubMed]

61. Laskowski, R.A.; Watson, J.D.; Thornton, J.M. ProFunc: A server for predicting protein function from 3D structure. *Nucleic Acids Res.* **2005**, *33*, W89–W93. [CrossRef]

62. Morris, G.M.; Huey, R.; Lindstrom, W.; Sanner, M.F.; Belew, R.K.; Goodsell, D.S.; Olson, A.J. AutoDock4 and AutoDockTools4: Automated docking with selective receptor flexibility. *J. Comput. Chem.* **2009**, *30*, 2785–2791. [CrossRef] [PubMed]

63. Trott, O.; Olson, A.J. AutoDock Vina: Improving the speed and accuracy of docking with a new scoring function, efficient optimization, and multithreading. *J. Comput. Chem.* **2010**, *31*, 455–461. [CrossRef] [PubMed]

64. Allen, W.J.; Balius, T.E.; Mukherjee, S.; Brozell, S.R.; Moustakas, D.T.; Lang, P.T.; Case, D.A.; Kuntz, I.D.; Rizzo, R.C. DOCK 6: Impact of new features and current docking performance. *J. Comput. Chem.* **2015**, *36*, 1132–1156. [CrossRef]

65. Verdonk, M.L.; Cole, J.C.; Hartshorn, M.J.; Murray, C.W.; Taylor, R.D. Improved protein–ligand docking using GOLD. *Proteins* **2003**, *52*, 609–623. [CrossRef]

66. Schrödinger. Glide. Available online: https://www.schrodinger.com/glide (accessed on 26 October 2019).

67. Schrödinger. Induced Fit. Available online: https://www.schrodinger.com/induced-fit (accessed on 26 October 2019).

68. Rarey, M.; Kramer, B.; Lengauer, T.; Klebe, G. A fast flexible docking method using an incremental construction algorithm. *J. Mol. Biol.* **1996**, *261*, 470–489. [CrossRef]

69. Davis, I.W.; Baker, D. RosettaLigand docking with full ligand and receptor flexibility. *J. Mol. Biol.* **2009**, *385*, 381–392. [CrossRef]

70. Wu, G.; Robertson, D.H.; Brooks, C.L., 3rd; Vieth, M. Detailed analysis of grid-based molecular docking: A case study of CDOCKER-A CHARMm-based MD docking algorithm. *J. Comput. Chem.* **2003**, *24*, 1549–1562. [CrossRef]

71. Bitencourt-Ferreira, G.; de Azevedo, W.F., Jr. Docking with SwissDock. *Methods Mol. Biol.* **2019**, *2053*, 189–202.

72. Grosdidier, A.; Zoete, V.; Michielin, O. SwissDock, a protein-small molecule docking web service based on EADock DSS. *Nucleic Acids Res.* **2011**, *39*, W270–W277. [CrossRef] [PubMed]

73. Koes, D.R.; Camacho, C.J. Pharmer: Efficient and exact pharmacophore search. *J. Chem. Inf. Model.* **2011**, *51*, 1307–1314. [CrossRef] [PubMed]

74. Dassault Systèmes BIOVIA. *Catalyst*. Available online: https://www.3dsbiovia.com/products/collaborative-science/biovia-discovery-studio/pharmacophore-and-ligand-based-design.html (accessed on 26 October 2019).

75. Schneidman-Duhovny, D.; Dror, O.; Inbar, Y.; Nussinov, R.; Wolfson, H.J. PharmaGist: A webserver for ligand-based pharmacophore detection. *Nucleic Acids Res.* **2008**, *36*, W223–W228. [CrossRef] [PubMed]

76. Inte:Ligand. LigandScout. Available online: http://www.inteligand.com/ligandscout/ (accessed on 26 October 2019).

77. Zoete, V.; Daina, A.; Bovigny, C.; Michielin, O. SwissSimilarity: A Web Tool for Low to Ultra High Throughput Ligand-Based Virtual Screening. *J. Chem. Inf. Model.* **2016**, *56*, 1399–1404. [CrossRef]

78. Douguet, D. e-LEA3D: A computational-aided drug design web server. *Nucleic Acids Res.* **2010**, *38*, W615–W621. [CrossRef]

79. Dallakyan, S.; Olson, A.J. Small-molecule library screening by docking with PyRx. *Methods Mol. Biol.* **2015**, *1263*, 243–250.

80. Schrödinger. PHASE. Available online: https://www.schrodinger.com/phase (accessed on 26 October 2019).

81. Case, D.A.; Cheatham, T.E., 3rd; Darden, T.; Gohlke, H.; Luo, R.; Merz, K.M., Jr.; Onufriev, A.; Simmerling, C.; Wang, B.; Woods, R.J. The Amber biomolecular simulation programs. *J. Comput. Chem.* **2005**, *26*, 1668–1688. [CrossRef]

82. Salomon-Ferrer, R.; Case, D.A.; Walker, R.C. An overview of the Amber biomolecular simulation package. *Wiley Interdiscip. Rev. Comput. Mol. Sci.* **2013**, *3*, 198–210. [CrossRef]

83. Brooks, B.R.; Brooks, C.L., 3rd; Mackerell, A.D., Jr.; Nilsson, L.; Petrella, R.J.; Roux, B.; Won, Y.; Archontis, G.; Bartels, C.; Boresch, S.; et al. CHARMM: The biomolecular simulation program. *J. Comput. Chem.* **2009**, *30*, 1545–1614. [CrossRef]

84. Miller, B.T.; Singh, R.P.; Klauda, J.B.; Hodoscek, M.; Brooks, B.R.; Woodcock, H.L., 3rd. CHARMMing: A new, flexible web portal for CHARMM. *J. Chem. Inf. Model.* **2008**, *48*, 1920–1929. [CrossRef]

85. Van Der Spoel, D.; Lindahl, E.; Hess, B.; Groenhof, G.; Mark, A.E.; Berendsen, H.J. GROMACS: Fast, flexible, and free. *J. Comput. Chem.* **2005**, *26*, 1701–1718. [CrossRef] [PubMed]

86. Abraham, M.J.; Murtola, T.; Schulz, R.; Páll, S.; Smith, J.C.; Hess, B.; Lindahl, E. GROMACS: High performance molecular simulations through multi-level parallelism from laptops to supercomputers. *SoftwareX* **2015**, *1–2*, 19–25. [CrossRef]

87. Phillips, J.C.; Braun, R.; Wang, W.; Gumbart, J.; Tajkhorshid, E.; Villa, E.; Chipot, C.; Skeel, R.D.; Kale, L.; Schulten, K. Scalable molecular dynamics with NAMD. *J. Comput. Chem.* **2005**, *26*, 1781–1802. [CrossRef] [PubMed]

88. Schrödinger. Desmond. Available online: https://www.schrodinger.com/desmond (accessed on 26 October 2019).

89. Zoete, V.; Cuendet, M.A.; Grosdidier, A.; Michielin, O. SwissParam: A fast force field generation tool for small organic molecules. *J. Comput. Chem.* **2011**, *32*, 2359–2368. [CrossRef] [PubMed]

90. Jo, S.; Kim, T.; Iyer, V.G.; Im, W. CHARMM-GUI: A web-based graphical user interface for CHARMM. *J. Comput. Chem.* **2008**, *29*, 1859–1865. [CrossRef] [PubMed]

91. Vanommeslaeghe, K.; Hatcher, E.; Acharya, C.; Kundu, S.; Zhong, S.; Shim, J.; Darian, E.; Guvench, O.; Lopes, P.; Vorobyov, I.; et al. CHARMM general force field: A force field for drug-like molecules compatible with the CHARMM all-atom additive biological force fields. *J. Comput. Chem.* **2010**, *31*, 671–690. [CrossRef] [PubMed]

92. Vanommeslaeghe, K.; MacKerell, A.D., Jr. Automation of the CHARMM General Force Field (CGenFF) I: Bond perception and atom typing. *J. Chem. Inf. Model.* **2012**, *52*, 3144–3154. [CrossRef] [PubMed]

93. Vanommeslaeghe, K.; Raman, E.P.; MacKerell, A.D., Jr. Automation of the CHARMM General Force Field (CGenFF) II: Assignment of bonded parameters and partial atomic charges. *J. Chem. Inf. Model.* **2012**, *52*, 3155–3168. [CrossRef]

94. Humphrey, W.; Dalke, A.; Schulten, K. VMD: Visual molecular dynamics. *J. Mol. Graph.* **1996**, *14*, 27–38. [CrossRef]

95. Helguera, A.M.; Combes, R.D.; Gonzalez, M.P.; Cordeiro, M.N. Applications of 2D descriptors in drug design: A DRAGON tale. *Curr. Top. Med. Chem.* **2008**, *8*, 1628–1655. [CrossRef]

96. Tetko, I.V.; Gasteiger, J.; Todeschini, R.; Mauri, A.; Livingstone, D.; Ertl, P.; Palyulin, V.A.; Radchenko, E.V.; Zefirov, N.S.; Makarenko, A.S.; et al. Virtual computational chemistry laboratory—Design and description. *J. Comput. Aided Mol. Des.* **2005**, *19*, 453–463. [CrossRef] [PubMed]

97. Schrödinger. Canvas. Available online: https://www.schrodinger.com/canvas (accessed on 26 October 2019).

98. Landrum, G. RDKit: Open-Source Cheminformatics. Available online: http://www.rdkit.org (accessed on 26 October 2019).

99.   Masand, V.H.; Rastija, V. PyDescriptor: A new PyMOL plugin for calculating thousands of easily understandable molecular descriptors. *Chemom. Intell. Lab. Syst.* **2017**, *169*, 12–18. [CrossRef]

100. Moriwaki, H.; Tian, Y.S.; Kawashita, N.; Takagi, T. Mordred: A molecular descriptor calculator. *J. Cheminform.* **2018**, *10*, 4. [CrossRef] [PubMed]

101. Tosco, P.; Balle, T. Open3DQSAR: A new open-source software aimed at high-throughput chemometric analysis of molecular interaction fields. *J. Mol. Model.* **2011**, *17*, 201–208. [CrossRef]

102. Dong, J.; Yao, Z.J.; Zhu, M.F.; Wang, N.N.; Lu, B.; Chen, A.F.; Lu, A.P.; Miao, H.; Zeng, W.B.; Cao, D.S. ChemSAR: An online pipelining platform for molecular SAR modeling. *J. Cheminform.* **2017**, *9*, 27. [CrossRef]

103. BioSolveIT. SeeSAR version 9.2. Available online: https://www.biosolveit.de/SeeSAR/ (accessed on 26 October 2019).

104. Schrödinger. QikProp. Available online: https://www.schrodinger.com/qikprop (accessed on 26 October 2019).

105. SimulationsPlus. ADMET Predictor. Available online: https://www.simulations-plus.com/software/admetpredictor/ (accessed on 26 October 2019).

106. ACD/Labs. Percepta Platform. Available online: https://www.acdlabs.com/products/percepta/ (accessed on 26 October 2019).

107. Miteva, M.A.; Violas, S.; Montes, M.; Gomez, D.; Tuffery, P.; Villoutreix, B.O. FAF-Drugs: Free ADME/tox filtering of compound collections. *Nucleic Acids Res.* **2006**, *34*, W738–W744. [CrossRef]

108. Rasolohery, I.; Moroy, G.; Guyon, F. PatchSearch: A Fast Computational Method for Off-Target Detection. *J. Chem. Inf. Model.* **2017**, *57*, 769–777. [CrossRef]

109. Dassault Systèmes BIOVIA. DS TOPKAT. Available online: https://www.3dsbiovia.com/products/collaborative-science/biovia-discovery-studio/qsar-admet-and-predictive-toxicology.html (accessed on 26 October 2019).

110. Dassault Systèmes BIOVIA. DS ADMET. Available online: https://www.3dsbiovia.com/products/collaborative-science/biovia-pipeline-pilot/component-collections/adme-tox.html (accessed on 26 October 2019).

111. Poroikov, V.; Filimonov, D.; Lagunin, A.; Gloriozova, T.; Zakharov, A. PASS: Identification of probable targets and mechanisms of toxicity. *SAR QSAR Environ. Res.* **2007**, *18*, 101–110. [CrossRef]

112. Daina, A.; Michielin, O.; Zoete, V. SwissADME: A free web tool to evaluate pharmacokinetics, drug-likeness and medicinal chemistry friendliness of small molecules. *Sci. Rep.* **2017**, *7*, 42717. [CrossRef]

113. Cruciani, G.; Carosati, E.; De Boeck, B.; Ethirajulu, K.; Mackie, C.; Howe, T.; Vianello, R. MetaSite: Understanding metabolism in human cytochromes from the perspective of the chemist. *J. Med. Chem.* **2005**, *48*, 6970–6979. [CrossRef]

114. Tcheremenskaia, O.; Benigni, R.; Nikolova, I.; Jeliazkova, N.; Escher, S.E.; Batke, M.; Baier, T.; Poroikov, V.; Lagunin, A.; Rautenberg, M.; et al. OpenTox predictive toxicology framework: Toxicological ontology and semantic media wiki-based OpenToxipedia. *J. Biomed. Semant.* **2012**, *3*, S7. [CrossRef] [PubMed]

115. Smiesko, M.; Vedani, A. VirtualToxLab: Exploring the Toxic Potential of Rejuvenating Substances Found in Traditional Medicines. *Methods Mol. Biol.* **2016**, *1425*, 121–137.

116. Vedani, A.; Dobler, M.; Smiesko, M. VirtualToxLab—A platform for estimating the toxic potential of drugs, chemicals and natural products. *Toxicol. Appl. Pharmacol.* **2012**, *261*, 142–153. [CrossRef] [PubMed]

117. Vedani, A.; Smiesko, M.; Spreafico, M.; Peristera, O.; Dobler, M. VirtualToxLab—In silico prediction of the toxic (endocrine-disrupting) potential of drugs, chemicals and natural products. Two years and 2000 compounds of experience: A progress report. *ALTEX* **2009**, *26*, 167–176. [CrossRef] [PubMed]

118. Vedani, A.; Dobler, M.; Spreafico, M.; Peristera, O.; Smiesko, M. VirtualToxLab—In silico prediction of the toxic potential of drugs and environmental chemicals: Evaluation status and internet access protocol. *ALTEX* **2007**, *24*, 153–161. [CrossRef]

119. Cheng, F.; Li, W.; Zhou, Y.; Shen, J.; Wu, Z.; Liu, G.; Lee, P.W.; Tang, Y. Correction to "admetSAR: A Comprehensive Source and Free Tool for Assessment of Chemical ADMET Properties". *J. Chem. Inf. Model.* **2019**. [CrossRef]

120. Yang, H.; Lou, C.; Sun, L.; Li, J.; Cai, Y.; Wang, Z.; Li, W.; Liu, G.; Tang, Y. admetSAR 2.0: Web-service for prediction and optimization of chemical ADMET properties. *Bioinformatics* **2019**, *35*, 1067–1069. [CrossRef]

121. Cheng, F.; Li, W.; Zhou, Y.; Shen, J.; Wu, Z.; Liu, G.; Lee, P.W.; Tang, Y. admetSAR: A comprehensive source and free tool for assessment of chemical ADMET properties. *J. Chem. Inf. Model.* **2012**, *52*, 3099–3105. [CrossRef]

122. Rudik, A.; Bezhentsev, V.; Dmitriev, A.; Lagunin, A.; Filimonov, D.; Poroikov, V. Metatox-Web application for generation of metabolic pathways and toxicity estimation. *J. Bioinform. Comput. Biol.* **2019**, *17*, 1940001. [CrossRef] [PubMed]

123. Rudik, A.V.; Bezhentsev, V.M.; Dmitriev, A.V.; Druzhilovskiy, D.S.; Lagunin, A.A.; Filimonov, D.A.; Poroikov, V.V. MetaTox: Web Application for Predicting Structure and Toxicity of Xenobiotics' Metabolites. *J. Chem. Inf. Model.* **2017**, *57*, 638–642. [CrossRef]

124. Cole, S.T.; Brosch, R.; Parkhill, J.; Garnier, T.; Churcher, C.; Harris, D.; Gordon, S.V.; Eiglmeier, K.; Gas, S.; Barry, C.E., 3rd; et al. Deciphering the biology of *Mycobacterium tuberculosis* from the complete genome sequence. *Nature* **1998**, *393*, 537–544. [CrossRef] [PubMed]

125. Rosenkrands, I.; King, A.; Weldingh, K.; Moniatte, M.; Moertz, E.; Andersen, P. Towards the proteome of *Mycobacterium tuberculosis*. *Electrophoresis* **2000**, *21*, 3740–3756. [CrossRef]

126. Jungblut, P.R.; Schaible, U.E.; Mollenkopf, H.J.; Zimny-Arndt, U.; Raupach, B.; Mattow, J.; Halada, P.; Lamer, S.; Hagens, K.; Kaufmann, S.H. Comparative proteome analysis of *Mycobacterium tuberculosis* and Mycobacterium bovis BCG strains: Towards functional genomics of microbial pathogens. *Mol. Microbiol.* **1999**, *33*, 1103–1117. [CrossRef] [PubMed]

127. Kruh, N.A.; Troudt, J.; Izzo, A.; Prenni, J.; Dobos, K.M. Portrait of a pathogen: The *Mycobacterium tuberculosis* proteome in vivo. *PLoS ONE* **2010**, *5*, e13938. [CrossRef]

128. Keren, I.; Minami, S.; Rubin, E.; Lewis, K. Characterization and transcriptome analysis of *Mycobacterium tuberculosis* persisters. *MBio* **2011**, *2*, e00100–e00111. [CrossRef]

129. Rachman, H.; Strong, M.; Ulrichs, T.; Grode, L.; Schuchhardt, J.; Mollenkopf, H.; Kosmiadi, G.A.; Eisenberg, D.; Kaufmann, S.H. Unique transcriptome signature of *Mycobacterium tuberculosis* in pulmonary tuberculosis. *Infect. Immun.* **2006**, *74*, 1233–1242. [CrossRef]

130. Xu, D. Protein databases on the internet. *Curr. Protoc. Protein Sci.* **2012**. [CrossRef]

131. Kim, S.; Thiessen, P.A.; Bolton, E.E.; Chen, J.; Fu, G.; Gindulyte, A.; Han, L.; He, J.; He, S.; Shoemaker, B.A.; et al. PubChem Substance and Compound databases. *Nucleic Acids Res.* **2016**, *44*, D1202–D1213. [CrossRef]

132. Williams, A.J. Public chemical compound databases. *Curr. Opin. Drug Discov. Dev.* **2008**, *11*, 393–404.

133. UniProt, C. Ongoing and future developments at the Universal Protein Resource. *Nucleic Acids Res.* **2011**, *39*, D214–D219.

134. Reddy, T.B.; Riley, R.; Wymore, F.; Montgomery, P.; DeCaprio, D.; Engels, R.; Gellesch, M.; Hubble, J.; Jen, D.; Jin, H.; et al. TB database: An integrated platform for tuberculosis research. *Nucleic Acids Res.* **2009**, *37*, D499–D508. [CrossRef] [PubMed]

135. Galagan, J.E.; Sisk, P.; Stolte, C.; Weiner, B.; Koehrsen, M.; Wymore, F.; Reddy, T.B.; Zucker, J.D.; Engels, R.; Gellesch, M.; et al. TB database 2010: Overview and update. *Tuberculosis* **2010**, *90*, 225–235. [CrossRef] [PubMed]

136. Kapopoulou, A.; Lew, J.M.; Cole, S.T. The MycoBrowser portal: A comprehensive and manually annotated resource for mycobacterial genomes. *Tuberculosis* **2011**, *91*, 8–13. [CrossRef]

137. Rosenthal, A.; Gabrielian, A.; Engle, E.; Hurt, D.E.; Alexandru, S.; Crudu, V.; Sergueev, E.; Kirichenko, V.; Lapitskii, V.; Snezhko, E.; et al. The TB Portals: An Open-Access, Web-Based Platform for Global Drug-Resistant-Tuberculosis Data Sharing and Analysis. *J. Clin. Microbiol.* **2017**, *55*, 3267–3282. [CrossRef] [PubMed]

138. Laskowski, R.A.; Jablonska, J.; Pravda, L.; Varekova, R.S.; Thornton, J.M. PDBsum: Structural summaries of PDB entries. *Protein Sci.* **2018**, *27*, 129–134. [CrossRef] [PubMed]

139. Liu, R.; Li, X.; Lam, K.S. Combinatorial chemistry in drug discovery. *Curr. Opin. Chem. Biol.* **2017**, *38*, 117–126. [CrossRef]

140. Ruddigkeit, L.; van Deursen, R.; Blum, L.C.; Reymond, J.L. Enumeration of 166 billion organic small molecules in the chemical universe database GDB-17. *J. Chem. Inf. Model.* **2012**, *52*, 2864–2875. [CrossRef]

141. Shivanyuk, A.; Ryabukhin, S.; Bogolyubsky, A.V.; Mykytenko, D.M.; Chuprina, A.; Heilman, W.; Kostyuk, A.N.; Tolmachev, A. Enamine real database: Making chemical diversity real. *Chem. Today* **2007**, *25*, 58–59.

142. Williams, A.J. ChemSpider: Integrating Structure-Based Resources Distributed across the Internet. In *Enhancing Learning with Online Resources, Social Networking, and Digital Libraries*; American Chemical Society: Washington, DC, USA, 2010; Volume 1060, pp. 23–39.

143. Sterling, T.; Irwin, J.J. ZINC 15—Ligand Discovery for Everyone. *J. Chem. Inf. Model.* **2015**, *55*, 2324–2337. [CrossRef]

144. Gaulton, A.; Hersey, A.; Nowotka, M.; Bento, A.P.; Chambers, J.; Mendez, D.; Mutowo, P.; Atkinson, F.; Bellis, L.J.; Cibrian-Uhalte, E.; et al. The ChEMBL database in 2017. *Nucleic Acids Res.* **2017**, *45*, D945–D954. [CrossRef]

145. Voigt, J.H.; Bienfait, B.; Wang, S.; Nicklaus, M.C. Comparison of the NCI open database with seven large chemical structural databases. *J. Chem. Inf. Comput. Sci.* **2001**, *41*, 702–712. [CrossRef]

146. Pierleoni, A.; Martelli, P.L.; Fariselli, P.; Casadio, R. eSLDB: Eukaryotic subcellular localization database. *Nucleic Acids Res.* **2007**, *35*, D208–D212. [CrossRef]

147. Sprenger, J.; Lynn Fink, J.; Karunaratne, S.; Hanson, K.; Hamilton, N.A.; Teasdale, R.D. LOCATE: A mammalian protein subcellular localization database. *Nucleic Acids Res.* **2008**, *36*, D230–D233. [CrossRef]

148. Peabody, M.A.; Laird, M.R.; Vlasschaert, C.; Lo, R.; Brinkman, F.S. PSORTdb: Expanding the bacteria and archaea protein subcellular localization database to better reflect diversity in cell envelope structures. *Nucleic Acids Res.* **2016**, *44*, D663–D668. [CrossRef]

149. Hendlich, M.; Bergner, A.; Gunther, J.; Klebe, G. Relibase: Design and development of a database for comprehensive analysis of protein-ligand interactions. *J. Mol. Biol.* **2003**, *326*, 607–620. [CrossRef]

150. Chen, X.; Lin, Y.; Liu, M.; Gilson, M.K. The Binding Database: Data management and interface design. *Bioinformatics* **2002**, *18*, 130–139. [CrossRef]

151. Liu, T.; Lin, Y.; Wen, X.; Jorissen, R.N.; Gilson, M.K. BindingDB: A web-accessible database of experimentally determined protein-ligand binding affinities. *Nucleic Acids Res.* **2007**, *35*, D198–D201. [CrossRef]

152. Salwinski, L.; Miller, C.S.; Smith, A.J.; Pettit, F.K.; Bowie, J.U.; Eisenberg, D. The Database of Interacting Proteins: 2004 update. *Nucleic Acids Res.* **2004**, *32*, D449–D451. [CrossRef]

153. Oughtred, R.; Stark, C.; Breitkreutz, B.J.; Rust, J.; Boucher, L.; Chang, C.; Kolas, N.; O'Donnell, L.; Leung, G.; McAdam, R.; et al. The BioGRID interaction database: 2019 update. *Nucleic Acids Res.* **2019**, *47*, D529–D541. [CrossRef]

154. Jensen, L.J.; Kuhn, M.; Stark, M.; Chaffron, S.; Creevey, C.; Muller, J.; Doerks, T.; Julien, P.; Roth, A.; Simonovic, M.; et al. STRING 8—A global view on proteins and their functional interactions in 630 organisms. *Nucleic Acids Res.* **2009**, *37*, D412–D416. [CrossRef]

155. Franca, T.C. Homology modeling: An important tool for the drug discovery. *J. Biomol. Struct. Dyn.* **2015**, *33*, 1780–1793. [CrossRef]

156. McGinnis, S.; Madden, T.L. BLAST: At the core of a powerful and diverse set of sequence analysis tools. *Nucleic Acids Res.* **2004**, *32*, W20–W25. [CrossRef]

157. Ye, J.; McGinnis, S.; Madden, T.L. BLAST: Improvements for better sequence analysis. *Nucleic Acids Res.* **2006**, *34*, W6–W9. [CrossRef]

158. Madden, T. The BLAST Sequence Analysis Tool. In *The NCBI Handbook [Internet]*, 2nd ed.; National Center for Biotechnology Information (US): Bethesda, MD, USA, 2013.

159. Papadopoulos, J.S.; Agarwala, R. COBALT: Constraint-based alignment tool for multiple protein sequences. *Bioinformatics* **2007**, *23*, 1073–1079. [CrossRef]

160. Sievers, F.; Higgins, D.G. Clustal Omega, Accurate Alignment of Very Large Numbers of Sequences. In *Multiple Sequence Alignment Methods*; Russell, D.J., Ed.; Humana Press: Totowa, NJ, USA, 2014; pp. 105–116.

161. Lassmann, T.; Sonnhammer, E.L. Kalign—An accurate and fast multiple sequence alignment algorithm. *BMC Bioinf.* **2005**, *6*, 298. [CrossRef]

162. Schmidtke, P.; Barril, X. Understanding and predicting druggability. A high-throughput method for detection of drug binding sites. *J. Med. Chem.* **2010**, *53*, 5858–5867. [CrossRef]

163. Kalyaanamoorthy, S.; Chen, Y.P. Structure-based drug design to augment hit discovery. *Drug Discov. Today* **2011**, *16*, 831–839. [CrossRef]

164. Hetenyi, C.; van der Spoel, D. Blind docking of drug-sized compounds to proteins with up to a thousand residues. *FEBS Lett.* **2006**, *580*, 1447–1450. [CrossRef]

165. Shi, Y.; Colombo, C.; Kuttiyatveetil, J.R.; Zalatar, N.; van Straaten, K.E.; Mohan, S.; Sanders, D.A.; Pinto, B.M. A Second, Druggable Binding Site in UDP-Galactopyranose Mutase from *Mycobacterium tuberculosis*? *Chembiochem* **2016**, *17*, 2264–2273. [CrossRef]

166. Stahura, F.L.; Bajorath, J. Virtual screening methods that complement HTS. *Comb. Chem. High Throughput Screen.* **2004**, *7*, 259–269. [CrossRef]

167. Steindl, T.M.; Schuster, D.; Wolber, G.; Laggner, C.; Langer, T. High-throughput structure-based pharmacophore modelling as a basis for successful parallel virtual screening. *J. Comput. Aided Mol. Des.* **2006**, *20*, 703–715. [CrossRef]

168. Halperin, I.; Ma, B.; Wolfson, H.; Nussinov, R. Principles of docking: An overview of search algorithms and a guide to scoring functions. *Proteins* **2002**, *47*, 409–443. [CrossRef] [PubMed]

169. Leach, A.R.; Gillet, V.J.; Lewis, R.A.; Taylor, R. Three-dimensional pharmacophore methods in drug discovery. *J. Med. Chem.* **2010**, *53*, 539–558. [CrossRef] [PubMed]

170. Wermuth, C.G.; Ganellin, C.R.; Lindberg, P.; Mitscher, L.A. Glossary of terms used in medicinal chemistry (IUPAC Recommendations 1998). *Pure Appl. Chem.* **1998**, *70*, 1129. [CrossRef]

171. Hein, M.; Zilian, D.; Sotriffer, C.A. Docking compared to 3D-pharmacophores: The scoring function challenge. *Drug Discov. Today Technol.* **2010**, *7*, e229–e236. [CrossRef]

172. Hessler, G.; Baringhaus, K. The scaffold hopping potential of pharmacophores. *Drug Discov. Today Technol.* **2010**, *7*, e263–e269. [CrossRef]

173. Pagadala, N.S.; Syed, K.; Tuszynski, J. Software for molecular docking: A review. *Biophys. Rev.* **2017**, *9*, 91–102. [CrossRef]

174. Cross, J.B.; Thompson, D.C.; Rai, B.K.; Baber, J.C.; Fan, K.Y.; Hu, Y.; Humblet, C. Comparison of several molecular docking programs: Pose prediction and virtual screening accuracy. *J. Chem. Inf. Model.* **2009**, *49*, 1455–1474. [CrossRef]

175. Cummings, M.D.; DesJarlais, R.L.; Gibbs, A.C.; Mohan, V.; Jaeger, E.P. Comparison of automated docking programs as virtual screening tools. *J. Med. Chem.* **2005**, *48*, 962–976. [CrossRef]

176. Annamala, M.K.; Inampudi, K.K.; Guruprasad, L. Docking of phosphonate and trehalose analog inhibitors into *M. tuberculosis* mycolyltransferase Ag85C: Comparison of the two scoring fitness functions GoldScore and ChemScore, in the GOLD software. *Bioinformation* **2007**, *1*, 339–350. [CrossRef]

177. Perola, E.; Walters, W.P.; Charifson, P.S. A detailed comparison of current docking and scoring methods on systems of pharmaceutical relevance. *Proteins* **2004**, *56*, 235–249. [CrossRef]

178. Xu, W.; Lucke, A.J.; Fairlie, D.P. Comparing sixteen scoring functions for predicting biological activities of ligands for protein targets. *J. Mol. Graph. Model.* **2015**, *57*, 76–88. [CrossRef]

179. Billones, J.B.; Carrillo, M.C.; Organo, V.G.; Sy, J.B.; Clavio, N.A.; Macalino, S.J.; Emnacen, I.A.; Lee, A.P.; Ko, P.K.; Concepcion, G.P. In silico discovery and in vitro activity of inhibitors against *Mycobacterium tuberculosis* 7,8-diaminopelargonic acid synthase (Mtb BioA). *Drug Des. Devel. Ther.* **2017**, *11*, 563–574. [CrossRef]

180. Dassault Systèmes BIOVIA. Discovery Studio. Available online: https://www.3dsbiovia.com/products/collaborative-science/biovia-discovery-studio/ (accessed on 26 October 2019).

181. Huang, S.Y.; Grinter, S.Z.; Zou, X. Scoring functions and their evaluation methods for protein-ligand docking: Recent advances and future directions. *Phys. Chem. Chem. Phys.* **2010**, *12*, 12899–12908. [CrossRef]

182. Ericksen, S.S.; Wu, H.; Zhang, H.; Michael, L.A.; Newton, M.A.; Hoffmann, F.M.; Wildman, S.A. Machine Learning Consensus Scoring Improves Performance Across Targets in Structure-Based Virtual Screening. *J. Chem. Inf. Model.* **2017**, *57*, 1579–1590. [CrossRef]

183. Li, D.D.; Meng, X.F.; Wang, Q.; Yu, P.; Zhao, L.G.; Zhang, Z.P.; Wang, Z.Z.; Xiao, W. Consensus scoring model for the molecular docking study of mTOR kinase inhibitor. *J. Mol. Graph. Model.* **2018**, *79*, 81–87. [CrossRef]

184. Charifson, P.S.; Corkery, J.J.; Murcko, M.A.; Walters, W.P. Consensus scoring: A method for obtaining improved hit rates from docking databases of three-dimensional structures into proteins. *J. Med. Chem.* **1999**, *42*, 5100–5109. [CrossRef]

185. Clark, R.D.; Strizhev, A.; Leonard, J.M.; Blake, J.F.; Matthew, J.B. Consensus scoring for ligand/protein interactions. *J. Mol. Graph. Model.* **2002**, *20*, 281–295. [CrossRef]

186. Harrison, A.J.; Yu, M.; Gardenborg, T.; Middleditch, M.; Ramsay, R.J.; Baker, E.N.; Lott, J.S. The structure of MbtI from *Mycobacterium tuberculosis*, the first enzyme in the biosynthesis of the siderophore mycobactin, reveals it to be a salicylate synthase. *J. Bacteriol.* **2006**, *188*, 6081–6091. [CrossRef]

187. Manos-Turvey, A.; Bulloch, E.M.; Rutledge, P.J.; Baker, E.N.; Lott, J.S.; Payne, R.J. Inhibition studies of *Mycobacterium tuberculosis* salicylate synthase (MbtI). *ChemMedChem* **2010**, *5*, 1067–1079. [CrossRef]

188. Vasan, M.; Neres, J.; Williams, J.; Wilson, D.J.; Teitelbaum, A.M.; Remmel, R.P.; Aldrich, C.C. Inhibitors of the salicylate synthase (MbtI) from *Mycobacterium tuberculosis* discovered by high-throughput screening. *ChemMedChem* **2010**, *5*, 2079–2087. [CrossRef]

189. Pini, E.; Poli, G.; Tuccinardi, T.; Chiarelli, L.R.; Mori, M.; Gelain, A.; Costantino, L.; Villa, S.; Meneghetti, F.; Barlocco, D. New Chromane-Based Derivatives as Inhibitors of *Mycobacterium tuberculosis* Salicylate Synthase (MbtI): Preliminary Biological Evaluation and Molecular Modeling Studies. *Molecules* **2018**, *23*, 1506. [CrossRef]

190. Chiarelli, L.R.; Mori, M.; Barlocco, D.; Beretta, G.; Gelain, A.; Pini, E.; Porcino, M.; Mori, G.; Stelitano, G.; Costantino, L.; et al. Discovery and development of novel salicylate synthase (MbtI) furanic inhibitors as antitubercular agents. *Eur. J. Med. Chem.* **2018**, *155*, 754–763. [CrossRef]

191. McGann, M. FRED pose prediction and virtual screening accuracy. *J. Chem. Inf. Model.* **2011**, *51*, 578–596. [CrossRef]

192. Korb, O.; Stützle, T.; Exner, T.E. PLANTS: Application of Ant Colony Optimization to Structure-Based Drug Design. In *ANTS 2006: Ant Colony Optimization and Swarm Intelligence*; Dorigo, M., Gambardella, L.M., Birattari, M., Martinoli, A., Poli, R., Stützle, T., Eds.; Springer: Heidelberg/Berlin, Germany, 2016; pp. 247–258.

193. Koshland, D.E. Application of a Theory of Enzyme Specificity to Protein Synthesis. *Proc. Natl. Acad. Sci. USA* **1958**, *44*, 98–104. [CrossRef]

194. Sotriffer, C.A. Accounting for induced-fit effects in docking: What is possible and what is not? *Curr. Top. Med. Chem.* **2011**, *11*, 179–191. [CrossRef]

195. Hartkoorn, R.C.; Sala, C.; Neres, J.; Pojer, F.; Magnet, S.; Mukherjee, R.; Uplekar, S.; Boy-Rottger, S.; Altmann, K.H.; Cole, S.T. Towards a new tuberculosis drug: Pyridomycin-nature's isoniazid. *EMBO Mol. Med.* **2012**, *4*, 1032–1042. [CrossRef]

196. Rozwarski, D.A.; Vilcheze, C.; Sugantino, M.; Bittman, R.; Sacchettini, J.C. Crystal structure of the *Mycobacterium tuberculosis* enoyl-ACP reductase, InhA, in complex with NAD+ and a C16 fatty acyl substrate. *J. Biol. Chem.* **1999**, *274*, 15582–15589. [CrossRef]

197. Rozwarski, D.A.; Grant, G.A.; Barton, D.H.; Jacobs, W.R., Jr.; Sacchettini, J.C. Modification of the NADH of the isoniazid target (InhA) from *Mycobacterium tuberculosis*. *Science* **1998**, *279*, 98–102. [CrossRef]

198. Amaro, R.E.; Li, W.W. Emerging methods for ensemble-based virtual screening. *Curr. Top. Med. Chem.* **2010**, *10*, 3–13. [CrossRef]

199. Brindha, S.; Sundaramurthi, J.C.; Velmurugan, D.; Vincent, S.; Gnanadoss, J.J. Docking-based virtual screening of known drugs against murE of *Mycobacterium tuberculosis* towards repurposing for TB. *Bioinformation* **2016**, *12*, 359–367. [CrossRef]

200. Wishart, D.S.; Knox, C.; Guo, A.C.; Cheng, D.; Shrivastava, S.; Tzur, D.; Gautam, B.; Hassanali, M. DrugBank: A knowledgebase for drugs, drug actions and drug targets. *Nucleic Acids Res.* **2008**, *36*, D901–D906. [CrossRef]

201. Schmidt, M.F.; Korb, O.; Howard, N.I.; Dias, M.V.B.; Blundell, T.L.; Abell, C. Discovery of Schaeffer's Acid Analogues as Lead Structures of *Mycobacterium tuberculosis* Type II Dehydroquinase Using a Rational Drug Design Approach. *ChemMedChem* **2013**, *8*, 54–58. [CrossRef]

202. Lovell, S.C.; Word, J.M.; Richardson, J.S.; Richardson, D.C. The penultimate rotamer library. *Proteins* **2000**, *40*, 389–408. [CrossRef]

203. Korb, O.; Stutzle, T.; Exner, T.E. Empirical scoring functions for advanced protein-ligand docking with PLANTS. *J. Chem. Inf. Model.* **2009**, *49*, 84–96. [CrossRef]

204. Bhabha, G.; Biel, J.T.; Fraser, J.S. Keep on moving: Discovering and perturbing the conformational dynamics of enzymes. *Acc. Chem. Res.* **2015**, *48*, 423–430. [CrossRef]

205. Goh, C.S.; Milburn, D.; Gerstein, M. Conformational changes associated with protein-protein interactions. *Curr. Opin. Struct. Biol.* **2004**, *14*, 104–109. [CrossRef]

206. Hospital, A.; Goni, J.R.; Orozco, M.; Gelpi, J.L. Molecular dynamics simulations: Advances and applications. *Adv. Appl. Bioinform. Chem.* **2015**, *8*, 37–47.

207. Lee, Y.; Jeong, L.S.; Choi, S.; Hyeon, C. Link between allosteric signal transduction and functional dynamics in a multisubunit enzyme: S-adenosylhomocysteine hydrolase. *J. Am. Chem. Soc.* **2011**, *133*, 19807–19815. [CrossRef]

208. McCammon, J.A.; Gelin, B.R.; Karplus, M. Dynamics of folded proteins. *Nature* **1977**, *267*, 585–590. [CrossRef]

209. Larsson, P.; Hess, B.; Lindahl, E. Algorithm improvements for molecular dynamics simulations. *WIREs Comput. Mol. Sci.* **2011**, *1*, 93–108. [CrossRef]

210. Orozco, M.; Orellana, L.; Hospital, A.; Naganathan, A.N.; Emperador, A.; Carrillo, O.; Gelpi, J.L. Coarse-grained representation of protein flexibility. Foundations, successes, and shortcomings. *Adv. Protein Chem. Struct. Biol.* **2011**, *85*, 183–215.

211. Linge, J.P.; Williams, M.A.; Spronk, C.A.; Bonvin, A.M.; Nilges, M. Refinement of protein structures in explicit solvent. *Proteins* **2003**, *50*, 496–506. [CrossRef]

212. Anandakrishnan, R.; Drozdetski, A.; Walker, R.C.; Onufriev, A.V. Speed of conformational change: Comparing explicit and implicit solvent molecular dynamics simulations. *Biophys. J.* **2015**, *108*, 1153–1164. [CrossRef]

213. MacKerell, A.D.; Bashford, D.; Bellott, M.; Dunbrack, R.L.; Evanseck, J.D.; Field, M.J.; Fischer, S.; Gao, J.; Guo, H.; Ha, S.; et al. All-atom empirical potential for molecular modeling and dynamics studies of proteins. *J. Phys. Chem. B* **1998**, *102*, 3586–3616. [CrossRef]

214. Cornell, W.D.; Cieplak, P.; Bayly, C.I.; Gould, I.R.; Merz, K.M.; Ferguson, D.M.; Spellmeyer, D.C.; Fox, T.; Caldwell, J.W.; Kollman, P.A. A Second Generation Force Field for the Simulation of Proteins, Nucleic Acids, and Organic Molecules. *J. Am. Chem. Soc.* **1995**, *117*, 5179–5197. [CrossRef]

215. Oostenbrink, C.; Villa, A.; Mark, A.E.; van Gunsteren, W.F. A biomolecular force field based on the free enthalpy of hydration and solvation: The GROMOS force-field parameter sets 53A5 and 53A6. *J. Comput. Chem.* **2004**, *25*, 1656–1676. [CrossRef]

216. Jorgensen, W.L.; Maxwell, D.S.; Tirado-Rives, J. Development and Testing of the OPLS All-Atom Force Field on Conformational Energetics and Properties of Organic Liquids. *J. Am. Chem. Soc.* **1996**, *118*, 11225–11236. [CrossRef]

217. Kaminski, G.A.; Friesner, R.A.; Tirado-Rives, J.; Jorgensen, W.L. Evaluation and Reparametrization of the OPLS-AA Force Field for Proteins via Comparison with Accurate Quantum Chemical Calculations on Peptides†. *J. Phys. Chem. B* **2001**, *105*, 6474–6487. [CrossRef]

218. Daggett, V.; Levitt, M. Protein Unfolding Pathways Explored Through Molecular Dynamics Simulations. *J. Mol. Biol.* **1993**, *232*, 600–619. [CrossRef]

219. Alonso, H.; Bliznyuk, A.A.; Gready, J.E. Combining docking and molecular dynamic simulations in drug design. *Med. Res. Rev.* **2006**, *26*, 531–568. [CrossRef]

220. Papaleo, E. Integrating atomistic molecular dynamics simulations, experiments, and network analysis to study protein dynamics: Strength in unity. *Front. Mol. Biosci.* **2015**, *2*, 28. [CrossRef]

221. Prada-Gracia, D.; Gomez-Gardenes, J.; Echenique, P.; Falo, F. Exploring the free energy landscape: From dynamics to networks and back. *PLoS Comput. Biol.* **2009**, *5*, e1000415. [CrossRef]

222. Wahab, H.A.; Choong, Y.S.; Ibrahim, P.; Sadikun, A.; Scior, T. Elucidating isoniazid resistance using molecular modeling. *J. Chem. Inf. Model.* **2009**, *49*, 97–107. [CrossRef]

223. Schroeder, E.K.; Basso, L.A.; Santos, D.S.; de Souza, O.N. Molecular dynamics simulation studies of the wild-type, I21V, and I16T mutants of isoniazid-resistant *Mycobacterium tuberculosis* enoyl reductase (InhA) in complex with NADH: Toward the understanding of NADH-InhA different affinities. *Biophys. J.* **2005**, *89*, 876–884. [CrossRef]

224. Cruz, J.N.; Costa, J.F.S.; Khayat, A.S.; Kuca, K.; Barros, C.A.L.; Neto, A. Molecular dynamics simulation and binding free energy studies of novel leads belonging to the benzofuran class inhibitors of *Mycobacterium tuberculosis* Polyketide Synthase 13. *J. Biomol. Struct. Dyn.* **2019**, *37*, 1616–1627. [CrossRef]

225. Aggarwal, A.; Parai, M.K.; Shetty, N.; Wallis, D.; Woolhiser, L.; Hastings, C.; Dutta, N.K.; Galaviz, S.; Dhakal, R.C.; Shrestha, R.; et al. Development of a Novel Lead that Targets, *M. tuberculosis* Polyketide Synthase 13. *Cell* **2017**, *170*, 249–259. [CrossRef]

226. Nikolova, N.; Jaworska, J. Approaches to Measure Chemical Similarity—A Review. *QSAR Comb. Sci.* **2003**, *22*, 1006–1026. [CrossRef]

227. Johnson, M.A.; Maggiora, G.M. American Chemical Society. In *Concepts and Applications of Molecular Similarity*; Wiley: New York, NY, USA, 1990.

228. Bacilieri, M.; Moro, S. Ligand-based drug design methodologies in drug discovery process: An overview. *Curr. Drug Discov. Technol.* **2006**, *3*, 155–165. [CrossRef]

229. Sukumar, N.; Das, S. Current trends in virtual high throughput screening using ligand-based and structure-based methods. *Comb. Chem. High Throughput Screen.* **2011**, *14*, 872–888. [CrossRef]

230. Cereto-Massague, A.; Ojeda, M.J.; Valls, C.; Mulero, M.; Garcia-Vallve, S.; Pujadas, G. Molecular fingerprint similarity search in virtual screening. *Methods* **2015**, *71*, 58–63. [CrossRef]

231. Bajusz, D.; Racz, A.; Heberger, K. Why is Tanimoto index an appropriate choice for fingerprint-based similarity calculations? *J. Cheminform.* **2015**, *7*, 20. [CrossRef]

232. Faulon, J.-L.; Bender, A. *Handbook of Chemoinformatics Algorithms*; Chapman & Hall/CRC: Boca Raton, FL, USA, 2010; 440p.

233. Haranczyk, M.; Holliday, J. Comparison of similarity coefficients for clustering and compound selection. *J. Chem. Inf. Model.* **2008**, *48*, 498–508. [CrossRef]

234. Al Khalifa, A.; Haranczyk, M.; Holliday, J. Comparison of nonbinary similarity coefficients for similarity searching, clustering and compound selection. *J. Chem. Inf. Model.* **2009**, *49*, 1193–1201. [CrossRef]

235. Ginn, C.M.R.; Willett, P.; Bradshaw, J. Combination of molecular similarity measures using data fusion. *Perspect. Drug Discov. Des.* **2000**, *20*, 1–16. [CrossRef]

236. Medina-Franco, J.L.; Maggiora, G.M.; Giulianotti, M.A.; Pinilla, C.; Houghten, R.A. A similarity-based data-fusion approach to the visual characterization and comparison of compound databases. *Chem. Biol. Drug Des.* **2007**, *70*, 393–412. [CrossRef]

237. Hu, G.; Kuang, G.; Xiao, W.; Li, W.; Liu, G.; Tang, Y. Performance evaluation of 2D fingerprint and 3D shape similarity methods in virtual screening. *J. Chem. Inf. Model.* **2012**, *52*, 1103–1113. [CrossRef]

238. Drwal, M.N.; Griffith, R. Combination of ligand- and structure-based methods in virtual screening. *Drug Discov. Today Technol.* **2013**, *10*, e395–e401. [CrossRef]

239. Hu, Y.; Bajorath, J. Extending the activity cliff concept: Structural categorization of activity cliffs and systematic identification of different types of cliffs in the ChEMBL database. *J. Chem. Inf. Model.* **2012**, *52*, 1806–1811. [CrossRef]

240. Hu, Y.; Stumpfe, D.; Bajorath, J. Advancing the activity cliff concept. *F1000Res* **2013**, *2*, 199. [CrossRef]

241. Stumpfe, D.; de la Vega de Leon, A.; Dimova, D.; Bajorath, J. Advancing the activity cliff concept, part II. *F1000Res* **2014**, *3*, 75. [CrossRef]

242. Verma, J.; Khedkar, V.M.; Coutinho, E.C. 3D-QSAR in drug design—A review. *Curr. Top. Med. Chem.* **2010**, *10*, 95–115. [CrossRef]

243. Exploring QSAR. *Environ. Sci. Technol.* **1995**, *29*, 444A. [CrossRef]

244. Bostrom, J.; Norrby, P.O.; Liljefors, T. Conformational energy penalties of protein-bound ligands. *J. Comput. Aided Mol. Des.* **1998**, *12*, 383–396. [CrossRef]

245. Perola, E.; Charifson, P.S. Conformational analysis of drug-like molecules bound to proteins: An extensive study of ligand reorganization upon binding. *J. Med. Chem.* **2004**, *47*, 2499–2510. [CrossRef]

246. Melo-Filho, C.C.; Braga, R.C.; Andrade, C.H. 3D-QSAR approaches in drug design: Perspectives to generate reliable CoMFA models. *Curr. Comput. Aided Drug Des.* **2014**, *10*, 148–159. [CrossRef]

247. Cherkasov, A.; Muratov, E.N.; Fourches, D.; Varnek, A.; Baskin, II; Cronin, M.; Dearden, J.; Gramatica, P.; Martin, Y.C.; Todeschini, R.; et al. QSAR modeling: Where have you been? Where are you going to? *J. Med. Chem.* **2014**, *57*, 4977–5010. [CrossRef]

248. Cramer, R.D.; Patterson, D.E.; Bunce, J.D. Comparative molecular field analysis (CoMFA). 1. Effect of shape on binding of steroids to carrier proteins. *J. Am. Chem. Soc.* **1988**, *110*, 5959–5967. [CrossRef]

249. Klebe, G.; Abraham, U.; Mietzner, T. Molecular similarity indices in a comparative analysis (CoMSIA) of drug molecules to correlate and predict their biological activity. *J. Med. Chem.* **1994**, *37*, 4130–4146. [CrossRef]

250. Bajpai, A.; Agarwal, N.; Gupta, S.P. A comparative 2D QSAR study on a series of hydroxamic acid-based histone deacetylase inhibitors vis-a-vis comparative molecular field analysis (CoMFA) and comparative molecular similarity indices analysis (CoMSIA). *Indian J. Biochem. Biophys.* **2014**, *51*, 244–252.

251. Chhatbar, D.M.; Chaube, U.J.; Vyas, V.K.; Bhatt, H.G. CoMFA, CoMSIA, Topomer CoMFA, HQSAR, molecular docking and molecular dynamics simulations study of triazine morpholino derivatives as mTOR inhibitors for the treatment of breast cancer. *Comput. Biol. Chem.* **2019**, *80*, 351–363. [CrossRef]

252. Singh, S.; Supuran, C.T. 3D-QSAR CoMFA studies on sulfonamide inhibitors of the Rv3588c beta-carbonic anhydrase from *Mycobacterium tuberculosis* and design of not yet synthesized new molecules. *J. Enzym. Inhib. Med. Chem.* **2014**, *29*, 449–455. [CrossRef]

253. Punkvang, A.; Hannongbua, S.; Saparpakorn, P.; Pungpo, P. Insight into the structural requirements of aminopyrimidine derivatives for good potency against both purified enzyme and whole cells of *M. tuberculosis*: Combination of HQSAR, CoMSIA, and MD simulation studies. *J. Biomol. Struct. Dyn.* **2016**, *34*, 1079–1091. [CrossRef] [PubMed]

254. Schuster, D. 3D pharmacophores as tools for activity profiling. *Drug Discov. Today Technol.* **2010**, *7*, e205–e211. [CrossRef] [PubMed]

255. Tawari, N.R.; Degani, M.S. Predictive models for nucleoside bisubstrate analogs as inhibitors of siderophore biosynthesis in *Mycobacterium tuberculosis*: Pharmacophore mapping and chemometric QSAR study. *Mol. Divers.* **2011**, *15*, 435–444. [CrossRef]

256. Hohenberg, P.; Kohn, W. Inhomogeneous Electron Gas. *Phys. Rev.* **1964**, *136*, B864–B871. [CrossRef]

257. Kohn, W.; Sham, L.J. Self-Consistent Equations Including Exchange and Correlation Effects. *Phys. Rev.* **1965**, *140*, A1133–A1138. [CrossRef]

258. Sharma, S. *Molecular Dynamics Simulation of Nanocomposites Using BIOVIA Materials Studio, Lammps and Gromacs*, 1st ed.; Elsevier: Waltham, MA, USA, 2019.

259. Fiolhais, C.; Nogueira, F.; Marques, M.A.L. *A Primer in Density Functional Theory*; Springer: Berlin, Germany; New York, NY, USA, 2003.

260. Becke, A.D. Perspective: Fifty years of density-functional theory in chemical physics. *J. Chem. Phys.* **2014**, *140*, 18A301. [CrossRef]

261. Rabi, S.; Patel, A.H.G.; Burger, S.K.; Verstraelen, T.; Ayers, P.W. Exploring the substrate selectivity of human sEH and *M. tuberculosis* EHB using QM/MM. *Struct. Chem.* **2017**, *28*, 1501–1511. [CrossRef]

262. Ramalho, T.C.; Caetano, M.S.; Josa, D.; Luz, G.P.; Freitas, E.A.; da Cunha, E.F. Molecular modeling of *Mycobacterium tuberculosis* dUTpase: Docking and catalytic mechanism studies. *J. Biomol. Struct. Dyn.* **2011**, *28*, 907–917. [CrossRef]

263. Oliveira, C.G.; da, S.M.P.I.; Souza, P.C.; Pavan, F.R.; Leite, C.Q.; Viana, R.B.; Batista, A.A.; Nascimento, O.R.; Deflon, V.M. Manganese(II) complexes with thiosemicarbazones as potential anti-*Mycobacterium tuberculosis* agents. *J. Inorg. Biochem.* **2014**, *132*, 21–29. [CrossRef]

264. Chi, G.; Manos-Turvey, A.; O'Connor, P.D.; Johnston, J.M.; Evans, G.L.; Baker, E.N.; Payne, R.J.; Lott, J.S.; Bulloch, E.M. Implications of binding mode and active site flexibility for inhibitor potency against the salicylate synthase from *Mycobacterium tuberculosis*. *Biochemistry* **2012**, *51*, 4868–4879. [CrossRef]

265. Frisch, M.J.; Trucks, G.W.; Schlegel, H.B.; Scuseria, G.E.; Robb, M.A.; Cheeseman, J.R.; Scalmani, G.; Barone, V.; Mennucci, B.; Petersson, G.A.; et al. *Gaussian 09, Revision E.01*; Gaussian, Inc.: Wallingford, CT, USA, 2009.

266. Stephens, P.J.; Devlin, F.J.; Chabalowski, C.F.; Frisch, M.J. Ab Initio Calculation of Vibrational Absorption and Circular Dichroism Spectra Using Density Functional Force Fields. *J. Phys. Chem.* **1994**, *98*, 11623–11627. [CrossRef]

267. Becke, A.D. Density-functional thermochemistry. III. The role of exact exchange. *J. Chem. Phys.* **1993**, *98*, 5648–5652. [CrossRef]

268. Indarto, A. *Theoretical Modelling and Mechanistic Study of the Formation and Atmospheric Transformations of Polycyclic Aromatic Compounds and Carbonaceous Particles*; Universal Publishers: Irvine, CA, USA, 2010.

269. Hamada, I. van der Waals density functional made accurate. *Phys. Rev. B* **2014**, *89*, 121103. [CrossRef]

270. Berland, K.; Cooper, V.R.; Lee, K.; Schroder, E.; Thonhauser, T.; Hyldgaard, P.; Lundqvist, B.I. van der Waals forces in density functional theory: A review of the vdW-DF method. *Rep. Prog. Phys.* **2015**, *78*, 066501. [CrossRef]

271. Grimme, S. Accurate description of van der Waals complexes by density functional theory including empirical corrections. *J. Comput. Chem.* **2004**, *25*, 1463–1473. [CrossRef]

272. Cohen, A.J.; Mori-Sanchez, P.; Yang, W. Insights into current limitations of density functional theory. *Science* **2008**, *321*, 792–794. [CrossRef]

273. Wilson, G.L.; Lill, M.A. Integrating structure-based and ligand-based approaches for computational drug design. *Future Med. Chem.* **2011**, *3*, 735–750. [CrossRef]

274. Polgar, T.; Keseru, G.M. Integration of virtual and high throughput screening in lead discovery settings. *Comb. Chem. High Throughput Screen.* **2011**, *14*, 889–897. [CrossRef]

275. Tanrikulu, Y.; Kruger, B.; Proschak, E. The holistic integration of virtual screening in drug discovery. *Drug Discov. Today* **2013**, *18*, 358–364. [CrossRef]

276. Tan, L.; Geppert, H.; Sisay, M.T.; Gutschow, M.; Bajorath, J. Integrating structure- and ligand-based virtual screening: Comparison of individual, parallel, and fused molecular docking and similarity search calculations on multiple targets. *ChemMedChem* **2008**, *3*, 1566–1571. [CrossRef]

277. Huang, S.Y.; Li, M.; Wang, J.; Pan, Y. HybridDock: A Hybrid Protein-Ligand Docking Protocol Integrating Protein- and Ligand-Based Approaches. *J. Chem. Inf. Model.* **2016**, *56*, 1078–1087. [CrossRef]

278. Lam, P.C.; Abagyan, R.; Totrov, M. Ligand-biased ensemble receptor docking (LigBEnD): A hybrid ligand/receptor structure-based approach. *J. Comput. Aided Mol. Des.* **2018**, *32*, 187–198. [CrossRef]

279. Mestres, J.; Knegtel, R.M.A. Similarity versus docking in 3D virtual screening. *Perspect. Drug Discov. Des.* **2000**, *20*, 191–207. [CrossRef]

280. Kruger, D.M.; Evers, A. Comparison of structure- and ligand-based virtual screening protocols considering hit list complementarity and enrichment factors. *ChemMedChem* **2010**, *5*, 148–158. [CrossRef]

281. Billones, J.B.; Carrillo, M.C.; Organo, V.G.; Macalino, S.J.; Sy, J.B.; Emnacen, I.A.; Clavio, N.A.; Concepcion, G.P. Toward antituberculosis drugs: In silico screening of synthetic compounds against *Mycobacterium tuberculosis*l,d-transpeptidase 2. *Drug Des. Devel Ther.* **2016**, *10*, 1147–1157. [CrossRef]

282. Fakhar, Z.; Govender, T.; Maguire, G.E.M.; Lamichhane, G.; Walker, R.C.; Kruger, H.G.; Honarparvar, B. Differential flap dynamics in l,d-transpeptidase2 from *Mycobacterium tuberculosis* revealed by molecular dynamics. *Mol. Biosyst.* **2017**, *13*, 1223–1234. [CrossRef]

283. Sandhu, P.; Akhter, Y. The drug binding sites and transport mechanism of the RND pumps from *Mycobacterium tuberculosis*: Insights from molecular dynamics simulations. *Arch. Biochem. Biophys.* **2016**, *592*, 38–49. [CrossRef]

284. Shah, P.; Mistry, J.; Reche, P.A.; Gatherer, D.; Flower, D.R. In silico design of *Mycobacterium tuberculosis* epitope ensemble vaccines. *Mol. Immunol.* **2018**, *97*, 56–62. [CrossRef]

285. Li, D.; Chi, B.; Wang, W.-W.; Gao, J.-M.; Wan, J. Exploring the possible binding mode of trisubstituted benzimidazoles analogues in silico for novel drug designtargeting Mtb FtsZ. *Med. Chem. Res.* **2017**, *26*, 153–169. [CrossRef]

286. Spitzer, R.; Jain, A.N. Surflex-Dock: Docking benchmarks and real-world application. *J. Comput. Aided Mol. Des.* **2012**, *26*, 687–699. [CrossRef]

287. Villoutreix, B.O.; Eudes, R.; Miteva, M.A. Structure-based virtual ligand screening: Recent success stories. *Comb. Chem. High Throughput Screen.* **2009**, *12*, 1000–1016. [CrossRef]

288. Talele, T.T.; Khedkar, S.A.; Rigby, A.C. Successful applications of computer aided drug discovery: Moving drugs from concept to the clinic. *Curr. Top. Med. Chem.* **2010**, *10*, 127–141. [CrossRef]

289. Clark, D.E. What has virtual screening ever done for drug discovery? *Expert Opin. Drug Discov.* **2008**, *3*, 841–851. [CrossRef]

290. Scior, T.; Bender, A.; Tresadern, G.; Medina-Franco, J.L.; Martinez-Mayorga, K.; Langer, T.; Cuanalo-Contreras, K.; Agrafiotis, D.K. Recognizing pitfalls in virtual screening: A critical review. *J. Chem. Inf. Model.* **2012**, *52*, 867–881. [CrossRef]

291. Baig, M.H.; Ahmad, K.; Roy, S.; Ashraf, J.M.; Adil, M.; Siddiqui, M.H.; Khan, S.; Kamal, M.A.; Provaznik, I.; Choi, I. Computer Aided Drug Design: Success and Limitations. *Curr. Pharm. Des.* **2016**, *22*, 572–581. [CrossRef]

292. Coupez, B.; Lewis, R.A. Docking and scoring—Theoretically easy, practically impossible? *Curr. Med. Chem.* **2006**, *13*, 2995–3003.

293. Geppert, H.; Vogt, M.; Bajorath, J. Current trends in ligand-based virtual screening: Molecular representations, data mining methods, new application areas, and performance evaluation. *J. Chem. Inf. Model.* **2010**, *50*, 205–216. [CrossRef]

294. Jain, A.N.; Nicholls, A. Recommendations for evaluation of computational methods. *J. Comput. Aided Mol. Des.* **2008**, *22*, 133–139. [CrossRef]

295. Lindorff-Larsen, K.; Maragakis, P.; Piana, S.; Shaw, D.E. Picosecond to Millisecond Structural Dynamics in Human Ubiquitin. *J. Phys. Chem. B* **2016**, *120*, 8313–8320. [CrossRef] [PubMed]

296. Noe, F. Beating the millisecond barrier in molecular dynamics simulations. *Biophys. J.* **2015**, *108*, 228–229. [CrossRef] [PubMed]

297. Shi, J.; Nobrega, R.P.; Schwantes, C.; Kathuria, S.V.; Bilsel, O.; Matthews, C.R.; Lane, T.J.; Pande, V.S. Atomistic structural ensemble refinement reveals non-native structure stabilizes a sub-millisecond folding intermediate of CheY. *Sci. Rep.* **2017**, *7*, 44116. [CrossRef]

298. Fujita, T. Recent Success Stories Leading to Commercializable Bioactive Compounds with the Aid of Traditional QSAR Procedures. *QSAR* **1997**, *16*, 107–112. [CrossRef]

299. Gao, Q.; Yang, L.; Zhu, Y. Pharmacophore based drug design approach as a practical process in drug discovery. *Curr. Comput. Aided Drug Des.* **2010**, *6*, 37–49. [CrossRef]

300. Sardari, S.; Dezfulian, M. Cheminformatics in anti-infective agents discovery. *Mini Rev. Med. Chem.* **2007**, *7*, 181–189. [CrossRef]

301. Topol, E.J. High-performance medicine: The convergence of human and artificial intelligence. *Nat. Med.* **2019**, *25*, 44–56. [CrossRef]

Role of Resultant Dipole Moment in Mechanical Dissociation of Biological Complexes

Maksim Kouza [1,2,*] (iD), **Anirban Banerji** [2,†], **Andrzej Kolinski** [1], **Irina Buhimschi** [3,4] (iD) and **Andrzej Kloczkowski** [2,4]

[1]   Faculty of Chemistry, University of Warsaw, Pasteura 1, 02-093 Warsaw, Poland; kolinski@chem.uw.edu.pl

[2]   Battelle Center for Mathematical Medicine, Nationwide Children's Hospital, Columbus, OH 43215, USA; andrzej.kloczkowski@nationwidechildrens.org

[3]   Center for Perinatal Research, Research Institute at Nationwide Children's Hospital, Columbus, OH 43215, USA; Irina.Buhimschi@nationwidechildrens.org

[4]   Department of Pediatrics, The Ohio State University College of Medicine, Columbus, OH 43215, USA

*   Correspondence: mkouza@chem.uw.edu.pl

†   Deceased 12 August 2015.

Abstract: Protein-peptide interactions play essential roles in many cellular processes and their structural characterization is the major focus of current experimental and theoretical research. Two decades ago, it was proposed to employ the steered molecular dynamics (SMD) to assess the strength of protein-peptide interactions. The idea behind using SMD simulations is that the mechanical stability can be used as a promising and an efficient alternative to computationally highly demanding estimation of binding affinity. However, mechanical stability defined as a peak in force-extension profile depends on the choice of the pulling direction. Here we propose an uncommon choice of the pulling direction along resultant dipole moment (RDM) vector, which has not been explored in SMD simulations so far. Using explicit solvent all-atom MD simulations, we apply SMD technique to probe mechanical resistance of ligand-receptor system pulled along two different vectors. A novel pulling direction—when ligand unbinds along the RDM vector—results in stronger forces compared to commonly used ligand unbinding along center of masses vector. Our observation that RDM is one of the factors influencing the mechanical stability of protein-peptide complex can be used to improve the ranking of binding affinities by using mechanical stability as an effective scoring function.

Keywords: steered molecular dynamics; all-atom molecular dynamics simulation; resultant dipole moment; mechanical stability; protein-peptide interactions

1. Introduction

Discovery of a new effective drug is a costly and time-consuming process. Billions of US dollars and years in research are spent to place an approved drug on the market. The cost of success is very high due to the fact that many drug candidates fail. One of the possibilities to reduce costs and improve efficiency in current drug discovery processes is to use computer-aided drug design. With the help of molecular modeling, one can predict the success of a potential new drug based on its ability to bind strongly to the target. One of the most popular computational approaches to estimate binding energy is molecular docking simulation by AutoDock [1], whereby the bound conformation of ligand-receptor complex is predicted followed by binding affinity estimation. AutoDock tool can be used for high-throughput virtual drug screening involving thousands to millions of drug candidates. However, it is worth noting that its high performance comes at the cost of accuracy. Limitations of AutoDock and other similar software packages that neglect entropic and solvation effects as well as dynamics properties of the receptor lead to lower accuracy compared to more sophisticated methods such as exact free energy perturbation calculations [2] and molecular mechanics Poisson–Boltzmann surface area (MMPBSA) approach [3]. The first method archived unprecedented level of accuracy establishing an astonishing agreement between experimental and computationally predicted values of binding affinities [2]. The later approach is an efficient method for the estimation of relative binding affinity for diverse biomolecular systems in reasonable time, however at present applicability of both methods for screening large compounds libraries is limited. Fast and simple methods based on a single or a minimal set of biomolecular structural features, which will be able to reveal latent details in quantitative terms about the strength of protein-peptide complex in a consistent and general manner, are still lacking. Consequently, further development of effective protein-peptide docking techniques [4–7] and finding an efficient alternative to binding affinity [8–15] have been a major focus of computational studies in recent years.

Recently, steered molecular dynamics (SMD) simulations have become popular to measure mechanical stability which could be used to assess the strength of the molecular interactions. The SMD approach was shown to be an efficient alternative to conventional MMPBSA method, but it can be few orders of magnitude faster [9], which enables screening of a correspondingly larger number of compounds. Such gain in performance is possible due to extreme conditions used in SMD simulations, e.g., the pulling speed in simulation is several orders of magnitude higher than that used in single molecule force spectroscopy experiments. Recent studies claim that mechanical unfolding pathways of some proteins are insensitive to pulling forces and speeds if all-atom explicit solvent simulations are employed [16–18]. Therefore, it is reasonable to assume that the mechanical stability measured as a force required to unbind a ligand from the receptor corresponds to the strength of interactions. In other words, mechanical stability computed in explicit solvent all-atom SMD simulations could be efficiently used to assess the strength of molecular interactions much faster than conventional methods like MMPBSA.

SMD simulations which mimic the Atomic Force Microscopy (AFM) experiment have been successfully used to study many processes including protein unfolding [19], enzyme-inhibitor unbinding [9] and disaggregation of beta-amyloid oligomers [20]. In our previous paper, we demonstrated that kinetic stability of the fibril state can be accessed via mechanical stability extracted from SMD simulation in such a way that the higher mechanical stability or kinetic stability the faster fibril formation [21]. A common strategy in SMD simulations applied to single molecules is to pull a protein by force ramped linearly with a time and monitor the mechanical stability as a function of the end-to-end displacement (or time). More than two decades ago, SMD simulations were utilized to measure the interaction strength of the streptavidin-biotin complex. The idea behind using SMD simulations is that the mechanical stability or rupture force required to unbind peptide from the receptor corresponds to the strength of the interactions or in other words peptide mechanical stability is proportional to its binding energy. The ligand was pulled along the vector connecting center of masses (COM) of receptor and ligand. Primarily due to easy implementation of COM's pulling, this direction has become a widely accepted option in MD studies of ligand unbinding. However, it should be pointed out that pulling in the direction connecting the COMs of the protein-peptide system does not necessary align the force vector with it. In this work, we attempt to identify the most prominent non-bonded interaction-based force which may act as a crucial determinant that influences the stability of the protein-peptide complex.

Recently it has been shown [22] that any protein molecule in solution can be represented by a set of polarizable dipoles embedded in a dielectric medium of solvent molecules. Taking a clue from this study, we investigated the role of the resultant dipole moment vector emerging out of the local stretch of protein backbone. In contrast to COMs pulling, the electrostatic force emerging out of the resultant dipole moment ensures the stability of any protein or biological complex. The resultant vector of the peptide-dipoles characterizing the local stretch of protein backbone in the peptide binding site may act as an important determinant of the mechanical strength of protein-peptide complex, especially because, the side-chain dipole moments may either neutralize itself, or, may become neutralized by the innumerable non-bonded interactions which dominate the interactional space involving disordered regions.

In this paper, we investigate the effect of the novel pulling direction on the mechanical stability of ligand-receptor complex using solvent all-atom SMD simulations. As follows from the studies of mechanical unfolding of proteins, the rupture force (or unfolding time in constant force experiments) is sensitive to the pulling direction [23,24]. To our best knowledge, the idea of the pulling ligand from the receptor along resultant dipole moment vector has not been previously explored. For the calmodulin N-lobe bound with ER alpha peptide complex (we will refer to it simply as 2LLO in the rest of the paper) studied here we show that pulling along RDM vector results in stronger forces compared to pulling along COMs vector. We conclude that resultant dipole moment is an important factor influencing the mechanical stability of biological complexes. This can be used to improve the ranking of binding affinities by using mechanical stability or its derivatives as effective scoring functions.

2. Results and Discussion

2.1. Assessing the Mechanical Stability of 2LLO Peptide-Protein Complex Using Steered Molecular Dynamics

The recent advancement of single-molecule force-spectroscopy (SMFS) techniques has allowed us to detect forces in the pico-newton range [25–28]. As a force necessary to unfold protein is in the order of piconewtons, SMFS techniques have become not only one of the most widely applied to study the mechanical unfolding and refolding of biomolecules [25–31] but also a powerful tool to probe the binding of ligand to receptor [9,10,12,32]. One of the strategies used in SMFS is to pull a protein by force ramped linearly with time, while monitoring the mechanical resistance as a function of distance between protein ends. The resulting force is computed for each time step to generate a force-extension profile, which has a peak(s) corresponding to the most mechanically stable region(s) in the protein. A typical force-extension profile obtained by constant velocity stretching experiments for a multi-domain construct of the I27 domain is shown in Figure 1a. Each peak of ~200 pN in the force-extension profile arises due to sequential unfolding of the individual domains [25]. This remarkable finding was subsequently reproduced by all-atom SMD simulations developed to mimic SMFS experiments [19].

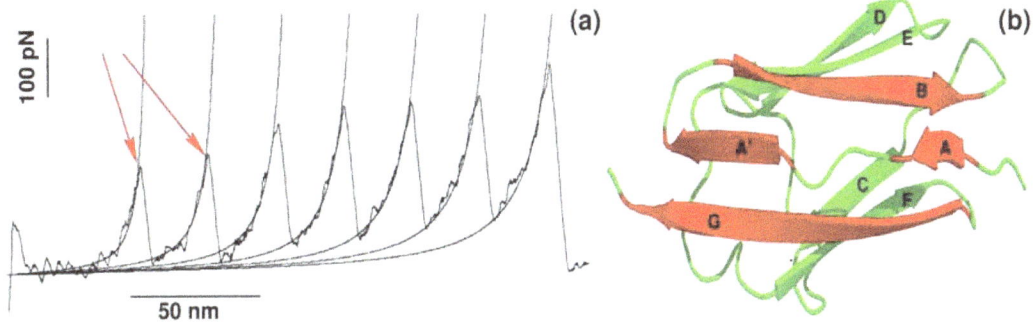

Figure 1. (**a**) Dependence of the force as a function of extension for stretching of multi-domain protein titin. The peaks correspond to the unfolding of individual domains with maximum resisting force to stretching, F_{max}. Figure adopted from Ref. [25]; (**b**) The 3D structure of titin (Brookhaven PDB databank; PDB ID 1TIT). Titin has eight β-strands: A (4–8), A′ (11–15), B (18–25), C (32–36), D (47–52), E (55–61), F (69–75), G (78–88). Each peak in force-extension profile corresponds to the breaking of hydrogen bonds between beta-strands marked by red color.

It is worth pointing out that apart from mechanical protein stability measured as F_{max} in the force-extension profile, SMD simulations can be used to investigate the molecular determinants of mechanical stability. Using all-atom explicit solvent SMD simulations [17,26,33–35], it was found that each peak in the force-extension profile is associated with breaking hydrogen bonds between strands A and B as well as A' and G in a single domain of a multi-domain construct (Figure 1b). Thus, not only protein mechanical stability, but also molecular interactions and the mechanism behind mechanical unfolding can be revealed using SMD simulations.

Using SMD simulations we undertook a detailed and systematic investigation to quantify the mechanical stability of 2LLO peptide-protein complex. The structure of 2LLO complex has been determined by NMR spectroscopy [36] and its structure in cartoon representation is shown on left of Figure 2 (marked by NS) with alpha-helical ligand colored red and four-helix receptor colored black. We regulated the local environment and applied dissociating force by employing exactly the same set of criteria. We employed the constant pulling speed method for SMD studies. Figure 2 shows the typical force-extension curves for pulling speed $v = 0.01$ nm/ps for 2LLO system. For the system studied, one distinct peak was consistently observed, which corresponded to the detachment of the peptide from the receptor. Application of a force leads to the external perturbation which drives the system away from equilibrium. At the beginning the force dependence on extension is almost linear obeying the Hooke law. The peak shows the most mechanically stable conformations of protein-ligand complex. Once the hydrogen bonds and van der Waals interactions are broken, the force drops drastically and ligand no longer resists force. The peak in force-time profiles appears to be similar to the two different pulling directions studied, but the height of the peaks is different. Typical conformations observed before and after the occurrence of this peak are shown as snapshots in Figure 2. The separating force was found to drop drastically, though expectedly, once the interactions between peptide and protein ruptured, that is because a ligand can no longer resist the applied force after detachment from the receptor.

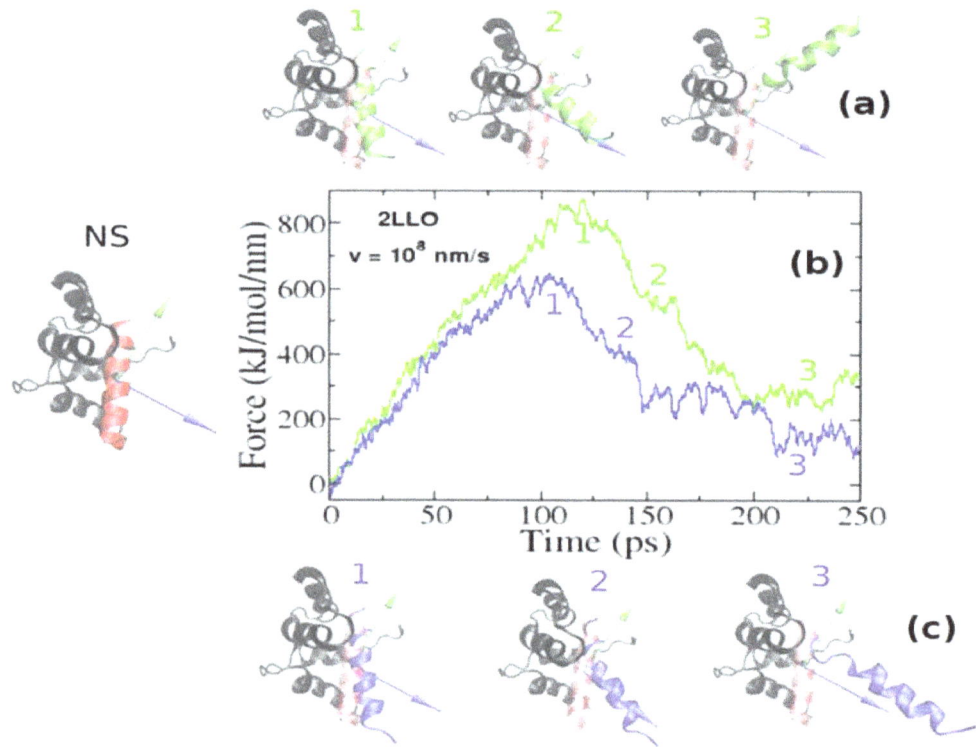

Figure 2. Examples of force-extension profiles for 2LLO complex pulled in different directions (**b**). Green and blue colors refer to the resultant dipole moment (RDM) and COMs pulling directions, respectively. The native conformation of 2LLO is shown on the left (marked by NS) with ligand colored red and receptor colored black. Representative snapshots of pathways for the mechanical unfolding along COMs and RDM directions are shown at the top (**a**) and bottom (**c**), respectively. In representative snapshots, we show the position of ligand in native conformation in transparent red.

2.2. Mechanical Stability Depends on the Pulling Direction of 2LLO Ligand-Receptor Complex

To elucidate the role of pulling direction on ligand mechanical stability, we have computed mechanical stability, F_{max}, for 2LLO protein-peptide complex. Figure 2 shows typical examples of force-extension curves for two different pulling directions and histograms of rupture forces computed from 50 trajectories are presented in Figure 3. The position of the peak corresponding to the most probable rupture force moves toward higher values for the RDM vector compared to the COM vector. The difference between F_{max} for RDM and COM pulling directions is around 180 kJ/mol/nm and indicates that pulling direction alters the mechanical stability of protein-peptide complex drastically. If we consider the averaged value of rupture force, F_{av}, we can see that the value of F_{av} is 828 ± 119 and 613 ± 57 kJ/mol/nm for RDM and COM pulling directions, respectively. Thus, regardless of whether the averaged or the most probable rupture force was used, we found that pulling the ligand along the resultant dipole moment vector results in stronger forces compared to the ligand unbinding along the center of masses vector.

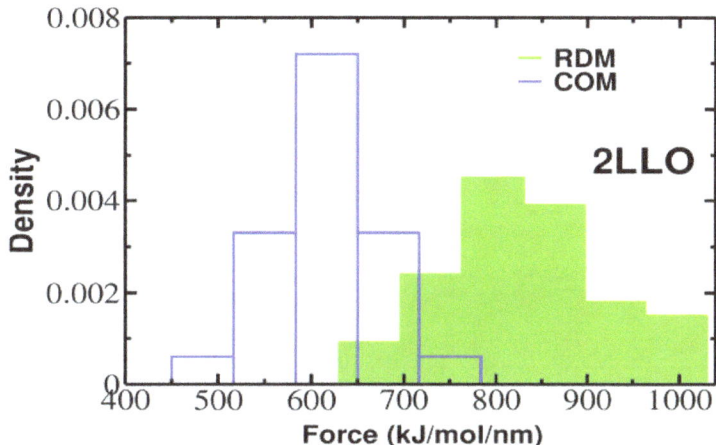

Figure 3. Histograms of rupture forces for 2LLO peptide-protein complex along RDM (green) and COM (blue) directions. The histograms clearly show that force peak moves towards higher values for RDM vector compared to COMs one.

It should be noted that the investigation of mechanical dissociation of a biological complex at lower pulling rates by explicit solvent all-atom MD simulations is still a challenge due to the enormous computation time required. However, the knowledge gathered from a two-decade-long spectrum of protein unfolding studies provides further evidence to support the claim that the difference in mechanical stability observed in high force regime will remain robust in low force regime. This seems to be supported by the Bell theory [37], which proposed that the most probable rupture force, F_{max}, decreases logarithmically as pulling rate, v, is lowered, e.g., $F_{max} \sim ln(v)$. The logarithmic dependence of F_{max} on the pulling speed v was confirmed by numerous experiments and simulations [17,38,39].

Our finding shows that pulling a ligand from the receptor along RDM vector results in stronger mechanical stability compared to pulling along COM vector. To make sure that our finding is valid for other protein-peptide systems, in the next subsection, we performed additional simulations on two different protein-peptide complexes.

2.3. Robustness of Results Against Different Protein-Peptide Complexes

So far, we have performed SMD simulations for 2LLO protein-peptide complex. The important question arises whether the effect of superior mechanical stability of RDM over COM vectors is universal and holds also for other protein-peptide complexes. In order to check whether our approach holds the same increase on dissociation force trend compared to COM, we performed additional SMD simulations for two different protein-peptide systems. Unlike the first protein-peptide complex that has an alpha-helical peptide, we chose two protein-peptide systems with different classes of peptidic ligands. We used Homer EVH1 domain with bound MGLUR peptide [40] where the bound peptide is unstructured (pdb code is 1DDV) and the inhibitor of apoptosis protein DIAP1 with bound N-terminal peptides from Hid and Grim [41] where the bound peptide takes a β-strand form (pdb code is 1JD5) (See Figure 4a,b). We generated 50 trajectories for each system at pulling rate of 0.01 nm/ns like before. Figure 4a,b show the native conformations of both complexes, while histograms of rupture forces are presented in Figure 4c,d. We computed the values of averaged rupture forces which are presented in Table 1. The difference between different pulling directions for both systems are easily identified, however it should be noted that the difference between RDM and COM for 2LLO is more noticeable compared to 1DDV and 1D5J systems. Intuitively, this can be explained by the difference in peptide size (Table 2). The size of the peptide in 2LLO is nearly 2–3 times larger compared to the peptides in 1DDV and 1D5J protein-peptide systems. Overall, our finding demonstrates that pulling a peptidic ligand from the receptor along RDM vector results in stronger mechanical stability compared to the pulling along COM for three diverse peptide-protein systems. Thus, regardless of the protein-peptide system used, pulling along RDM vector results in higher mechanical stability compared to the commonly used COM pulling.

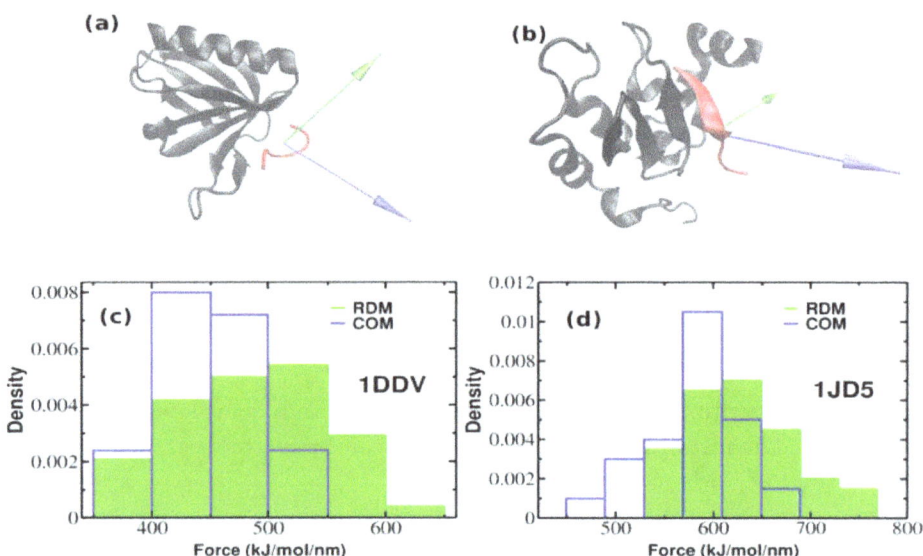

Figure 4. The native conformations of 1DDV (**a**) and 1JD5 (**b**) complexes with ligand colored red and receptor colored black. Green and blue colors refer to the resultant dipole moment (RDM) and COMs pulling directions, respectively. Histograms of rupture forces for 1DDV (**c**) and 1JD5 (**d**) peptide-protein complexes along RDM (green) and COMs (blue) directions.

Table 1. List of three protein-peptide complexes used in all-atom SMD simulations. Amino acids of protein involved in protein-peptide interactions for each system are shown. The pulling vectors used in the all-atom simulations (RDM and COM) and the averaged rupture forces and standard deviations obtained along RDM and COM pulling are also shown. Data are averaged over 50 trajectories.

PDB Code of the Protein-Peptide Complex	Identified Protein Residues Involved in Protein-Peptide Interactions	RDM Vector	COM Vector	Force (kJ/mol/nm) RDM COM	
2LLO	7–21, 25–29, 31–40, 43, 47–58, 61–65, 67–80	$0.144i + 0.983j - 0.11k$	$0.193i + 0.068j + 0.979k$	828.5 ± 118.7	613.5 ± 56.4
1DDV	10–16, 22–26, 30–31, 69–76, 87–92, 96, 109	$-0.636i + 0.111j + 0.764k$	$-0.799i + 0.454j - 0.395k$	486.1 ± 53.6	432.4 ± 42.1
1JD5	219–220, 242, 252–257, 259–263, 265–279, 282–283, 285–290, 311, 314–315, 317–318	$0.226i + 0.182j - 0.957k$	$-0.77i - 0.416j - 0.484k$	773.9 ± 149.4	595.9 ± 55.1

Table 2. List of 3 complexes used in the all atom SMD simulations. The bound structures of the complexes are obtained from the structures deposited in Protein Data Bank. The lengths and structural classes of both the proteins and the peptides are provided.

PDB ID of The Protein-Peptide Complex	Protein		Peptide	
	Length	Class	Length	Class
2LLO	80	α/β (34/2%)	19	α (84%)
1DDV	104	α/β (13/45%)	6	unstructured
1JD5	105	α/β (41/7%)	8	β (40%)

3. Materials and Methods

We used GROMOS43a1 [42] force-field [43] to describe the peptides and SPC [44] water model for solvent. All-atom MD simulations have been carried out using Gromacs program suite [45] which was previously successfully employed by our group for studying protein folding, unfolding and aggregation [46–49]. We use periodic boundary conditions and calculate the electrostatic interactions by the particle mesh Ewald method [50]. The non-bonded interaction pair-lists are updated every 10 fs, using a cutoff of 1.4 nm. All bond lengths are constrained with the linear constraint solver LINCS [51], allowing us to integrate the equations of motion with a time step of 2 fs.

To avoid improper structures, the whole system was minimized with the steepest-descent method, before being equilibrated at 310 K with two successive molecular dynamics runs of length 1ns each; the first one at constant volume, the second at constant pressure (1 atm). Initial velocities of the atoms were generated from the Maxwell distribution at 310 K. The temperature was kept close to 310 K using the v-rescale thermostat. Data analysis was done using the corresponding Gromacs programs and snapshots of all peptides were created with Visual Molecular Dynamics molecular graphics software [52]. Resultant dipole moment was defined as net dipole moment of those receptor backbone atoms which interacts with the bonded ligand.

During the steered molecular dynamics (SMD) simulations, the spring constant was chosen as $k = 1000$ kJ/(mol·nm^2) ≈ 1700 pN/nm, which corresponds to the upper limit of k of cantilever used in AFM experiments. We applied an external force to the center of mass (COM) of the ligand and pulled it along two different vectors. The first vector is drawn between COM of pulled peptide and COM of the receptor. The second vector is the resultant dipole moment defined as net dipole moment of those receptor backbone atoms which interacts with the bonded ligand. Pulled movement of the peptide under external force caused its dissociation from the receptor and the total force needed to bring about this dissociation was measured by $F = k(vt - x)$, where x denoted the displacement of the pulled peptide from its initial position. The resulting force was computed for each time step to generate a force-extension profile, which recorded a single peak showing the most mechanically resisting conformation in our system. Once the critical interactions were disrupted, the pulled peptide was found to no longer resist the applied force. Overall, the simulation procedure could be described similarly to those followed during the AFM experiments, except that the pulling speeds in our SMD simulations were fixed at several orders of magnitude higher than those used in AFM experiments [53]. We performed simulations at room temperature (T = 310 K) for $v = 10^7$ nm/s and generated 50 trajectories for each pulling direction. The 50 peak forces extracted were subsequently used to construct histogram of most probable rupture forces.

4. Conclusions

In the reported here studies we have tested the influence of RDM pulling direction on mechanical stability of three peptide-protein complexes. Unlike in widely-used COMs pulling simulations where COM does not talk about the forces that contribute to the stability of the complex, RDM vector retains information about the electrostatic forces associated with the resultant dipole moment. Pulling along COMs vector turns out to lead to a weaker resistance compared to RDM direction which has a significant electrostatic force aligned with it. Thus, together with other geometric and dynamics properties of protein binding pockets [54], RDM is one of the important factors influencing stability of biological complexes. Consequently, we hypothesize that peptide ligand binding affinity might be more accurately predicted using mechanical stability obtained by a computational approach that incorporates RDM factor in SMD studies. Our finding can provide a basis, through qualitative, for improvement of the computationally predicted mechanical stability. We believe that this should lead to development of new strategies that employ the mechanical stability as an effective scoring function for ranking binding affinities and/or for the quick testing of peptide ligands that might eventually block formation of pathological aggregates.

Author Contributions: M.K. conceived research, M.K. and A.B. performed simulations and analyzed data, and all authors wrote manuscript.

Acknowledgments: The second author, Anirban Banerji, passed away in Columbus, OH, USA on 12 August 2015 at the age of 39. This research was supported in part by the High Performance Computing Facility at The Research Institute at Nationwide Children's Hospital.

References

1.   Trott, O.; Olson, A.J. Software news and update autodock vina: Improving the speed and accuracy of docking with a new scoring function, efficient optimization, and multithreading. *J. Comput. Chem.* **2010**, *31*, 455–461. [PubMed]

2.   Wang, L.; Wu, Y.J.; Deng, Y.Q.; Kim, B.; Pierce, L.; Krilov, G.; Lupyan, D.; Robinson, S.; Dahlgren, M.K.; Greenwood, J.; et al. Accurate and reliable prediction of relative ligand binding potency in prospective drug discovery by way of a modern free-energy calculation protocol and force field. *J. Am. Chem. Soc.* **2015**, *137*, 2695–2703. [CrossRef] [PubMed]

3.   Kollman, P.A.; Massova, I.; Reyes, C.; Kuhn, B.; Huo, S.H.; Chong, L.; Lee, M.; Lee, T.; Duan, Y.; Wang, W.; et al. Calculating structures and free energies of complex molecules: Combining molecular mechanics and continuum models. *Acc. Chem. Res.* **2000**, *33*, 889–897. [CrossRef] [PubMed]

4.   Blaszczyk, M.; Kurcinski, M.; Kouza, M.; Wieteska, L.; Debinski, A.; Kolinski, A.; Kmiecik, S. Modeling of protein-peptide interactions using the cabs-dock web server for binding site search and flexible docking. *Methods* **2016**, *93*, 72–83. [CrossRef] [PubMed]

5.   Kozakov, D.; Hall, D.R.; Xia, B.; Porter, K.A.; Padhorny, D.; Yueh, C.; Beglov, D.; Vajda, S. The cluspro web server for protein-protein docking. *Nat. Protoc.* **2017**, *12*, 255–278. [CrossRef] [PubMed]

6.   Kurcinski, M.; Jamroz, M.; Blaszczyk, M.; Kolinski, A.; Kmiecik, S. Cabs-dock web server for the flexible docking of peptides to proteins without prior knowledge of the binding site. *Nucleic Acids Res.* **2015**, *43*, W419–W424. [CrossRef] [PubMed]

7.   London, N.; Raveh, B.; Cohen, E.; Fathi, G.; Schueler-Furman, O. Rosetta flexpepdock web server-high resolution modeling of peptide-protein interactions. *Nucleic Acids Res.* **2011**, *39*, W249–W253. [CrossRef] [PubMed]

8.   Bruce, N.J.; Ganotra, G.K.; Kokh, D.B.; Sadiq, S.K.; Wade, R.C. New approaches for computing ligand-receptor binding kinetics. *Curr. Opin. Struct. Biol.* **2018**, *49*, 1–10. [CrossRef] [PubMed]

9.   Li, M.S.; Mai, B.K. Steered molecular dynamics-a promising tool for drug design. *Curr. Bioinform.* **2012**, *7*, 342–351.

10. Rydzewski, J.; Nowak, W. Ligand diffusion in proteins via enhanced sampling in molecular dynamics. *Phys. Life Rev.* **2017**, *22–23*, 58–74. [CrossRef] [PubMed]

11. Van Vuong, Q.; Nguyen, T.T.; Li, M.S. A new method for navigating optimal direction for pulling ligand from binding pocket: Application to ranking binding affinity by steered molecular dynamics. *J. Chem. Inf. Model.* **2015**, *55*, 2731–2738. [CrossRef] [PubMed]

12. Grubmuller, H.; Heymann, B.; Tavan, P. Ligand binding: Molecular mechanics calculation of the streptavidin biotin rupture force. *Science* **1996**, *271*, 997–999. [CrossRef] [PubMed]

13. Bernetti, M.; Cavalli, A.; Mollica, L. Protein-ligand (un)binding kinetics as a new paradigm for drug discovery at the crossroad between experiments and modelling. *MedChemComm* **2017**, *8*, 534–550. [CrossRef]

14. Colizzi, F.; Perozzo, R.; Scapozza, L.; Recanatini, M.; Cavalli, A. Single-molecule pulling simulations can discern active from inactive enzyme inhibitors. *J. Am. Chem. Soc.* **2010**, *132*, 7361–7371. [CrossRef] [PubMed]

15. Gu, J.F.; Li, H.X.; Wang, X.C. A self-adaptive steered molecular dynamics method based on minimization of stretching force reveals the binding affinity of protein-ligand complexes. *Molecules* **2015**, *20*, 19236–19251. [CrossRef] [PubMed]

16. Lee, E.H.; Hsin, J.; Sotomayor, M.; Comellas, G.; Schulten, K. Discovery through the computational microscope. *Structure* **2009**, *17*, 1295–1306. [CrossRef] [PubMed]

17. Kouza, M.; Hu, C.K.; Li, M.S.; Kolinski, A. A structure-based model fails to probe the mechanical unfolding pathways of the titin i27 domain. *J. Chem. Phys.* **2013**, *139*, 065103. [CrossRef] [PubMed]

18. Lichter, S.; Rafferty, B.; Flohr, Z.; Martini, A. Protein high-force pulling simulations yield low-force results. *PLoS ONE* **2012**, *7*, e34781. [CrossRef] [PubMed]

19. Lu, H.; Isralewitz, B.; Krammer, A.; Vogel, V.; Schulten, K. Unfolding of titin immunoglobulin domains by steered molecular dynamics simulation. *Biophys. J.* **1998**, *75*, 662–671. [CrossRef]

20. Lemkul, J.A.; Bevan, D.R. Assessing the stability of alzheimer's amyloid protofibrils using molecular dynamics. *J. Phys. Chem. B* **2010**, *114*, 1652–1660. [CrossRef] [PubMed]

21. Kouza, M.; Co, N.T.; Li, M.S.; Kmiecik, S.; Kolinski, A.; Kloczkowski, A.; Buhimschi, I.A. Kinetics and mechanical stability of the fibril state control fibril formation time of polypeptide chains: A computational study. *J. Chem. Phys.* **2018**, *148*, 215106. [CrossRef] [PubMed]

22. Song, X.Y. An inhomogeneous model of protein dielectric properties: Intrinsic polarizabilities of amino acids. *J. Chem. Phys.* **2002**, *116*, 9359–9363. [CrossRef]

23. Brockwell, D.J.; Paci, E.; Zinober, R.C.; Beddard, G.S.; Olmsted, P.D.; Smith, D.A.; Perham, R.N.; Radford, S.E. Pulling geometry defines the mechanical resistance of a beta-sheet protein. *Nat. Struct. Biol.* **2003**, *10*, 731–737. [CrossRef] [PubMed]

24. Best, R.B.; Paci, E.; Hummer, G.; Dudko, O.K. Pulling direction as a reaction coordinate for the mechanical unfolding of single molecules. *J. Phys. Chem. B* **2008**, *112*, 5968–5976. [CrossRef] [PubMed]

25. Rief, M.; Gautel, M.; Oesterhelt, F.; Fernandez, J.M.; Gaub, H.E. Reversible unfolding of individual titin immunoglobulin domains by afm. *Science* **1997**, *276*, 1109–1112. [CrossRef] [PubMed]

26. Marszalek, P.E.; Lu, H.; Li, H.B.; Carrion-Vazquez, M.; Oberhauser, A.F.; Schulten, K.; Fernandez, J.M. Mechanical unfolding intermediates in titin modules. *Nature* **1999**, *402*, 100–103. [CrossRef] [PubMed]

27. Kotamarthi, H.C.; Sharma, R.; Ainavarapu, R.K. Single-molecule studies on polysumo proteins reveal their mechanical flexibility. *Biophys. J.* **2013**, *104*, 2273–2281. [CrossRef] [PubMed]

28. Kumar, S.; Li, M.S. Biomolecules under mechanical force. *Phys. Rep.* **2010**, *486*, 1–74. [CrossRef]

29. Kouza, M.; Lan, P.D.; Gabovich, A.M.; Kolinski, A.; Li, M.S. Switch from thermal to force-driven pathways of protein refolding. *J. Chem. Phys.* **2017**, *146*, 135101. [CrossRef] [PubMed]

30. Glyakina, A.V.; Likhachev, I.V.; Balabaev, N.K.; Galzitskaya, O.V. Right- and left-handed three-helix proteins. Ii. Similarity and differences in mechanical unfolding of proteins. *Proteins* **2014**, *82*, 90–102. [CrossRef] [PubMed]

31. Kouza, M.; Hu, C.K.; Zung, H.; Li, M.S. Protein mechanical unfolding: Importance of non-native interactions. *J. Chem. Phys.* **2009**, *131*, 215103. [CrossRef] [PubMed]

32. Li, M.S. Ligand migration and steered molecular dynamics in drug discovery: Comment on "ligand diffusion in proteins via enhanced sampling in molecular dynamic" by jakub rydzewski and wieslaw nowak. *Phys. Life Rev.* **2017**. [CrossRef] [PubMed]

33. MacKerell, A.D.; Bashford, D.; Bellott, M.; Dunbrack, R.L.; Evanseck, J.D.; Field, M.J.; Fischer, S.; Gao, J.; Guo, H.; Ha, S.; et al. All-atom empirical potential for molecular modeling and dynamics studies of proteins. *J. Phys. Chem. B* **1998**, *102*, 3586–3616. [CrossRef] [PubMed]

34. Lu, H.; Schulten, K. Steered molecular dynamics simulation of conformational changes of immunoglobulin domain i27 interpret atomic force microscopy observations. *Chem. Phys.* **1999**, *247*, 141–153. [CrossRef]

35. Fowler, S.B.; Best, R.B.; Herrera, J.L.T.; Rutherford, T.J.; Steward, A.; Paci, E.; Karplus, M.; Clarke, J. Mechanical unfolding of a titin Ig domain: Structure of unfolding intermediate revealed by combining AFM, molecular dynamics simulations, NMR and protein engineering. *J. Mol. Biol.* **2002**, *322*, 841–849. [CrossRef]

36. Zhang, Y.H.; Li, Z.G.; Sacks, D.B.; Ames, J.B. Structural basis for Ca^{2+}-induced activation and dimerization of estrogen receptor alpha by calmodulin. *J. Biol. Chem.* **2012**, *287*, 9336–9344. [CrossRef] [PubMed]

37. Bell, G.I. Models for the specific adhesion of cells to cells. *Science* **1978**, *200*, 618–627. [CrossRef] [PubMed]

38. Arad-Haase, G.; Chuartzman, S.G.; Dagan, S.; Nevo, R.; Kouza, M.; Binh, K.M.; Hung, T.N.; Li, M.S.; Reich, Z. Mechanical unfolding of acylphosphatase studied by single-molecule force spectroscopy and md simulations. *Biophys. J.* **2010**, *99*, 238–247. [CrossRef] [PubMed]

39. Klimov, D.K.; Thirumalai, D. Stretching single-domain proteins: Phase diagram and kinetics of force-induced unfolding. *Proc. Natl. Acad. Sci. USA* **1999**, *96*, 6166–6170. [CrossRef] [PubMed]

40. Irie, K.; Nakatsu, T.; Mitsuoka, K.; Miyazawa, A.; Sobue, K.; Hiroaki, Y.; Doi, T.; Fujiyoshi, Y.; Kato, H. Crystal structure of the homer 1 family conserved region reveals the interaction between the evh1 domain and own proline-rich motif. *J. Mol. Biol.* **2002**, *318*, 1117–1126. [CrossRef]

41. Wu, J.W.; Cocina, A.E.; Chai, J.J.; Hay, B.A.; Shi, Y.G. Structural analysis of a functional diap1 fragment bound to grim and hid peptides. *Mol. Cell.* **2001**, *8*, 95–104. [CrossRef]

42. Scott, W.R.P.; Hunenberger, P.H.; Tironi, I.G.; Mark, A.E.; Billeter, S.R.; Fennen, J.; Torda, A.E.; Huber, T.; Kruger, P.; van Gunsteren, W.F. The gromos biomolecular simulation program package. *J. Phys. Chem. A* **1999**, *103*, 3596–3607. [CrossRef]

43. Cornell, W.D.; Cieplak, P.; Bayly, C.I.; Gould, I.R.; Merz, K.M.; Ferguson, D.M.; Spellmeyer, D.C.; Fox, T.; Caldwell, J.W.; Kollman, P.A. A second generation force field for the simulation of proteins, nucleic acids, and organic molecules (vol 117, pg 5179, 1995). *J. Am. Chem. Soc.* **1996**, *118*, 2309. [CrossRef]

44. Berendsen, H.J.C.; Postma, J.P.M.; van Gunsteren, W.F.; Hermans, J. Interaction models for water in relation to protein hydration. *Intermol. Forces* **1981**, *14*, 331–442.

45. Hess, B.; Kutzner, C.; van der Spoel, D.; Lindahl, E. Gromacs 4: Algorithms for highly efficient, load-balanced, and scalable molecular simulation. *J. Chem. Theory Comput.* **2008**, *4*, 435–447. [CrossRef] [PubMed]

46. Wabik, J.; Kmiecik, S.; Gront, D.; Kouza, M.; Kolinski, A. Combining coarse-grained protein models with replica-exchange all-atom molecular dynamics. *Int. J. Mol. Sci.* **2013**, *14*, 9893–9905. [CrossRef] [PubMed]

47. Kouza, M.; Banerji, A.; Kolinski, A.; Buhimschi, I.A.; Kloczkowski, A. Oligomerization of fvflm peptides and their ability to inhibit beta amyloid peptides aggregation: Consideration as a possible model. *Phys. Chem. Chem. Phys.* **2017**, *19*, 2990–2999. [CrossRef] [PubMed]

48. Kouza, M.; Hansmann, U.H.E. Velocity scaling for optimizing replica exchange molecular dynamics. *J. Chem. Phys.* **2011**, *134*, 044124. [CrossRef] [PubMed]

49. Kouza, M.; Co, N.T.; Nguyen, P.H.; Kolinski, A.; Li, M.S. Preformed template fluctuations promote fibril formation: Insights from lattice and all-atom models. *J. Chem. Phys.* **2015**, *142*, 145104. [CrossRef] [PubMed]

50. Darden, T.; York, D.; Pedersen, L. Particle mesh ewald—An n.Log(n) method for ewald sums in large systems. *J. Chem. Phys.* **1993**, *98*, 10089–10092. [CrossRef]

51. Hess, B.; Bekker, H.; Berendsen, H.J.C.; Fraaije, J.G.E.M. Lincs: A linear constraint solver for molecular simulations. *J. Comput. Chem.* **1997**, *18*, 1463–1472. [CrossRef]

52. Humphrey, W.; Dalke, A.; Schulten, K. Vmd: Visual molecular dynamics. *J. Mol. Graph. Model.* **1996**, *14*, 33–38. [CrossRef]

53. Peplowski, L.; Sikora, M.; Nowak, W.; Cieplak, M. Molecular jamming-the cystine slipknot mechanical clamp in all-atom simulations. *J. Chem. Phys.* **2011**, *134*. [CrossRef] [PubMed]

54. Stank, A.; Kokh, D.B.; Fuller, J.C.; Wade, R.C. Protein binding pocket dynamics. *Accounts Chem. Res.* **2016**, *49*, 809–815. [CrossRef] [PubMed]

3

Artificial Intelligence in Drug Design

Gerhard Hessler [1,*] and **Karl-Heinz Baringhaus** [2]

[1]   R&D, Integrated Drug Discovery, Industriepark Hoechst, 65926 Frankfurt am Main, Germany
[2]   R&D, Industriepark Hoechst, 65926 Frankfurt am Main, Germany; karl-heinz.baringhaus@sanofi.com
*   Correspondence: gerhard.hessler@sanofi.com

Abstract: Artificial Intelligence (AI) plays a pivotal role in drug discovery. In particular artificial neural networks such as deep neural networks or recurrent networks drive this area. Numerous applications in property or activity predictions like physicochemical and ADMET properties have recently appeared and underpin the strength of this technology in quantitative structure-property relationships (QSPR) or quantitative structure-activity relationships (QSAR). Artificial intelligence in de novo design drives the generation of meaningful new biologically active molecules towards desired properties. Several examples establish the strength of artificial intelligence in this field. Combination with synthesis planning and ease of synthesis is feasible and more and more automated drug discovery by computers is expected in the near future.

Keywords: artificial intelligence; deep learning; neural networks; property prediction; quantitative structure-activity relationship (QSAR); quantitative structure-property prediction (QSPR); de novo design

1. Introduction

Artificial intelligence (AI) plays an important role in daily life. Significant achievements in numerous different areas such as image and speech recognition, natural language processing etc. have emerged [1–3]. Some of the progress in the field is highlighted by computers beating world class players in chess and in Go. While Deep Blue, beating world chess champion Kasparov in 1997, used a set of hard-coded rules and brute force computing power, Alpha Go has learned from playing against itself and won against the world strongest Go player [4,5].

Artificial intelligence is considered as intelligence demonstrated by machines. This term is used, when a machine shows cognitive behavior associated with humans, such as learning or problem solving [6]. AI comprises technologies like machine learning, which are well established for learning and prediction of novel properties. In particular, artificial neural networks, such as deep neural networks (DNN) or recurrent neural networks (RNN) drive the evolution of artificial intelligence.

In pharmaceutical research, novel artificial intelligence technologies received wide interest, when deep learning architectures demonstrated superior results in property prediction. In the Merck Kaggle [7] and the NIH Tox21 challenge [8], deep neural networks showed improved predictivity in comparison to baseline machine learning methods. In the meantime, the scope of AI applications for early drug discovery has been widely increased, for example to de novo design of chemical compounds and peptides as well as to synthesis planning.

Recently, numerous reviews have been published comprising good introductions into the field [9–18]. Here, we want to focus on recent developments of artificial intelligence in the field of property or activity prediction, de novo design and retrosynthetic approaches.

2. Artificial Intelligence in Property Prediction

In drug discovery, clinical candidate molecules must meet a set of different criteria. Next to the right potency for the biological target, the compound should be rather selective against undesired targets and also exhibit good physicochemical as well as ADMET properties (absorption, distribution, metabolism, excretion and toxicity properties). Therefore, compound optimization is a multidimensional challenge. Numerous in-silico prediction methods are applied along the optimization process for efficient compound design. In particular, several machine learning technologies have been successfully used, such as support vector machines (SVM) [19], Random Forests (RF) [20,21] or Bayesian learning [22,23].

One important aspect of the success of machine learning for property prediction is access to large datasets, which is a prerequisite for applying AI. In pharmaceutical industry, large datasets are collected during compound optimization for many different properties. Such large datasets for targets and antitargets are available across different chemical series and are systematically used for training machine learning models to drive compound optimization.

Prediction of activities against different kinases is an illustrative example. Selectivity profiling in different kinase projects generates larger datasets, which have been systematically used for model generation. For Profiling-QSAR [24], binary Bayesian QSAR models were generated from a large, but sparsely populated data matrix of 130,000 compounds on 92 different kinases. These models are applied to novel compounds to generate an affinity fingerprint, which is used to train models for prediction of biological activity against new kinases with relatively few data points. Models are iteratively refined with new experimental data. Thus, machine learning has become part of an iterative approach to discover novel kinase inhibitors.

In another example of predicting kinase activities Random Forest models could be successfully derived for ~200 different kinases combining publically available datasets with in-house datasets [25]. Random Forest models showed a better performance than other machine learning technologies. Only a DNN showed comparable performance with better sensitivity but worse specificity. Nevertheless, the authors preferred the Random Forest models since they are easier to train. Several recent reviews summarize numerous different additional aspects of machine learning [26–29].

In the public domain large datasets are available and can be used to derive machine learning models for the prediction of cross target activities [30–34]. These models can be applied to drug repurposing, the identification of new targets for an existing drug. Successful applications for repurposing of compounds have been shown using the SEA (Similarity Ensemble Approach) methodology [35]. SEA is a similarity based method, in which ensembles of ligands for each target are compared with each other. Similarities are compared to a distribution obtained from random comparisons to judge the significance of the observed similarities against a random distribution. For repurposing of a ligand, the analysis can also be done with a single molecule queried against an ensemble of ligands for each protein target.

Stimulated by the success of the Kaggle competition, deep neural networks have been used in numerous property prediction problems. Deep neural networks belong to the class of artificial neural networks, which are brain-inspired systems. Multiple nodes, also called neurons, are interconnected like the neurons in the brain. Signals coming in from different nodes are transformed and cascaded to the neurons of the next layer as illustrated in Figure 1. Layers between the input and output layer are called hidden layers. During training of a neural network, weights and biases at the different nodes are adjusted. Deep neural networks are using a significantly larger number of hidden layers and nodes than shallow architectures. Thus, a large number of parameters have to be fitted during the training of the neural network. Therefore, increases in compute power as well as a number of algorithmic improvements were necessary to address the overfitting problem such as dropout [36] or use of rectified linear units to address the vanishing gradient problem [37].

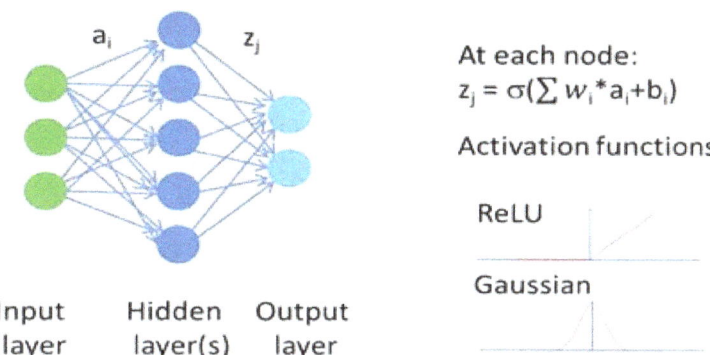

Figure 1. Neurons are connected to each other. Incoming signals are multiplied by a weight. The output signal z_j is given by the sum of this products plus a bias transformed by an activation function. Examples of activation functions are graphically shown, like the rectified linear unit (ReLU) or the Gaussian function. For each neuron in the neural net, weights and biases need to be trained. Deep neural networks have several hidden layers with many neurons. The number of neurons typically varies between different layers.

DNNs have been used in numerous examples for property prediction. In many of these studies a comparison to other machine learning approaches has been performed indicating, that DNNs show comparable or better performance than other machine learning approaches, e.g., for different properties ranging from biological activity prediction, ADMET properties to physicochemical parameters. For example, in the Kaggle competition, the DNN shows a better performance for 13 of the 15 assays than a Random Forest approach using 2D topological descriptors [7]. The study revealed that the performance of the DNN is variable, depending on the hyperparameters used, such as the architecture of the network (number of hidden layers as well as the number of neurons in each layer) and the activation function. Definition of a reasonable parameter set is crucial to achieve good performance.

In another study, a broad dataset from ChEMBL [38] was used comprising more than 5000 different assays and almost 750,000 compounds using Extend Connectivity Fingerprint (ECFP4) [39]. Again, DNNs outperformed several other machine learning methods used for comparison with respect to the area under the ROC curve.

Lenselik et al. performed a large benchmark study on a dataset from ChEMBL coming to a similar conclusion of better performance of the DNN methodology [40]. In this study temporal validation was used for performance comparison where training and test data are split according to publication date. This way of performance measurement is more stringent [41]. In temporal validation performance measures are significantly smaller than in the random split approach, which is probably closer to real life predictivity.

Korotcov et al. compared DNNs to other machine learning algorithms for diverse endpoints comprising biological activity, solubility and ADME properties [42]. In this study, Functional Class Fingerprint (FCFP6) fingerprints were used. The DNN performed better than the SVM approach, which in turn was superior to other machine learning technologies tested. Another interesting aspect of that study revealed that the performance and sensitivity rankings depend on the applied metrices.

Deep learning has also been applied to prediction of toxicity. Results from the Tox21 competition showed, that DNN shows good predictivity on 12 different toxic endpoints [8]. In this study, some emphasis was given to the selection of the molecular descriptors. Absence or presence of known toxicophores was included as one descriptor set in addition to physicochemical descriptors and ECFP type fingerprints. The authors demonstrate, that the DNN is capable of extracting molecular features, which are supposedly associated with known toxicophoric elements, illustrating, that such networks appear to learn more abstract representations in the different hidden layers. Figure 2 gives examples of such features detected by the network. While it is promising, that relevant structural elements can be derived from a DNN, the shown fragments are certainly too generic to be applied to drug discovery without human expertise in the field of toxicology. Additionally, the composition of the training dataset has a strong influence on predictivity and applicability domain of the model as well as the representation learnt by the network, creating a high barrier to such automated learnings. The DeepTox pipeline uses an ensemble of different models, but is dominated by DNN predictions. It outperformed other machine learning approaches in 9 out of 12 toxic endpoints.

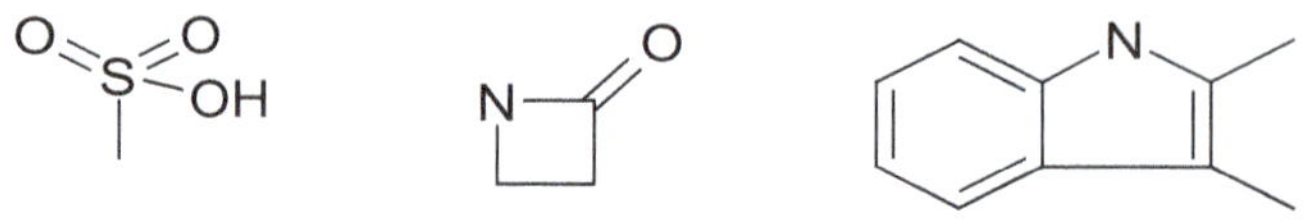

Figure 2. Toxicophoric features identified from the Tox21 dataset by the neural network [8].

Another example for the prediction of toxic endpoints has been given for the prediction of drug-induced liver injury (DILI) [43]. In this example, the network was trained on 475 compounds and performance was tested on 198 compounds. Good statistical parameters could be achieved for the predicition of drug-induced liver toxicity with accuracy of 86.9%, sensitivity of 82.5%, specificity of 92.9%, and AUC of 0.955. Molecular descriptors from PaDEL [44] and Mold [45] were used as well as a molecular description derived from the UG-RNN method for structural encoding [46] in combination with a line bisection method [47]. In the UG-RNN method, the descriptor is derived from the chemical structures captured as undirected graphs (UGs). Heavy atoms are represented as nodes and bonds as edges. The graph is fed into a recursive neuronal network (RNN) (Figure 3).

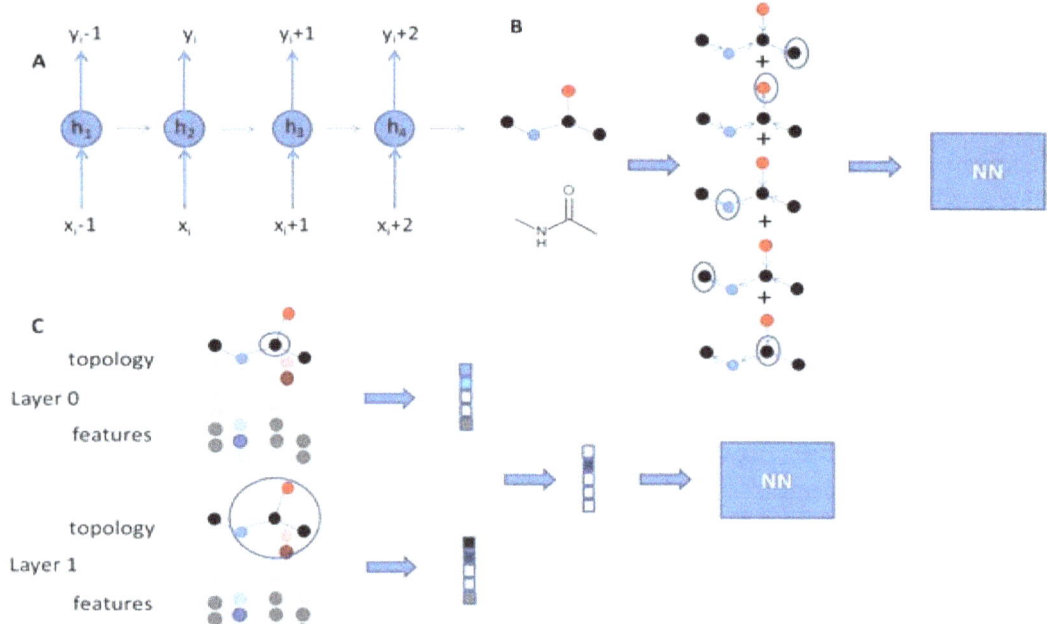

Figure 3. (**A**) Recurrent neural networks (RNNs) use sequential data. The output for the next element depends on the previous element. Thus, RNNs have a memory. h_i represent the hidden state at each neuron. They are updated based on the input x and the hidden state from the previous neuron. (**B**) In the UG-RNN approach, molecules are described as undirected graphs and fed into a RNN. Each vertex of a molecular graph is selected as a root node and becomes the endpoint of a directed graph. Output for all nodes is traversed along the graph until the root node is reached. All signals are summed to give the final output of the RNN, which enters into the NN for property training. (**C**) Graph convolutional models use the molecular graph. For each atom a feature vector is defined and used to pass on information for the neighboring atoms. In analogy to circular fingerprints different layers of neighboring atoms are passed through convolutional networks. Summation of the different atomic layers for all atoms results in the final vector entering the neural network for training.

All edges are directed towards the selected root node along the shortest path to cascade vector-encoded atom information to the root node. Every atom becomes the root node. The final output is summed over the different iterations. The UG-RNN derived descriptors show significantly better performance than the other two descriptor sets.

Using neural networks for encoding of molecular structures is a novel development in the cheminformatics field. While most of the examples described so far, use classical descriptors, more and more implementations allow selection of the chemical descriptor by the neural net. The idea is that the neural network can learn the representation which is best suited for the actual problem in question. Several ways have been described so far. Some of these approaches are shortly described in Wu et al. [48]. Graph convolutional (GC) models are derived from the concept of circular fingerprints [49]. Information is added by adding information from distant atoms in growing out along certain bond distances. These iterations are done for each atom and finally merged into a fixed length vector, which enters a neural network for property prediction. In graph convolutional models the molecular description layer is part of the differentiable network (Figure 3). Thus, training of the neuronal net also optimizes a useful representation of molecules suited to the task at hand.

In several examples, it was shown, that this training approach indeed improves predictivity for several properties. Duvenaud et al. showed improved performance for a solubility dataset and photovoltaic efficiency, while a biological activity prediction did not benefit from this approach. Additionally, the authors could identify molecular descriptors which are relevant for the different properties [50]. Li et al. introduced a dummy supernode as a new layer for molecular representation and could show good results on datasets from the MoleculeNet [48] dataset [51]. Other versions of convolutional networks have also been introduced. Graph convolution models using a simple molecular graph description show good results already. Kearnes et al. conclude that current graph convolutional models do not consistently outperform classical descriptors, but are a valuable extension of method repertoire, which provide novel flexibility, since the model can pick relevant molecular features and thus give access to a large descriptor space [52]. Related to work on image recognition, molecular structures have also been captured as images and fed into the network. This representation slightly outperformed a network trained on ECFP fingerprints on solubility and activity datasets, but performed slightly worse in toxicity prediction [53].

QSAR and machine learning models are usually trained for one endpoint, although multiple endpoints can be used. DNNs offer the possibility to systematically combine predictions for several endpoints as multitask learning. Multitask learning can improve prediction quality as has been shown by several studies, which compared the performance of singletask vs. multitask models. Ramsundar et al. analysed the benefit of multitask learning for a dataset containing up to 200 assays [54]. Overall, an increase of the performance of the models is observed with multitask learning, while it appears to be stronger for certain tasks. A dataset appears to show improved performance when it shares many active compounds with other tasks. In addition, both, the amount of data and the number of tasks were described to beneficially influence multitask learning. In another study on industry sized ADME datasets beneficial effects for multitask learning could be identified as well, although the improvement appears to be highly dataset dependent [55].

Conclusions about the best performance were also observed to be dependent on temporal or random split type validation. Simply adding massive amount of data does not guarantee a positive effect on predictivity. While multitask learning appears to have beneficial effects on a wide variety of different datasets, there are also examples of a drop in predictivity for some endpoints [56]. Xu et al. showed, that in multitask learning some information is "borrowed" from other endpoints, which leads to improved predictions [57]. According to the authors, an improved r^2 can be observed, when compounds in the training data for one endpoint are similar to compounds from the test data for a second endpoint and activities are correlated (positively or negatively). If activities are uncorrelated, a tendency for a decrease of r^2 was observed. If molecules between two endpoints are different from each other, no significant effect on r^2 can be expected from multitask learning.

Bajorath et al. used a set of about 100,000 compounds to develop a model prediction panel against 53 different targets [58]. Overall, good predictivity was achieved. Interestingly, the comparison between DNNs and other machine learning technologies does not yield any superior performance of the deep learning approach. The authors discuss, that the dataset is relatively small and thus might not be suited to demonstrate the full potential of DNNs.

Deep learning has also been used to predict potential energies of small organic molecules replacing a computational demanding quantum chemical calculation by a fast machine learning method. For large datasets, quantum chemically derived DFT potential energies have been calculated and used to train deep neuronal nets. The network was possible to predict the potential energy, called ANI-1, even for test molecules with higher molecular weight than the training set molecules [59].

Deep learning has been extensively validated for a number of different datasets and learning tasks. In a number of comparisons, DNNs show an improvement compared to well established machine learning technologies. This has also been demonstrated in a recent large-scale comparison of different methods, in which the performance of DNNs was described as comparable to in-vitro assays [60]. Nevertheless, many of the studies are performed retrospectively to show the applicability of deep learning architectures for property prediction and to compare the method to established machine learning algorithms. Often, public datasets like ChEMBL are used. In ChEMBL, biological data are often only available for one target resulting in a sparsely populated matrix, making cross-target learnings a significant challenge. Thus, it still remains to be seen, in which scenarios, DNNs clearly outperform other machine learning approaches, in particular since training and parameter optimization is less demanding for many other machine learning methods. A promising development is the self-encoding of the compound description by the learning engine, which will offer problem-dependent optimized compound descriptions.

3. Artificial Intelligence for de novo Design

De novo design aiming to generate new active molecules without reference compounds was developed approximately 25 years ago. Numerous approaches and software solutions have been introduced [61,62]. But de novo design has not seen a widespread use in drug discovery. This is at least partially related to the generation of compounds, which are synthetically difficult to access. The field has seen some revival recently due to developments in the field of artificial intelligence. An interesting approach is the variational autoencoder (Figure 4), which consists of two neural networks, an encoder network and a decoder network [63]. The encoder network translates the chemical structures defined by SMILES representation into a real-value continuous vector as a latent space. The decoder part is capable to translate vectors from that latent space into chemical structures. This feature was used to search for optimal solutions in latent space by an in-silico model and to back translate these vectors into real molecules by the decoder network. For most back translations one molecule dominates, but slight structural modifications exist with smaller probability. The authors used the latent space representation to train a model based on the QED drug-likeness score [64] and the synthetic accessibility score SAS [65]. A path of molecules with improved target properties could be obtained. In another publication, the performance of such a variational autoencoder was compared to an adversarial autoencoder [66]. The adversarial autoencoder consists of a generative model producing novel chemical structures. A second discriminative adversarial model is trained to tell apart real molecules from generated ones, while the generative model tries to fool the discriminative one. The adversarial autoencoder produced significantly more valid structures than the variational autoencoder in generation mode. In combination with an in-silico model novel structures predicted to be active against the dopamine receptor type 2 could be obtained. Kadurin et al. used a generative adversarial network (GAN) to suggest compounds with putative anticancer properties [67].

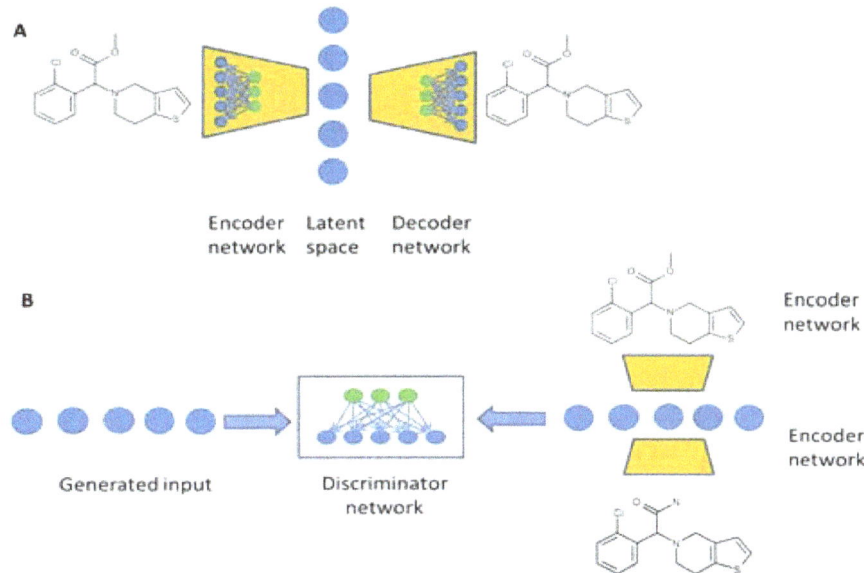

Figure 4. (**A**) A variational autoencoder consists of two neural networks. The encoder network transforms the molecular description into a description vector, the latent space, while the decoder network is trained to translate a latent space vector into a molecule. (**B**) The adversarial autoencoder comprises a standard autoencoder, which learns to generate chemical structures. The discriminator network compares descriptions from a defined distribution to structures generated from the autoencoder.

Recursive neural networks (RNN) have also been successfully used for de novo design. Originally, they have been established in the area of natural language processing [68]. RNNs take sequential information as input. Since SMILES strings encode chemical structures in a sequence of letters, RNNs have been used for generation of chemical structures. To teach the neural network the grammar of SMILES strings, RNNs are trained with a large set of chemical compounds taken from existing compound collections such as ChEMBL or commercially available compounds. It was shown, that RNNs are capable of producing a large fraction of valid SMILES strings [69,70]. The same approach was also successfully used for the generation of novel peptide structures [71]. Reinforcement learning was successfully applied to bias the generated compounds towards desired properties [72].

Transfer learning was used as another strategy to generate novel chemical structures with a desired biological activity. In the first step, the network is trained to learn the SMILES grammar with a large training set. In the second step, the training is continued with compounds having the desired activity. Few additionally epochs of training are sufficient to bias the generation of novel compounds into a chemical space occupied by active molecules [69]. Based on such an approach five molecules were synthesized and the design activity could be confirmed for four molecules against nuclear hormone receptors [73].

Several different architectures have been implemented, which are capable of generating valid, meaningful novel structures. The novel chemical space can be explored by these methods with the property distribution of the generated molecules being similar to the training space. The first prospective application for this methodology was successful with 4 out of 5 molecules showing the desired activity. Nevertheless, more experience need to be gained with respect to the size of the chemical space sampled and chemical feasibility of the proposed molecules.

4. Artificial Intelligence for Synthesis Planning

Organic synthesis is a critical part of any small molecule drug discovery program. New molecules are synthesized to progress along the compound optimization path and to identify molecules with improved properties. In certain situations, synthetic challenges restrict the available chemical space accessible for design molecules. Therefore synthesis planning is a key discipline in drug discovery. Accordingly, numerous computational approaches have been developed to assist synthesis planning. Three different aspects can be distinguished: prediction of the outcome of a reaction with a given set of educts, prediction of the yield of a chemical reactions as well as retrosynthetic planning. In particular, retrosynthetic planning is dominated by knowledge-based systems, which are built on expert-derived rules or automatically extracted rules from reaction databases [74–77].

Recently, a number of machine learning based approaches have been described for forward synthesis prediction. Forward synthesis prediction offers the ranking of synthetic routes from retrosynthetic analysis. In one type of approaches, quantum chemical descriptors have been combined with manual encoded rules and machine learning to predict a reaction and its product(s). [78–80]. The methodology has recently been extended to predict multi-step reactions [81]. In another approach [82], a deep neural network has been trained with a set of millions of reactions extracted from Reaxys [83]. The described network outperforms an expert system used for comparison. For reactions in the automatically derived rule set of 8720 templates, the authors report 78% accuracy for the best network.

Public reaction databases do not contain examples of failed chemical reactions, which is a clear limitation for machine learning approaches. Therefore, in another example, the dataset was augmented with chemical plausible negative examples [84]. At first possible reactions are selected and ranked by the neural network. Based on a training set of 15,000 reactions from granted US patents, the major reaction product was correctly identified top ranked in 71.8% in a 5-fold cross-validation experiment. In a subsequent publication, the authors use a template-free approach to improve coverage of chemical reactions [85]. Forward prediction of chemical reactions based on machine learning shows good performance in published validation studies. Nevertheless, some aspects need further consideration in future developments, such as the inclusion of reaction conditions, used catalysts etc.

Artificial intelligence has also been described for retrosynthetic analysis. Liu et al. used a sequence-to-sequence based model for retrosynthetic reaction prediction. Reactants and products are coded by SMILES strings for RNNs and coupled to each other in an encoder-decoder architecture. The training set spans 10 broad reaction types such as C-C bond formation, reductions, oxidations, heteroatom alkylation etc. and comprises 50,000 reactions from US patent literature [86]. The performance of the technology overall was comparable to rule-based expert systems, but large differences have been observed over different reaction classes. In a different approach recommender systems have been used to identify reactants yielding a desired product in combination with a chemical reaction graph [87]. Nevertheless, AUCs obtained in the validation indicated, that further improvement needs to be done.

The combination of three deep neural networks with a Monte Carlo tree search for retrosynthetic prediction yielded an excellent performance [88]. Training and test dataset were extracted from the entire Reaxys database and were split in time. For a test set of 497 diverse molecules, synthesized after 2015, over 80% correct synthetic routes were proposed. According to a blind test, medicinal chemists prefer the route proposed by this methodology over proposals from rule-based approaches.

Machine learning-based approaches can mine large datasets humans cannot handle in an unbiased manner. For synthesis planning, the combination of knowledge-based and machine learning approaches for prediction of chemical reactions turned out to be quite powerful. On the other hand, the purely machine-based approach capitalizing on a large reaction database shows excellent performance. Nevertheless, one limitation remains for in-silico tools, the capability to propose and develop novel chemical reactions. Here, a detailed analysis is necessary and will rely on the use of quantum chemical methods in the future [89].

5. Conclusions and Outlook

Artificial intelligence has received much attention recently and also has entered the field of drug discovery successfully. Many machine learning methods, such as QSAR methods, SVMs or

Random Forests are well-established in the drug discovery process. Novel algorithms based on neural networks, such as deep neural networks, offer further improvements for property predictions, as has been shown in numerous benchmark studies comparing deep learning to classical machine learning. The applicability of these novel algorithms for a number of different applications has been demonstrated including physicochemical properties as well as biological activities, toxicity etc. Some benefit from multitask learning has also been shown, where the prediction of related properties appears to benefit from joint learning. Future improvement can be expected from the capability of learning a chemical representation which is adapted to the problem at hand. First efforts have been taken, to identify relevant chemical features from such representations, which also points to one major challenge of these algorithms, which is their "black box" character. It is very difficult to extract from deep neural networks, why certain compounds are predicted to be good. This becomes relevant, if synthesis resources are more and more guided by artificial intelligence.

On the other hand, the effort of training such models will be increasing compared to established machine learning technologies. A large number of hyperparameters need to be tuned and optimized for good performance, although some studies indicate, that some reasonable parameter set can be used for starting the optimization.

The application of artificial intelligence for drug discovery benefits strongly from open source implementations, which provide access to software libraries allowing implementation of complex neural networks. Accordingly, open source libraries like Tensorflow [90] or Keras [91] are frequently used to implement different neural network architectures in drug discovery. Additionally, the Deepchem library provides a wrapper around Tensorflow that simplifies processing of chemical structures [92].

The scope of applications of artificial intelligence systems has been largely increased over recent years, now also comprising de novo design or retrosynthetic analysis, highlighting, that we will see more and more applications in areas where large datasets are available. With progress in these different areas, we can expect a tendency towards more and more automated drug discovery done by computers. In particular, large progress in robotics will accelerate this development. Nevertheless, artificial intelligence is far from being perfect. Other technologies with sound theoretical background will remain important, in particular, since they also benefit from increase in compute power, thus larger systems can be simulated with more accurate methods. Furthermore, there are still missing areas, novel ideas, which cannot be learned from data, giving a combination of human and machine intelligence a good perspective.

Author Contributions: Both authors contributed to the writing of the manuscript.

References and Note

1. Howard, J. The business impact of deep learning. In Proceedings of the 19th ACM SIGKDD International Conference on Knowledge Discovery and Data Mining, Chicago, IL, USA, 11–14 August 2013; p. 1135.
2. Impact Analysis: Buisiness Impact of Deep Learning. Available online: https://www.kaleidoinsights.com/impact-analysis-businedd-impacts-of-deep-learning/ (accessed on 10 August 2018).
3. Deep Learning, with Massive Amounts of Computational Power, Machines Can Now Recognize Objects and Translate Speech in Real Time. Artificial Intelligence Is Finally Getting Smart. Available online: https://www.technologyreview.com/s/513696/deep-learning/ (accessed on 10 August 2018).
4. Silver, D.; Huang, A.; Maddison, C.J.; Guez, A.; Sifre, L.; van den Driessche, G.; Schrittieser, J.; Antonoglou, I.; Panneershelvam, V.; Lactot, M.; et al. Mastering the game of Go with deep neural networks and tree search. *Nature* **2016**, *529*, 484. [CrossRef] [PubMed]
5. Hassabis, D. Artificial intelligence: Chess match of the century. *Nature* **2017**, *544*, 413–414. [CrossRef]
6. Artificial Intelligence. Available online: https://en.wikipedia.org/wiki/Artificial_intelligence (accessed on 16 June 2018).
7. Ma, J.; Sheridan, R.P.; Liaw, A.; Dahl, G.E.; Svetnik, V. Deep neural nets as a method for quantitative structure–activity relationships. *J. Chem. Inf. Model.* **2015**, *55*, 263–274. [CrossRef] [PubMed]

8.	Mayr, A.; Klambauer, G.; Unterthiner, T.; Hochreither, S. Deep Tox: Toxicity prediction using Deep Learning. *Front. Environ. Sci.* **2016**, 3. [CrossRef]

9.	Jørgensen, P.B.; Schmidt, M.N.; Winther, O. Deep Generative Models for Molecular. *Sci. Mol. Inf.* **2018**, *37*, 1700133. [CrossRef] [PubMed]

10.	Goh, G.B.; Hodas, N.O.; Vishnu, A. Deep Learning for Computational Chemistry. *J. Comp. Chem.* **2017**, *38*, 1291–1307. [CrossRef] [PubMed]

11.	Jing, Y.; Bian, Y.; Hu, Z.; Wang, L.; Xie, X.S. Deep Learning for Drug Design: An Artificial Intelligence Paradigm for Drug Discovery in the Big Data Era. *AAPS J.* **2018**, *20*, 58. [CrossRef] [PubMed]

12.	Gawehn, E.; Hiss, J.A.; Schneider, G. Deep Learning in Drug Discovery. *Mol. Inf.* **2016**, *35*, 3–14. [CrossRef] [PubMed]

13.	Gawehn, E.; Hiss, J.A.; Brown, J.B.; Schneider, G. Advancing drug discovery via GPU-based deep learning. *Expert Opin. Drug Discov.* **2018**, *13*, 579–582. [CrossRef] [PubMed]

14.	Colwell, L.J. Statistical and machine learning approaches to predicting protein–ligand interactions. *Curr. Opin. Struct. Biol.* **2018**, *49*, 123–128. [CrossRef] [PubMed]

15.	Zhang, L.; Tan, J.; Han, D.; Zhu, H. From machine learning to deep learning: Progress in machine intelligence for rational drug discovery. *Drug Discov. Today* **2017**, *22*, 1680–1685. [CrossRef] [PubMed]

16.	Chen, H.; Engkvist, O.; Wang, Y.; Olivecrona, M.; Blaschke, T. The rise of deep learning in drug discovery. *Drug Discov. Today* **2018**, *23*, 1241–1250. [CrossRef] [PubMed]

17.	Panteleeva, J.; Gaoa, H.; Jiab, L. Recent applications of machine learning in medicinal chemistry. *Bioorg. Med. Chem. Lett.* **2018**, in press. [CrossRef] [PubMed]

18.	Bajorath, J. Data analytics and deep learning in medicinal chemistry. *Future Med. Chem.* **2018**, *10*, 1541–1543. [CrossRef] [PubMed]

19.	Cortes, C.; Vapnik, V. Support vector networks. *Mach. Learn.* **1995**, *20*, 273–297. [CrossRef]

20.	Breiman, L. Random forests. *Mach. Learn.* **2001**, *45*, 5–32. [CrossRef]

21.	Svetnik, V.; Liaw, A.; Tong, C.; Culberson, J.C.; Sheridan, R.P.; Feuston, B.P. Random forest: A classification and regression tool for compound classification and QSAR modeling. *J. Chem. Inf. Model.* **2003**, *43*, 1947–1958. [CrossRef] [PubMed]

22.	Duda, R.O.; Hart, P.E.; Stork, G.E. *Pattern Classification*, 2nd ed.; John Wiley & Sons, Inc.: New York, NY, USA, 2001; pp. 20–83. ISBN 0-471-05669-3.

23.	Rogers, D.; Brown, R.D.; Hahn, M. Using Extended-Connectivity Fingerprints with Laplacian-Modified Bayesian Analysis in High-Throughput Screening Follow-Up. *J. Biomol. Screen.* **2005**, *10*, 682–686. [CrossRef] [PubMed]

24.	Martin, E.; Mukherjee, P.; Sullivan, D.; Jansen, J. Profile-QSAR: A novel meta-QSAR method that combines activities across the kinase family to accurately predict affinity, selectivity, and cellular activity. *J. Chem. Inf. Model.* **2011**, *51*, 1942–1956. [CrossRef] [PubMed]

25.	Merget, B.; Turk, S.; Eid, S.; Rippmann, F.; Fulle, S. Profiling Prediction of Kinase Inhibitors: Toward the Virtual Assay. *J. Med. Chem.* **2017**, *60*, 474–485. [CrossRef] [PubMed]

26.	Varnek, A.; Baskin, I. Machine Learning Methods for Property Prediction in Cheminformatics: Quo Vadis? *J. Chem. Inf. Model.* **2012**, *52*, 1413–1437. [CrossRef] [PubMed]

27.	Lo, Y.C.; Rensi, S.E.; Torng, W.; Altman, R.B. Machine learning in chemoinformatics and drug discovery. *Drug Discov. Today* **2018**, *23*, 1538–1546. [CrossRef]

28.	Lima, A.N.; Philot, E.A.; Trossini, G.H.G.; Scott, L.P.B.; Maltarollo, V.G.; Honorio, K.M. Use of machine learning approaches for novel drug discovery. *Expert Opin. Drug Discov.* **2016**, *2016 11*, 225–239. [CrossRef]

29.	Ghasemi, F.; Mehridehnavi, A.; Pérez-Garrido, A.; Pérez-Sánchez, H. Neural network and deep-learning algorithms used in QSAR studies: Merits and drawbacks. *Drug Discov. Today* **2018**, in press. [CrossRef] [PubMed]

30.	Keiser, M.J.; Roth, B.L.; Armbruster, B.N.; Ernsberger, P.; Irwin, J.J.; Shoichet, B.K. Relating protein pharmacology by ligand chemistry. *Nat. Biotechnol.* **2007**, *25*, 197–206. [CrossRef] [PubMed]

31.	Pogodin, P.V.; Lagunin, A.A.; Filimonov, D.A.; Poroikov, V.V. PASS Targets: Ligand-based multi-target computational system based on a public data and naïve Bayes approach. *SAR QSAR Environ. Res.* **2015**, *26*, 783–793. [CrossRef] [PubMed]

32.	Mervin, L.H.; Afzal, A.M.; Drakakis, G.; Lewis, R.; Engkvist, O.; Bender, A. Target prediction utilising negative bioactivity data covering large chemical space. *J. Cheminform.* **2015**, *7*, 51. [CrossRef] [PubMed]

33. Vidal, D.; Garcia-Serna, R.; Mestres, J. Ligand-based approaches to in silico pharmacology. *Methods Mol. Biol.* **2011**, *672*, 489–502. [CrossRef] [PubMed]

34. Steindl, T.M.; Schuster, D.; Laggner, C.; Langer, T. Parallel Screening: A Novel Concept in Pharmacophore Modelling and Virtual Screening. *J. Chem. Inf. Model.* **2006**, *46*, 2146–2157. [CrossRef] [PubMed]

35. Laggner, C.; Abbas, A.I.; Hufeisen, S.J.; Jensen, N.H.; Kuijer, M.B.; Matos, R.C.; Tran, T.B.; Whaley, R.; Glennon, R.A.; Hert, J.; et al. Predicting new molecular targets for known drugs. *Nature* **2009**, *462*, 175–181. [CrossRef]

36. Srivastava, N.; Hinton, G.; Krizhevsky, A.; Sutskever, I.; Salakhutdinov, R. Dropout: A simple way to prevent neural networks from overfitting. *J. Mach. Learn. Res.* **2014**, *15*, 1929–1958.

37. Nair, V.; Hinton, G.E. Rectified linear units improve restricted boltzmann machines. In Proceedings of the 27th International Conference on International Conference on Machine Learning, Haifa, Israel, 21–24 June 2010; pp. 807–814.

38. ChEMBL. Available online: https://www.ebi.ac.uk/chembl (accessed on 15 September 2018).

39. Unterthiner, T.; Mayr, A.; Klambauer, G.; Steijaert, M.; Ceulemans, H.; Wegner, J.; Hochreiter, S. Deep Learning as an Opportunity in Virtual Screening. In Proceedings of the NIPS Workshop on Deep Learning and Representation Learning, Montreal, QC, Canada, 8–13 December 2014; pp. 1058–1066. Available online: http://www.bioinf.at/publications/2014/NIPS2014a.pdf (accessed on 15 September 2018).

40. Lenselink, E.B.; ten Dijke, N.; Bongers, B.; Papadatos, G.; van Vlijmen, H.W.T.; Kowalczyk, W.; IJzerman, A.P.; van Westen, G.J.P. Beyond the hype: Deep neural networks outperform established methods using a ChEMBL bioactivity benchmark set. *J. Cheminform.* **2017**, *9*, 45. [CrossRef] [PubMed]

41. Sheridan, R.P. Time-split cross-validation as a method for estimating the goodness of prospective prediction. *J. Chem. Inf. Model.* **2013**, *53*, 783–790. [CrossRef] [PubMed]

42. Korotcov, A.; Tkachenko, V.; Russo, D.P.; Ekins, S. Comparison of Deep Learning with Multiple Machine Learning Methods and Metrics Using Diverse Drug Discovery Data Sets. *Mol. Pharm.* **2017**, *14*, 4462–4475. [CrossRef] [PubMed]

43. Xu, Y.; Dai, Z.; Chen, F.; Gao, S.; Pei, J.; Lai, L. Deep Learning for Drug-Induced Liver Injury. *J. Chem. Inf. Model.* **2015**, *55*, 2085–2093. [CrossRef] [PubMed]

44. Yap, C.W. PaDEL-descriptor: An open source software to calculate molecular descriptors and fingerprints. *J. Comput. Chem.* **2011**, *32*, 1466–1474. [CrossRef] [PubMed]

45. Hong, H.; Xie, Q.; Ge, W.; Qian, F.; Fang, H.; Shi, L.; Su, Z.; Perkins, R.; Tong, W. Mold^2, Molecular Descriptors from 2D Structures for Chemoinformatics and Toxicoinformatics. *J. Chem. Inf. Model.* **2008**, *48*, 1337–1344. [CrossRef] [PubMed]

46. Lusci, A.; Pollastri, G.; Baldi, P. Deep architectures and deep learning in chemoinformatics: The prediction of aqueous solubility for drug-like molecules. *J. Chem. Inf. Model.* **2013**, *53*, 1563–1574. [CrossRef] [PubMed]

47. Schenkenberg, T.; Bradford, D.; Ajax, E. Line Bisection and Unilateral Visual Neglect in Patients with Neurologic Impairment. *Neurology* **1980**, *30*, 509. [CrossRef] [PubMed]

48. Wu, Z.; Ramsundar, B.; Feinberg, E.N.; Gomes, J.; Geniesse, C.; Pappu, A.S.; Leswing, K.; Pande, V. MoleculeNet: A benchmark for molecular machine learning. *Chem. Sci.* **2018**, *9*, 513–530. [CrossRef] [PubMed]

49. Glen, R.C.; Bender, A.; Arnby, C.H.; Carlsson, L.; Boyer, S.; Smith, J. Circular fingerprints: Flexible molecular descriptors with applications from physical chemistry to ADME. *IDrugs Investig. Drugs J.* **2006**, *9*, 199–204.

50. Duvenaud, D.; Maclaurin, D.; Aguilera-Iparraguirre, J.; Gomez-Bombarelli, R.; Hirzel, T.; Aspuru-Guzik, A.; Adams, R.P. Convolutional Networks on Graphs for Learning Molecular Fingerprints. In Proceedings of the 28th International Conference on Neural Information Processing Systems, Montreal, QC, Canada, 7–12 December 2015; Volume 2, pp. 2224–2232. Available online: http://arxiv.org/abs/1509.09292 (accessed on 15 September 2018).

51. Li, J.; Cai, D.; He, X. Learning Graph-Level Representation for Drug Discovery. *arXiv* 2017, arXiv:1709.03741v2. Available online: http://arxiv.org/abs/1709.03741 (accessed on 15 September 2018).

52. Kearnes, S.; McCloskey, K.; Berndl, M.; Pande, V.; Riley, P. Molecular graph convolutions: Moving beyond fingerprints. *J. Comput. Aided Mol. Des.* **2016**, *30*, 595–608. [CrossRef] [PubMed]

53. Goh, G.B.; Siegel, C.; Vishnu, A.; Hodas, N.O.; Baker, N. A Deep Neural Network with Minimal Chemistry Knowledge Matches the Performance of Expert-developed QSAR/QSPR Models. arXiv:1706.06689.

54.   Ramsundar, B.; Kearnes, S.; Riley, P.; Webster, D.; Konerding, D.; Pandey, V. Massively Multitask Networks for Drug Discovery. *arXiv* 2015, arXiv:1502.02072v1. Available online: http://arxiv.org/abs/1502.02072 (accessed on 15 September 2018).

55.   Kearnes, S.; Goldman, B.; Pande, V. Modeling Industrial ADMET Data with Multitask Networks. *arXiv* 2016, arXiv:1606.08793v3. Available online: http://arxiv.org/abs/1606.08793v3 (accessed on 15 September 2018).

56.   Ramsundar, B.; Liu, B.; Wu, Z.; Verras, A.; Tudor, M.; Sheridan, R.P.; Pande, V. Is Multitask Deep Learning Practical for Pharma? *J. Chem. Inf. Model.* **2017**, *57*, 2068–2076. [CrossRef] [PubMed]

57.   Xu, Y.; Ma, J.; Liaw, A.; Sheridan, R.P.; Svetnik, V. Demystifying Multitask Deep Neural Networks for Quantitative Structure−Activity Relationships. *J. Chem. Inf. Model.* **2017**, *57*, 2490–2504. [CrossRef] [PubMed]

58.   Vogt, M.; Jasial, S.; Bajorath, J. Extracting Compound Profiling Matrices from Screening Data. *ACS Omega* **2018**, *3*, 4713–4723. [CrossRef] [PubMed]

59.   Smith, J.S.; Isayev, O.; Roitberg, A.E. ANI-1, A data set of 20 million calculated off-equilibrium conformations for organic molecules. *Sci. Data* **2017**, *4*, 170193. [CrossRef] [PubMed]

60.   Mayr, A.; Klambauer, G.; Unterthiner, T.; Steijaert, M.; Wegner, J.K.; Ceulemans, H.; Clevert, D.-A.; Hochreiter, S. Large-scale comparison of machine learning methods for drug prediction on ChEMBL. *Chem. Sci.* **2018**, *9*, 5441–5451. [CrossRef] [PubMed]

61.   Hartenfeller, M.; Schneider, G. Enabling future drug discovery by de novo design. *WIREs Comput. Mol. Sci.* **2011**, *1*, 742–759. [CrossRef]

62.   Schneider, P.; Schneider, G. De Novo Design at the Edge of Chaos. *J. Med. Chem.* **2016**, *59*, 4077–4086. [CrossRef] [PubMed]

63.   Gómez-Bombarelli, R.; Wei, J.N.; Duvenaud, D.; Hernández-Lobato, J.M.; Sánchez-Lengeling, B.; Sheberla, D.; Aguilera-Iparraguirre, J.; Hirzel, T.D.; Adamsk, P.; Aspuru-Guzik, A. Automatic Chemical Design Using a Data-Driven Continuous Representation of Molecules. *arXiv*, 2016; arXiv:1610.02415v3. Available online: http://arxiv.org/abs/1610.02415(accessed on 15 September 2018).

64.   Bickerton, G.R.; Paolini, G.V.; Besnard, J.; Muresan, S.; Hopkins, A.L. Quantifying the chemical beauty of drugs. *Nat. Chem.* **2012**, *4*, 90–98. [CrossRef] [PubMed]

65.   Ertl, P.; Schuffenhauer, A. Estimation of synthetic accessibility score of drug-like molecules based on molecular complexity and fragment contributions. *J. Cheminf.* **2009**, *1*. [CrossRef] [PubMed]

66.   Blaschke, T.; Olivecrona, M.; Engkvist, O.; Bajorath, J.; Chen, H. Application of Generative Autoencoder in De Novo Molecular Design. *Mol. Inf.* **2018**, *37*, 1700123. [CrossRef] [PubMed]

67.   Kadurin, A.; Aliper, A.; Kazennov, A.; Mamoshina, P.; Vanhaelen, Q.; Kuzma, K.; Zhavoronkov, A. The cornucopia of meaningful leads: Applying deep adversarial autoencoders for new molecule development in oncology. *Oncotarget* **2017**, *8*, 10883–10890. [CrossRef] [PubMed]

68.   Bengio, Y. Learning Deep Architectures for AI. *Found. Trends Mach. Learn.* **2009**, *2*, 1–127. [CrossRef]

69.   Segler, M.H.S.; Kogej, T.; Tyrchan, C.; Waller, M.P. Generating Focused Molecule Libraries for Drug Discovery with Recurrent Neural Networks. *ACS Cent. Sci.* **2018**, *4*, 120–131. [CrossRef] [PubMed]

70.   Gupta, A.; Müller, A.T.; Huisman, B.J.H.; Fuchs, J.A.; Schneider, P.; Schneider, G. Generative Recurrent Networks for De Novo Drug Design. *Mol. Inf.* **2018**, *37*, 1700111. [CrossRef] [PubMed]

71.   Muller, A.T.; Hiss, J.A.; Schneider, G. Recurrent Neural Network Model for Constructive Peptide Design. *J. Chem. Inf. Model.* **2018**, *58*, 472–479. [CrossRef] [PubMed]

72.   Olivecrona, M.; Blaschke, T.; Engkvist, O.; Chen, H. Molecular de-novo design through deep reinforcement learning. *J. Cheminform.* **2017**, *9*, 48. [CrossRef] [PubMed]

73.   Merk, D.; Friedrich, L.; Grisoni, F.; Schneider, G. De Novo Design of Bioactive Small Molecules by Artificial Intelligence Daniel. *Mol. Inf.* **2018**, *37*, 1700153. [CrossRef] [PubMed]

74.   Engkvist, O.; Norrby, P.-O.; Selmi, N.; Lam, Y.; Peng, Z.; Sherer, C.E.; Amberg, W.; Erhard, T.; Smyth, L.A. Computational prediction of chemical reactions: Current status and outlook. *Drug Discov. Today* **2018**, *23*, 1203–1218. [CrossRef] [PubMed]

75.   Szymkuc, S.; Gajewska, E.P.; Klucznik, T.; Molga, K.; Dittwald, P.; Startek, M.; Bajczyk, M.; Grzybowski, B.A. Computer-Assisted Synthetic Planning: The End of the Beginning. *Angew. Chem. Int. Ed.* **2016**, *55*, 5904–5937. [CrossRef] [PubMed]

76.   Chen, J.H.; Baldi, P. Synthesis Explorer: A Chemical Reaction Tutorial System for Organic Synthesis Design and Mechanism Prediction. *J. Chem. Educ.* **2008**, *85*, 1699–1703. [CrossRef]

77. Law, J.; Zsoldos, Z.; Simon, A.; Reid, D.; Liu, Y.; Khew, S.Y.; Johnson, A.P.; Major, S.; Wade, R.A.; Ando, H.Y. Route Designer: A Retrosynthetic Analysis Tool Utilizing Automated Retrosynthetic Rule Generation. *J. Chem. Inf. Model.* **2009**, *49*, 593–602. [CrossRef] [PubMed]

78. Kayala, M.A.; Azencott, C.A.; Chen, J.H.; Baldi, P. Learning to predict chemical reactions. *J. Chem. Inf. Model.* **2011**, *51*, 2209–2222. [CrossRef] [PubMed]

79. Kayala, M.A.; Baldi, P. ReactionPredictor: Prediction of complex chemical reactions at the mechanistic level using machine learning. *J. Chem. Inf. Model.* **2012**, *52*, 2526–2540. [CrossRef] [PubMed]

80. Sadowski, P.; Fooshee, D.; Subrahmanya, N.; Baldi, P. Synergies between quantum mechanics and machine learning in reaction prediction. *J. Chem. Inf. Model.* **2016**, *56*, 2125–2128. [CrossRef] [PubMed]

81. Fooshee, D.; Mood, A.; Gutman, E.; Tavakoli, M.; Urban, G.; Liu, F.; Huynh, N.; Van Vranken, D.; Baldi, P. Deep learning for chemical reaction prediction. *Mol. Syst. Des. Eng.* **2018**, *3*, 442–452. [CrossRef]

82. Segler, M.H.S.; Waller, M.P. Neural-Symbolic Machine Learning for Retrosynthesis and Reaction Prediction. *Chem. Eur. J.* **2017**, *23*, 5966–5971. [CrossRef] [PubMed]

83. http://www.reaxys.com, Reaxys is a registered trademark of RELX Intellectual Properties SA used under license.

84. Coley, W.; Barzilay, R.; Jaakkola, T.S.; Green, W.H.; Jensen, K.F. Prediction of Organic Reaction Outcomes Using Machine Learning. *ACS Cent. Sci.* **2017**, *3*, 434–443. [CrossRef] [PubMed]

85. Jin, W.; Coley, C.W.; Barzilay, R.; Jaakkola, T. Predicting Organic Reaction Outcomes with Weisfeiler-Lehman Network. *arXiv*, 2017; arXiv:1709.04555.

86. Liu, B.; Ramsundar, B.; Kawthekar, P.; Shi, J.; Gomes, J.; Luu Nguyen, Q.; Ho, S.; Sloane, J.; Wender, P.; Pande, V. Retrosynthetic reaction prediction using neural sequence-to-sequence models. *ACS Cent. Sci.* **2017**, *3*, 103–1113. [CrossRef] [PubMed]

87. Savage, J.; Kishimoto, A.; Buesser, B.; Diaz-Aviles, E.; Alzate, C. Chemical Reactant Recommendation Using a Network of Organic Chemistry. Available online: https://cseweb.ucsd.edu/classes/fa17/cse291-b/reading/p210-savage.pdf (accessed on 18 September 2018).

88. Segler, M.H.S.; Preuss, M.; Waller, M.P. Planning chemical syntheses with deep neural networks and symbolic AI. *Nature* **2018**, *555*, 604–610. [CrossRef] [PubMed]

89. Grimme, S.; Schreiner, P.R. Computational Chemistry: The Fate of Current Methods and Future Challenges. *Angew. Chem. Int. Ed.* **2018**, *57*, 4170–4176. [CrossRef] [PubMed]

90. Abadi, M.; Barham, P.; Chen, J.; Chen, Z.; Davis, A.; Dean, J.; Devin, M.; Ghemawat, S.; Irving, G.; Isard, M.; et al. TensorFlow: A System for Large-Scale Machine Learning. *arXiv* 2016, arXiv:1605.08695v2.

91. Keras: The Python Deep Learning library. Available online: https://keras.io (accessed on 18 September 2018).

92. Deepchem. Available online: https://deepchem.io (accessed on 18 September 2018).

4

Solvents to Fragments to Drugs: MD Applications in Drug Design

Lucas A. Defelipe [1,2,†], Juan Pablo Arcon [1,2,†], Carlos P. Modenutti [1,2], Marcelo A. Marti [1,2,*], Adrián G. Turjanski [1,2,*] and Xavier Barril [3,4,*]

[1] Departamento de Química Biológica, Facultad de Ciencias Exactas y Naturales, Universidad de Buenos Aires, Buenos Aires 1428, Argentina; ldefelipe@gmail.com (L.A.D.); juanarcon@gmail.com (J.P.A.); cpmode@gmail.com (C.P.M.)

[2] IQUIBICEN/UBA-CONICET, Facultad de Ciencias Exactas y Naturales, Universidad de Buenos Aires, Buenos Aires 1428, Argentina

[3] Catalan Institution for Research and Advanced Studies (ICREA), Passeig Lluís Companys 23, 08010 Barcelona, Spain

[4] Faculty of Pharmacy and Institute of Biomedicine (IBUB), University of Barcelona, Avgda. Diagonal 643, 08028 Barcelona, Spain

* Correspondence: marti.marcelo@gmail.com (M.A.M.); adrian@qb.fcen.uba.ar (A.G.T.); xbarril@ub.edu (X.B.);

† Both authors contributed equally.

Academic Editors: Rebecca C. Wade and Outi Salo-Ahen

Abstract: Simulations of molecular dynamics (MD) are playing an increasingly important role in structure-based drug discovery (SBDD). Here we review the use of MD for proteins in aqueous solvation, organic/aqueous mixed solvents (MDmix) and with small ligands, to the classic SBDD problems: Binding mode and binding free energy predictions. The simulation of proteins in their condensed state reveals solvent structures and preferential interaction sites (hot spots) on the protein surface. The information provided by water and its cosolvents can be used very effectively to understand protein ligand recognition and to improve the predictive capability of well-established methods such as molecular docking. The application of MD simulations to the study of the association of proteins with drug-like compounds is currently only possible for specific cases, as it remains computationally very expensive and labor intensive. MDmix simulations on the other hand, can be used systematically to address some of the common tasks in SBDD. With the advent of new tools and faster computers we expect to see an increase in the application of mixed solvent MD simulations to a plethora of protein targets to identify new drug candidates.

Keywords: molecular dynamics; cosolvent molecular dynamics; drug design; fragment screening; docking

1. Introduction

The first revolution in structural biology, in the early 1990's, increased the available structural information by 20-fold in a decade, creating a high expectation for computational methods that could turn this information into drug candidates. A large body of methods emerged, and some drugs owe their existence—at least in part—to them [1,2]. But it is obvious that the impact of structure-based drug design (SBDD) has not met these expectations. For instance, out of 66 clinical candidates, published by the Journal of Medicinal Chemistry in the 2016–2017 period, none originated as a virtual screening hit [3]. It is a fact that predicting binding affinities ($K_A = 1/K_D = \exp(-\Delta G_{BIND}/RT)$) is terribly difficult, and one of the main compounding factors is the solvent's effect. Contrary to many expectations, designing a good ligand is not a simple matter of finding a molecule that offers a good shape, or electrostatic and chemical complementarity to its protein target. Binding that occurs in the presence of solvent and related predictions will always fall short if this is not fully accounted for. Accurate predictions will unavoidably consider the protein and the ligand embedded in the solvent as part of a condensed state with a great number of configurational possibilities. Molecular dynamics (MD) is uniquely suited to simulate such systems by identifying true ensembles that can be related to macroscopic observables [4]. Here we will review how MD can be used to understand the behavior of water, the universal biological solvent, in relation to the protein surface, and to accurately predict its molecular association properties. We will then discuss how MD simulations of proteins in water and mixed solvents can be used to identify key interactions on their surface, and how these can be incorporated into computational docking, to identify better drug candidates.

We will start by showing that, far from being empty space, a protein's binding sites in the unbound state are occupied mainly by water (but also ions and metabolites) that does not behave as a homogeneous solvent. Rather, there are well-defined hydration spots and also regions where water density is much lower than in bulk solvents [5]. This determines binding in ways that were not initially expected. Solvation also affects the bound state and binding pathways, thus the gold standard for computational methods is to recapitulate the binding process of a ligand to its target by means of molecular simulations that consider the solvent explicitly. As the timescale of the binding/unbinding events has an exponential relationship with molecular size [6], observing binding on a 'computational microscope' [7] is greatly facilitated when the ligand has only a few atoms, particularly if it can be simulated at high concentrations. In Section 2, we will discuss applications and practical aspects of this approach (termed MD simulations with mixed solvents, or MDmix for short). The use of simple ligands as probes to elucidate interaction preferences of protein binding sites has a long history in SBDD. Except for the crucial difference of including explicit solvation in all the computational procedures, MDmix-type simulations can trace their roots to Goodford's GRID [8], Karplus' MCSS [9] or the more recent FTmap [10]. All such methods assume that the behavior of the probe is transferable to bigger molecules. Their documented ability to locate binding hot spots confirm that this is at least partially true. But binding free energy is clearly a non-additive property, [11,12] thus it becomes necessary to consider the actual molecules of interest to obtain quantitative predictions. Once considered a dream, major advances in the field of molecular dynamics (See [13–15] and references therein) have finally made it possible to directly simulate the binding and unbinding process of actual ligands to their targets. In Sections 3 and 4, we will review applications with small ligands (fragments) and actual drugs, respectively. We will conclude by discussing the practical limitations and future perspectives for the application of these methods in drug discovery.

2. Solvent Structure as a Predictor of Protein-Ligand Interaction Sites

Among the most relevant processes underlying the formation of protein-ligand complexes is the associated solvent reorganization at the contact surfaces, particularly that of the protein receptor. Water molecules bound to the ligand binding regions must either be displaced to allow direct protein-ligand contact [16–19] or be retained, bridging specific protein-ligand interactions, as is sometimes observed [20–25]. The thermodynamics of this solvent reorganization process is a key contribution to the complex formation free energy and thus to the ligand binding affinity. Initial attention was paid to the role of tightly bound or ordered waters, as revealed by X-ray structures [26], which after displacement by the incoming ligand, were proposed to contribute favorably to ligand affinity. [21,22]. This observation can be further extended to other proteins even if waters are not resolved in diffraction experiments due to lack of resolution, by the use of molecular dynamics in the explicit solvent.

Explicit solvent molecular dynamics allows studying the structure and dynamics of water molecules, which as a consequence of the shape and charge distribution of protein surfaces, are distributed inhomogeneously in the solvation shell, giving rise to space regions where the probability of finding water molecules is significantly higher (or lower) than that of the bulk solvent, and where rotational and translational motions of each molecule vary significantly. Wiesner et al. for example [27], found that confined waters can have residence times in the range of 1 ns to 106 ns, while for the more mobile waters residence times were only 10–50 ps. Further thermodynamic characterization of these surface waters can be achieved by means of the inhomogeneous fluid solvation theory (IFST), developed by Lazaridis et. al. [5], through the identification of the so-called water sites (WS) [28].

Water sites (sometimes also called hydration sites) are defined as confined space regions close to the protein surface, and internal cavities or packing effects, showing a high probability of finding a water molecule inside them (water finding probability, WFP). They can be evidenced by the presence of crystallographic water molecules, or from MD simulations as defined by their position (whose coordinates correspond to the center of mass of all oxygen atoms, from those water molecules that visit the site during the simulation timescale), their WFP, and their size (characterized by the R90 values, which describes in Angstrom the radius of the WS that contains a water molecule 90% of the time). WS are usually identified by applying a clustering algorithm to a collection of snapshots derived from MD simulations, and despite some special cases, good convergence is achieved in 20–50 ns [29].

In addition to their application as detailed descriptors of the solvent structure, the relevance of WS determination stems from their capacity to reveal key hydrophilic protein-ligand interaction sites, such as those established by ligand hydroxyl, carbonyl and carboxylate groups, among others. This is nicely exemplified by hydrophilic ligands such as carbohydrates, where several groups reported that the solvent structure in the receptor carbohydrate recognition domain, as revealed by the WS, mimics the framework of the sugar -OH groups, as shown in Figure 1. Moreover, detailed analysis of WS properties showed that those WS that are replaced by the incoming ligand-OH group tend to be those with higher WFP and establishing more interactions with the protein.

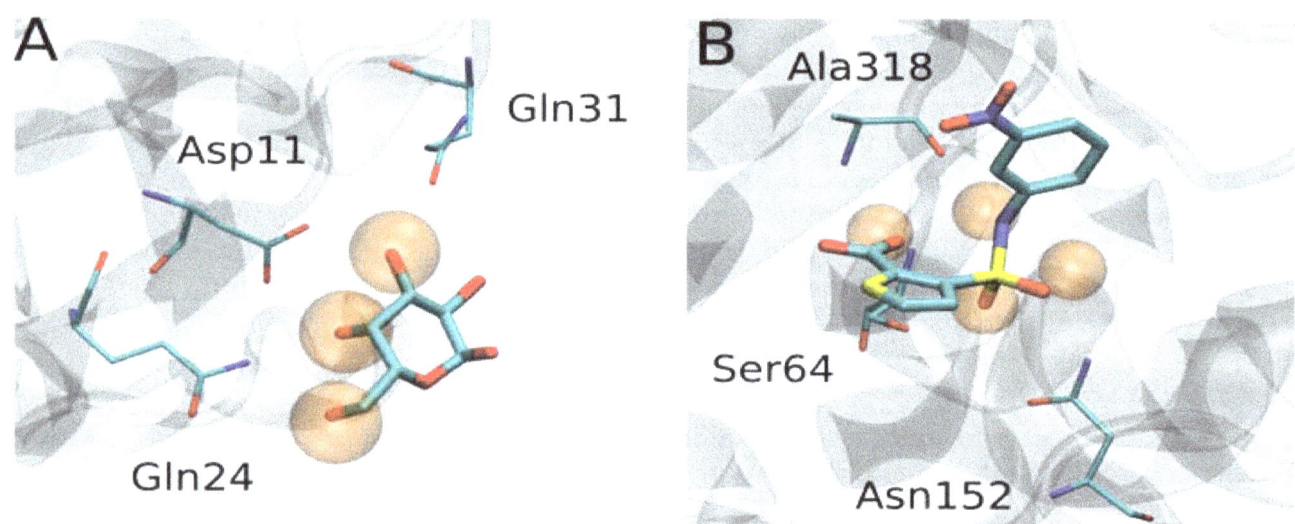

Figure 1. (**A**) Superposition of *Sambucus nigra* agglutinin II in complex with Lactose (PDB ID 3CA4) showing how the Water Sites (Orange transparent spheres) mimic the ligand -OH framework. (**B**) *Escherichia coli* AmpC beta-lactamase (PDB ID 1XGI) WS.

More recently, the role WS as predictors of protein-ligand interactions was extended beyond the sugars, again showing that WS, particularly those with high probe finding probability (PFP), tend to be replaced by ligand hydrophilic groups that establish key interactions with the protein receptor, as shown in Figure 1B, for AMPc beta-lactamase.

Having established the tight relationship between WS and protein-ligand interactions, the next logical move was to apply this knowledge in the context of protein-ligand complex structure prediction (i.e., docking methods) and determination of ligand binding free energies. However, before moving to this topic we will present the use of other solvents as tools for the prediction of protein-ligand interactions.

3. Mixed Solvents Simulations in Drug Design

While water is the universal biological solvent, organic solvents are ubiquitous in laboratories. Some exceptional proteins remain active in neat organic solvents and have been explored as catalysts in industrial applications [30]. More frequently, the buffers used in chemical and structural biology contain small concentrations of organic solvents. Most proteins preserve their structure and function in the presence of 1–5% of DMSO and other common organic molecules [31]. This fact led to the independent observation by NMR and X-ray crystallography that solvents bind preferentially to the active sites of proteins [32,33]. Systematic studies on proteins crystals showed an increasing number of solvent interaction sites as the solvent concentration was increased, and some degree of selectivity for various solvents [34,35]. The most frequently occupied regions coincided with key interaction sites for the substrates, which agreed with the recently postulated notion of 'hot spots', i.e., regions on the protein surface that provide most of the binding affinity. [36] Interestingly, the same authors also showed that the computational methods available at the time, GRID [8] and MCSS [9,37] did a mediocre job at predicting binding sites due to the use of implicit solvation and neglecting entropic contributions [34,35]. While the possibility of detecting binding sites by crystallography or NMR with mixed solvents was enticing, the method had limited practical impact because proteins and their crystals rarely withstand high solvent concentrations. Retrospectively, it may seem surprising that it took more than 20 years to perform analogous experiments using molecular dynamics, but it wasn't until the late 2000's that MD simulations could routinely explore sufficiently long timescales to ensure meaningful results. In 2009 the Barril's lab published the first MD application of mixed solvents. In this work, the probe solvent was isopropanol to capture in a single molecule, the hydrophobic and hydrogen bond donor, and acceptor moieties that are common in drug-like molecules. The aim was to detect binding sites and quantify their potential to bind drug-like molecules [38]. This property, often referred to as 'druggability' (but note the parallelism with the term 'ligandability' [39]), is crucial to predicting the probabilities of successful development of a drug candidate tackling a particular site [40]. The authors noted that "in addition to a prediction for the (druggability of the) whole site, one also obtains a map of the interaction preferences". Independently and almost simultaneously, the MacKerell's lab described another mixed solvent approach that focused precisely on this application [41]. In this case, the solvents used were propane as an aliphatic probe, benzene as aromatic probe and water itself was used as a polar probe. Probe interaction maps (called FragMaps) showed an excellent correlation with the binding modes of existing ligands. Since then, a large number of contributions have emerged. Besides druggability [42–44] and binding site mapping [45,46], mixed solvents have also been used to predict water displaceability [47,48], to probe protein flexibility and the detection of more druggable conformations [49], or cryptic pockets [50–52], or used to re-score docking poses [53,54]. As the diverse implementations and applications of mixed-solvent MD have been extensively reviewed by Ghanakota and Carlson [55], we will place emphasis on the issue of convergence, which is essential for correct predictions.

Convergence of a mixed solvents MD is determined by three interrelated aspects that merit individual discussion: Simulation time, solvent concentration, and protein flexibility.

(1) Simulation time should be sufficient to observe multiple binding and unbinding events. Naturally, the accuracy of the predictions increases and variability decreases as the number of observations (N) increases. Ns as low as 5 are sufficient for qualitative applications but must reach hundreds to be truly quantitative [6]. The other factor determining the total simulation time is the residence time of the solvent ($t_{1/2} = \ln 2/k_{off}$; $k_{off} \propto \exp(-\Delta G^{\ddagger}/RT)$ [56]. For barrierless dissociation ($\Delta G_{BIND} = -\Delta G^{TS}$) $t_{1/2}$ depends on the binding free energy, which can increase almost linearly with the number of atoms [57]. Thus, simulation times should increase exponentially with the size of the solvent. But the pathways leading to and from particular binding sites may be hindered, particularly for large

ligands, resulting in $\Delta G_{BIND} << -\Delta G^{TS}$ and, in consequence, much larger $t_{1/2}$. Conventional recipes suggest running several replicas of 10–40 ns each, for a total timeframe of 50–100 ns. This is sufficient to ensure qualitative convergence of the published solvents on the surface of the protein. But direct counting of the number of binding/unbinding events or other forms of measuring convergence should always be used (Figure 2).

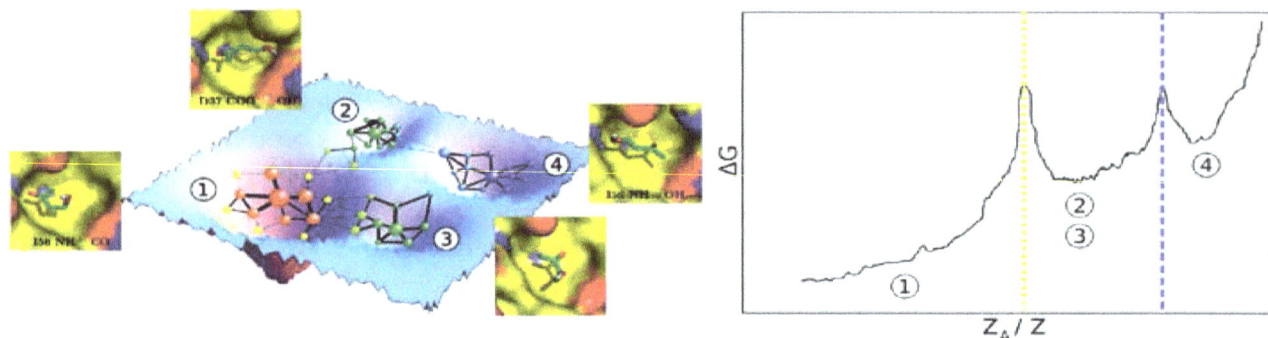

Figure 2. Exhaustive sampling of solvent-protein configurational space enables quantification of binding free energies. The figure is taken from [58].

(2) Solvent concentration increases sampling effectiveness. Not only due to the increase in effective on-rate (i.e., the number of binding events), but also because multiple binding sites can be sampled simultaneously. The behavior of the organic solvent should remain ideal (i.e., as in infinite dilution) to avoid artifacts caused by solvent-solvent interactions in the unbound state (e.g., inhomogeneous dilution and phase separation). Particular solutions to this problem include the introduction of repulsive terms between solvent molecules [41], or the use of amphiphilic molecules that are highly soluble and do not self-aggregate [59]. Additionally, protein dynamics should not be excessively perturbed by the solvent [60]. Considering that most solvents are denaturants at high concentrations, concentrations should be kept relatively low (<5%), as the protein could be artificially constrained, or simulation times could be much shorter than the denaturation time.

(3) Protein flexibility also determines convergence. Ideally, proteins should be allowed complete conformational freedom, but sampling the configurational space of regular proteins requires excessively long timeframes. Not only that, but it also complicates interpretation of results, as many hotspots are conformation-specific and not representative of the whole ensemble [61]. Constraining the mobility of protein atoms, on the other hand, is a straightforward way of increasing convergence. But this can lead to the overestimation of some hot spots and missing others. As a compromise, for many applications, it is useful and correct to use weak restraints that prevent conformational drift but allow sampling of the local conformational space [61]. Contrarily, if the goal is to induce conformational changes in the protein, such as the opening of cryptic pockets, simulations should be extended to the μs scale [50–52].

4. Small Ligands and Fragment Screening

Midway between solvent-sized and drug-like molecules, we find the so-called fragments. Fragment-based drug discovery initial hits are small molecules (roughly 10 to 20 non-hydrogen atoms) that are then grown and optimized to become standard drugs (30–40 atoms) [62]. Considering the industrial interest and the small size of these molecules, the use of MD as a screening technique raises considerable interest. In this approach, each compound in the virtual screening collection would be considered a probe that would be subjected to long MD simulations in the presence of the target protein. Probes that bind would then be considered fragment hits.

At present, molecular docking is the tool of choice for virtual fragment screening. Pioneering work by the Shoichet's Lab in this area led to the conclusion that although virtual fragment screening is adequate, with hit rates of 14.5% [63] and correct pose prediction, it mostly finds low specificity molecules. The effectiveness of this method for screening and de novo design are well documented in the literature [64–68]. Docking is particularly well suited for fragment screening since the molecules used as fragments are small and not very flexible (less than three rotatable bonds). Nevertheless, if the binding site is not known, it can lead to many false positives. Consensus strategies, like the ones used in FTMap [10], have been used to identify new binding sites. However, in shallow interfaces, as seen in many protein-protein interactions (PPI) sites, the lack of proper treatment for the receptor flexibility can be a drawback for these strategies [69].

MD is an essential tool to include receptor flexibility and therefore to compute the binding free energy [1,70]. Both Free Energy Perturbation (FEP) and kinetic parameter estimation methods have been used for fragment discovery, while FEP has been successful for rescoring [71,72] as well as predicting absolute free energies (but not routinely due to high computational cost) [73–75]. On the other hand, recent works have focused on the determination of the binding kinetics of small molecules and fragments from MD simulations [76–79]. Many methods rely on an intelligent design of the analysis strategy to predict the kinetic binding parameters k_{off}, mainly using Markov state models [80]. Although most of the reports use molecular simulations to characterize the binding kinetics of known fragments/small molecules [81–83], there are some reports on fragment-based screening from "first principles" using molecular simulations [84]. The De Fabritiis' Lab [85] recently presented a proof of concept of fragment-based screening using MD. They screened a library of 129 fragments (6 to 16 heavy atoms) using short simulations (100 ns), applying a bias and analyzing the trajectories with Markov state models (MSMs). Although the authors found promising fragments binding (8 mM) to the receptor surface (CXCL12), the computational expense is still prohibitive (380,000 GPU hours).

Work at Shaw D.E. Research sets the bar for quantitative prediction for fragment-based drug discovery [6]. They explored the binding thermodynamics and kinetics of 7 molecules of 4 to 10 heavy atoms to FKBP protein. After hundreds of direct observations of binding and unbinding events, they computed the k_{on}, k_{off} and binding affinities. They showed a perfect agreement with FEP simulations, demonstrating that when convergence is ensured, direct simulation of the binding equilibrium by molecular dynamics, can be a quantitative tool. Unfortunately, the RMSE of the computed binding free energy with experimental values was 2.1 kcal/mol, which illustrates the challenges that still lie ahead.

There is significant scope for cross-fertilization between mixed solvent MD and fragment-based drug discovery that has not been extensively explored. For instance, fragments often bind to multiple binding sites on the protein surface [86] which could potentially be identified by cosolvent MD. Fragments can also induce opening of new cavities (cryptic pockets). Gervasio's research on an exciting tool to address this topic, which combines co-solvent MD and advanced sampling (SWISH), helped to discover cryptic pockets [50,51]. Specifically, simulations on NPC2, p38α, LfrR, and hPNMT were performed, and due to the combined nature of SWISH and CoSolvent, MD was able to find all the cryptic pockets. Once in the binding site, the information derived from the cosolvent MD simulations could potentially be used to predict binding modes and affinities, or to guide the fragment evolution process. Work in this area has been done by MacKerell's Group with the SILCS methodology. They used the information derived from cosolvent MD to derive so-called FragMaps [41]. These grids were used to rank different ligands and to determine the free energy of binding.

5. Molecular Dynamics Simulations of Drugs or Drug-Like Compounds

Molecular Dynamics simulations could be used to study the free energy of binding of a drug or a drug-like molecule (30–40 heavy atoms) to a protein. This would require the sampling of several binding and unbinding events and therefore unbiased MD runs of at least hundreds of microseconds. Direct observation of drugs binding to their target has been an outstanding achievement of MD applications. Unbiased simulations have revealed the binding pathways of dasatinib to Src kinase [87] and alprenolol binding to the β2 adrenergic G protein-coupled receptor [76]. However, due to the long timescale involved in the dissociation of a drug from its target, direct observation of several unbinding events is not possible. Massive short unbiased simulations in conjunction with Markok

State analysis has been used to study benzamidine binding to trypsin [80]. On the other hand, biased simulations can be used to study the Potential of Mean Force of drug binding. Cavalli published the first study of its kind, showing that it was possible to discern active from inactive compounds of the beta-hydroxyacyl-ACP dehydratase of *Plasmodium falciparum* using steered MD [88]. Since then, a large variety of biasing potentials have been investigated and applied to the problem [89]. Even so, the problem remains computationally prohibitive. For instance, the study of a single inhibitor of p38 MAP kinase, that is a fragment of Doramapimod (BIRB 796) and dissociates 4 orders of magnitude faster than the parent compound, took 6.8 μs of production simulations and a total CPU time of 2.5 million core-hours [90]. In addition, identification of the reaction coordinate is often a trial and error process that takes considerable human time and is difficult to automatize [89]. Intriguingly the initial steps of the dissociation may already provide useful information [91], but full reconstruction of the process and quantitative binding affinity estimates remain a major challenge that is only applicable to particularly relevant protein-drug pairs.

For higher throughput applications, docking is widely used to predict protein-ligand interactions and has become extremely useful for virtual screening of huge collections of small molecules [92–94]. Most popular docking methods show that success rates are highly system-dependent, with an overall good performance for pose prediction with binding free energy errors of 2–3 kcal/mol for small drug-like molecules and in the absence of significant receptor conformational adjustment [95]. However, it is well known that better results can be obtained by adjusting the docking protocol using previous knowledge for a particular system, such as binding sites or crucial molecular interactions [96,97].

The term "biased docking" (or "guided docking") refers to the use of additional, experimental or in silico, information to influence the outcome of a docking experiment, e.g., the use of chemical information to favor a certain orientation and conformation of a ligand inside the binding site. The source of this information can be either the protein target structure or its known ligands. A protein-derived bias extracts the information directly from the protein surface and its available molecular interactions and generates a chemically complementary representation of the surface with more weight on particularly important residues, e.g., those confirmed to be essential for the activity by point mutations. As we discussed before, the use of probe atoms, functional groups, small molecules (e.g., mixed solvents) or molecular fragments is another approach to detect important interaction sites or hotspots without involving actual ligands. In this way, a protein-derived pharmacophore is obtained and defines energetically favorable binding site locations for docked compounds. A currently common technique for obtaining these hotspots is to run molecular dynamics simulations with small probes (see **Mixed Solvents** section). The hotspots can then be used to adjust the docking protocol, e.g., by adding a restriction towards the formation of a given protein-probe interaction. Recently, Arcon et al. showed that determination of water and/or ethanol sites derived from molecular dynamics simulations in mixed solvents allowed identification of over 79% of all protein-ligand interactions, especially those that were most important for the binding [54]. They also stated how this knowledge could be used to improve docking. On the other hand, a ligand-derived bias extracts the information from the known ligands for a particular protein target, for example, a particular substructure such as the core of a congeneric series. Several protein-ligand complex structures are available and the conserved interactions of the co-crystallized ligands (ligand-derived pharmacophore) can be inferred and used to improve docking accuracy [97,98].

The improved performance of knowledge-based biased docking is highlighted by the different options available in the most common docking programs. For example, Glide [99] and GOLD [100] allow hydrogen bonds and substructure-based constraints, while Glide also permits metal restraints to enforce coordination geometries. On the other hand, rDock [94] and MOE [101] are able to constrain generated poses to satisfy pharmacophores, and thus bias the results towards important interactions, and also perform knowledge-based template guided (or tethered) docking. DOCK6 [102] has a conformational search option to bias the sampling towards poses in accordance with a defined set of known ligand structures. AutoDock [92] and DOCK3 [103] were subjected, by us [29] and others [104,105], to implementations considering the energy accounted from water displacement through inhomogeneous solvation theory for guiding the docking. Lopez et al. have proposed a scheme to add a bias to AutoDock [29] that has been recently implemented for performing biased docking with AutoDock4 (AutoDock-Bias, in preparation Arcon et al., 2018). The versatile definition of the different types of biases in AutoDock-Bias accounts for all of the above cases. It allows guided docking towards pharmacophoric interactions in a straightforward way for hydrogen bond and hydrophobic/aromatic interactions. Furthermore, it allows researchers to get ideal interaction patterns for a specific protein structure, thus easily defining interactor locations. In addition, the capability of modifying any specific energy map and assigning any bias potential strength permits the precise localization of any desired atom (e.g., metal) or group (e.g., substructure core of a congeneric ligand series or for fragment growth) in a defined region space relative to the target protein. Finally, the specific energy map modification may also be used as an anchor for covalent docking studies. Since we addressed the problem of incorporating single target information, in the present discussion, we omitted potentials used for docking scoring functions [106–108] generally derived for diverse protein-ligand complexes.

In summary, mixed solvents simulations can lead to the identification of hot spots that can then be used in biased docking. The bias may affect the conformational search and/or scoring of the obtained poses.

6. Conclusions and Perspectives

Simulation of molecular dynamics in an explicit solvent are needed for accurate drug design. As the thermodynamics of the solvent reorganization upon drug binding is a key contribution to the complex formation free energy and thus to the ligand binding affinity. Therefore, accurate predictions have to consider the protein and the ligand embedded in the solvent, as part of a condensed state and have to account for a great number of configurational possibilities. On the other hand, explicit water MD allows studying the structure and dynamics of water molecules, and therefore the identification of water sites, that are relevant for their capacity to reveal key hydrophilic protein-ligand interaction sites. Water provides useful information for drug design, like guiding thermodynamic integration computations for compound optimization by allowing researchers to predict where it is favorable to grow the compound by displacing waters [109,110]. Another recent use of water molecules is to design specific inhibitors between a protein family, like the bromodomain proteins where structural water position determines drug selectivity [111]. The MD application of mixed solvents allows researchers to detect binding sites and quantify their potential to bind drug-like molecules. In turn, the identified hot spots can then be used as a bias in docking simulations to better identify drug candidates.

Mixed solvent MD with a cosolvent of no more than 10 heavy atoms is feasible and as we have described in this review, can clearly contribute to drug design, but has not yet been fully exploited. With the advent of new web services and user-friendly software, good algorithms to analyze the simulations and faster computers, we expect to see an increase in the application of these techniques to a plethora of protein targets. Docking simulations have not increased accuracy for drug-protein conformational predictions in the last decade, but most probably will get better in the near future, with the increased use of knowledge-based algorithms. MDMix will also help to obtain more accurate binding free energy estimations, but much effort in the community is needed in order to derive new algorithms that are not only able to estimate the free energy contribution of drug-protein interactions, but also the free energy of protein and drug desolvation.

Author Contributions: Conceptualization: L.A.D., J.P.A., C.P.M., M.A.M., A.G.T., X.B. Writing-Original Draft Preparation, L.A.D., J.P.A., M.A.M., A.G.T., X.B.; Writing-Review & Editing, L.A.D., M.A.M., A.G.T., X.B.; Funding Acquisition, A.G.T., X.B.

References

1. Jorgensen, W.L. The many roles of computation in drug discovery. *Science* **2004**, *303*, 1813–1818. [CrossRef] [PubMed]

2. Sliwoski, G.; Kothiwale, S.; Meiler, J.; Lowe, E.W., Jr. Computational methods in drug discovery. *Pharmacol. Rev.* **2014**, *66*, 334–395. [CrossRef] [PubMed]

3. Brown, D.G.; Boström, J. Where Do Recent Small Molecule Clinical Development Candidates Come From? *J. Med. Chem.* **2018**, *61*, 9442–9468. [CrossRef] [PubMed]

4. Bottaro, S.; Lindorff-Larsen, K. Biophysical experiments and biomolecular simulations: A perfect match? *Science* **2018**, *361*, 355–360. [CrossRef] [PubMed]

5. Lazaridis, T. Inhomogeneous Fluid Approach to Solvation Thermodynamics. 1. Theory. *J. Phys. Chem. B* **1998**, *102*, 3531–3541. [CrossRef]

6. Pan, A.C.; Xu, H.; Palpant, T.; Shaw, D.E. Quantitative Characterization of the Binding and Unbinding of Millimolar Drug Fragments with Molecular Dynamics Simulations. *J. Chem. Theory Comput.* **2017**, *13*, 3372–3377. [CrossRef] [PubMed]

7. Lee, E.H.; Hsin, J.; Sotomayor, M.; Comellas, G.; Schulten, K. Discovery through the computational microscope. *Structure* **2009**, *17*, 1295–1306. [CrossRef] [PubMed]

8. Goodford, P.J. A computational procedure for determining energetically favorable binding sites on biologically important macromolecules. *J. Med. Chem.* **1985**, *28*, 849–857. [CrossRef]

9. Miranker, A.; Karplus, M. Functionality maps of binding sites: A multiple copy simultaneous search method. *Proteins* **1991**, *11*, 29–34. [CrossRef]

10. Brenke, R.; Kozakov, D.; Chuang, G.-Y.; Beglov, D.; Hall, D.; Landon, M.R.; Mattos, C.; Vajda, S. Fragment-based identification of druggable "hot spots" of proteins using Fourier domain correlation techniques. *Bioinformatics* **2009**, *25*, 621–627. [CrossRef]

11. Baum, B.; Muley, L.; Smolinski, M.; Heine, A.; Hangauer, D.; Klebe, G. Non-additivity of functional group contributions in protein-ligand binding: A comprehensive study by crystallography and isothermal titration calorimetry. *J. Mol. Biol.* **2010**, *397*, 1042–1054. [CrossRef] [PubMed]

12. Biela, A.; Betz, M.; Heine, A.; Klebe, G. Water Makes the Difference: Rearrangement of Water Solvation Layer Triggers Non-additivity of Functional Group Contributions in Protein–Ligand Binding. *ChemMedChem.* **2012**, *7*, 1423–1434. [CrossRef] [PubMed]

13. Eastman, P.; Swails, J.; Chodera, J.D.; McGibbon, R.T.; Zhao, Y.; Beauchamp, K.A.; Wang, L.-P.; Simmonett, A.C.; Harrigan, M.P.; Stern, C.D.; et al. OpenMM 7: Rapid development of high performance algorithms for molecular dynamics. *PLoS Comput. Biol.* **2017**, *13*, e1005659. [CrossRef] [PubMed]

14. Kutzner, C.; Páll, S.; Fechner, M.; Esztermann, A.; de Groot, B.L.; Grubmüller, H. Best bang for your buck: GPU nodes for GROMACS biomolecular simulations. *J. Comput. Chem.* **2015**, *36*, 1990–2008. [CrossRef] [PubMed]

15. Lee, T.-S.; Cerutti, D.S.; Mermelstein, D.; Lin, C.; LeGrand, S.; Giese, T.J.; Roitberg, A.; Case, D.A.; Walker, R.C.; York, D.M. GPU-Accelerated Molecular Dynamics and Free Energy Methods in Amber18: Performance Enhancements and New Features. *J. Chem. Inf. Model.* **2018**. [CrossRef] [PubMed]

16. Li, Z.; Lazaridis, T. The effect of water displacement on binding thermodynamics: Concanavalin A. *J. Phys. Chem. B* **2005**, *109*, 662–670. [CrossRef] [PubMed]

17. Englert, L.; Biela, A.; Zayed, M.; Heine, A.; Hangauer, D.; Klebe, G. Displacement of disordered water molecules from hydrophobic pocket creates enthalpic signature: Binding of phosphonamidate to the S1'-pocket of thermolysin. *BBA—Gen. Subj.* **2010**, *1800*, 1192–1202. [CrossRef]

18. Michel, J.; Tirado-Rives, J.; Jorgensen, W.L. Energetics of displacing water molecules from protein binding sites: Consequences for ligand optimization. *J. Am. Chem. Soc.* **2009**, *131*, 15403–15411. [CrossRef]

19. García-Sosa, A.T.; Mancera, R.L.; Dean, P.M. WaterScore: A novel method for distinguishing between bound and displaceable water molecules in the crystal structure of the binding site of protein-ligand complexes. *J. Mol. Model.* **2003**, *9*, 172–182. [CrossRef]

20. Crawford, T.D.; Tsui, V.; Flynn, E.M.; Wang, S.; Taylor, A.M.; Côté, A.; Audia, J.E.; Beresini, M.H.; Burdick, D.J.; Cummings, R.; et al. Diving into the Water: Inducible Binding Conformations for BRD4, TAF1(2), BRD9, and CECR2 Bromodomains. *J. Med. Chem.* **2016**, *59*, 5391–5402. [CrossRef]

21. Ladbury, J.E. Just add water! The effect of water on the specificity of protein-ligand binding sites and its potential application to drug design. *Chem. Biol.* **1996**, *3*, 973–980. [CrossRef]

22. Poornima, C.S.; Dean, P.M. Hydration in drug design. 1. Multiple hydrogen-bonding features of water molecules in mediating protein-ligand interactions. *J. Comput. Aided Mol. Des.* **1995**, *9*, 500–512. [CrossRef] [PubMed]

23. Levinson, N.M.; Boxer, S.G. A conserved water-mediated hydrogen bond network defines bosutinib's kinase selectivity. *Nat. Chem. Biol.* **2014**, *10*, 127–132. [CrossRef] [PubMed]

24. García-Sosa, A.T. Hydration properties of ligands and drugs in protein binding sites: Tightly-bound, bridging water molecules and their effects and consequences on molecular design strategies. *J. Chem. Inf. Model.* **2013**, *53*, 1388–1405. [CrossRef] [PubMed]

25. Sridhar, A.; Ross, G.A.; Biggin, P.C. Waterdock 2.0: Water placement prediction for Holo-structures with a pymol plugin. *PLoS ONE* **2017**, *12*, e0172743. [CrossRef] [PubMed]

26. Bissantz, C.; Kuhn, B.; Stahl, M. A Medicinal Chemist's Guide to Molecular Interactions. *J. Med. Chem.* **2010**, *53*, 5061–5084. [CrossRef] [PubMed]

27. Wiesner, S.; Kurian, E.; Prendergast, F.G.; Halle, B. Water molecules in the binding cavity of intestinal fatty acid binding protein: Dynamic characterization by water 17O and 2H magnetic relaxation dispersion. *J. Mol. Biol.* **1999**, *286*, 233–246. [CrossRef] [PubMed]

28. Gauto, D.F.; Di Lella, S.; Guardia, C.M.A.; Estrin, D.A.; Martí, M.A. Carbohydrate-binding proteins: Dissecting ligand structures through solvent environment occupancy. *J. Phys. Chem. B* **2009**, *113*, 8717–8724. [CrossRef] [PubMed]

29. López, E.D.; Arcon, J.P.; Gauto, D.F.; Petruk, A.A.; Modenutti, C.P.; Dumas, V.G.; Marti, M.A.; Turjanski, A.G. WATCLUST: A tool for improving the design of drugs based on protein-water interactions. *Bioinformatics* **2015**, *31*, 3697–3699. [CrossRef] [PubMed]

30. Klibanov, A.M. Improving enzymes by using them in organic solvents. *Nature* **2001**, *409*, 241–246. [CrossRef] [PubMed]

31. Halling, P.J. What can we learn by studying enzymes in non-aqueous media? *Philos. Trans. R. Soc. Lond. B Biol. Sci.* **2004**, *359*, 1287–1296. [CrossRef] [PubMed]

32. Allen, K.N.; Bellamacina, C.R.; Ding, X.; Jeffery, C.J.; Mattos, C.; Petsko, G.A.; Ringe, D. An Experimental Approach to Mapping the Binding Surfaces of Crystalline Proteins. *J. Phys. Chem.* **1996**, *100*, 2605–2611. [CrossRef]

33. Liepinsh, E.; Otting, G. Organic solvents identify specific ligand binding sites on protein surfaces. *Nat. Biotechnol.* **1997**, *15*, 264–268. [CrossRef] [PubMed]

34. English, A.C.; Done, S.H.; Caves, L.S.; Groom, C.R.; Hubbard, R.E. Locating interaction sites on proteins: The crystal structure of thermolysin soaked in 2% to 100% isopropanol. *Proteins* **1999**, *37*, 628–640. [CrossRef]

35. English, A.C.; Groom, C.R.; Hubbard, R.E. Experimental and computational mapping of the binding surface of a crystalline protein. *Protein Eng.* **2001**, *14*, 47–59. [CrossRef] [PubMed]

36. Clackson, T.; Wells, J.A. A hot spot of binding energy in a hormone-receptor interface. *Science* **1995**, *267*, 383–386. [CrossRef] [PubMed]

37. Caflisch, A. Computational combinatorial ligand design: Application to human alpha-thrombin. *J. Comput. Aided Mol. Des.* **1996**, *10*, 372–396. [CrossRef]

38. Seco, J.; Luque, F.J.; Barril, X. Binding site detection and druggability index from first principles. *J. Med. Chem.* **2009**, *52*, 2363–2371. [CrossRef]

39. Vukovic, S.; Huggins, D.J. Quantitative metrics for drug-target ligandability. *Drug Discov. Today* **2018**, *23*, 1258–1266. [CrossRef]

40. Barril, X. Druggability predictions: Methods, limitations, and applications. *Wiley Interdiscip. Rev. Comput. Mol. Sci.* **2013**, *3*, 327–338. [CrossRef]

41. Guvench, O.; MacKerell, A.D., Jr. Computational fragment-based binding site identification by ligand competitive saturation. *PLoS Comput. Biol.* **2009**, *5*, e1000435. [CrossRef] [PubMed]

42. Bakan, A.; Nevins, N.; Lakdawala, A.S.; Bahar, I. Druggability Assessment of Allosteric Proteins by Dynamics Simulations in the Presence of Probe Molecules. *J. Chem. Theory Comput.* **2012**, *8*, 2435–2447. [CrossRef] [PubMed]

43. Ghanakota, P.; Carlson, H.A. Moving Beyond Active-Site Detection: MixMD Applied to Allosteric Systems. *J. Phys. Chem. B* **2016**, *120*, 8685–8695. [CrossRef] [PubMed]

44. Sayyed-Ahmad, A.; Gorfe, A.A. Mixed-Probe Simulation and Probe-Derived Surface Topography Map Analysis for Ligand Binding Site Identification. *J. Chem. Theory Comput.* **2017**, *13*, 1851–1861. [CrossRef] [PubMed]

45. Yu, W.; Lakkaraju, S.K.; Raman, E.P.; Fang, L.; MacKerell, A.D., Jr. Pharmacophore modeling using site-identification by ligand competitive saturation (SILCS) with multiple probe molecules. *J. Chem. Inf. Model.* **2015**, *55*, 407–420. [CrossRef] [PubMed]

46. Ghanakota, P.; van Vlijmen, H.; Sherman, W.; Beuming, T. Large-Scale Validation of Mixed-Solvent Simulations to Assess Hotspots at Protein-Protein Interaction Interfaces. *J. Chem. Inf. Model.* **2018**, *58*, 784–793. [CrossRef] [PubMed]

47. Alvarez-Garcia, D.; Barril, X. Molecular simulations with solvent competition quantify water displaceability and provide accurate interaction maps of protein binding sites. *J. Med. Chem.* **2014**, *57*, 8530–8539. [CrossRef]

48. Graham, S.E.; Smith, R.D.; Carlson, H.A. Predicting Displaceable Water Sites Using Mixed-Solvent Molecular Dynamics. *J. Chem. Inf. Model.* **2018**, *58*, 305–314. [CrossRef]

49. Uehara, S.; Tanaka, S. Cosolvent-Based Molecular Dynamics for Ensemble Docking: Practical Method for Generating Druggable Protein Conformations. *J. Chem. Inf. Model.* **2017**, *57*, 742–756. [CrossRef]

50. Oleinikovas, V.; Saladino, G.; Cossins, B.P.; Gervasio, F.L. Understanding Cryptic Pocket Formation in Protein Targets by Enhanced Sampling Simulations. *J. Am. Chem. Soc.* **2016**, *138*, 14257–14263. [CrossRef]

51. Comitani, F.; Gervasio, F.L. Exploring Cryptic Pockets Formation in Targets of Pharmaceutical Interest with SWISH. *J. Chem. Theory Comput.* **2018**, *14*, 3321–3331. [CrossRef] [PubMed]

52. Kimura, S.R.; Hu, H.P.; Ruvinsky, A.M.; Sherman, W.; Favia, A.D. Deciphering Cryptic Binding Sites on Proteins by Mixed-Solvent Molecular Dynamics. *J. Chem. Inf. Model.* **2017**, *57*, 1388–1401. [CrossRef]

53. Raman, E.P.; Yu, W.; Guvench, O.; Mackerell, A.D. Reproducing crystal binding modes of ligand functional groups using Site-Identification by Ligand Competitive Saturation (SILCS) simulations. *J. Chem. Inf. Model.* **2011**, *51*, 877–896. [CrossRef] [PubMed]

54. Arcon, J.P.; Defelipe, L.A.; Modenutti, C.P.; López, E.D.; Alvarez-Garcia, D.; Barril, X.; Turjanski, A.G.; Martí, M.A. Molecular Dynamics in Mixed Solvents Reveals Protein-Ligand Interactions, Improves Docking, and Allows Accurate Binding Free Energy Predictions. *J. Chem. Inf. Model.* **2017**, *57*, 846–863. [CrossRef] [PubMed]

55. Ghanakota, P.; Carlson, H.A. Driving Structure-Based Drug Discovery through Cosolvent Molecular Dynamics. *J. Med. Chem.* **2016**, *59*, 10383–10399. [CrossRef] [PubMed]

56. Pan, A.C.; Borhani, D.W.; Dror, R.O.; Shaw, D.E. Molecular determinants of drug-receptor binding kinetics. *Drug Discov. Today* **2013**, *18*, 667–673. [CrossRef] [PubMed]

57. Kuntz, I.D.; Chen, K.; Sharp, K.A.; Kollman, P.A. The maximal affinity of ligands. *Proc. Natl. Acad. Sci. USA* **1999**, *96*, 9997–10002. [CrossRef]

58. Huang, D.; Caflisch, A. The free energy landscape of small molecule unbinding. *PLoS Comput. Biol.* **2011**, *7*, e1002002. [CrossRef]

59. Lexa, K.W.; Goh, G.B.; Carlson, H.A. Parameter choice matters: Validating probe parameters for use in mixed-solvent simulations. *J. Chem. Inf. Model.* **2014**, *54*, 2190–2199. [CrossRef]

60. Foster, T.J.; MacKerell, A.D., Jr.; Guvench, O. Balancing target flexibility and target denaturation in computational fragment-based inhibitor discovery. *J. Comput. Chem.* **2012**, *33*, 1880–1891. [CrossRef]

61. Alvarez-Garcia, D.; Barril, X. Relationship between Protein Flexibility and Binding: Lessons for Structure-Based Drug Design. *J. Chem. Theory Comput.* **2014**, *10*, 2608–2614. [CrossRef] [PubMed]

62. Erlanson, D.A.; Fesik, S.W.; Hubbard, R.E.; Jahnke, W.; Jhoti, H. Twenty years on: The impact of fragments on drug discovery. *Nat. Rev. Drug Discov.* **2016**, *15*, 605–619. [CrossRef] [PubMed]

63. Chen, Y.; Shoichet, B.K. Molecular docking and ligand specificity in fragment-based inhibitor discovery. *Nat. Chem. Biol.* **2009**, *5*, 358–364. [CrossRef] [PubMed]

64. Giannetti, A.M.; Shoichet, B.K. Docking for fragment inhibitors of AmpC β-lactamase. *Proc. Natl. Acad. Sci. USA* **2009**, *106*, 7455–7460.

65. Zhao, H.; Gartenmann, L.; Dong, J.; Spiliotopoulos, D.; Caflisch, A. Discovery of BRD4 bromodomain inhibitors by fragment-based high-throughput docking. *Bioorg. Med. Chem. Lett.* **2014**, *24*, 2493–2496. [CrossRef] [PubMed]

66. Spiliotopoulos, D.; Zhu, J.; Wamhoff, E.-C.; Deerain, N.; Marchand, J.-R.; Aretz, J.; Rademacher, C.; Caflisch, A. Virtual screen to NMR (VS2NMR): Discovery of fragment hits for the CBP bromodomain. *Bioorg. Med. Chem. Lett.* **2017**, *27*, 2472–2478. [CrossRef] [PubMed]

67. Vass, M.; Agai-Csongor, E.; Horti, F.; Keserű, G.M. Multiple fragment docking and linking in primary and secondary pockets of dopamine receptors. *ACS Med. Chem. Lett.* **2014**, *5*, 1010–1014. [CrossRef]

68. Marchand, J.-R.; Dalle Vedove, A.; Lolli, G.; Caflisch, A. Discovery of Inhibitors of Four Bromodomains by Fragment-Anchored Ligand Docking. *J. Chem. Inf. Model.* **2017**, *57*, 2584–2597. [CrossRef]

69. Jubb, H.; Blundell, T.L.; Ascher, D.B. Flexibility and small pockets at protein-protein interfaces: New insights into druggability. *Prog. Biophys. Mol. Biol.* **2015**, *119*, 2–9. [CrossRef]

70. Chipot, C.; Pohorille, A. *Free Energy Calculations: Theory and Applications in Chemistry and Biology*; Springer Science & Business Media: Berlin, Germany, 2007; ISBN 9783540384472.

71. Chen, D.; Ranganathan, A.; IJzerman, A.P.; Siegal, G.; Carlsson, J. Complementarity between in silico and biophysical screening approaches in fragment-based lead discovery against the A(2A) adenosine receptor. *J. Chem. Inf. Model.* **2013**, *53*, 2701–2714. [CrossRef]

72. Steinbrecher, T.B.; Dahlgren, M.; Cappel, D.; Lin, T.; Wang, L.; Krilov, G.; Abel, R.; Friesner, R.; Sherman, W. Accurate Binding Free Energy Predictions in Fragment Optimization. *J. Chem. Inf. Model.* **2015**, *55*, 2411–2420. [CrossRef] [PubMed]

73. Jiang, W.; Roux, B. Free Energy Perturbation Hamiltonian Replica-Exchange Molecular Dynamics (FEP/H-REMD) for Absolute Ligand Binding Free Energy Calculations. *J. Chem. Theory Comput.* **2010**, *6*, 2559–2565. [CrossRef] [PubMed]

74. Aldeghi, M.; Heifetz, A.; Bodkin, M.J.; Knapp, S.; Biggin, P.C. Accurate calculation of the absolute free energy of binding for drug molecules. *Chem. Sci.* **2016**, *7*, 207–218. [CrossRef] [PubMed]

75. Lin, Y.-L.; Meng, Y.; Jiang, W.; Roux, B. Explaining why Gleevec is a specific and potent inhibitor of Abl kinase. *Proc. Natl. Acad. Sci. USA* **2013**, *110*, 1664–1669. [CrossRef] [PubMed]

76. Dror, R.O.; Pan, A.C.; Arlow, D.H.; Borhani, D.W.; Maragakis, P.; Shan, Y.; Xu, H.; Shaw, D.E. Pathway and mechanism of drug binding to G-protein-coupled receptors. *Proc. Natl. Acad. Sci. USA* **2011**, *108*, 13118–13123. [CrossRef] [PubMed]

77. Mondal, J.; Friesner, R.A.; Berne, B.J. Role of Desolvation in Thermodynamics and Kinetics of Ligand Binding to a Kinase. *J. Chem. Theory Comput.* **2014**, *10*, 5696–5705. [CrossRef] [PubMed]

78. Tiwary, P.; Mondal, J.; Berne, B.J. How and when does an anticancer drug leave its binding site? *Sci. Adv.* **2017**, *3*, e1700014. [CrossRef] [PubMed]

79. Lotz, S.D.; Dickson, A. Unbiased Molecular Dynamics of 11 min Timescale Drug Unbinding Reveals Transition State Stabilizing Interactions. *J. Am. Chem. Soc.* **2018**, *140*, 618–628. [CrossRef] [PubMed]

80. Plattner, N.; Noé, F. Protein conformational plasticity and complex ligand-binding kinetics explored by atomistic simulations and Markov models. *Nat. Commun.* **2015**, *6*, 7653. [CrossRef] [PubMed]

81. Bisignano, P.; Doerr, S.; Harvey, M.J.; Favia, A.D.; Cavalli, A.; De Fabritiis, G. Kinetic characterization of fragment binding in AmpC β-lactamase by high-throughput molecular simulations. *J. Chem. Inf. Model.* **2014**, *54*, 362–366. [CrossRef]

82. Ferruz, N.; Harvey, M.J.; Mestres, J.; De Fabritiis, G. Insights from Fragment Hit Binding Assays by Molecular Simulations. *J. Chem. Inf. Model.* **2015**, *55*, 2200–2205. [CrossRef] [PubMed]

83. Buch, I.; Giorgino, T.; De Fabritiis, G. Complete reconstruction of an enzyme-inhibitor binding process by molecular dynamics simulations. *Proc. Natl. Acad. Sci. USA* **2011**, *108*, 10184–10189. [CrossRef] [PubMed]

84. Rathi, P.C.; Ludlow, R.F.; Hall, R.J.; Murray, C.W.; Mortenson, P.N.; Verdonk, M.L. Predicting "Hot" and "Warm" Spots for Fragment Binding. *J. Med. Chem.* **2017**, *60*, 4036–4046. [CrossRef] [PubMed]

85. Martinez-Rosell, G.; Harvey, M.J.; De Fabritiis, G. Molecular-Simulation-Driven Fragment Screening for the Discovery of New CXCL12 Inhibitors. *J. Chem. Inf. Model.* **2018**, *58*, 683–691. [CrossRef] [PubMed]

86. Ludlow, R.F.; Verdonk, M.L.; Saini, H.K.; Tickle, I.J.; Jhoti, H. Detection of secondary binding sites in proteins using fragment screening. *Proc. Natl. Acad. Sci. USA* **2015**, *112*, 15910–15915. [CrossRef] [PubMed]

87. Shan, Y.; Kim, E.T.; Eastwood, M.P.; Dror, R.O.; Seeliger, M.A.; Shaw, D.E. How does a drug molecule find its target binding site? *J. Am. Chem. Soc.* **2011**, *133*, 9181–9183. [CrossRef]

88. Colizzi, F.; Perozzo, R.; Scapozza, L.; Recanatini, M.; Cavalli, A. Single-molecule pulling simulations can discern active from inactive enzyme inhibitors. *J. Am. Chem. Soc.* **2010**, *132*, 7361–7371. [CrossRef]

89. Gioia, D.; Bertazzo, M.; Recanatini, M.; Masetti, M.; Cavalli, A. Dynamic Docking: A Paradigm Shift in Computational Drug Discovery. *Molecules* **2017**, *22*, 2029. [CrossRef]

90. Casasnovas, R.; Limongelli, V.; Tiwary, P.; Carloni, P.; Parrinello, M. Unbinding Kinetics of a p38 MAP Kinase Type II Inhibitor from Metadynamics Simulations. *J. Am. Chem. Soc.* **2017**, *139*, 4780–4788. [CrossRef]

91. Ruiz-Carmona, S.; Schmidtke, P.; Luque, F.J.; Baker, L.; Matassova, N.; Davis, B.; Roughley, S.; Murray, J.; Hubbard, R.; Barril, X. Dynamic undocking and the quasi-bound state as tools for drug discovery. *Nat. Chem.* **2017**, *9*, 201–206. [CrossRef]

92. Morris, G.M.; Huey, R.; Lindstrom, W.; Sanner, M.F.; Belew, R.K.; Goodsell, D.S.; Olson, A.J. AutoDock4 and AutoDockTools4: Automated docking with selective receptor flexibility. *J. Comput. Chem.* **2009**, *30*, 2785–2791. [CrossRef] [PubMed]

93. Forli, S.; Huey, R.; Pique, M.E.; Sanner, M.F.; Goodsell, D.S.; Olson, A.J. Computational protein-ligand docking and virtual drug screening with the AutoDock suite. *Nat. Protoc.* **2016**, *11*, 905–919. [CrossRef] [PubMed]

94. Ruiz-Carmona, S.; Alvarez-Garcia, D.; Foloppe, N.; Garmendia-Doval, A.B.; Juhos, S.; Schmidtke, P.; Barril, X.; Hubbard, R.E.; Morley, S.D. rDock: A fast, versatile and open source program for docking ligands to proteins and nucleic acids. *PLoS Comput. Biol.* **2014**, *10*, e1003571. [CrossRef] [PubMed]

95. Sousa, S.F.; Ribeiro, A.J.M.; Coimbra, J.T.S.; Neves, R.P.P.; Martins, S.A.; Moorthy, N.S.; Fernandes, P.A.; Ramos, M.J. Protein-Ligand Docking in the New Millennium—A Retrospective of 10 Years in the Field. *Curr. Med. Chem.* **2013**, *20*, 2296–2314. [CrossRef] [PubMed]

96. Cleves, A.E.; Jain, A.N. Knowledge-guided docking: Accurate prospective prediction of bound configurations of novel ligands using Surflex-Dock. *J. Comput. Aided Mol. Des.* **2015**, *29*, 485–509. [CrossRef] [PubMed]

97. Hu, B.; Lill, M.A. PharmDock: A pharmacophore-based docking program. *J. Cheminform.* **2014**, *6*, 14. [CrossRef] [PubMed]

98. Perryman, A.L.; Santiago, D.N.; Forli, S.; Martins, D.S.; Olson, A.J. Virtual screening with AutoDock Vina and the common pharmacophore engine of a low diversity library of fragments and hits against the three allosteric sites of HIV integrase: Participation in the SAMPL4 protein-ligand binding challenge. *J. Comput. Aided Mol. Des.* **2014**, *28*, 429–441. [CrossRef]

99. Friesner, R.A.; Banks, J.L.; Murphy, R.B.; Halgren, T.A.; Klicic, J.J.; Mainz, D.T.; Repasky, M.P.; Knoll, E.H.; Shelley, M.; Perry, J.K.; et al. Glide: A new approach for rapid, accurate docking and scoring. 1. Method and assessment of docking accuracy. *J. Med. Chem.* **2004**, *47*, 1739–1749. [CrossRef]

100. Jones, G.; Willett, P.; Glen, R.C.; Leach, A.R.; Taylor, R. Development and validation of a genetic algorithm for flexible docking. *J. Mol. Biol.* **1997**, *267*, 727–748. [CrossRef] [PubMed]

101. Corbeil, C.R.; Williams, C.I.; Labute, P. Variability in docking success rates due to dataset preparation. *J. Comput. Aided Mol. Des.* **2012**, *26*, 775–786. [CrossRef]

102. Allen, W.J.; Balius, T.E.; Mukherjee, S.; Brozell, S.R.; Moustakas, D.T.; Lang, P.T.; Case, D.A.; Kuntz, I.D.; Rizzo, R.C. DOCK 6: Impact of new features and current docking performance. *J. Comput. Chem.* **2015**, *36*, 1132–1156. [CrossRef] [PubMed]

103. Coleman, R.G.; Carchia, M.; Sterling, T.; Irwin, J.J.; Shoichet, B.K. Ligand pose and orientational sampling in molecular docking. *PLoS ONE* **2013**, *8*, e75992. [CrossRef] [PubMed]

104. Balius, T.E.; Fischer, M.; Stein, R.M.; Adler, T.B.; Nguyen, C.N.; Cruz, A.; Gilson, M.K.; Kurtzman, T.; Shoichet, B.K. Testing inhomogeneous solvation theory in structure-based ligand discovery. *Proc. Natl. Acad. Sci. USA* **2017**, *114*, E6839–E6846. [CrossRef] [PubMed]

105. Uehara, S.; Tanaka, S. AutoDock-GIST: Incorporating Thermodynamics of Active-Site Water into Scoring Function for Accurate Protein-Ligand Docking. *Molecules* **2016**, *21*, 1604. [CrossRef] [PubMed]

106. Gohlke, H.; Hendlich, M.; Klebe, G. Knowledge-based scoring function to predict protein-ligand interactions. *J. Mol. Biol.* **2000**, *295*, 337–356. [CrossRef] [PubMed]

107. Muegge, I.; Martin, Y.C. A general and fast scoring function for protein-ligand interactions: A simplified potential approach. *J. Med. Chem.* **1999**, *42*, 791–804. [CrossRef] [PubMed]

108. Zheng, Z.; Merz, K.M. Development of the Knowledge-Based and Empirical Combined Scoring Algorithm (KECSA) To Score Protein–Ligand Interactions. *J. Chem. Inf. Model.* **2013**, *53*, 1073–1083. [CrossRef] [PubMed]

109. Wang, L.; Wu, Y.; Deng, Y.; Kim, B.; Pierce, L.; Krilov, G.; Lupyan, D.; Robinson, S.; Dahlgren, M.K.; Greenwood, J.; et al. Accurate and reliable prediction of relative ligand binding potency in prospective drug discovery by way of a modern free-energy calculation protocol and force field. *J. Am. Chem. Soc.* **2015**, *137*, 2695–2703. [CrossRef]

110. García-Sosa, A.T.; Mancera, R.L. Free Energy Calculations of Mutations Involving a Tightly Bound Water Molecule and Ligand Substitutions in a Ligand-Protein Complex. *Mol. Inform.* **2010**, *29*, 589–600. [CrossRef]

111. Aldeghi, M.; Ross, G.A.; Bodkin, M.J.; Essex, J.W.; Knapp, S.; Biggin, P.C. Large-scale analysis of water stability in bromodomain binding pockets with grand canonical Monte Carlo. *Commun. Chem.* **2018**, *1*, 19. [CrossRef]

5

Targeting Dynamical Binding Processes in the Design of Non-Antibiotic Anti-Adhesives by Molecular Simulation—The Example of FimH

Eva-Maria Krammer *[iD], Jerome de Ruyck[iD], Goedele Roos, Julie Bouckaert[iD] and Marc F. Lensink *[iD]

Unite de Glycobiologie Structurale et Fonctionnelle, UMR 8576 of the Centre National de la Recherche Scientifique and the University of Lille, 50 Avenue de Halley, 59658 Villeneuve d'Ascq, France; jerome.de-ruyck@univ-lille.fr (J.d.R.); goedele.roos@univ-lille.fr (G.R.) julie.bouckaert@univ-lille1.fr (J.B.)
* Correspondence: eva-maria.krammer@univ-lille1.fr (E.-M.K.); marc.lensink@univ-lille.fr (M.F.L.)

Academic Editors: Rebecca C. Wade and Outi Salo-Ahen

Abstract: Located at the tip of type I fimbria of *Escherichia coli*, the bacterial adhesin FimH is responsible for the attachment of the bacteria to the (human) host by specifically binding to highly-mannosylated glycoproteins located on the exterior of the host cell wall. Adhesion represents a necessary early step in bacterial infection and specific inhibition of this process represents a valuable alternative pathway to antibiotic treatments, as such anti-adhesive drugs are non-intrusive and are therefore unlikely to induce bacterial resistance. The currently available anti-adhesives with the highest affinities for FimH still feature affinities in the nanomolar range. A prerequisite to develop higher-affinity FimH inhibitors is a molecular understanding of the FimH-inhibitor complex formation. The latest insights in the formation process are achieved by combining several molecular simulation and traditional experimental techniques. This review summarizes how molecular simulation contributed to the current knowledge of the molecular function of FimH and the importance of dynamics in the inhibitor binding process, and highlights the importance of the incorporation of dynamical aspects in (future) drug-design studies.

Keywords: adhesion; FimH; rational drug design; molecular dynamics; molecular docking; ligand binding

1. Introduction

Although commensal *Escherichia coli* bacteria live in symbiosis with their human hosts as part of the gut flora, several *E. coli* strains are pathogenic to humans [1]. These pathogens are at the origin of a wide variety of diseases including intestinal (enteritis, and diarrhea) and extra-intestinal diseases (urinary tract infections (UTIs), sepsis, and meningitis). Uropathogenic *E. coli* (UPEC) for example are the primary cause of a large majority of UTIs (up to 70–95% of community acquired UTIs) [2,3]. UTIs are often recurrent or relapsing and although they are common infections they cause serious morbidity [4] and account for substantial medical costs worldwide [2]. The standard treatment for uncomplicated UTIs is a short course of antibiotics, which are highly effective against sensitive UPECs. However, antibiotic-resistant UPEC strains are on the rise as evidenced in urine cultures of UTI patients [5–7] and highlighted in 2016 by the first case of an US UTI patient carrying a pan-drug resistant *E. coli* strain [8]. The emergence of multi- and pan-drug resistance bacteria as well as the latency in the development of new antibiotics highlight the need for new non-antibiotic treatment alternatives against UPEC and other pathogenic *E. coli* infections [9,10]. A promising target for such a drug development is the FimH adhesin [11,12]. Drugs targeting FimH are unlikely to induce bacterial resistance as they do not interfere with the bacterial metabolism. Furthermore, it has been shown in mice and primate studies that vaccination with FimH leads to protection against bacterial infection [13].

FimH is located at the tip of the *E. coli* type I fimbria and used by the bacteria to adhere to their host cells. Extensive research performed on murine cystitis models evidenced that type 1 pili and FimH-mediated adhesion are essential for bacterial invasion [14–17]. UPEC (and most other *E. coli* strains) express a few hundreds of these about 1 μm-long rod-shaped organelles on their cell surface to adhere in a multivalent fashion to the superficial bladder cells. Adhesion is mediated at the molecular level by FimH binding to highly-mannosylated glycoproteins (MGP). In the case of UTIs, the primary partner for FimH adhesion is Uroplakin Ia (UPIa), a MGP present on the surface of epithelial umbrella cells of the urinary tract [18].

More recently, another class of pathogenic *E. coli* strains, the adherent and invasive *E. coli* (AIEC) strains have been evidenced to be of central importance in the development of Crohn's disease (CD) [19–21]. In CD, chronic inflammation of the ileal epithelium leads to the over-expression and the display of the MGP carcinoembryonic antigen-related cell adhesion molecule 6 (CEACAM6) on epithelial cell surfaces. The adhesion of these AIEC bacteria via FimH-CEACAM6 binding leads to further bacterial invasion of the gut mucosa [21,22]. Current results show that FimH antagonists can decrease the AIEC population in-vivo [23]. An anti-adhesive mannosidic compound named EB8018 (Enterome; licensed in early 2016) treating CD is currently in the human testing phase [24]. EB8018 is a divalent compound allowing for the binding of two FimH proteins at the same time.

The FimH proteins of UPEC and AIEC have been used in the last two decades as a target in the development of precision antimicrobial drugs [25]. Such drugs have several advantages over the more traditional antibiotic drugs: (1) they are specific for a certain type of process or bacterial species (2) they do not disturb the host microbiota and (3) they are not likely to induce bacterial resistance as they interfere with the pathogen without killing it. Most of the currently known FimH inhibitors (e.g., heptyl α-D-mannopyranoside (HM), K_D = 5 nM) [26] have been rationally designed on the basis of structural information obtained by X-ray crystallography [24,26,27]. A new route for drug design is to include the dynamical aspects of the binding process. This review summarizes how the inclusion of dynamical information from molecular dynamics (MD) studies as well as other molecular simulation techniques can be used to gain further insight into the interaction between the anti-adhesive compound and its receptor FimH and how this information is incorporated into rational drug design to further improve the efficiency of the anti-adhesive compounds.

2. The Molecular Binding Mechanism of Small Mannosidic Compounds to the FimH Binding Site

2.1. The FimH Mannose-Binding Site

The first crystal structure of an α-D-mannose molecule bound FimH was reported in 2002 [14], disclosing that FimH is composed of two structurally similar domains, both with an immunoglobulin (Ig)-like fold (11-stranded β-barrel) connected through a flexible linker (amino acids (aa.) 154–160) (see Figure 1A). The N-terminal, lectin domain (aa. 1–153) carries the mannose-binding site, whereas the C-terminal, pilin domain (aa. 161–276) mediates the connection with the other proteins of the type 1 pili. The co-crystallized α-D-mannose molecule is located in a polar pocket (see Figure 1B, Asn46, Asp47, Asp54, Gln133, Asn135, Asn138 and Asp140) of the lectin domain. Its tight binding is achieved predominantly through hydrogen (H) bonding (direct and water-mediated) and other electrostatic interactions. The binding pocket is surrounded by a collar of hydrophobic residues (see Figure 1B, Phe1, Ile13, Tyr48, Ile52, Tyr137 and Phe142).

The lectin binding site is highly specific for α-D-mannose (K_D = 2.3 μM) as evidenced by surface-plasmon resonance (SPR) measurements [26]. Minor changes in the chemical structure of the sugar as for example the change of the 2-hydroxyl group position (D-glucose, K_D = 9240 μM) or its complete removal (2-deoxy-α-D-mannose, K_D = 300 μM) results in compounds only very poorly recognized by FimH [26]. Only the sugar fructose, a five membered ring with the 2-hydroxyl group being axial, shows an affinity for FimH binding that is near the one of α-D-mannose, albeit 15-fold less (K_D = 31 μM) [26]. Most of the key residues (Phe1, Asn46, Asp47, Asp54, Gln133, Asn135, Asp140 and Phe142) shaping the FimH mannose-binding pocket are invariant throughout all known strains of *E. coli*. The mutation of any of these residues led to a loss of mannose binding and diminished virulence [14,28]. These observations are in line with the high specificity of FimH for α-D-mannose.

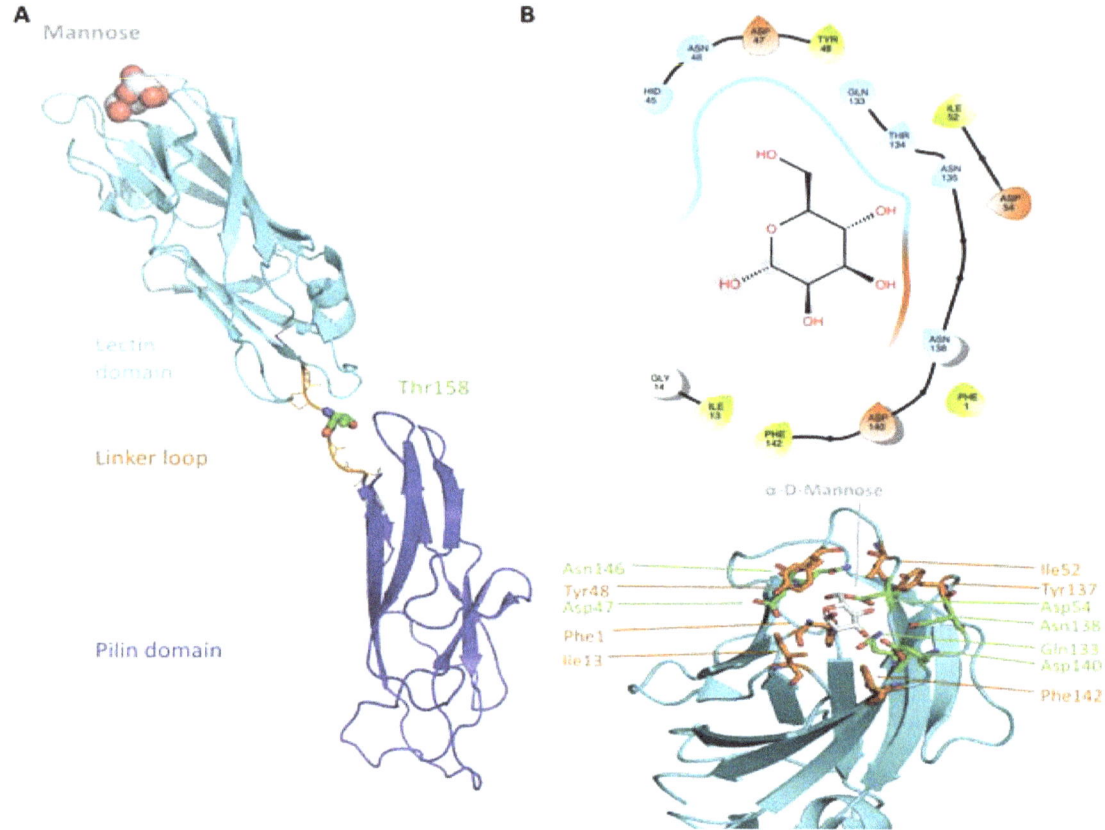

Figure 1. The FimH structure and organization (**A**) An elongated linker (orange) connects the pilin (blue) and the lectin (cyan) domain of FimH (PDB code 1KLF [14]). The protein is shown in cartoon and the bound α-D-mannose molecule is depicted as atom-colored (grey for carbon) van-der-Waals spheres. Additionally, the position of T158 is shown as atom-colored sticks (green for carbon). (**B**) The mannose-binding site of the FimH lectin domain. On the top the 2D diagram of the binding site is depicted (prepared with Maestro using a cutoff of 5 Å) and on the bottom the 3D representation of the same site. The mannose molecule is highlighted in gray. The polar (green) binding site residues as well as the hydrophobic rim residues (orange) are additionally depicted. The 3D protein representations in this and the following figures were prepared using Pymol [29].

As the FimH lectin domain is highly specific for mannose, and no other site was exploited so far in anti-FimH drug design, most so-far developed FimH inhibitors contain a mannose compound (see Section 3.1). In the more than 50 crystal structures of FimH in an inhibitor-bound state, that can be accessed today in the PDB database, the mannosidic moiety binds in the same way, independent of the chemical nature of the aglycon moiety (see Figure 2A). Very recently, however, a series of FimH inhibitors were designed featuring instead of a α-D-mannose ring a seven-membered ring analog (septanose rings). One among them, the 2-*n*-heptyl-1-deoxyseptanose (HS), is very promising as it features only an about 10-fold reduced affinity (K_D = 0.26 μM) compared to HM (K_D = 0.029 μM in the same isothermal titration calorimetry (ITC) measurement) [30]. Furthermore, the crystal structure of

FimH in complex with this HS compound highlights that the septanose ring is very similarly bound as the mannose ring of HM sharing the same H bond partners (Phe1, Asp47, Gln133, Asn137 and Asp140) [30]. Further optimization of this HS might lead to a new class of potent FimH inhibitors.

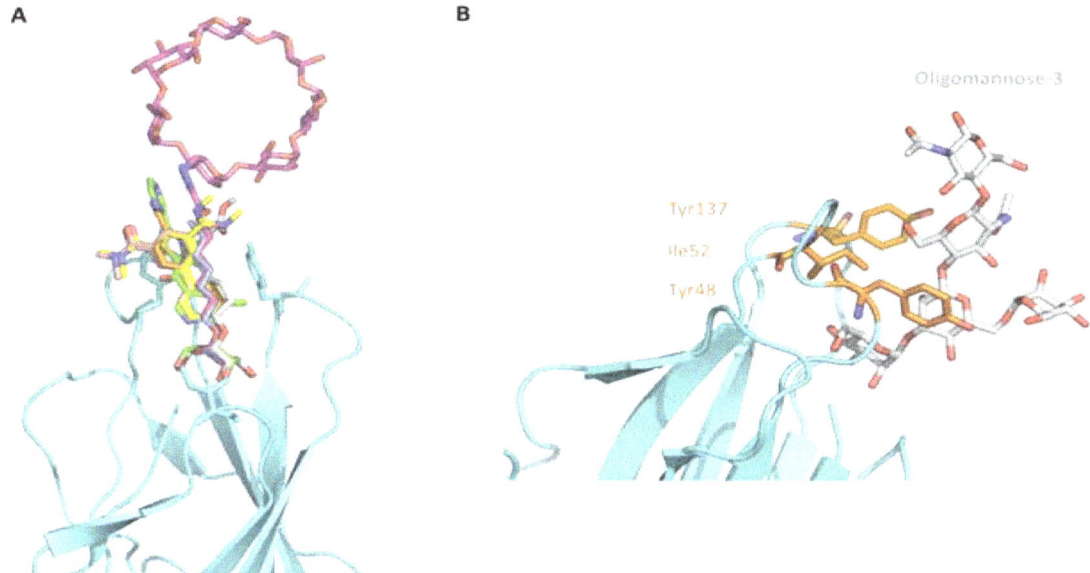

Figure 2. FimH in complex with different inhibitors. (**A**) The mannose ring of different recent high-affinity inhibitors is bound similarly to FimH. Shown are the following inhibitors: HM (lilac, PDB ID: 4BUQ [31]), thiazolylaminomannoside (green; PDB code 5MTS [32], β-cyclodextrin-α-D-mannoside (purple; PDB code 5AB1 [23]), para-biphenyl-2-methyl-3′,5′di-methylamide-α-D-mannoside (yellow; PDB 5F2F [33]), 8-(Methoxycarbonyl)octyl-α-D-mannoside (grey, PDB code 4AVI [34]), 3′-Chloro-4′-(α-D-mannopyranosyloxy)biphenyl-4-carbonitrile (orange, PDB code 4CST [35]), *para*-biphenyl-2-methyl-3′-methylamidemannoside (rose, PDB code 5F3F [36]). (**B**) Oligomannose-3 bound to FimH. The mannoside is highlighted in grey and the tyrosine gate residues in yellow. The FimH lectin domain is shown in cyan in the cartoon (PDB code 2VCO [37]).

2.2. The Tyrosine Gate and Its Impact on Mannoside Binding

The crystal structure of FimH with the branched oligomannose-3 [37] highlights the particular importance of the tyrosine gate, formed by Ile52, Tyr48 and Tyr137, for the binding of the mannose rings adjacent to the first mannose ring bound in the pocket (see Figure 2B). The tyrosine gate is located at the entry of the binding pocket and forms part of the hydrophobic collar (see Figure 1B). It is at the level of the tyrosine gate, that the isolated FimH lectin domain differentiates between different high-mannosidic glycans, mainly based on their capability to form hydrophobic interactions with this gate. A recent combined molecular simulation and experimental study highlighted the coupling of the motion of the two tyrosine residues via Ile52 (see Table 1) [38]. The tyrosine gate has attracted large interest because of its potential to generate nanomolar affinities for mannosides conjugated to hydrophobic aglycons through the formation of favorable van der Waals and stacking interactions within the gate. Based on crystallographic data, different inhibitor interaction modes have been evidenced: the non-glycon substituents either travel through the gate and interact in multiple stacking modes either (1) with Tyr48 (Tyr48-loving) or (2) with Tyr137 (Tyr137-loving), or (3) bypass the tyrosine gate and interact with either one or both the tyrosine residues from the outside [31,34,35,37,39–41].

The interaction of the aglycon moiety of the anti-adhesives with one or several tyrosine gate residues has been shown to impact the affinity of the inhibitor. Furthermore both Tyr48 and Tyr137 have been evidenced to be highly dynamic [34,38]. A detailed molecular understanding of the mode of action of the tyrosine gate, including its dynamical behavior, is therefore required in order to design more efficient inhibitors.

Table 1. Residues from the FimH lectin domain important for its function (and discussed in this review) are listed. These residues are either (i) involved in binding of the algycon moiety in the FimH mannose-binding pocket, (ii) important for the conformational change of FimH or (iii) have been shown to be involved in promising alternative binding positions. For each residue the available experimental evidence as well as the insight gained from molecular simulation shortly summarized. The most promising residues are highlight by an asterisk. The sequence from the UPEC strain UTI89 was used.

Residue	Important Due to	Exp. Evidence	Insight from Molecular Simulation
Ile13	Located in the clamp loop (changes conformation due to shear force) Possibly involved in alternative binding position	Ile13 forms van der Waals interactions with the C1–C2 bond of mannose [42] Crystal structures of the HA and LA state highlight the movement of the clamp loop [43]	The aglycon moiety of the C117 and of biantenarry mannosides orients towards Ile13 [39,44].
Glu50	Part of a possible new binding site for anti-adhesives	EDTA binding site [38] Implied in the shear-force dependent conformational change [45] Less adhesion of the E50A mutant under shear [45]	
Ile52	Belongs to the tyrosine gate	Attributed to the tyrosine gate on the basis of crystal structures [42]	Mediates coupled motion of Tyr48 and Tyr137 [38]
Thr53	Part of a possible new binding site for anti-adhesives	EDTA binding site [38] Implied in the shear-force dependent conformational change [45] Less adhesion of the T53A mutant under shear [45]	
Asn136	Part of a possible new binding site for anti-adhesives	EDTA binding site [38]	
Tyr137	Belongs to the tyrosine gate Binding of the aglycon part in the mannose-binding moiety	Y137A mutation significantly reduces FimH affinity towards f HM [38]	The flexibility of the bound HM is increased in the Y137A mutant; The apo mutant already is in a quasi-bound configuration [38]
Thr158	Implicated in the shear-force dependent conformational change	Natural variation leads to bacteria with different stress responses [22,46,47]	A force was applied to this residue in the sMD simulation [48]

We recently generated single-residue FimH mutants in which one of the two tyrosine-gate residues was mutated to alanine (Y48A and Y137A). The effect of these mutations on the binding of three synthetic ligands (1,5-anhydro-D-mannitol, HM, and 4-biphenyl-α-D-mannose) was tested by X-ray crystallography, affinity measurements and molecular simulation studies [38]. The experimentally determined affinity data highlight the importance of Tyr137, as its mutation clearly alters the binding properties of the FimH lectin independently from the ligand used (Table 1). No major structural changes were evidenced in the mutant by X-ray crystallography and CD measurements. Only the combination of quantum mechanics (QM) calculations and MD simulations revealed why the FimH Tyr137Ala (Y137A) mutant shows such a dramatic loss of affinity without being in direct contact with its mannose ligand: in the ligand-free state of the FimH Y137A mutants, several of the binding site residues (48, 136, and 137) exhibit backbone dihedral angles that are normally only found after binding of the mannose. This is because the Y137A mutation disrupts a dynamic coupling between Tyr137 and Tyr48 via the inner Ile52 residue and holds the binding cavity in a highly energetic mannose binding conformation [38]. In addition, the ligand retained a higher flexibility in the binding site of the Y137A mutant compared to the wild-type. In contrast to this, the in-silico mutation of Y48A only minimally affects the binding affinity of the different ligands as shown by smaller observed effects on the flexibility of the ligand and on the protein local dynamics. This is in good agreement with entropy-enthalpy compensation effects seen in ITC measurements performed within our study [38] as well as in an earlier study of the Y48A mutant [49]. The latter study also showed, using NMR,

X-ray crystallography and SAXS measurements that the Y48A mutation does not affect FimH structure and function.

2.3. The Conformational States of FimH

All inhibitors discussed so far in this review target FimH in its high affinity (HA) state, however, FimH can also exist in a low affinity (LA) state (see Figure 3A), which is at least 100 times less efficient in mannose binding [50]. In the absence of any force FimH is in its LA state, most likely loosely adhering to its receptor, allowing thereby the UPEC or the AIEC bacteria to change their position and move along the tissue [51,52]. Shear force can be observed in the human body in the form of the flow of fluids such as mucosal secretions used as natural body defenses against bacterial colonization [53]. Furthermore, shear forces can also act on UPEC in the form of urine flow. Under laboratory conditions, force application triggers the conversion of FimH from its LA state to its HA state. The conversion most likely allows the bacteria to withstand the vigorous shear stress imposed by the (human) host. The combination of steered MD (sMD) simulations and atomic force microscopy (AFM) measurements allowed to get insight into the molecular origin of the conformational change [54–56]. Moreover, using the fimbrial tip (consisting of FimH, FimG followed by one FimF) structure [57], a coarse-grained lattice Boltzmann MD simulation study showed that the application of fluid flow leads to a drastic alternation of the complex conformation [58]. In these simulations, the chain stretched according to the fluid velocity drag in accordance with a shear-force dependent conformational change. In the sMD simulations of the isolated FimH tensile forces were applied between residues in the mannose-binding pocket or the mannose and residues at the end of the interdomain linker chain [54]. In the study of the fimbrial tip tensile force were applied between the binding site residues and the donor-strand of the second FimF molecule [55]. Independent of the used sMD approach the interdomain linker loop extended under the applied force. In line with the observed importance of the linker, SPR experiments highlight that natural variants of FimH with different amino acids in the position 158 (located in the linker loop region, see Figure 1A) show different responses to stress [22,46,47]: the adhesion strength of the uropathogenic UTI89 E. coli strain with a threonine at position 158 of FimH shows an optimum at higher shear, whereas in strains (AIEC7082 and LF82) carrying an alanine or a proline respectively at aa. position 158, bacterial binding is less or not enhanced with increasing fluid shear.

The crystal structure of FimH as part of the multi-protein fimbrial tip (FimH followed by one FimG and two FimF molecules, the last one stabilized by the FimC chaperone) highlight the FimH in its LA state [57]. Even in absence of the FimG proteins, the LA state can be stabilized by the co-crystallization of FimH with a DsG peptide, which fills the place of the donor-strand of FimG complementing the missing β-strand of the Ig-fold of the FimH pilin domain [59]. In the FimH LA structure (see Figure 3A), the anchoring (pilin) domain of FimH interacts with the mannose-binding (lectin) domain and causes a twist in the β-sandwich fold of the latter. This loosens the mannose-binding pocket on the opposite end of lectin domain. The HA was observed in the isolated lectin domain structures [26,37], the FimH-FimC structure [14,60], and the HM-bound FimH-DsF peptide structure [59]. In these structures the lectin domain is untwisted and elongated compared to the LA state resulting in a tight, high-affinity mannose-binding pocket. In the FimH-FimC structures the HA is most likely stabilized by the FimC chaperon that is wedged between the FimH lectin and pilin domain thereby separating the two domains. Three flexible loops, the so-called swing (aa. 27–33), linker (aa. 154–160) and insertion (aa. 112–118) loop, have been identified to mediate contact between the pilin and lecin domain in the LA state (Figure 3) [57]. During the conformational change of the LA to the HA state these loops are rearranged leading to the rupture of the inter-domain connections and elongation of the linker loop. SPR experiments with natural variants of FimH (e.g., aa. 158; Table 1) found in different E. coli strains corroborate the importance of the linker loop in the shear-dependent conformational change [22,46,47]. The analysis of sMD simulations combined with experimental data highlighted that below a force of about 60 pN, the unbinding of the mannose is mainly observed from the LA state (with a rate of $6\ \text{s}^{-1}$ [61]. Above that force, the conformational change to the HA state is the main occurring event (with a rate of $0.00125\ \text{s}^{-1}$) [55,62].

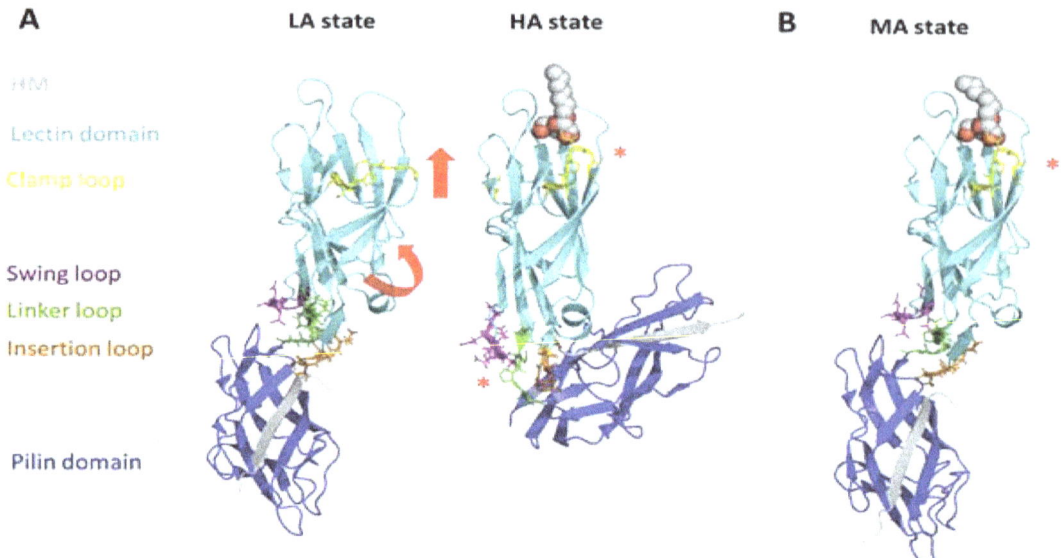

Figure 3. The conformational flexibility of FimH. The LA (**left**; PDB code 4XOD), HA (**middle**; PDB code 4XOB), and MA (**right**; PDB code 4XOE, chain G and H) state are depicted. Following a β-sheet twisting mechanism, the lectin domain is elongated and straightened in the HA (MA) state (red arrows) leading to a local conformational change in the mannose-binding site (red star) locking it in its high affinity conformation. In the HA state the pilin domain is elongated and the link between the pilin and the lectin domain is weakened. The lectin (cyan) and the pilin (blue) domain as well as the clamp (yellow), swing (purple), insertion (orange), and linker (green) loops are shown in cartoon (domains, loops) and in lines (loops). The co-crystallized peptide is shown in grey cartoon. The HM bound to the HA and MA state is shown in van-der-Waals spheres (grey).

The mechanical activation of FimH (the switch of FimH from its LA to its HA state) was proposed to be due to allosteric coupling of its two domains: the pilin domain functions as an allosteric autoinhibitor of the lectin domain, which is pulled away by the mechanical force (as described above). The interdomain loops (see Figure 3) have been identified as structural elements important for the allosteric activation of FimH [48,54,57]. Based on computational structural analysis (using the Rosetta Design tool) [63], MD simulations, site-directed mutagenesis and enzyme-linked immunosorbent assay (ELISA), a β-bulge (aa. 59 to 63) and α-switch (aa. 64 to 71) region have been pinpointed to be also tightly coupled to the pilin domain and playing an important role in the allosteric change [64]. These regions and the clamp loop (aa. 10 to 15), which closes the binding site in response to mannose binding (see Figure 3), have also been identified as significantly changing their conformation in an extended MD study of the FimH HA/LA change using the Anton supercomputer [65]. Based on crystallographic data [57] and MD simulations [65], a large β-strand (aa. 16 to 22) connecting the clamp loop was identified as mediator to propagate the allosteric signal from the binding site to the pilin domain. Thus a possible treatment alternative to antibiotic treatment could be to develop allosteric inhibitors or antibodies against FimH [66]. Several anti-FimH antibodies have been identified so far [66–68]. The antibodies carry out inhibition using either an allosteric (mAb21, [66]), competitive, orthosteric (mAb475, [68]) or non-competitive, parasteric (mAb926, [67]) binding mechanism. Whereas the mAb21 was found to significantly enhance adhesion, most likely by stabilization of the HA state [58,60], the mAb926 and mAb475 antibodies are strongly inhibiting adhesion, the latter via blocking the switch from the LA to the HA state [67]. In the context of antibody design, MD simulations are a helpful method as they allow to verify if epitopes, identified from static structures, are also accessible to the antibody considering the dynamical nature of the protein [69].

Recently, the crystal structure of a HM-bound FimH-DsG peptide complex highlighted a third possible conformational state, named the middle affinity (MA) state [59]. In this state, the interdomain loops are in the same conformation as in the LA state, whereas the clamp loop already closed upon the mannose-binding site (see Figure 3B). MD simulations highlighted that the MA state is stabilized by ligand binding as after in silico removal of the HM led to a spontaneous relaxation of the clamp loop relaxes back to the LA state [59].

Although the conformational flexibility of FimH is the largest contributing factor in the shear-force-dependent binding strength, the other fimbrial tip proteins (FimG and FimF) also play a crucial function in the adhesion process. The quaternary structure of the multi-protein fimbrial tip was proposed to be highly flexible to optimize the binding rate [51] in contrast to the rod which was shown to be rigid [70]. Indeed both MD and NMR studies of FimG-FimF and FimF-FimF dimers highlight high levels of mobility [55,71]. The rigid rod has been shown to be able to recoil under increased force conditions to prevent breakage of the high-affinity mannose bond in the FimH lectin domain [72]. The fimbria and in particular FimH thus allow the bacteria to hold firmly onto the cells in the presence of a shear stress (such as the urine flow) as well as in the absence of this stress in order to detach and change location. The phenomena of sustained FimH binding (slower off rates) under stress conditions was descripted by "catch bonds" [73]. Catch bonds were also observed in other adhesive proteins and are thus likely to be a common phenomenon for proteins involved in various adhesion processes [61,74].

3. Rational Drug Design of FimH Inhibitors

3.1. Monovalent FimH Inhibitors Targeting the Mannose-Binding Pocket of the HA FimH State

As historically only the HA state of FimH was available, several classes of mannosidic inhibitors have been rationally designed based on structural information targeting the FimH mannose-binding site of the HA state [42,75]. These compounds can be subdivided into the following chemotypes: alkyl/aryl mannosides, biaryl mannosides, mannose ring modifications including *O*-glycosidic bond replacement leading to *N*-, *S*-, or *C*-linked compounds (Table 2) [24,39]. Several of these compounds have been crystallized in complex with FimH recently, highlighting the fact that the mannoside moiety binds in the same fashion, independently of the identity of the atom type at the glycosidic bond [36,40,76–78] (Figure 1C) and that the non-sugar moiety interacts with the tyrosine gate in one of the three modes described in Section 2.2. The added advantage of non-*O*-glycosidic linked compounds is that they are less sensible to host glycosidases and thus might be better suited for therapeutic use [24]. An extensive overview of all physiochemical properties of the currently known FimH inhibitors has been published elsewhere (see for example Reference [24]). Of particular interest are the recently developed thiazolylaminomannosides (TazMans), as they have been identified as potent anti-adhesives of different *E. coli* strains isolated from patients with CD, cystitis or osteoarticular infections [77–79].

As the search for new FimH inhibitors is largely structure-driven, several examples exist in which structural and affinity data were combined with molecular docking in a rational drug design approach [27,36,80,81]. In 2006, shortly after the first X-ray structures of FimH became accessible, a first combined experimental/molecular docking study was published [80]. This study highlights that for most of the tested compounds (alkyl and squaric glycans) the computed docking score is related to the affinity data obtained by ELISA measurements. The combination of docking with bioassays allowed to determine the binding mode of squaric acid monoamine mannosidic compounds to FimH and their affinities [82]. Based on docking poses with biphenyl derivatives with nanomolar affinities for FimH, new biphenyl inhibitors were designed, some of which showed higher affinities, increased solubility and slightly improved pharmacokinetic properties as the original compounds [81]. In 2013, over 100 mannoside compounds were also used to develop a multi-dimensional quantitative structure-activity relationship (mQSAR) and to develop an automatized tool box for in-silico rational drug design and MD simulation [83]. Docking models of *C*-linked mannosides in *R*- and *S*-isomer highlighted that the binding of the *R*-isomer to FimH is energetically favored due to a water-mediated H-bond with Asn138 and Asp140, which is only observed in the former [36]. The induced-fit docking of *C*-, *O*-, *N*- and *S*-glycosidic compounds to FimH further indicated that the position of this water and thus the distance to the linkage depends on the identity of the exocyclic atom (distance water O-exocyclic atom 2.9 Å for O and N, 3.5 Å for S and 3.6 Å for C) [27]. The dependence of the distance of the water to the glycosidic linkage was also highlighted in a crystallographic study of C-, N- and O-linked compounds [40]. The position of the water in the different bound states might influence the affinity of FimH for the different compounds.

Table 2. Classification of FimH mannosidic inhibitors. For each compound type one or more examples with their affinities are listed. PDB codes for wild-type structures of FimH in complex with the corresponding inhibitors are given. The following abbreviations are used: FDA for fluorescence polarization assay; ELLSA for Enzyme-linked lectinosorbent assay; HAI for hemagglutination inhibition.

(A) Different O-Linked Mannosidic Compounds					
Compound Type	**Example (s) (R=)**	**Measure (Technique)**	**Value [nM]**	**Ref.**	**PDB Code**	**Ref.**
Mannose	H	K_D (ITC) K_D (SPR) EC_{90} (HAI)	1672.2 2300.0 >1 mM	[34] [24] [24]	1KEF	[14]
Alkyl mannosides		K_D (SPR) K_D (ITC) K_D (ITC) K_D (FDA) EC_{90} (HAI) EC_{90} (HAI) IC_{50} (ELLSA)	5.0 28.9 7.3 28.3 1500.0 6300.0 160.0	[26] [38] [34] [35] [24] [39] [24]	4BUQ 4LOV 4XOE 4XOB	[31] [49] [59] [59]
		K_D (ITC)	23.6	[34]		
Aryl mannosides		K_D (ITC)	18.3	[34]		
		IC_{50} (Bioassay)	1730.0	[82]		
Biaryl mannosides		K_D (ITC) K_D (FPA)	17.7 15.1	[38] [35]	5FWR	[38]
		K_D (ITC)	3.5	[81]		

Table 2. *Cont.*

(B) Mannose Ring Modifications

Ring modification	Example(s) (R=)	Measure (Technique)	Value [nM]	Ref.	PDB Code	Ref.
N-linked compounds X = N		IC$_{50}$ (ELLSA)	70.0	[32]	5MTS	[32]
		IC$_{50}$ (ELLSA)	205.0	[32]	3LZ2	[77]
C-linked compounds X = C		EC$_{90}$ (HAI)	3.1	[36]		
		IC$_{50}$ (ELLSA)	194.0	[32]		
S-linked compounds X = S		IC$_{50}$ (ELLSA)	146.0	[23]		

Only a few examples exist in which the flexibility of the ligand and the very dynamical behavior of the protein were taken into account into the drug development and/or the understanding of the underlying mechanism(s). One such example is the determination of an alternative binding position of a C-linked *ortho*-subsituted biphenyl mannose derivative (C117) in the FimH binding site in its HA state [39]. After overlaying the sugar of the NMR-solution C117 with the position of the mannose ring of other FimH inhibitors in the binding site indicated that the C117 first phenyl moiety of C117 interacts with Tyr48 and the second one points towards Ile13 (Table 1), which is part of the clamp loop (see Figure 2). In the absence of structural information of the C117-FimH complex, the existence of such a secondary binding position for C117 could only be evidenced by combining molecular docking and MD simulations. Indeed, following the dynamics of the generated C117-FimH complex the Ile13-oriented binding mode could be identified as a minor binding mode (see Figure 4A) [76]. The Ile13-oriented binding mode was also identified as the secondary binding mode for biantennary mannosides using molecular docking [44]. The identified minor binding mode is of particular interest as the clamp loop (see Section 2.3) undergoes a major conformational change when FimH forms high-affinity catch bonds with mannosides [76] and changes from the LA to the MA and eventually to the HA state. Mannosides targeting and stabilizing this secondary binding site are in the focus of further inhibitor development as they might alter the kinetics of the FimH conformational change.

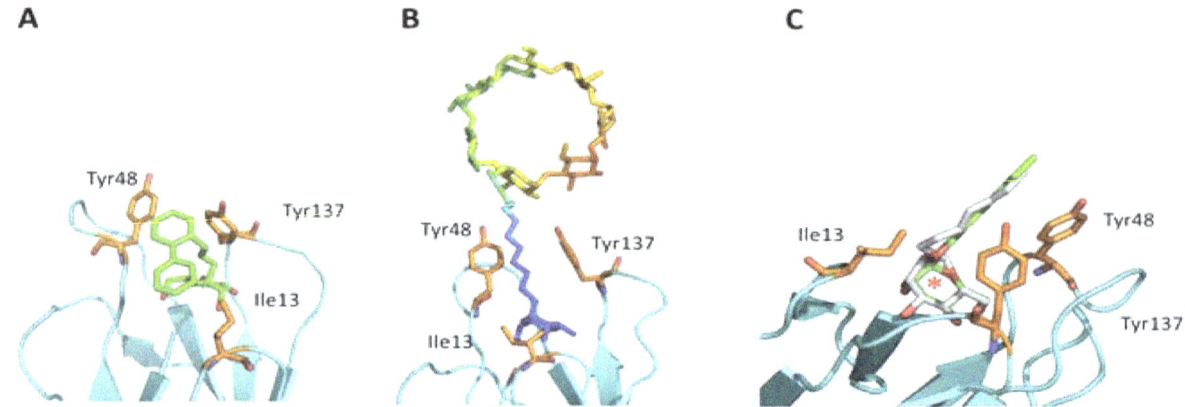

Figure 4. Recent results including the dynamics aspect of the FimH-ligand complex (formation). (**A**) A minor MD conformation (populated with 11%) shows the C117 second phenyl ring orientated towards C117 [76]. (**B**) The position of bCD in the binding pocket (PDB code: 5AB1). The ligand is colored according to the structure factor (from blue: rigid to red: highly flexible). (**C**) Comparison of the septanoside position (green; HS; PDB code: 5CGB) compared to a mannoside compound (grey; HM; PDB code: 4BUQ). The difference in the sugar ring is highlighted by a red asterisk.

Another example of how molecular simulation can contribute to understand the drug properties is captured in the case of a β-cyclodextrin (bCD)-containing HM (bCD-1HM; Figure 4B). This inhibitor was recently shown to disrupt the attachment of *E. coli* to the bladder or gut of mice models of cystitis and CD, respectively [84,85]. The crystal structure of the bCD-1HM HA FimH complex highlighted that the HM part of this compound adopts a similar conformation as the isolated HM antagonist in the FimH binding site [23]. According to the obtained electron density and MD simulations, the bCD moiety does not form any significant interactions with the protein and moves freely in solution. Thus, it does not seem to impact bCD-1HM binding. Surprisingly however, bCD-1HM has a much lower effect as HM on the capability of *E. coli* LF82 strains to adhere to intestinal epithelial cells (T84) [23]. MD simulations of the bCD-1HM compound alone in water allowed to provide a possible explanation of the observed effect difference: the bCD moiety of bCD-1HM seems to fold back and interact with the HM moiety of the compound thereby locking it in a state unfavorable to FimH binding as the mannose part is shielded by the interaction. The addition of the bCD moiety to HM is thus likely to modulate the pharmacokinetics of the compound but not the in-vivo affinity.

Metadynamics simulations of the inhibitors alone in solution also allowed to understand why the change of the mannose sugar to a septanose ring led to an increased entropic penalty in ITC measurements even so they show similar binding to FimH (Figure 4C) [30]: the HM (with a mannose) ring had only a single energy minimum for the considered O1–C1–C4 angle and the O1–C1–O5–C5 of dihedral torsion, which represents the HM conformation in the FimH bound state, whereas for the corresponding septanose (HS) compound a more shallow energy landscape was observed with two energy minima for the corresponding angle/dihedral, one of which does not correspond to the bound state of the compound. Thus, the binding of the septanose derivative leads to a higher reduction in conformational flexibility of the sugar and thus accounts for the higher entropic cost.

As shown in the examples, the incorporation of data on the dynamics behavior of the protein, the complex and the ligand in water, often originating from MD simulations, allows to describe and therefore understand the dynamics of the binding process. It thus complements the more traditional approaches such as crystallography and affinity measurements from which static pictures are obtained.

3.2. Multivalent FimH Inhibitors Targeting the Mannose-Binding Pocket

Adhesion of pathogenic E. coli is mediated by multiple type 1 fimbria and thus FimH binding to MGP displayed on the host cells (for example UPIa in the case of UPEC [18] or CEACAM6 in the case of AIEC [86]). It is well known that so-called glycoclusters can improve the affinity for lectins to a large extent [11,87]. Moreover, rather high concentrations of HM are needed to obtain 90% reduction of the bacterial load in a mouse model [42,85], making the design of better binding inhibitors needed. Therefore, multivalent antagonists were designed mimicking the clusters of glycans on the host cells [11,12]. These multivalent versions have the advantage over the monovalent counterpart that they could induce FimH aggregation leading to fimbrial entanglement followed by the formation of large bacterial aggregates that are less prone to adhering to human epithelial host cells. For example, a multivalent version of bCD-1HM carrying seven HM on the bCD ring (bCD-7HM) was shown to interact with different FimH molecules simultaneously and to induce FimH aggregation and precipitation [85,88]. Interestingly bCD-7HM, highlighted a 100-fold reduction in the effective dose in CEACAM6-expressing mice compared to its monovalent version [84]. Further development of more efficient multivalent inhibitors will largely benefit from the incorporation of the dynamical behavior, as assessable by MD simulations, of the complex and the inhibitor alone in water.

3.3. Alternative Binding Positions for Inhibitors

An alternative route to develop higher affinity FimH antagonists would be to target other off-site positions instead of improving the binding affinity of FimH inhibitors that target the mannose-binding pocket. Targeting such off-site binding positions might have the advantage that such an inhibitor might block the protein in a state non-accessible to MGP binding. Recently a promising off-site binding pocket has been serendipitously discovered in the ligand-free Y137A FimH mutant crystal structure [38]. In this structure a single ethylenediaminetetraacetic acid (EDTA) molecule was observed to be bound in several orientations near to Glu50, Thr53, and Asn136 (see Figure 5).

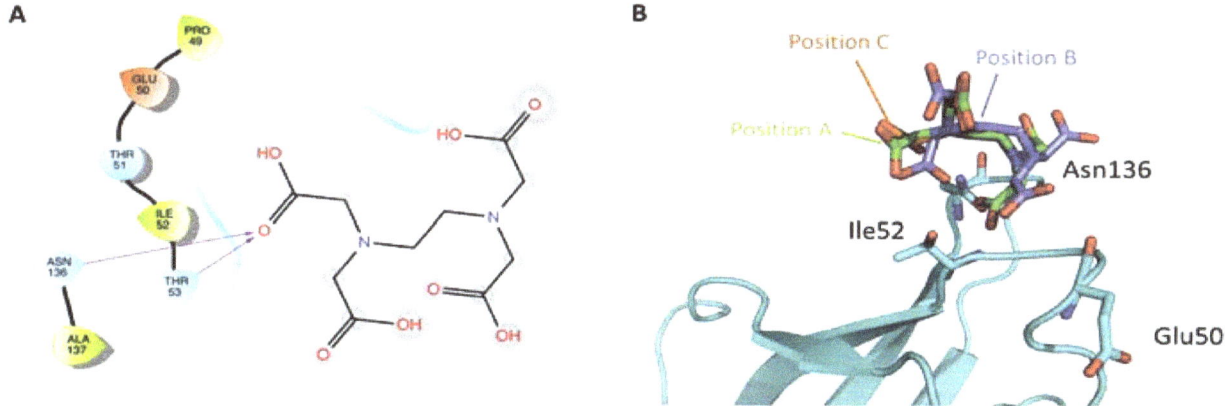

Figure 5. EDTA-binding site in the FimH lectin domain. (**A**) The 2D diagram of the binding site is depicted (prepared with maestro using a cutoff of 5 Å) (**B**) on the right the location of the binding site in the crystal structure (PDB code: 5FX3 [38]) is shown. Slightly different, alternative positions (**A–C**) of EDTA have been observed in the crystal structure and are colored differently (green, lilac, and orange).

Most of these residues have been shown to be important for the shear-force enhanced E.coli adhesion to vascular epithelium cells (Table 1) [45]. The EDTA molecule was not found in the mannoside-occupied wild-type, Y48A or Y137A mutant FimH structures in spite of identical extraction, purification and crystallization protocols [38]. The relaxation of the FimH mannose-binding pocket due to the Y137A mutation (see Section 2.2) might have allowed for the binding of EDTA. The newly discovered EDTA-binding site is close to a belt of positively charged residues (aa: Arg60, Arg92, and Arg132), which are moreover strictly conserved within E. coli. This might indicate a protein-docking pocket in the continuation of the mannose-binding site of FimH [27]. Investigation of the flexibility of the bound EDTA as well as the design of specific compounds targeting this site will allow in the future to judge the importance of this site in the FimH adhesion process.

Also the allosteric inhibition of FimH (see Section 2.3) by side-specific antibodies could be an alternative route for treatment of *E. coli* infections such as UTI. Promisingly, a vaccination study with FimH highlighted protection against bacterial infection in the case of mice and primates [13] and antibodies inhibiting the conformational change from the LA to the HA state have been described [67]. In the design of more efficient antibodies, the incorporation of dynamical data, such as provided by MD, will prove helpful. In more general terms, the investigation of the conformational flexibility of the FimH protein and its different affinity states might open new avenues for non-mannose-binding site inhibitors of FimH.

4. Molecular Simulation as a Tool to Study FimH Function and Inhibition

A wide variety of different molecular simulation techniques as protein-ligand docking, MD and QM calculations were applied to FimH in order to study its conformational change, the binding of its substrates and inhibitors and to design new and more efficient anti-adhesive molecules targeting FimH. The different methods range from the investigation of model systems consisting of a few atoms like in QM, to a few thousands of atoms representing the entire protein as in molecular docking or to several tens of thousands of atoms, describing the solvated protein in an explicit solvation sphere in MD simulations. These methods representing both a static (as in QM and docking) and a dynamic modelling (as in MD) were applied together with different experimental techniques including SPR, ITC, X-ray crystallography or NMR.

All different molecular simulation approaches have their own advantages and problems. Force-field based methods such as MD for example are computationally very efficient and can handle large systems, giving an (almost) correct representation of the biological reality. In contrast, QM calculations are very time consuming and can handle only a fraction of the real system. However, force-field based methods suffer from serious shortcomings in e.g., the description of charge transfer, halogen bonding and polarizability (see for example [89] and references herein). As such, it is expected that QM calculations outperform the force-field based approaches in accuracy of (relative) energies (see for example [90] and references herein). When used for specific purposes, QM calculations on small models can thus add to the classic (MD) description. An example of a combined QM and MD approach is the study of the Y48A and Y137A mutation impact on FimH inhibitor binding [38] (see Section 2.2).

In most studies on FimH, molecular simulation was used to understand functional details of the protein not decipherable by the *priori* performed experiments. Example are the study of the molecular origin of the increased entropic penalty in ITC measurements if the HM mannose sugar is replaced by a septanose leading to HS [30] or the determination of the molecular reason for the Y137A mutation effect on the FimH HM affinity [38]. However, recent studies use molecular simulation as tool to predict effects on FimH function and regulation, which are afterwards proven experimentally. Examples hereof are the generation of a recombinant fusion proteins as a possible UTI vaccine [91,92]: the three-dimensional structural models of FimH fused to either flagellin [91] or MrpH [92] produced using molecular modelling were tested in-silico for their binding affinity towards Toll-like receptors. The fusion proteins with the best binding affinities also showed immune responses in an cell-based assay [92] and in mice experiments [91]. A similar combined molecular simulation experimental approach was followed to generate dimeric and trimeric fusion proteins of FimH with CsgA, and PapG adhesins [93].

It is a necessity for the development of UTI or CD vaccines as well as for the development of treatment alternatives to antibiotics for these diseases to combine experimental and theoretical approaches in future research. Molecular simulation can help to predict possible conformations and interactions as well as affinities of ligands and can thus explain experimental results and/or allow for the design of new experiments.

5. Conclusions and Outlook

The use of molecular dynamics simulation in complement to crystallographic assays offers a powerful combination to study ligand binding. For instance, the application of molecular dynamics simulation in combination with quantum-chemical calculations have allowed to understand the molecular importance of the FimH tyrosine gate and the impact of mutation of its residues on binding affinity. The incorporation of dynamical information on the wild-type FimH lectin domain in the ligand-free and ligand-bound state into a structure-based rational drug design approach allowed for the identification of a previously unidentified and promising binding mode of the ligand in the FimH binding site, in which it is oriented towards the clamp loop. These two examples among others highlight the added value of molecular simulation in the drug design of inhibitor molecules, here targeting FimH.

Further applications of molecular simulations techniques could be the identification of alternative binding positions for anti-adhesives on the FimH lectin domain, such as the recently identified EDTA binding site, or of compounds binding tightly to these positions. Such compounds might fix FimH in an off-path conformation and could thus abolish FimH receptor binding. In a similar fashion, simulation techniques can be expected to lead to the identification of key residues in the conformational change which could be then be targeted.

Molecular simulation will allow to rationally design new anti-adhesives either being specific to a single FimH conformational state or to all states and will give further insight into the molecular details of FimH binding. Therefore, it will provide the necessary details allowing experimentalists to design and perform new and more precise experiments proving and complementing the concepts provided in-silico.

It is our expectation that molecular simulation integrated with experimental techniques will lead to new routes for drug development not only for bacterial adhesins but for a variety of proteins involved in bacterial infection. In contrast to the currently used antibiotics, precision antimicrobial drugs will allow to specifically target selected bacterial strains and will thus constitute a valuable non-antibiotic alternative treatment.

References

1. Leimbach, A.; Hacker, J.; Dobrindt, U. *E. coli* as an All-Rounder: The thin line between commensalism and pathogenicity. *Curr. Top. Microbiol. Immunol.* **2013**, *358*, 3–32. [PubMed]
2. Foxman, B. Epidemiology of Urinary tract infections: Incidence, morbidity, and economic costs. *Dis. Mon.* **2003**, *49*, 53–70. [CrossRef] [PubMed]
3. Ronald, A.R.; Nicolle, L.E.; Stamm, E.; Krieger, J.; Warren, J.; Schaeffer, A.; Naber, K.G.; Hooton, T.M.; Johnson, J.; Chambers, S.; et al. Urinary tract infection in adults: Research priorities and strategies. *Int. J. Antimicrob. Agents* **2001**, *17*, 343–348. [CrossRef]
4. Foxman, B. Urinary tract infection syndromes. *Infect. Dis. Clin. N. Am.* **2014**, *28*, 1–13. [CrossRef] [PubMed]
5. Sanchez, G.V.; Master, R.N.; Bordon, J. Trimethoprim-sulfamethoxazole may no longer be acceptable for the treatment of acute uncomplicated cystitis in the United States. *Clin. Infect. Dis.* **2011**, *53*, 316–317. [CrossRef] [PubMed]
6. Karlowsky, J.A.; Hoban, D.J.; DeCorby, M.R.; Laing, N.M.; Zhanel, G.G. Fluoroquinolone-resistant urinary isolates of *Escherichia coli* from outpatients are frequently multidrug resistant: Results from the north american urinary tract infection collaborative alliance-quinolone resistance study. *Antimicrob. Agents Chemother.* **2006**, *50*, 2251–2254. [CrossRef] [PubMed]
7. Zhanel, G.G.; Hisanaga, T.L.; Laing, N.M.; DeCorby, M.R.; Nichol, K.A.; Palatnick, L.P.; Johnson, J.; Noreddin, A.; Harding, G.K.M.; Nicolle, L.E.; et al. Antibiotic resistance in outpatient urinary isolates: Final results from the North American Urinary Tract Infection Collaborative Alliance (NAUTICA). *Int. J. Antimicrob. Agents* **2005**, *26*, 380–388. [CrossRef] [PubMed]
8. McGann, P.; Snesrud, E.; Maybank, R.; Corey, B.; Ong, A.C.; Clifford, R.; Hinkle, M.; Whitman, T.; Lesho, E.; Schaecher, K.E. *Escherichia coli* Harboring mcr-1 and bla CTX-M on a novel incf plasmid: First report of mcr-1 in the United States. *Antimicrob. Agents Chemother.* **2016**, *60*, 4420–4421. [CrossRef] [PubMed]

9. Cole, S.T. Who will develop new antibacterial agents? *Philos. Trans. R. Soc. B Biol. Sci.* **2014**, *369*, 20130430. [CrossRef] [PubMed]

10. Nathan, C. Fresh approaches to anti-infective therapies. *Sci. Transl. Med.* **2012**, *4*, 140sr2. [CrossRef] [PubMed]

11. Hartmann, M.; Lindhorst, T.K. The bacterial lectin fimh, a target for drug discovery–carbohydrate inhibitors of type 1 fimbriae-mediated bacterial adhesion. *Eur. J. Org. Chem.* **2011**, *2011*, 3583–3609. [CrossRef]

12. Pieters, R.J. Intervention with bacterial adhesion by multivalent carbohydrates. *Med. Res. Rev.* **2007**, *27*, 796–816. [CrossRef] [PubMed]

13. Langermann, S. Prevention of mucosal *Escherichia coli* infection by fimh-adhesin-based systemic vaccination. *Science* **1997**, *276*, 607–611. [CrossRef] [PubMed]

14. Hung, C.-S.; Bouckaert, J.; Hung, D.; Pinkner, J.; Widberg, C.; DeFusco, A.; Auguste, C.G.; Strouse, R.; Langermann, S.; Waksman, G.; et al. Structural basis of tropism of *escherichia coli* to the bladder during urinary tract infection. *Mol. Microbiol.* **2002**, *44*, 903–915. [CrossRef] [PubMed]

15. Eto, D.S.; Jones, T.A.; Sundsbak, J.L.; Mulvey, M.A. Integrin-mediated host cell invasion by type 1–piliated uropathogenic *Escherichia coli*. *PLoS Pathog.* **2007**, *3*, e100. [CrossRef] [PubMed]

16. Rosen, D.A.; Hooton, T.M.; Stamm, W.E.; Humphrey, P.A.; Hultgren, S.J. Detection of intracellular bacterial communities in human urinary tract infection. *PLoS Med.* **2007**, *4*, e329. [CrossRef] [PubMed]

17. Martinez, J.J. Type 1 Pilus-mediated bacterial invasion of bladder epithelial cells. *EMBO J.* **2000**, *19*, 2803–2812. [CrossRef] [PubMed]

18. Zhou, G.; Mo, W.J.; Sebbel, P.; Min, G.; Neubert, T.A.; Glockshuber, R.; Wu, X.R.; Sun, T.T.; Kong, X.P. Uroplakin Ia is the urothelial receptor for uropathogenic *Escherichia coli*: Evidence from in vitro FimH binding. *J. Cell Sci.* **2001**, *114*, 4095–4103. [PubMed]

19. Baumgart, M.; Dogan, B.; Rishniw, M.; Weitzman, G.; Bosworth, B.; Yantiss, R.; Orsi, R.H.; Wiedmann, M.; McDonough, P.; Kim, S.G.; et al. Culture independent analysis of ileal mucosa reveals a selective increase in invasive *Escherichia coli* of novel phylogeny relative to depletion of clostridiales in crohn's disease involving the ileum. *ISME J.* **2007**, *1*, 403–418. [CrossRef] [PubMed]

20. DeFilippis, E.M.; Longman, R.; Harbus, M.; Dannenberg, K.; Scherl, E.J. Crohn's Disease: Evolution, epigenetics, and the emerging role of microbiome-targeted therapies. *Curr. Gastroenterol. Rep.* **2016**, *18*, 13. [CrossRef] [PubMed]

21. Sivignon, A.; Bouckaert, J.; Bernard, J.; Gouin, S.G.; Barnich, N. The potential of fimh as a novel therapeutic target for the treatment of crohn's disease. *Expert Opin. Ther. Targets* **2017**, *21*, 837–847. [CrossRef] [PubMed]

22. Barnich, N.; Carvalho, F.A.; Glasser, A.-L.; Darcha, C.; Jantscheff, P.; Allez, M.; Peeters, H.; Bommelaer, G.; Desreumaux, P.; Colombel, J.-F.; et al. CEACAM6 Acts as a receptor for adherent-invasive *E. coli*, supporting ileal mucosa colonization in crohn disease. *J. Clin. Investig.* **2007**, *117*, 1566–1574. [CrossRef] [PubMed]

23. Alvarez Dorta, D.; Sivignon, A.; Chalopin, T.; Dumych, T.I.; Roos, G.; Bilyy, R.O.; Deniaud, D.; Krammer, E.-M.; De Ruyck, J.; Lensink, M.F.; et al. The Antiadhesive strategy in crohn's disease: Orally active mannosides to decolonize pathogenic *Escherichia coli* from the gut. *ChemBioChem* **2016**, *17*, 936–952. [CrossRef] [PubMed]

24. Mydock-McGrane, L.K.; Hannan, T.J.; Janetka, J.W. Rational design strategies for fimh antagonists: New drugs on the horizon for urinary tract infection and crohn's sisease. *Expert Opin. Drug Discov.* **2017**, *12*, 711–731. [CrossRef] [PubMed]

25. Spaulding, C.N.; Klein, R.D.; Schreiber, H.L.; Janetka, J.W.; Hultgren, S.J. Precision antimicrobial therapeutics: The path of least resistance? *NPJ Biofilms Microbiomes* **2018**, *4*, 4. [CrossRef] [PubMed]

26. Bouckaert, J.; Berglund, J.; Schembri, M.; De Genst, E.; Cools, L.; Wuhrer, M.; Hung, C.-S.; Pinkner, J.; Slättegård, R.; Zavialov, A.; et al. Receptor binding studies disclose a novel class of high-affinity inhibitors of the *escherichia coli* fimh adhesin. *Mol. Microbiol.* **2004**, *55*, 441–455. [CrossRef] [PubMed]

27. De Ruyck, J.; Roos, G.; Krammer, E.-M.; Prévost, M.; Lensink, M.F.; Bouckaert, J. Molecular mechanisms of drug action: X-ray crystallography at the basis of structure-based and ligand-based drug design. In *Biophysical Techniques in Drug Discovery*; The Royal Society of Chemstry: London, UK, 2017; Chapter 4; pp. 67–86. ISBN 9781788010016.

28. Chen, S.L.; Hung, C.S.; Pinkner, J.S.; Walker, J.N.; Cusumano, C.K.; Li, Z.; Bouckaert, J.; Gordon, J.I.; Hultgren, S.J. Positive selection identifies an in vivo role for fimh during urinary tract infection in addition to mannose binding. *Proc. Natl. Acad. Sci. USA* **2009**, *106*, 22439–22444. [CrossRef] [PubMed]

29. *The PyMOL Molecular Graphics System*; Version 2.0; Schrödinger LLC: New York, NY, USA, 2018.

30. Sager, C.P.; Fiege, B.; Zihlmann, P.; Vannam, R.; Rabbani, S.; Jakob, R.P.; Preston, R.C.; Zalewski, A.; Maier, T.; Peczuh, M.W.; et al. The Price of flexibility—A case study on septanoses as pyranose mimetics. *Chem. Sci.* **2018**, *9*, 646–654. [CrossRef] [PubMed]

31. Roos, G.; Wellens, A.; Touaibia, M.; Yamakawa, N.; Geerlings, P.; Roy, R.; Wyns, L.; Bouckaert, J. Validation of reactivity descriptors to assess the aromatic stacking within the tyrosine gate of FimH. *ACS Med. Chem. Lett.* **2013**, *4*, 1085–1090. [CrossRef] [PubMed]

32. Alvarez Dorta, D.; Chalopin, T.; Sivignon, A.; de Ruyck, J.; Dumych, T.I.; Bilyy, R.O.; Deniaud, D.; Barnich, N.; Bouckaert, J.; Gouin, S.G. Physiochemical tuning of potent *Escherichia coli* anti-adhesives by microencapsulation and methylene homologation. *ChemMedChem* **2017**, *12*, 986–998. [CrossRef] [PubMed]

33. Jarvis, C.; Han, Z.; Kalas, V.; Klein, R.; Pinkner, J.S.; Ford, B.; Binkley, J.; Cusumano, C.K.; Cusumano, Z.; Mydock-McGrane, L.; et al. Antivirulence isoquinolone mannosides: Optimization of the biaryl aglycone for fimh lectin binding affinity and efficacy in the treatment of chronic UTI. *Chem. Med. Chem.* **2016**, *11*, 367–373. [CrossRef] [PubMed]

34. Wellens, A.; Lahmann, M.; Touaibia, M.; Vaucher, J.; Oscarson, S.; Roy, R.; Remaut, H.; Bouckaert, J. The tyrosine gate as a potential entropic lever in the receptor-binding site of the bacterial adhesin FimH. *Biochemistry* **2012**, *51*, 4790–4799. [CrossRef] [PubMed]

35. Kleeb, S.; Pang, L.; Mayer, K.; Eris, D.; Sigl, A.; Preston, R.C.; Zihlmann, P.; Sharpe, T.; Jakob, R.P.; Abgottspon, D.; et al. FimH antagonists: Bioisosteres to improve the in vitro and in vivo pk/pd profile. *J. Med. Chem.* **2015**, *58*, 2221–2239. [CrossRef] [PubMed]

36. Mydock-McGrane, L.; Cusumano, Z.; Han, Z.; Binkley, J.; Kostakioti, M.; Hannan, T.; Pinkner, J.S.; Klein, R.; Kalas, V.; Crowley, J.; et al. Antivirulence C-mannosides as antibiotic-sparing, oral therapeutics for urinary tract infections. *J. Med. Chem.* **2016**, *59*, 9390–9408. [CrossRef] [PubMed]

37. Wellens, A.; Garofalo, C.; Nguyen, H.; Van Gerven, N.; Slättegård, R.; Hernalsteens, J.-P.; Wyns, L.; Oscarson, S.; De Greve, H.; Hultgren, S.; et al. Intervening with urinary tract infections using anti-adhesives based on the crystal structure of the Fimh–oligomannose-3 complex. *PLoS ONE* **2008**, *3*, e2040. [CrossRef]

38. Rabbani, S.; Krammer, E.-M.; Roos, G.; Zalewski, A.; Preston, R.; Eid, S.; Zihlmann, P.; Prévost, M.; Lensink, M.F.; Thompson, A.; et al. Mutation of Tyr137 of the Universal *Escherichia coli* Fimbrial Adhesin FimH relaxes the tyrosine gate prior to mannose binding. *IUCrJ* **2017**, *4*, 7–23. [CrossRef] [PubMed]

39. Touaibia, M.; Krammer, E.-M.; Shiao, T.; Yamakawa, N.; Wang, Q.; Glinschert, A.; Papadopoulos, A.; Mousavifar, L.; Maes, E.; Oscarson, S.; et al. Sites for dynamic protein-carbohydrate interactions of O- and C-linked mannosides on the *E. coli* FimH Adhesin. *Molecules* **2017**, *22*, 1101. [CrossRef] [PubMed]

40. de Ruyck, J.; Lensink, M.F.; Bouckaert, J. Structures of C-mannosylated anti-adhesives bound to the type 1 fimbrial fimh adhesin. *IUCrJ* **2016**, *3*, 163–167. [CrossRef] [PubMed]

41. Han, Z.; Pinkner, J.S.; Ford, B.; Obermann, R.; Nolan, W.; Wildman, S.A.; Hobbs, D.; Ellenberger, T.; Cusumano, C.K.; Hultgren, S.J.; et al. Structure-based drug design and optimization of mannoside bacterial fimh antagonists. *J. Med. Chem.* **2010**, *53*, 4779–4792. [CrossRef] [PubMed]

42. Gouin, S.G.; Roos, G.; Bouckaert, J. Discovery and Application of FimH Antagonists. In *Carbohydrates as Drugs*; Topics in Medicinal Chemistry; Seeberger, P., Rademacher, C., Eds.; Springer: Berlin, Germany, 2014; Volume 12, pp. 123–168.

43. Le Trong, I.; Aprikian, P.; Kidd, B.A.; Thomas, W.E.; Sokurenko, E.V.; Stenkamp, R.E. Donor strand exchange and conformational changes during *E. coli* fimbrial formation. *J. Struct. Biol.* **2010**, *172*, 380–388. [CrossRef] [PubMed]

44. Tomašić, T.; Rabbani, S.; Gobec, M.; Raščan, I.M.; Podlipnik, Č.; Ernst, B.; Anderluh, M. Branched α-d-mannopyranosides: A new class of potent fimh antagonists. *Med. Chem. Commun.* **2014**, *5*, 1247–1253. [CrossRef]

45. Feenstra, T.; Thøgersen, M.S.; Wieser, E.; Peschel, A.; Ball, M.J.; Brandes, R.; Satchell, S.C.; Stockner, T.; Aarestrup, F.M.; Rees, A.J.; et al. Adhesion of *Escherichia coli* under flow conditions reveals potential novel effects of Fimh mutations. *Eur. J. Clin. Microbiol. Infect. Dis.* **2017**, *36*, 467–478. [CrossRef] [PubMed]

46. Szunerits, S.; Zagorodko, O.; Cogez, V.; Dumych, T.; Chalopin, T.; Alvarez Dorta, D.; Sivignon, A.; Barnich, N.; Harduin-Lepers, A.; Larroulet, I.; et al. Differentiation of Crohn's disease-associated isolates from other pathogenic *Escherichia coli* by Fimbrial adhesion under shear force. *Biology (Basel)* **2016**, *5*, 14. [CrossRef] [PubMed]

47. Sokurenko, E.V.; Chesnokova, V.; Dykhuizen, D.E.; Ofek, I.; Wu, X.-R.; Krogfelt, K.A.; Struve, C.; Schembri, M.A.; Hasty, D.L. Pathogenic adaptation of *Escherichia coli* by natural variation of the fimh adhesin. *Proc. Natl. Acad. Sci. USA* **1998**, *95*, 8922–8926. [CrossRef] [PubMed]

48. Thomas, W.E.; Trintchina, E.; Forero, M.; Vogel, V.; Sokurenko, E. V Bacterial adhesion to target cells enhanced by shear force. *Cell* **2002**, *109*, 913–923. [CrossRef]

49. Vanwetswinkel, S.; Volkov, A.N.; Sterckx, Y.G.J.; Garcia-Pino, A.; Buts, L.; Vranken, W.F.; Bouckaert, J.; Roy, R.; Wyns, L.; van Nuland, N.A.J. Study of the structural and dynamic effects in the FimH adhesin upon α-D-Heptyl Mannose Binding. *J. Med. Chem.* **2014**, *57*, 1416–1427. [CrossRef] [PubMed]

50. Aprikian, P.; Tchesnokova, V.; Kidd, B.; Yakovenko, O.; Yarov-Yarovoy, V.; Trinchina, E.; Vogel, V.; Thomas, W.; Sokurenko, E. Interdomain interaction in the FimH adhesin of *Escherichia coli* regulates the affinity to mannose. *J. Biol. Chem.* **2007**, *282*, 23437–23446. [CrossRef] [PubMed]

51. Nilsson, L.M.; Thomas, W.E.; Sokurenko, E.V.; Vogel, V. Elevated shear stress protects *Escherichia coli* cells adhering to surfaces via catch bonds from detachment by soluble inhibitors. *Appl. Environ. Microbiol.* **2006**, *72*, 3005–3010. [CrossRef] [PubMed]

52. Anderson, B.N.; Ding, A.M.; Nilsson, L.M.; Kusuma, K.; Tchesnokova, V.; Vogel, V.; Sokurenko, E.V.; Thomas, W.E. Weak rolling adhesion enhances bacterial surface colonization. *J. Bacteriol.* **2007**, *189*, 1794–1802. [CrossRef] [PubMed]

53. Anderson, G.G.; Dodson, K.W.; Hooton, T.M.; Hultgren, S.J. Intracellular Bacterial communities of uropathogenic *Escherichia coli* in urinary tract pathogenesis. *Trends Microbiol.* **2004**, *12*, 424–430. [CrossRef] [PubMed]

54. Thomas, W.E.; Nilsson, L.M.; Forero, M.; Sokurenko, E.V.; Vogel, V. Shear-dependent 'Stick-and-Eoll' adhesion of type 1 Fimbriated *Escherichia coli*. *Mol. Microbiol.* **2004**, *53*, 1545–1557. [CrossRef] [PubMed]

55. Aprikian, P.; Interlandi, G.; Kidd, B.A.; Le Trong, I.; Tchesnokova, V.; Yakovenko, O.; Whitfield, M.J.; Bullitt, E.; Stenkamp, R.E.; Thomas, W.E.; et al. The bacterial fimbrial tip acts as a mechanical force sensor. *PLoS Biol.* **2011**, *9*. [CrossRef] [PubMed]

56. Nilsson, L.M.; Thomas, W.E.; Sokurenko, E.V.; Vogel, V. Beyond Induced-fit receptor-ligand interactions: Structural changes that can significantly extend bond lifetimes. *Structure* **2008**, *16*, 1047–1058. [CrossRef] [PubMed]

57. Le Trong, I.; Aprikian, P.; Kidd, B.A.; Forero-Shelton, M.; Tchesnokova, V.; Rajagopal, P.; Rodriguez, V.; Interlandi, G.; Klevit, R.; Vogel, V.; et al. Structural basis for mechanical force regulation of the adhesin fimh via finger trap-like β sheet twisting. *Cell* **2010**, *141*, 645–655. [CrossRef] [PubMed]

58. Sterpone, F.; Doutreligne, S.; Tran, T.T.; Melchionna, S.; Baaden, M.; Nguyen, P.H.; Derreumaux, P. Multi-Scale Simulations of biological systems using the opep coarse-grained model. *Biochem. Biophys. Res. Commun.* **2018**, *498*, 296–304. [CrossRef] [PubMed]

59. Sauer, M.M.; Jakob, R.P.; Eras, J.; Baday, S.; Eriş, D.; Navarra, G.; Bernèche, S.; Ernst, B.; Maier, T.; Glockshuber, R. Catch-bond mechanism of the bacterial adhesin FimH. *Nat. Commun.* **2016**, *7*, 10738. [CrossRef] [PubMed]

60. Choudhury, D. X-ray Structure of the fimc-fimh chaperone-adhesin complex from uropathogenic *Escherichia coli*. *Science* **1999**, *285*, 1061–1066. [CrossRef] [PubMed]

61. Sokurenko, E.V.; Vogel, V.; Thomas, W.E. Catch-Bond mechanism of force-enhanced adhesion: Counterintuitive, elusive, but . . . widespread? *Cell Host Microbe* **2008**, *4*, 314–323. [CrossRef] [PubMed]

62. Whitfield, M.; Ghose, T.; Thomas, W. Shear-stabilized rolling behavior of *E. coli* examined with simulations. *Biophys. J.* **2010**, *99*, 2470–2478. [CrossRef] [PubMed]

63. Rohl, C.A.; Strauss, C.E.M.; Misura, K.M.S.; Baker, D. Protein Structure Prediction Using Rosetta. In *Methods in Enzymology*; Academic Press: Cambridge, MA, USA, 2004; Volume 383, pp. 66–93. ISBN 9780121827885.

64. Rodriguez, V.B.; Kidd, B.A.; Interlandi, G.; Tchesnokova, V.; Sokurenko, E.V.; Thomas, W.E. Allosteric coupling in the bacterial adhesive protein FimH. *J. Biol. Chem.* **2013**, *288*, 24128–24139. [CrossRef] [PubMed]

65. Interlandi, G.; Thomas, W.E. Mechanism of allosteric propagation across a β-sheet structure investigated by molecular dynamics simulations. *Proteins Struct. Funct. Bioinf.* **2016**, *84*, 990–1008. [CrossRef] [PubMed]

66. Tchesnokova, V.; Aprikian, P.; Kisiela, D.; Gowey, S.; Korotkova, N.; Thomas, W.; Sokurenko, E. Type 1 fimbrial adhesin Fimh elicits an immune response that enhances cell adhesion of *Escherichia coli*. *Infect. Immun.* **2011**, *79*, 3895–3904. [CrossRef] [PubMed]

67. Kisiela, D.I.; Avagyan, H.; Friend, D.; Jalan, A.; Gupta, S.; Interlandi, G.; Liu, Y.; Tchesnokova, V.; Rodriguez, V.B.; Sumida, J.P.; et al. Inhibition and reversal of microbial attachment by an antibody with parasteric activity against the FimH adhesin of uropathogenic *E. coli*. *PLoS Pathog.* **2015**, *11*, e1004857. [CrossRef] [PubMed]

68. Kisiela, D.I.; Rodriguez, V.B.; Tchesnokova, V.; Avagyan, H.; Aprikian, P.; Liu, Y.; Wu, X.-R.; Thomas, W.E.; Sokurenko, E.V. Conformational inactivation induces immunogenicity of the receptor-binding pocket of a bacterial adhesin. *Proc. Natl. Acad. Sci. USA* **2013**, *110*, 19089–19094. [CrossRef] [PubMed]

69. Singaravelu, M.; Selvan, A.; Anishetty, S. Molecular dynamics simulations of lectin somain of FimH and immunoinformatics for the design of potential vaccine candidates. *Comput. Biol. Chem.* **2014**, *52*, 18–24. [CrossRef] [PubMed]

70. Hahn, E.; Wild, P.; Hermanns, U.; Sebbel, P.; Glockshuber, R.; Häner, M.; Taschner, N.; Burkhard, P.; Aebi, U.; Müller, S.A. Exploring the 3D molecular architecture of *Escherichia coli* type 1 Pili. *J. Mol. Biol.* **2002**, *323*, 845–857. [CrossRef]

71. Gossert, A.D.; Bettendorff, P.; Puorger, C.; Vetsch, M.; Herrmann, T.; Glockshuber, R.; Wüthrich, K. NMR structure of the *Escherichia coli* type 1 pilus subunit fimf and its interactions with other pilus subunits. *J. Mol. Biol.* **2008**, *375*, 752–763. [CrossRef] [PubMed]

72. Forero, M.; Yakovenko, O.; Sokurenko, E.V.; Thomas, W.E.; Vogel, V. Uncoiling mechanics of *Escherichia coli* type i fimbriae are optimized for catch bonds. *PLoS Biol.* **2006**, *4*, e298. [CrossRef] [PubMed]

73. Yakovenko, O.; Sharma, S.; Forero, M.; Tchesnokova, V.; Aprikian, P.; Kidd, B.; Mach, A.; Vogel, V.; Sokurenko, E.; Thomas, W.E. FimH forms catch bonds that are enhanced by mechanical force due to allosteric regulation. *J. Biol. Chem.* **2008**, *283*, 11596–11605. [CrossRef] [PubMed]

74. Rakshit, S.; Sivasankar, S. Biomechanics of cell adhesion: How force regulates the lifetime of adhesive bonds at the single molecule level. *Phys. Chem. Chem. Phys.* **2014**, *16*, 2211. [CrossRef] [PubMed]

75. Mydock-McGrane, L.K.; Cusumano, Z.T.; Janetka, J.W. Mannose-derived FimH antagonists: A promising anti-virulence therapeutic strategy for urinary tract infections and Crohn's disease. *Expert Opin. Ther. Pat.* **2016**, *26*, 175–197. [CrossRef] [PubMed]

76. Fiege, B.; Rabbani, S.; Preston, R.C.; Jakob, R.P.; Zihlmann, P.; Schwardt, O.; Jiang, X.; Maier, T.; Ernst, B. The tyrosine gate of the bacterial lectin FimH: A conformational analysis by NMR spectroscopy and X-ray crystallography. *Chem. Bio. Chem.* **2015**, *16*, 1235–1246. [CrossRef] [PubMed]

77. Brument, S.; Sivignon, A.; Dumych, T.I.; Moreau, N.; Roos, G.; Guérardel, Y.; Chalopin, T.; Deniaud, D.; Bilyy, R.O.; Darfeuille-Michaud, A.; et al. Thiazolylaminomannosides As potent antiadhesives of Type 1 piliated escherichia coli isolated from Crohn's disease patients. *J. Med. Chem.* **2013**, *56*, 5395–5406. [CrossRef] [PubMed]

78. Chalopin, T.; Brissonnet, Y.; Sivignon, A.; Deniaud, D.; Cremet, L.; Barnich, N.; Bouckaert, J.; Gouin, S.G. Inhibition profiles of mono- and polyvalent FimH antagonists against 10 different *Escherichia coli* strains. *Org. Biomol. Chem.* **2015**, *13*, 11369–11375. [CrossRef] [PubMed]

79. Chalopin, T.; Alvarez Dorta, D.; Sivignon, A.; Caudan, M.; Dumych, T.I.; Bilyy, R.O.; Deniaud, D.; Barnich, N.; Bouckaert, J.; Gouin, S.G. Second Generation of thiazolylmannosides, FimH antagonists for *E. coli*-induced Crohn's disease. *Org. Biomol. Chem.* **2016**, *14*, 3913–3925. [CrossRef] [PubMed]

80. Sperling, O.; Fuchs, A.; Lindhorst, T.K. Evaluation of the carbohydrate recognition domain of the bacterial adhesin FimH: Design, synthesis and binding properties of mannoside ligands. *Org. Biomol. Chem.* **2006**, *4*, 3913. [CrossRef] [PubMed]

81. Pang, L.; Kleeb, S.; Lemme, K.; Rabbani, S.; Scharenberg, M.; Zalewski, A.; Schädler, F.; Schwardt, O.; Ernst, B. FimH antagonists: Structure-activity and structure-property relationships for biphenyl α-D-mannopyranosides. *ChemMedChem* **2012**, *7*, 1404–1422. [CrossRef] [PubMed]

82. Grabosch, C.; Hartmann, M.; Schmidt-Lassen, J.; Lindhorst, T.K. Squaric Acid Monoamide Mannosides as Ligands for the Bacterial Lectin FimH: Covalent Inhibition or Not? *ChemBioChem* **2011**, *12*, 1066–1074. [CrossRef] [PubMed]

83. Eid, S.; Zalewski, A.; Smieško, M.; Ernst, B.; Vedani, A. A Molecular-Modeling Toolbox Aimed at Bridging the gap between medicinal chemistry and computational sciences. *Int. J. Mol. Sci.* **2013**, *14*, 684–700. [CrossRef] [PubMed]

84. Sivignon, A.; Yan, X.; Alvarez Dorta, D.; Bonnet, R.; Bouckaert, J.; Fleury, E.; Bernard, J.; Gouin, S.G.; Darfeuille-Michaud, A.; Barnich, N. Development of heptylmannoside-based glycoconjugate antiadhesive compounds against adherent-invasive *Escherichia coli* bacteria associated with Crohn's disease. *MBio* **2015**, *6*, e01298-15. [CrossRef] [PubMed]

85. Bouckaert, J.; Li, Z.; Xavier, C.; Almant, M.; Caveliers, V.; Lahoutte, T.; Weeks, S.D.; Kovensky, J.; Gouin, S.G. Heptyl α-D-Mannosides grafted on a β-Cyclodextrin core to interfere with *Escherichia coli* adhesion: An in vivo multivalent effect. *Chem. Eur. J.* **2013**, *19*, 7847–7855. [CrossRef] [PubMed]

86. Dumych, T.; Yamakawa, N.; Sivignon, A.; Garenaux, E.; Robakiewicz, S.; Coddeville, B.; Bongiovanni, A.; Bray, F.; Barnich, N.; Szunerits, S.; et al. Oligomannose-rich membranes of dying intestinal epithelial cells promote host colonization by adherent-invasive *E. coli*. *Front. Microbiol.* **2018**, *9*, 742. [CrossRef] [PubMed]

87. Lonardi, E.; Moonens, K.; Buts, L.; de Boer, A.; Olsson, J.; Weiss, M.; Fabre, E.; Guérardel, Y.; Deelder, A.; Oscarson, S.; et al. Structural sampling of glycan interaction profiles reveals mucosal receptors for fimbrial adhesins of enterotoxigenic *Escherichia coli*. *Biology* **2013**, *2*, 894–917. [CrossRef] [PubMed]

88. Almant, M.; Moreau, V.; Kovensky, J.; Bouckaert, J.; Gouin, S.G. Clustering of *Escherichia coli* Type-1 fimbrial adhesins by using multimeric heptyl α-D-mannoside probes with a carbohydrate core. *Chem. Eur. J.* **2011**, *17*, 10029–10038. [CrossRef] [PubMed]

89. Fanfrlík, J.; Bronowska, A.K.; Řezáč, J.; Přenosil, O.; Konvalinka, J.; Hobza, P. A Reliable Docking/scoring scheme based on the semiempirical quantum mechanical pm6-dh2 method accurately covering dispersion and H-bonding: HIV-1 protease with 22 ligands. *J. Phys. Chem. B* **2010**, *114*, 12666–12678. [CrossRef] [PubMed]

90. Lepšík, M.; Řezáč, J.; Kolář, M.; Pecina, A.; Hobza, P.; Fanfrlík, J. The Semiempirical quantum mechanical scoring function for in silico drug design. *Chempluschem* **2013**, *78*, 921–931. [CrossRef]

91. Savar, N.S.; Jahanian-Najafabadi, A.; Mahdavi, M.; Shokrgozar, M.A.; Jafari, A.; Bouzari, S. In silico and in vivo studies of Truncated Forms of Flagellin (FliC) of enteroaggregative *Escherichia coli* fused to FimH from uropathogenic *Escherichia coli* as a Vaccine candidate against urinary tract infections. *J. Biotechnol.* **2014**, *175*, 31–37. [CrossRef] [PubMed]

92. Habibi, M.; Reza, M.; Karam, A.; Bouzari, S. In silico design of fusion protein of fimh from uropathogenic *Escherichia coli* and mrph from proteus mirabilis against urinary tract infections. *Adv. Biomed. Res.* **2015**, *4*, 217. [CrossRef] [PubMed]

93. Luna-Pineda, V.M.; Reyes-Grajeda, J.P.; Cruz-Córdova, A.; Saldaña-Ahuactzi, Z.; Ochoa, S.A.; Maldonado-Bernal, C.; Cázares-Domínguez, V.; Moreno-Fierros, L.; Arellano-Galindo, J.; Hernández-Castro, R.; et al. Dimeric and Trimeric fusion proteins generated with fimbrial adhesins of uropathogenic *Escherichia coli*. *Front. Cell. Infect. Microbiol.* **2016**, *6*, 135. [CrossRef] [PubMed]

6

Theoretical Model of EphA2-Ephrin A1 Inhibition

Wiktoria Jedwabny [1] ⓘ, **Alessio Lodola** [2] **and Edyta Dyguda-Kazimierowicz** [1,*] ⓘ

[1] Department of Chemistry, Wrocław University of Science and Technology, 50370 Wrocław, Poland; wiktoria.jedwabny@pwr.edu.pl
[2] Department of Food and Drug, University of Parma, 43100 Parma, Italy; alessio.lodola@unipr.it
*   Correspondence: Edyta.Dyguda@pwr.edu.pl

Abstract: This work aims at the theoretical description of EphA2-ephrin A1 inhibition by small molecules. Recently proposed ab initio-based scoring models, comprising long-range components of interaction energy, is tested on lithocholic acid class inhibitors of this protein–protein interaction (PPI) against common empirical descriptors. We show that, although limited to compounds with similar solvation energy, the ab initio model is able to rank the set of selected inhibitors more effectively than empirical scoring functions, aiding the design of novel compounds.

Keywords: EphA2-ephrin A1; PPI inhibition; interaction energy

1. Introduction

The erythropoietin-producing hepatocellular carcinoma (Eph) receptors are probably the largest family of receptor tyrosine kinases (RTKs) and includes 14 members [1] divided into class A (EphA) and class B (EphB), based on the binding affinity for their ligands (ephrins, also divided into classes A and B), and sequence homology [2]. Ephrins are membrane proteins with the A class connected to the membrane by a phosphatidylinositol (GPI) linker, and the B class linked via a hydrophobic domain. While interclass binding has been reported [3,4], ephrin A-type ligands generally bind to EphA receptors, whereas ephrin B-type ligands interact with EphB receptors.

The Eph-ephrin signaling system is known to play important and diverse biological functions that involve cell–cell interactions both during embryonic development and for maintaining homeostasis in adult cells. For instance, in embryos, the Eph-ephrin system finely tunes tissue boundary formation, including central nervous system patterning [5], while in adults it controls bone and intestinal homeostasis, immune system functions and angiogenic processes. The Eph-ephrin system is currently gaining interest in the context of drug discovery as it has been found hyperactivated in several cancers [6]. Among the cloned Eph receptor subtype, EphA2 has been studied the most in the oncology field since the overexpression and/or the hyperactivation of this receptor has been linked to the insurgence and progression of several cancer types, including brain, lung, breast, ovarian and prostate [7]. Moreover, the abnormal activity of this receptor has been associated with poor prognosis [8]. Due to its increasing recognition as a tumorigenic protein, the EphA2 receptor has gained interest as a target protein for novel cancer therapies [9].

One of the available approaches targeting Eph-ephrin system (and EphA2 with its physiological ligand, ephrin-A1, in particular) involves small molecule inhibitors [1] able to prevent ephrin-A1 binding to EphA2. Several classes of inhibitors of this specific protein–protein interaction (PPI) have been recently identified [10–12]. The most promising class is represented by lithocholic acid (LCA) and its α-amino acid conjugates [7,13]. It has been demonstrated by surface plasmon resonance (SPR) analysis that this class of compounds prevents ephrin-A1 binding to EphA2 by targeting a conserved region of the ligand-binding domain of EphA2 [14,15].

Molecular modeling investigations performed with classical force fields have identified a likely binding mode for these inhibitors consistent with available structure–activity relationship (SAR) data, i.e., proposing a reasonable role for the terminal carboxylic group and the amino acid side-chain of the inhibitors during their docking within EphA2 [13,14]. However, attempts to build quantitative models correlating experimental activities to docking energies led to modest results [13], suggesting that classical methods may not be able to properly describe accommodation of amino acid conjugates of LCA within EphA2 ligand binding domain (LBD).

Ligand-receptor binding is often examined using empirical or semi-empirical methods with a diverse level of success [16–19], particularly in terms of the virtual screening campaigns. A way to improve the quality of the results could involve ab initio calculations, but due to the computational time required, these are rather impractical in the screening of potential drug candidates. On the other hand, quantum chemical calculations are able to provide insight into the physical nature of the receptor–ligand interactions. Studying small-molecule PPI inhibition is usually more challenging than evaluation of interactions in regular protein–ligand complexes [20]. For instance, binding cavities for inhibitors targeting PPIs are flat and often featured by the presence of aromatic residues, such as Phe, Tyr or Trp residues [21]. Empirical scoring functions, commonly used for scoring of receptor–ligand interactions, are not really suited for PPIs [22,23]. Despite the fact that some empirical and semi-empirical approaches have been applied to score PPI inhibitors with moderate success [24–27], ab initio derived models appear to be better suited for studying PPI recognition by small molecules, since they offer a detailed insight into the physical basis of such interactions.

When polar or charged systems are investigated, the computationally inexpensive non-empirical electrostatic term is sufficient to model the experimental data [28,29]. However, accounting for the dispersive interactions is required for a general description targeting any receptor–ligand complex, irrespectively of the physical nature of binding within such a system [30]. While non-empirical evaluation of the multipole electrostatic term conveniently scales with the size of the complex under study as the squared number of atoms, ab initio calculations of dispersion energy are computationally demanding, scaling with at least the fifth power of the number of atomic orbitals. However, dispersion interactions could be approximated, for instance, by the E_{Das} function, which successfully describes non-covalent interactions with atom–atom potentials fitted to reproduce the results of high-level quantum chemical calculations [31,32]. Recently developed non-empirical model comprising long-range terms of interaction energy, i.e., multipole electrostatic moment and dispersion contribution approximated by E_{Das} function [31,32], which offers a great enhancement in the computational time, was already tested on several systems, including essentially non-polar complexes of fatty acid amide hydrolase (FAAH) [33], pteridine reductase 1 (TbPTR1) featuring both dispersive and electrostatic interactions [34], and menin-mixed lineage leukemia (MLL) system [35], in which electrostatic interactions are dominant.

Such an approach neglects, among other entropic contributions, the influence of solvation effects. To include the latter, one would need a much more time-consuming method, for instance free-energy perturbation (FEP), Molecular Mechanics/Poisson-Boltzmann, Molecular Mechanics/Generalized Born Surface Area (MM/GBSA and MM/PBSA, respectively) [36] or Fragment Molecular Orbital (FMO) approach [37]. The quantum chemical methods (like DFT or MP2) are rather not combined with empirical solvation or ligand entropy estimates [36], and therefore they should work only if the neglected contributions to the energy of binding are similar within the studied set of complexes.

In the work presented herein, we attempt to reproduce the experimental ranking of a congeneric series of EphA2-ephrin A1 inhibitors [38] (shown in Table 1) with a recently developed simple ab initio model comprising multipole electrostatic and dispersion contributions, $E_{EL,MTP}^{(10)} + E_{Das}$. Such a model was previously validated on another set of protein–protein inhibitors [35], and not only the inhibitory activity ranking was reproduced, but novel inhibitors (i.e., not present in the training set) were successfully scored. We show here that if we limit our analysis to a set of EphA2-ephrin A1 inhibitors featuring similar solvation energy, ab initio modeling of the interactions provides computational results which parallel experimental potency data well. Moreover, such a model is able to outperform several commonly used empirical scoring functions.

Table 1. The structures and experimental activity [a] of inhibitors targeting EphA2-ephrin A1 interaction. The numbering of the structures is consistent with Table 1 from [13].

Inhibitor	X Substituent	$p\mathrm{IC}_{50}$
2 (Gly)		4.31
4 (L-Ala)		4.70
5 (D-Ala)		4.51
6 (L-Val)		4.62
7 (D-Val)		4.76
8 (L-Ser)		4.48
9 (D-Ser)		4.22
14 (L-Met)		4.56
15 (D-Met)		4.56
16 (L-Phe)		5.18
17 (D-Phe)		5.12

Table 1. *Cont.*

Inhibitor	X Substituent	pIC_{50}
18 (L-Tyr)		4.30
19 (D-Tyr)		4.00
20 (L-Trp)		5.69
21 (D-Trp)		4.69

[a] pIC_{50} values are taken from [13].

2. Results and Discussion

EphA2 binding site representation, shown in Figure 1, comprises six amino acid residues: Cys70, Cys188, Phe108, Arg103, Val72 and Met73 (more details regarding the model are given in the Materials and Methods section). All 15 analyzed inhibitors (Table 1) shared a similar binding mode [13], with a –COOH group facing Arg103 residue, in agreement with SAR data. Moreover, their common LCA scaffold was positioned almost identically. Thus, this steroidal moiety was excluded from the analysis and the compounds were cut in a way indicated by the red line in Table 1. Accordingly, the inhibitors were represented by smaller entities corresponding to the variable part of the inhibitor structure. Binding poses of models of two inhibitors, **20** (L-Trp derivative) and **19** (D-Tyr derivative), i.e., the most and least potent compounds, respectively, are presented in Figure 1.

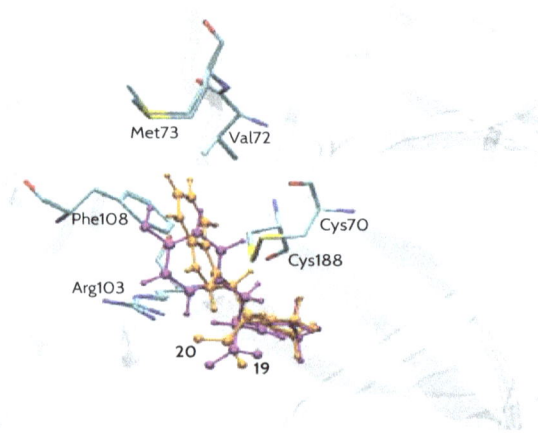

Figure 1. EphA2 binding site representation with bound inhibitors **19** (D-Tyr) and **20** (L-Trp).

2.1. Theoretical Models

Total binding energy values of EphA2 inhibitors for consecutive levels of Hybrid Variation–Perturbation Theory (HVPT) [39,40] and, in addition, $E_{EL,MTP}^{(10)} + E_{Das}$ energy results, are provided in Table 2. Pairwise interaction energy values between each inhibitor and a given amino acid residue are given in Table S1 in Supplementary Materials. Apparently, the main contribution to the total interaction energy calculated at the MP2 level of theory is due to the electrostatic energy. As a result, $E_{EL}^{(10)}$ and E_{MP2} energy values are comparable in magnitude (Table 2).

Table 2. Total EphA2-inhibitor interaction energy [a] at the consecutive levels of theory.

Inhibitor	pIC_{50} [b]	$E_{EL,MTP}^{(10)}$	$E_{EL}^{(10)}$	$E^{(10)}$	E_{SCF}	E_{MP2}	$E_{EL,MTP}^{(10)} + E_{Das}$
20 (L-Trp)	5.69	−89.2	−101.3	−66.5	−83.5	−102.7	−118.0
16 (L-Phe)	5.18	−90.7	−102.5	−65.6	−86.1	−100.5	−115.3
17 (D-Phe)	5.12	−98.5	−111.4	−70.1	−92.6	−109.6	−127.0
7 (D-Val)	4.76	−75.2	−83.3	−65.7	−77.4	−87.7	−91.3
4 (L-Ala)	4.70	−97.1	−108.5	−73.7	−94.1	−103.5	−116.5
21 (D-Trp)	4.69	−72.8	−82.3	−57.9	−70.9	−90.8	−99.4
6 (L-Val)	4.62	−99.3	−110.0	−71.9	−94.4	−104.4	−120.4
14 (L-Met)	4.56	−89.9	−101.1	−69.1	−87.7	−100.7	−112.3
15 (D-Met)	4.56	−80.5	−89.5	−67.3	−80.6	−94.2	−101.5
5 (D-Ala)	4.51	−75.1	−82.2	−66.7	−76.9	−85.6	−88.9
8 (L-Ser)	4.48	−85.9	−96.6	−70.4	−86.2	−95.5	−103.7
2 (Gly)	4.31	−64.6	−69.3	−56.2	−65.0	−72.5	−75.7
18 (L-Tyr)	4.30	−65.9	−73.2	−55.3	−65.3	−79.4	−85.3
9 (D-Ser)	4.22	−69.0	−74.7	−62.6	−71.4	−81.1	−83.2
19 (D-Tyr)	4.00	−65.3	−74.1	−55.8	−66.5	−81.9	−85.7
R [c]		−0.63	−0.65	−0.44	−0.55	−0.69	−0.72
N_{pred} [d]		75.0	76.9	65.4	69.2	75.0	77.9
SE [e]		10.1	11.5	5.6	9.0	8.2	11.5

[a] In units of $kcal \cdot mol^{-1}$; [b] pIC_{50} values are taken from [13]; [c] Correlation coefficient between the energy obtained at a given level of theory and the experimental inhibitory activity; [d] Percentage of successful predictions [%]; [e] Standard error of estimate, in units of $kcal \cdot mol^{-1}$.

The dominant electrostatic effects appear to arise from the interaction between counter-charged inhibitors and Arg103 residue (charges of −1 and +1, respectively). Indeed, as shown in Figure 2, which presents the electrostatic contribution to the binding energy of each amino acid residue, Arg103–inhibitor interaction has the major impact on the total $E_{EL}^{(10)}$ energy. Compared to Arg103, the remaining residues are of minor contribution. All inhibitors are directed towards Arg103 residue with their common –COOH group. Thus, any positional inaccuracy of the docked compounds related to Arg103 residue could mask the subtle interactions with other residues.

In general, more potent inhibitors are characterized by higher absolute values of the interaction energy (Table 2). To assess the relationship between the total binding energy and the inhibitory activity, interaction energy terms evaluated within HVPT energy decomposition scheme were correlated with pIC_{50} values established experimentally [13]. It can be seen in Table 2 that the interaction energy results computed at the electrostatic and MP2 levels of theory are comparable in terms of the correlation with the experimental inhibitory activity ($R = -0.65$ and -0.69, respectively). Correlation coefficient of the multipole electrostatic model of inhibitory activity is slightly lower ($R = -0.63$), but the values of the statistical predictor N_{pred} (the success rate of prediction of relative affinities, explained further in the Materials and Methods section) are comparable for all three levels of theory and remain within the range between 75.0% ($E_{EL,MTP}^{(10)}$, E_{MP2}) and 76.9% ($E_{EL}^{(10)}$). The first order Heitler–London energy ($E^{(10)}$) is characterized by the weakest relationship with the experimental inhibitory activity ($R = -0.44$, Table 2), which is due to the repulsive $E_{EX}^{(10)}$ term of the interaction energy. Apparently, the short-range exchange term of the interaction energy has contributed to the greatest extent to the binding of inhibitors with higher affinity to the EphA2 LBD, resulting in the drop of the R value at the $E^{(10)}$ level of theory. It has already been observed for other complexes [29,34] that structures obtained with empirical docking protocols and further evaluated with ab initio methods appear to suffer from the presence of artificially shortened intermolecular distances. Due to the sensitivity of short-range interaction energy components to any structural deficiencies, long-range binding energy terms seem to be more suitable for the determination of the relative ligand binging affinities [41]. Thus, the following E_{SCF} level of theory, which accounts for short-range delocalization contribution ($E_{DEL}^{(R0)}$), is only slightly improved compared to $E^{(10)}$ in terms of the correlation ($R = -0.55$, Table 2). Nevertheless, only the introduction of the correlation term $E_{CORR}^{(2)}$, that is present in E_{MP2} energy, is able to recover the predictive abilities of the inhibitory activity model, as the corresponding correlation coefficient amounts to -0.69. Similarly to values of the Pearson correlation coefficient, N_{pred} values associated with $E^{(10)}$ and E_{SCF} are also lower compared to the statistical outcome obtained for the remaining levels of theory.

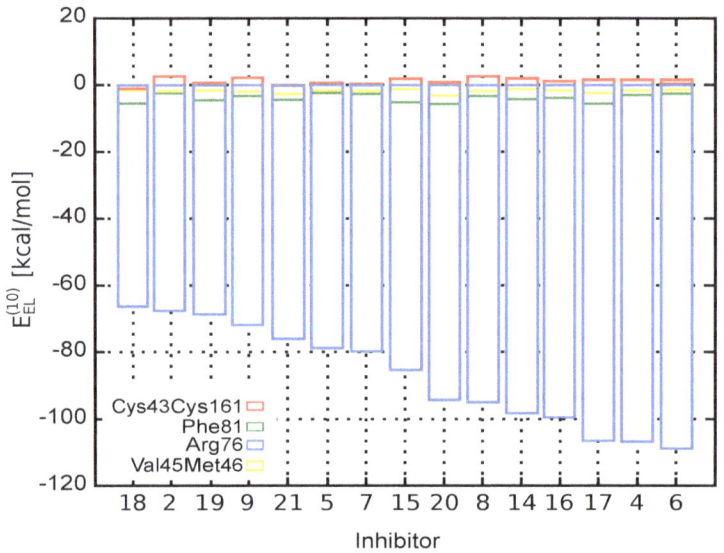

Figure 2. Contribution of EphA2 amino acid residues to the EphA2-inhibitor binding energy represented by the electrostatic term, $E_{EL}^{(10)}$.

Among all presented levels of theory, $E_{EL,MTP}^{(10)} + E_{Das}$ model offers the best performance ($R = -0.72$ or $R^2 = 0.52$, $N_{pred} = 77.9\%$). Reasonable agreement with experimental binding potency yielded by $E_{EL,MTP}^{(10)} + E_{Das}$ model indicates that accounting only for long-range interaction energy terms could compete with the computationally expensive MP2 level of theory. Still, its predictive abilities for EphA2-ephrin A1 inhibitors appear to be rather limited. Therefore, the impact of solvation was further analyzed to check whether it might be significant in this particular system.

2.2. Solvation Energy of Inhibitors

PPI contact surfaces are large [42], and the targeted EphA2 receptor fits into this description. Therefore, with a small molecule inhibitor bound, the EphA2 binding site remains relatively solvent exposed. As a result, solvation effects could possibly affect the interaction energy and influence the correlation between the latter and the experimental binding affinities. On the other hand, in the case of inhibition of another PPI system, i.e., menin-MLL complex [35], the nonempirical model accounting for the gas phase interaction only was sufficient to reproduce the experimental data. This could arise from the fact that substantially more amino acid residues surround menin ligands than in the case of

EphA2 receptor. To determine the importance of solvation effects for binding of EphA2-ephrin A1 inhibitors, solvation free energy was calculated for all compounds analyzed herein.

The solvation free energy, ΔG_{solv}, along with its electrostatic and non-electrostatic contributions ($\Delta G_{solv,el}$ and $\Delta G_{solv,non-el}$, respectively), is given in Table 3 for each EphA2 inhibitor. It can be concluded from the analysis of the correlation coefficient values provided in Table 3 that ΔG_{solv} energy values do not explicitly correlate with the experimental binding potency. Nonempirical models of inhibitory activity applied herein operate under the assumption that the enthalpic contribution to the binding free energy is responsible for the observed differences in ligand binding affinity. Accordingly, applicability of the interaction energy-based nonempirical approaches is limited to the set of ligands characterized by similar solvation free energy. Considering the suboptimal performance of $E_{EL,MTP}^{(10)} + E_{Das}$ model in predicting the inhibitory activity of EphA2 ligands ($R = -0.72$, see Table 2), compared to more significant correlation obtained previously for, e.g., menin-MLL inhibitors ($R = -0.87$ [35]), the possible influence of the solvation effects was further investigated by calculating ΔG of solvation for FAAH [33], TbPTR1 [34] and menin-MLL [35] inhibitors. In all cases, ΔG_{solv} is calculated at the MP2 level of theory, but the basis sets used depend on the system (FAAH: 6-31G(d), menin-MLL: 6-31G(d), TbPTR1: 6-311G(d) with diffuse functions on S and P orbitals of chlorine atoms; the choice of basis set was made to match the remaining ab initio interaction energy calculations performed for each of these systems). The solvation free energies of FAAH, TbPTR1 and menin-MLL inhibitors (22, 6, and 18 inhibitors in each system, respectively) are given in Supplementary Materials in Tables S2–S4. Comparison of the corresponding ΔG_{solv} standard deviation is provided in Table 4 for all abovementioned inhibitors.

Table 3. Solvation free energy (ΔG_{solv}) of inhibitors of EphA2-ephrin A1 interaction with its electrostatic, $\Delta G_{solv,el}$, and non-electrostatic, $\Delta G_{solv,non-el}$, contributions [a].

Inhibitor	pIC$_{50}$ [b]	ΔG_{solv}	$\Delta G_{solv,el}$	$\Delta G_{solv,non-el}$
20 (L-Trp)	5.69	−73.6	−81.2	7.6
16 (L-Phe)	5.18	−66.4	−73.5	7.2
17 (D-Phe)	5.12	−67.9	−75.3	7.4
7 (D-Val)	4.76	−63.2	−70.0	6.8
4 (L-Ala)	4.70	−70.9	−77.0	6.0
21 (D-Trp)	4.69	−67.5	−75.2	7.7
6 (L-Val)	4.62	−68.6	−75.5	7.0
14 (L-Met)	4.56	−69.0	−75.9	6.9
15 (D-Met)	4.56	−66.3	−73.5	7.2
5 (D-Ala)	4.51	−67.2	−73.4	6.2
8 (L-Ser)	4.48	−66.2	−72.0	5.8
2 (Gly)	4.31	−62.8	−68.1	5.3
18 (L-Tyr)	4.30	−71.7	−78.9	7.2
9 (D-Ser)	4.22	−64.2	−70.2	6.0
19 (D-Tyr)	4.00	−67.2	−74.9	7.7
R [c]		−0.43	−0.46	0.37

[a] In units of kcal · mol^{-1}; [b] pIC$_{50}$ values are taken from [13]; [c] Correlation coefficient between the solvation free energy and the experimental inhibitory activity.

Among the ligand sets presented in Table 4, EphA2-ephrin A1 inhibitors are characterized by the largest value of standard deviation of solvation free energy (3.0 kcal · mol^{-1}). Since the linear relationship between interaction energy and experimental affinities assumes, among other factors, that the solvation effects are comparable for all inhibitors within the set, this could indicate that this expectation is not met in the case of EphA2-ephrin A1 inhibitors. Considering that PCM results can be obtained easily, ΔG_{solv} standard deviation could be used as an initial predictor of the applicability of $E_{EL,MTP}^{(10)} + E_{Das}$ model.

Table 4. Performance of E_{MP2} and $E_{EL,MTP}^{(10)} + E_{Das}$ models and differences in ligand solvation free energy for EphA2-ephrin A1, menin-MLL [35], FAAH [33], and *Tb*PTR1 [34] inhibitors.

	EphA2-Ephrin A1	Menin-MLL	FAAH	*Tb*PTR1
R_{MP2} [a]	−0.69	−0.55	−0.83	−0.89
$R_{E_{EL,MTP}^{(10)}+E_{Das}}$ [b]	−0.72	−0.87	−0.67	−0.96
SD [c]	3.0	2.5	1.5	1.1

[a] Correlation coefficient between the energy obtained at MP2 level of theory and the experimental inhibitory activity; [b] Correlation coefficient between the energy obtained with $E_{EL,MTP}^{(10)} + E_{Das}$ model and the experimental inhibitory activity; [c] ΔG_{solv} standard deviation within a given set of inhibitors. In units of $kcal \cdot mol^{-1}$.

Compared to FAAH and *Tb*PTR1 ligand sets, characterized by significantly lower values of ΔG_{solv} standard deviation (Table 4), E_{MP2} model provides less accurate inhibitory activity predictions in the case of both EphA2-ephrin A1 and menin-MLL systems. On the other hand, the best performing $E_{EL,MTP}^{(10)} + E_{Das}$ model is not able to predict the EphA2-ephrin A1 inhibitory activity to the extent observed for menin-MLL or *Tb*PTR1 inhibitors. Therefore, it seemed interesting if omitting the inhibitors that differ the most in terms of ΔG_{solv} values (compounds **20**, **7**, **2** and **18**, all marked in white in Figure 3) would improve the results. The standard deviation of solvation free energy associated with the resulting reduced set of EphA2 inhibitors is equal to 1.8 $kcal \cdot mol^{-1}$. The correlation coefficients obtained for the full and reduced ligand sets are compared in Figure 4 for $E_{EL,MTP}^{(10)}$, E_{MP2}, and $E_{EL,MTP}^{(10)} + E_{Das}$ models. The corresponding correlation coefficients and N_{pred} values for all the nonempirical models of inhibitory activity, as applied to the full and reduced ligands sets, are provided in the Supplementary Materials (Table S5). Indeed, the reduced set of EphA2 inhibitors, obtained by selecting the compounds with essentially similar solvation free energies (Figure 3) features improved values of correlation coefficients. In particular, $E_{EL,MTP}^{(10)} + E_{Das}$ model provides the most accurate predictions (Figure 4), as the corresponding correlation coefficient R amounts to -0.79 ($R^2 = 0.62$).

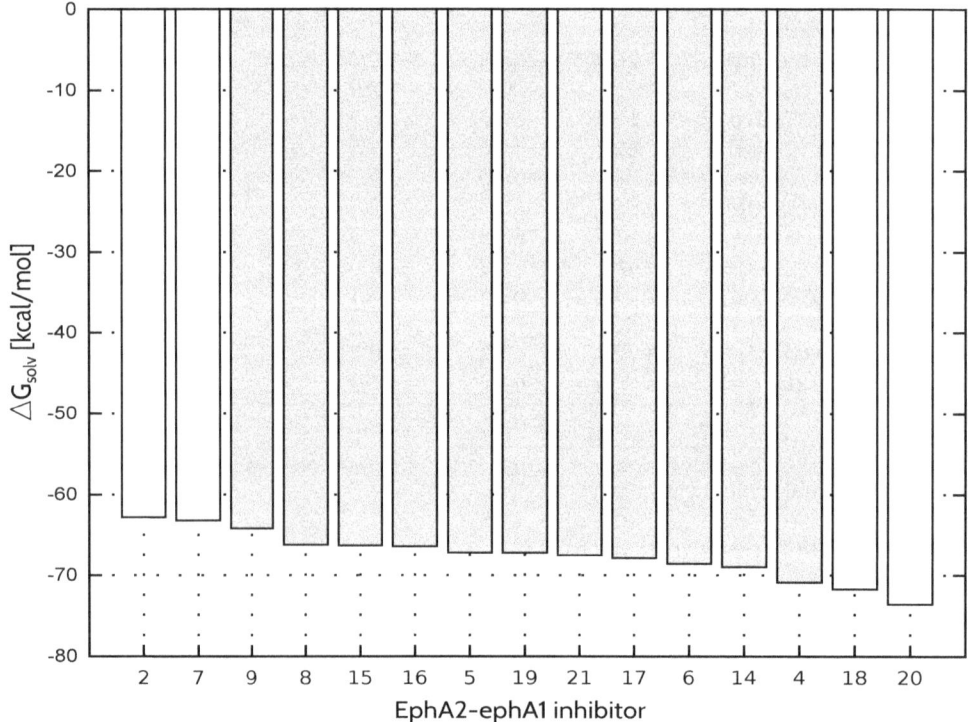

Figure 3. Solvation free energy of EphA2-ephA1 inhibitors. Compounds indicated in white were not included in the reduced ligand set.

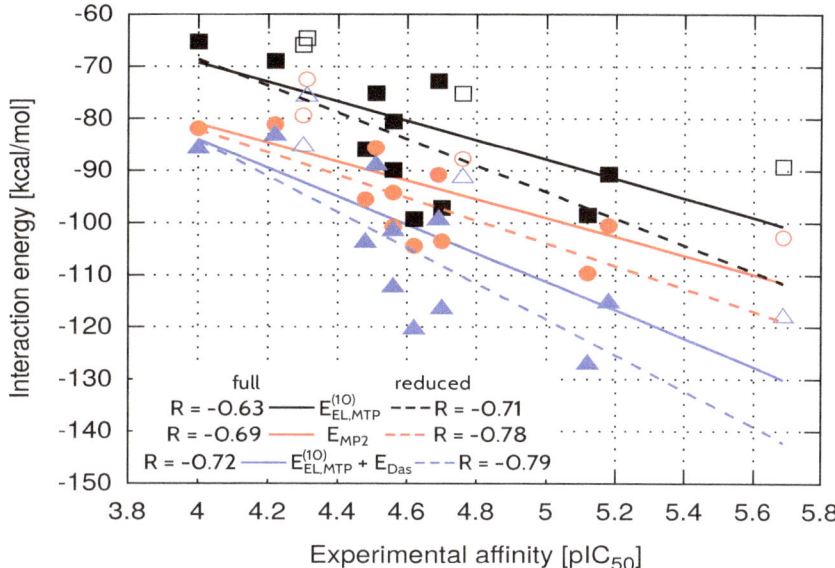

Figure 4. Total EphA2-inhibitor interaction energy at the selected levels of theory within the full (solid line) and reduced (dashed line) ligand sets. The reduced set of EphA2 inhibitors consists of the compounds shown with full symbols.

Overall, selection of ligands to be excluded based on their ΔG_{solv} differences is rather an arbitrary approach, as one could iteratively select inhibitors to reach even lower standard deviation values and, presumably, better predictive abilities of the nonempirical approach. On the other hand, a more extensive elimination of compounds does not necessarily improve the correlation coefficient between the given interaction energy model and the experimental binding potency. It can be seen in Figure 3 that ligands **4** and **9** feature ΔG_{solv} values similar to compounds **2**, **7**, **18** and **20**, already exluded from the initial set due to solvation free energy differing the most in comparison with the majority of EphA2 inhibitors considered herein. However, further limiting the size of the test set by removal of compounds **4** and **9** results in no improvement in the correlation coefficient values ($R = -0.75$ and -0.76 for E_{MP2} and $E_{EL,MTP}^{(10)} + E_{Das}$, respectively), despite substantial drop in the ΔG_{solv} standard deviation equal to 1.0 kcal·mol^{-1}. It should be noted that since the models of receptor–ligand complexes are developed with certain approximations due to the lack of experimental structures, they cannot be expected to provide perfect agreement with the experimental binding potency. Therefore, the ligand elimination based on the ΔG_{solv} differences also appears to be a limited approach. Nevertheless, it provides a reasonable basis for the exclusion of the ΔG_{solv} outliers with simultaneous improvement in the performance of nonempirical models applied herein.

2.3. Empirical Evaluation of EphA2-Ephrin A1 Inhibitors

To further evaluate the predictive potential of various empirical descriptors related to receptor–ligand binding, Solvent Accessible Surface Area (SASA) and Molecular Hydrophobicity Potential (MHP) were calculated for each EphA2-ephrin A1 inhibitor. Both lipophilic ($S_{L/L}$) and hydrophilic match surfaces ($S_{H/H}$) obtained with MHP calculation could help to assess the hydrophobic/hydrophilic complementarity of the analyzed ligands to the receptor binding site, which is based on the surface area of favorable (hydrophilic-hydrophilic) and unfavorable (hydrophilic-hydrophobic) contacts [43]. A number of scoring functions were also used for comparison, namely LigScore1 [44], PLP2 [45,46], Jain [47], PMF [48], PMF04 [49], Ludi1 [50], and Ludi3 [51] (available in Discovery Studio 2017 [52]), GoldScore, ChemScore and ASP (implemented in Gold 4.0 program [53]), AutoDock Vina [54], CHEMPLP (PLANTS program [55]), and Glide SP [56]. Correlation coefficients associated with all these empirical approaches are compared in Figure 5 for both full and reduced ligand sets. The numerical data reflecting each empirical score obtained for

EphA2 inhibitors alongwith the corresponding correlation coefficients and N_{pred} values are provided in Table S6 in Supplementary Materials.

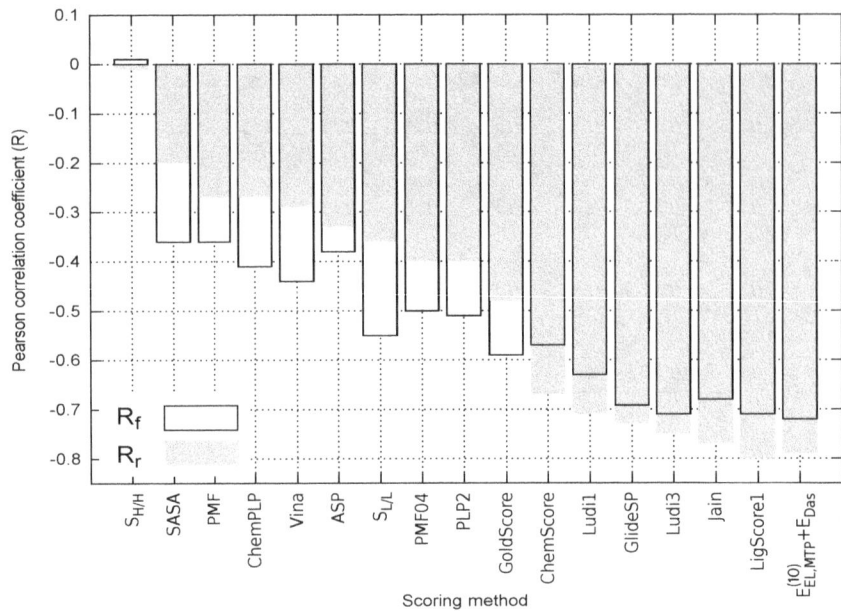

Figure 5. Pearson correlation coefficients obtained for the empirical scoring methods and $E_{EL,MTP}^{(10)} + E_{Das}$ model applied to the full (R_f) and reduced (R_r) ligand sets.

The best performing empirical descriptors for both full and reduced ligand sets include LigScore1, Jain, Ludi3, GlideSP and Ludi1 (Figure 5). In fact, the related correlation coefficients are comparable with the corresponding value characterizing $E_{EL,MTP}^{(10)} + E_{Das}$ model, e.g., in the case of full ligand set $R = -0.71$ ($R^2 = 0.50$) and -0.72 ($R^2 = 0.52$) for LigScore1 and $E_{EL,MTP}^{(10)} + E_{Das}$, respectively (see Table S6 in Supplementary Materials). Nevertheless, the majority of the analyzed empirical scoring functions yield unsatisfactory results and are outperformed by most of the nonempirical models, including $E_{EL,MTP}^{(10)} + E_{Das}$. As it has been pointed out by Li et al. [57], SASA appears to perform better as a scoring method than a number of popular scoring functions [57,58]. Accordingly, outperforming the SASA predictions might be viewed as a necessary condition, allowing for distinguishing between the scoring functions providing reasonable results and those failing to reflect the experimental binding affinity. In this particular case, most of the scoring approaches presented in Figure 5 seem to pass this test; however, only some of the empirical approaches, and $E_{EL,MTP}^{(10)} + E_{Das}$ model in particular, appear to provide at least semi-quantitative agreement with the experimental inhibitory activity.

In contrast to the theoretical models considered herein (Table S5 in Supplementary Materials), the correlation between the empirical scoring functions and experimental inhibitory activity values do not always improve when the reduced model is considered (Figure 5). This could arise from the fact that solvation effects might be implicitly included in the empirical description by parametrization performed with experimental binding potency. Depending on the ability of a given scoring function to account for the influence of solvent, limiting the test ligand set to the inhibitors featuring similar solvation energy might either decrease the performance of the method (PLP2 and PMF04) or improve the predictions, as can be seen for (e.g., LigScore1 and Jain; see Figure 5).

It seemed interesting to check whether there is some consistency in top scoring empirical functions throughout the systems tested so far in our group. Since some scoring functions implemented in Discovery Studio have been used also in the case of FAAH [33] and menin-MLL [35], comparison was made for these methods. The performance of LigScore1, PLP2, Jain, PMF, and Ludi1, described by correlation coefficients and percentage of successful predictions (N_{pred}) is presented in Table 5 for FAAH, menin-MLL and EphA2-ephA1 systems. In the latter case, comparison was made based on the results associated with the reduced set of ligands featuring similar ΔG_{solv} values. As demonstrated in Tables S5 and S6 in Supplementary Materials, selecting EphA2 inhibitors with relatively similar values of solvation free energy improves the performance of both nonempirical $E_{EL,MTP}^{(10)} + E_{Das}$ model and most of the scoring functions included in this comparison.

Table 5. Performance of empirical scoring for FAAH, menin-MLL and EphA2-ephrin A1 systems. The results obtained for nonempirical $E_{EL,MTP}^{(10)} + E_{Das}$ model are provided for comparison.

Scoring Function	FAAH [a]		menin-MLL [b]		EphA2-ephrin A1 [c]	
	R [d]	N_{pred} [e]	R	N_{pred}	R	N_{pred}
LigScore1	+0.25	44.6	−0.81	75.2	−0.80	79.6
Jain	−0.48	71.4	−0.80	77.8	−0.77	83.3
PLP2	−0.51	65.8	−0.79	80.4	−0.40	72.2
Ludi1	−0.62	73.2	−0.40	58.8	−0.71	75.9
PMF	−0.72	77.1	+0.24	41.2	−0.27	66.7
$E_{EL,MTP}^{(10)} + E_{Das}$	−0.67	74.9	−0.87	81.1	−0.79	79.6

[a] The results are taken from [33]; [b] The results are taken from [35]; [c] The results refer to the reduced set of EphA2 inhibitors; [d] Correlation coefficient between the score obtained with a given empirical function or $E_{EL,MTP}^{(10)} + E_{Das}$ energy and the experimental inhibitory activity; [e] Percentage of successful predictions [%].

It can be seen in Table 5 that both LigScore1 and Jain provide the best prediction for menin-MLL and EphA2-ephrin A1 systems. On the contrary, the performance of these scoring functions is unsatisfactory in the case of FAAH inhibitors. Entirely different predictive abilities seem to be associated with PMF function, that performs the best for FAAH system, yet it fails in the case of both menin and EphA2 inhibitors. As for the remaning empirical scoring functions compared herein, PLP2 appears to provide valid predictions only for menin-MLL system, whereas Ludi1 yields rather satisfactory agreement with the experimental data for both FAAH and EphA2 inhibitors. The interactions in menin-MLL [35] and EphA2-ephrin A1 system are predominantly electrostatic in nature, and it seems that LigScore1 or Jain functions might be better suited in such a case. On the other hand, for dispersion-dominated systems like FAAH [33], PMF could be a better choice. Nevertheless, the performance of $E_{EL,MTP}^{(10)} + E_{Das}$ model is comparable (or superior, as demonstrated in the case of menin-MLL system) to the best empirical scoring functions in each system analyzed so far. Considering that the physical nature of interactions for novel receptor–ligand complexes can hardly be determined without time-consuming ab initio calculations and the resulting choice of a reliable empirical scoring function might not be clear, the nonempirical $E_{EL,MTP}^{(10)} + E_{Das}$ model appears to be a preferable method, capable of providing the predictions with a reasonable quality at the computational cost comparable to that of empirical scoring functions.

3. Materials and Methods

3.1. Preparation of the Structures

From the LCA-based structures reported by Incerti et al. [13], all active α-amino acid LCA conjugates were selected. An LCA compound was not included in this analysis on account of the likely multiple binding modes within EphA2 [14]. In contrast, LCA amino acid conjugates studied herein presumably possess a single binding mode due to the interaction between the carboxylate group and Arg103 residue of EphA2 receptor. The structures of the selected inhibitors and the corresponding pIC$_{50}$ vales (taken from [13]) are given in Table 1.

The geometries of EphA2-inhibitor complexes, obtained from molecular docking simulation [13], were provided by Incerti et al. [13]. Since the goal of the analysis was to investigate the influence of amino acid substituent on the activity of the inhibitors, the common LCA scaffold, positioned similarly in all complexes, was not included in the analysis. In particular, the inhibitors were cut as indicated by the red line in the scaffold representation in Table 1.

To obtain more reliable positions of amino acid residues, all EphA2-inhibitor complexes were solvated with the TIP3 water model [59] and re-optimized in the CHARMM program [60] (version c36b1, Harvard University, Cambridge, MA, USA). Hydrogen atoms were built with HBUILD command. Both CHARMM General Force Field v. 2b7 [61] and CHARMM22 All-Hydrogen Force Field [62–64] parameter files were used. Missing parameters for inhibitor structures were generated with CGenFF interface at http://cgenff.paramchem.org [61,65–67] (interface version 1.0.0). LCA scaffold and all amino acid residues further than 4 Å from each inhibitor were kept frozen throughout 1000 steps of steepest descent followed by conjugate gradient optimization until a root mean squared deviation of the gradient (GRMS) of 0.01 kcal $\cdot$ mol$^{-1}\cdot$Å was reached.

The model of EphA2 binding site included all residues in the vicinity of 4 Å of the interchangeable fragment of the inhibitors, i.e., Cys70, Cys188, Phe108, Arg103, Val72 and Met73 (Figure 1). Dangling bonds resulting from cutting the amino acid residues from protein structure were filled with hydrogen atoms minimized in the Schrödinger Maestro [68] program (Maestro version 9.3, Schrödinger, LLC, New York, NY, USA) using OPLS 2005 force field [69].

3.2. Interaction Energy Calculations

Interaction energy between EphA2 receptor and each inhibitor was calculated within Hybrid Variation–Perturbation Theory (HVPT) [39,40] decomposition scheme as the sum of interaction energy components obtained for each residue-inhibitor dimer. Counterpoise correction was applied in the treatment of the basis set superposition error [70]. The calculations were carried out with a modified version [40] of GAMESS program [71] using 6-311+G(d) basis set [72–74]. HVPT introduces the partitioning of the Møller–Plesset second-order interaction energy (E_{MP2}) into the multipole electrostatic ($E_{EL,MTP}^{(10)}$), penetration ($E_{EL,PEN}^{(10)}$), exchange ($E_{EX}^{(10)}$), delocalization ($E_{DEL}^{(R0)}$) and correlation ($E_{CORR}^{(2)}$) terms:

$$
\begin{aligned}
&\quad\quad\quad\;\; \overbrace{R^{-n}}\quad\quad\quad\quad\overbrace{exp(-\gamma R)}\quad\quad\quad\overbrace{R^{-n}}\\
E_{MP2} &= E_{EL,MTP}^{(10)} + E_{EL,PEN}^{(10)} + E_{EX}^{(10)} + E_{DEL}^{(R0)} + E_{CORR}^{(2)}\\
O(N^5) &\qquad\underbrace{\hspace{7cm}}_{E_{MP2}}\\
O(N^4) &\qquad\underbrace{\hspace{5.5cm}}_{E_{SCF}}\\
O(N^4) &\qquad\underbrace{\hspace{4cm}}_{E^{(10)}}\\
O(N^4) &\qquad\underbrace{\hspace{2.5cm}}_{E_{EL}^{(10)}}\\
O(A^2) &\qquad\underbrace{\hspace{1cm}}_{E_{EL,MTP}^{(10)}}
\end{aligned}
\tag{1}
$$

which could be divided into the long- and short-range contributions that vary with the intermolecular distance R as R^{-n} and $exp(-\gamma R)$, respectively. $E_{EL,MTP}^{(10)}$ term from Equation (1) is the electrostatic multipole component of the binding energy. Herein, it was estimated from Cumulative Atomic Multipole Moment (CAMM) expansion (implemented in GAMESS), truncated at the R^{-4} term. The first-order electrostatic energy ($E_{EL}^{(10)}$) is obtained by adding the penetration term, $E_{EL,PEN}^{(10)}$, to the $E_{EL,MTP}^{(10)}$ energy. The first-order Heitler–London energy, $E^{(10)}$, is the sum of first-order electrostatic energy and the exchange component $E_{EX}^{(10)}$. The higher order delocalization energy,

$E_{DEL}^{(R0)}$, comprising classical induction and charge transfer terms, is defined as the difference between the counterpoise-corrected self-consistent field (SCF) variational energy, E_{SCF}, and the first-order Heitler–London energy, $E^{(10)}$. The correlation term $E_{CORR}^{(2)}$ is calculated as the difference of the second-order Møller–Plesset interaction energy, E_{MP2}, and converged SCF energy, E_{SCF}. $E_{CORR}^{(2)}$ consists mostly of intramolecular correlation contributions, dispersion and exchange-dispersion interaction energy terms. The zero value of the second superscript accompanying some energy terms in Equation (1) represents uncorrelated interaction energy contributions. $O(X)$ in Equation (1) denotes the scaling of the computational cost, where N and A indicate the number of atomic orbitals and atoms, respectively.

On account of the considerable computational cost of $E_{CORR}^{(2)}$ term, containing the dispersion contribution, atom–atom potential function E_{Das} [31,32] was calculated to obtain the approximate dispersion energy at a far more affordable computational expense. In contrast to $E_{CORR}^{(2)}$, computation scaling with at least the fifth power of the number of atomic orbitals, $O(N^5)$, E_{Das} calculation scales with the square number of atoms, $O(A^2)$.

Among amino acid residues in the close proximity of a varying fragment of the LCA derivatives, only Arg103 residue is not neutral, bearing +1 charge. Except for Arg103 residue and two polar cysteine residues, linked by disulfide bond, the remaining residues in the model of EphA2 receptor are nonpolar. The negatively charged (-1) ligands could be considered solvent exposed, as their large fragments face water environment. Since Cys70 and Cys188 residues constitute a disulfide bridge, these residues were considered as Cys70-Cys188 dimer interacting with inhibitors. Similarly, the subsequent Val72 and Met73 residues were not separated but treated as Val72-Met73 dimer to interact with all inhibitors. The remaining residues (Arg103 and Phe108) were included separately.

3.3. Solvation Energy Calculations

ΔG_{solv} for each inhibitor was computed at the MP2/6-311+G(d) level of theory in Gaussian09 [75]. The calculations involved Polarizable Continuum Model (PCM) using the integral equation formalism variant (IEFPCM) [76–78] and ExternalIteration [79,80], DoVacuum, and SMD [81] options.

3.4. Empirical Scoring

Empirical scoring with a variety of methods was performed for EphA2-inhibitor complexes. As scoring in the presence of water molecules appears to have little influence on the quality of predictions [82], solvent molecules were removed from protein–ligand complexes. Solvent Accessible Surface Area (SASA) [83,84] of each inhibitor was calculated in VMD [85,86] (http://www.ks.uiuc.edu/Research/vmd/) with SASA.TCL script [87] and the sphere radius set to 1.4 Å. Molecular Hydrophobicity Potential (MHP) was calculated in the PLATINUM program (version 1.0, Laboratory of Biomolecular Modeling, Russian Academy of Sciences, Moscow, Russia) [43]. GoldScore, ChemScore, and Astex Statistical Potential (ASP) were obtained using GOLD 4.0 (The Cambridge Crystallographic Data Centre, Cambridge, United Kingdom) [53] with a spherical grid centered at the alpha carbon of Arg103, comprising amino acid residues within 10 Å radius from the point of origin. PLANTS [55] docking program with its CHEMPLP scoring function was employed with a 10 Å radius sphere. PyMOL [88] and PyMOL AutoDock/Vina plugin [89] were used for preparation of the receptor and inhibitors for scoring in AutoDock Vina (version 1.1.2, Molecular Graphics Lab at The Scripps Research Institute, La Jolla, CA, USA). The latter was carried out with 22.5 Å cubic grid. Glide SP [56] (standard precision), implemented in Schrödinger Glide [90], was applied with a 15 Å grid centered on the ligand. The following scoring functions implemented in Discovery Studio 2017 [52] were used: LigScore1 [44], Piecewise Linear Potential, PLP2 [45,46], Jain [47], Potential of Mean Force, PMF [48] and PMF04 [49], Ludi1 [50] and Ludi3 [51]. In all cases, the scoring performed with Discovery Studio 2017 (Dassault Systèmes BIOVIA, San Diego, CA, USA) suite was carried out with a 10 Å radius sphere centered on the ligand. Calculations performed with AutoDock Vina, PLANTS, GOLD, Glide, and Discovery Studio 2017 involved only scoring of the available compounds' poses to avoid their re-docking, as this would affect the results. While using all these docking programs, the full protein structures were employed. In each case, standard settings were employed, as further described in Supplementary Materials.

3.5. Evaluation of the Results

To assess the performance of each scoring model, the Pearson correlation coefficients were calculated with respect to the experimentally determined inhibitory activity values [13]. The scoring functions with higher score indicating the greater binding potency were assigned the opposite values of the calculated correlation coefficient to facilitate the comparison with the non-empirical interaction energy results, assigning lower binding energy values to more potent inhibitors. Another performance measure applied herein involved the statistical predictor N_{pred}, constituting the success rate of prediction of relative affinities, and defined as the percentage of concordant pairs with relative stability of the same sign as in the reference experimentally measured activities, evaluated among all pairs of inhibitors [91]. Here, a special case has occurred as two of the examined inhibitors were reported with an identical experimental affinity ($pIC_{50} = 4.56$ for compounds **14** (L-Met) and **15** (D-Met) [13]). This particular pair of inhibitors was not taken into account while evaluating N_{pred} values.

4. Conclusions

The binding of inhibitors of EphA2-ephrin A1 system appears to be dominated by electrostatic interactions. Interaction due to the positively charged Arg103 residue constitutes the major contribution to the interaction energy between the receptor and the negatively charged inhibitors. Nevertheless, accounting for dispersion improves the predictive abilities of the theoretical models applied herein. Among the proposed nonempirical approaches characterizing EphA2-ephrin A1 inhibition, $E_{EL,MTP}^{(10)} + E_{Das}$ model, comprising solely long-range multipole electrostatic and approximate dispersion interactions, appears to be the best performing ($R = -0.72$, $N_{pred} = 77.9\%$) in terms of the agreement with the experimental data.

Furthermore, solvation effects are probably significant in the case of binding of the presented class of EphA2 inhibitors. Rather limited predictive abilities of $E_{EL,MTP}^{(10)} + E_{Das}$ model could be related to a relatively large standard deviation of solvation free energy of EphA2-ephrin A1 inhibitors. Compared to ΔG_{solv} standard deviation obtained for ligands in other systems previously studied in our group, this value is higher and thus could indicate the limited applicability of $E_{EL,MTP}^{(10)} + E_{Das}$ model for this particular case. In fact, once the set of EphA2 inhibitors is restricted to the ligands featuring essentially similar solvation free energy (i.e., without the compounds **2, 7, 18, 20**), the correlation of the theoretical models with the experimental results is improved, with the performance of $E_{EL,MTP}^{(10)} + E_{Das}$ model characterized by $R = -0.79$ and $N_{pred} = 79.6\%$.

Despite the limitations discussed above, $E_{EL,MTP}^{(10)} + E_{Das}$ model is able to outperform essentially all of the empirical descriptors tested herein, including the scoring functions implemented in popular docking programs, such as GOLD, AutoDock Vina or PLANTS. Among the empirical approaches tested herein for EphA2 inhibitors, the only scoring functions that perform comparably to $E_{EL,MTP}^{(10)} + E_{Das}$ model in this particular case involve LigScore1, Jain and Ludi. However, the scoring performance of these functions is hardly general, as it was not satisfactory in some of the systems studied in our group [33,35]. Based on the comparison encompassing FAAH [33], menin-MLL [35] and EphA2-ephrin A1 cases, it could be tentatively stated that LigScore1 or Jain functions might be better suited for systems with predominant electrostatic interactions (e.g., menin-MLL and EphA2-ephrin A1). In contrast, PMF is probably more appropriate for dispersion-dominated systems (FAAH). Irrespectively of the physical nature of the receptor–ligand binding, the nonempirical $E_{EL,MTP}^{(10)} + E_{Das}$ model yields the inhibitory activity predictions comparable or outperforming the best empirical scoring function in each of these cases, at similar computational cost. While more tests are required to validate the usefulness and general applicability of $E_{EL,MTP}^{(10)} + E_{Das}$ model, it appears to constitute an advantageous alternative to commonly used empirical scoring approaches.

Author Contributions: A.L. provided the data; W.J. performed the calculations and analyzed the data, together with E.D.-K.; all authors worked on the paper.

Acknowledgments: We thank W. Andrzej Sokalski for thoughtful reading and suggestions. Calculations were carried out using resources provided by the Wrocław Centre for Networking and Supercomputing (http://wcss.pl) and the Biovia Polish Academic Country Wide license.

References

1. Lisle, J.E.; Mertens-Walker, I.; Rutkowski, R.; Herington, A.C.; Stephenson, S.A. Eph receptors and their ligands: Promising molecular biomarkers and therapeutic targets in prostate cancer. *BBA-Rev. Cancer* **2013**, *1835*, 243–257, doi:10.1016/j.bbcan.2013.01.003. [CrossRef] [PubMed]

2. Tognolini, M.; Hassan-Mohamed, I.; Giorgio, C.; Zanotti, I.; Lodola, A. Therapeutic perspectives of Eph-ephrin system modulation. *Drug Discov. Today* **2014**, *19*, 661–669, doi:10.1016/j.drudis.2013.11.017. [CrossRef] [PubMed]

3. Himanen, J.P.; Chumley, M.J.; Lackmann, M.; Li, C.; Barton, W.A.; Jeffrey, P.D.; Vearing, C.; Geleick, D.; Feldheim, D.A.; Boyd, A.W.; et al. Repelling class discrimination: Ephrin-A5 binds to and activates EphB2 receptor signaling. *Nat. Neurosci.* **2004**, *7*, 501–509, doi:10.1038/nn1237. [CrossRef] [PubMed]

4. Qin, H.; Noberini, R.; Huan, X.; Shi, J.; Pasquale, E.B.; Song, J. Structural Characterization of the EphA4-Ephrin-B2 Complex Reveals New Features Enabling Eph-Ephrin Binding Promiscuity. *J. Biol. Chem.* **2010**, *285*, 644–654, doi:10.1074/jbc.M109.064824. [CrossRef] [PubMed]

5. Park, J.E.; Son, A.I.; Zhou, R. Roles of EphA2 in Development and Disease. *Genes* **2013**, *4*, 334–357, doi:10.3390/genes4030334. [CrossRef] [PubMed]

6. Chavent, M.; Seiradake, E.; Jones, E.Y.; Sansom, M.S.P. Structures of the EphA2 Receptor at the Membrane: Role of Lipid Interactions. *Structure* **2016**, *24*, 337–347, doi:10.1016/j.str.2015.11.008. [CrossRef] [PubMed]

7. Giorgio, C.; Mohamed, I.H.; Flammini, L.; Barocelli, E.; Incerti, M.; Lodola, A.; Tognolini, M. Lithocholic Acid Is an Eph-ephrin Ligand Interfering with Eph-kinase Activation. *PLoS ONE* **2011**, *6*, e18128, doi:10.1371/journal.pone.0018128. [CrossRef] [PubMed]

8. Miyazaki, T.; Kato, H.; Fukuchi, M.; Nakajima, M.; Kuwano, H. EphA2 overexpression correlates with poor prognosis in esophageal squamous cell carcinoma. *Int. J. Cancer* **2003**, *103*, 657–663, doi:10.1002/ijc.10860. [CrossRef] [PubMed]

9. Nikolov, D.B.; Xu, K.; Himanen, J.P. Eph/ephrin recognition and the role of Eph/ephrin clusters in signaling initiation. *BBA-Proteins Proteom.* **2013**, *1834*, 2160–2165, doi:10.1016/j.bbapap.2013.04.020. [CrossRef] [PubMed]

10. Russo, S.; Incerti, M.; Tognolini, M.; Castelli, R.; Pala, D.; Hassan-Mohamed, I.; Giorgio, C.; De Franco, F.; Gioiello, A.; Vicini, P.; et al. Synthesis and Structure-Activity Relationships of Amino Acid Conjugates of Cholanic Acid as Antagonists of the EphA2 Receptor. *Molecules* **2013**, *18*, 13043–13060, doi:10.3390/molecules181013043. [CrossRef] [PubMed]

11. Noberini, R.; Koolpe, M.; Peddibhotla, S.; Dahl, R.; Su, Y.; Cosford, N.D.P.; Roth, G.P.; Pasquale, E.B. Small Molecules Can Selectively Inhibit Ephrin Binding to the EphA4 and EphA2 Receptors. *J. Biol. Chem.* **2008**, *283*, 29461–29472, doi:10.1074/jbc.M804103200. [CrossRef] [PubMed]

12. Petty, A.; Myshkin, E.; Qin, H.; Guo, H.; Miao, H.; Tochtrop, G.P.; Hsieh, J.T.; Page, P.; Liu, L.; Lindner, D.J.; et al. A Small Molecule Agonist of EphA2 Receptor Tyrosine Kinase Inhibits Tumor Cell Migration In Vitro and Prostate Cancer Metastasis In Vivo. *PLoS ONE* **2012**, *7*, e42120, doi:10.1371/journal.pone.0042120. [CrossRef] [PubMed]

13. Incerti, M.; Tognolini, M.; Russo, S.; Pala, D.; Giorgio, C.; Hassan-Mohamed, I.; Noberini, R.; Pasquale, E.B.; Vicini, P.; Piersanti, S.; et al. Amino Acid Conjugates of Lithocholic Acid as Antagonists of the EphA2 Receptor. *J. Med. Chem.* **2013**, *56*, 2936–2947, doi:10.1021/jm301890k. [CrossRef] [PubMed]

14. Russo, S.; Callegari, D.; Incerti, M.; Pala, D.; Giorgio, C.; Brunetti, J.; Bracci, L.; Vicini, P.; Barocelli, E.; Capoferri, L.; et al. Exploiting Free-Energy Minima to Design Novel EphA2 Protein-Protein Antagonists: From Simulation to Experiment and Return. *Chem. Eur. J.* **2016**, *22*, 8048–8052, doi:10.1002/chem.201600993. [CrossRef] [PubMed]

15. Tognolini, M.; Incerti, M.; Hassan-Mohamed, I.; Giorgio, C.; Russo, S.; Bruni, R.; Lelli, B.; Bracci, L.; Noberini, R.; Pasquale, E.B.; et al. Structure-Activity Relationships and Mechanism of Action of Eph-ephrin Antagonists: Interaction of Cholanic Acid with the EphA2 Receptor. *ChemMedChem* **2012**, *7*, 1071–1083, doi:10.1002/cmdc.201200102. [CrossRef] [PubMed]

16. Ortiz, A.R.; Pisabarro, M.T.; Gago, F.; Wade, R.C. Prediction of drug binding affinities by computer binding analysis. *J. Med. Chem.* **1995**, *38*, 2681–2691. [CrossRef] [PubMed]

17. Leach, A.R.; Shoichet, B.K.; Peishoff, C.E. Prediction of Protein-Ligand Interactions. Docking and Scoring: Successes and Gaps. *J. Med. Chem.* **2006**, *49*, 5851–5855. [CrossRef] [PubMed]

18. Doweyko, A.M. 3D-QSAR Illusions. *J. Comput.-Aided Mol. Des.* **2004**, *18*, 587–596. [CrossRef] [PubMed]

19. Yilmazer, N.D.; Korth, M. Comparison of Molecular Mechanics, Semi-Empirical Quantum Mechanical, and Density Functional Theory Methods for Scoring Protein-Ligand Interactions. *J. Phys. Chem. B* **2013**, *117*, 8075–8084. [CrossRef] [PubMed]

20. Arkin, M.R.; Tang, Y.; Wells, J.A. Small-Molecule Inhibitors of Protein-Protein Interactions: Progressing toward the Reality. *Chem. Biol.* **2014**, *21*, 1102–1114. [CrossRef] [PubMed]

21. Bienstock, R.J. Computational drug design targeting protein–protein interactions. *Curr. Pharm. Des.* **2012**, *18*, 1240–1254. [CrossRef] [PubMed]

22. Laraia, L.; McKenzie, G.; Spring, D.R.; Venkitaraman, A.R.; Huggins, D.J. Overcoming Chemical, Biological, and Computational Challenges in the Development of Inhibitors Targeting Protein-Protein Interactions. *Chem. Biol.* **2015**, *22*, 689–703. [CrossRef] [PubMed]

23. Kuenemann, M.A.; Sperandio, O.; Labbé, C.M.; Lagorce, D.; Miteva, M.A.; Villoutreix, B.O. In silico design of low molecular weight protein–protein interaction inhibitors: Overall concept and recent advances. *Prog. Biophys. Mol. Biol.* **2015**, *119*, 20–32, doi:10.1016/j.pbiomolbio.2015.02.006. [CrossRef] [PubMed]

24. Jiang, Z.Y.; Lu, M.C.; Xu, L.L.; Yang, T.T.; Xi, M.Y.; Xu, X.L.; Guo, X.K.; Zhang, X.J.; You, Q.D.; Sun, H.P. Discovery of Potent Keap1-Nrf2 Protein-Protein Interaction Inhibitor Based on Molecular Binding Determinants Analysis. *J. Med. Chem.* **2014**, *57*, 2736–2745. [CrossRef] [PubMed]

25. Li, H.; Xiao, H.; Lin, L.; Jou, D.; Kumari, V.; Lin, J.; Li, C. Drug design targeting protein–protein interactions (PPIs) using multiple ligand simultaneous docking (MLSD) and drug repositioning: Discovery of raloxifene and bazedoxifene as novel inhibitors of IL-6/GP130 interface. *J. Med. Chem.* **2014**, *57*, 632–641. [CrossRef] [PubMed]

26. Chen, J.; Wang, J.; Zhang, Q.; Chen, K.; Zhu, W. Probing Origin of Binding Difference of inhibitors to MDM2 and MDMX by Polarizable Molecular Dynamics Simulation and QM/MM-GBSA Calculation. *Sci. Rep.* **2015**, *5*, 17421. [CrossRef] [PubMed]

27. Huang, W.; Cai, L.; Chen, C.; Xie, X.; Zhao, Q.; Zhao, X.; Zhou, H.Y.; Han, B.; Peng, C. Computational analysis of spiro-oxindole inhibitors of the MDM2-p53 interaction: Insights and selection of novel inhibitors. *J. Biomol. Struct. Dyn.* **2016**, *34*, 341–351. [CrossRef] [PubMed]

28. Dyguda, E.; Grembecka, J.; Sokalski, W.A.; Leszczyński, J. Origins of the activity of PAL and LAP enzyme inhibitors: Towards ab initio binding affinity prediction. *J. Am. Chem. Soc.* **2005**, *127*, 1658–1659. [CrossRef] [PubMed]

29. Grzywa, R.; Dyguda-Kazimierowicz, E.; Sieńczyk, M.; Feliks, M.; Sokalski, W.A.; Oleksyszyn, J. The molecular basis of urokinase inhibition: From the nonempirical analysis of intermolecular interactions to the prediction of binding affinity. *J. Mol. Model.* **2007**, *13*, 677–683, doi:10.1007/s00894-007-0193-8. [CrossRef] [PubMed]

30. Wagner, J.P.; Schreiner, P.R. London Dispersion in Molecular Chemistry—Reconsidering Steric Effects. *Angew. Chem. Int. Ed.* **2015**, *54*, 12274–12296, doi:10.1002/anie.201503476. [CrossRef] [PubMed]

31. Podeszwa, R.; Pernal, K.; Patkowski, K.; Szalewicz, K. Extension of the Hartree-Fock Plus Dispersion Method by First-Order Correlation Effects. *J. Phys. Chem. Lett.* **2010**, *1*, 550–555. [CrossRef]

32. Pernal, K.; Podeszwa, R.; Patkowski, K.; Szalewicz, K. Dispersionless Density Functional Theory. *Phys. Rev. Lett.* **2009**, *103*, 263201. [CrossRef] [PubMed]

33. Giedroyć-Piasecka, W.; Dyguda-Kazimierowicz, E.; Beker, W.; Mor, M.; Lodola, A.; Sokalski, W.A. Physical Nature of Fatty Acid Amide Hydrolase Interactions with Its Inhibitors: Testing a Simple Nonempirical Scoring Model. *J. Phys. Chem. B* **2014**, *118*, 14727–14736, doi:10.1021/jp5059287. [CrossRef] [PubMed]

34. Jedwabny, W.; Panecka-Hofman, J.; Dyguda-Kazimierowicz, E.; Wade, R.C.; Sokalski, W.A. Application of a simple quantum chemical approach to ligand fragment scoring for Trypanosoma brucei pteridine reductase 1 inhibition. *J. Comput.-Aided Mol. Des.* **2017**, *31*, 715–728. [CrossRef] [PubMed]

35. Jedwabny, W.; Kłossowski, S.; Purohit, T.; Cierpicki, T.; Grembecka, J.; Dyguda-Kazimierowicz, E. Theoretical models of inhibitory activity for inhibitors of protein–protein interactions: Targeting menin-mixed lineage leukemia with small molecules. *Med. Chem. Commun.* **2017**, *8*, 2216–2227. [CrossRef] [PubMed]

36. Ryde, U.; Söderhjelm, P. Ligand-Binding Affinity Estimates Supported by Quantum-Mechanical Methods. *Chem. Rev.* **2016**, *116*, 5520–5566, doi:10.1021/acs.chemrev.5b00630. [CrossRef] [PubMed]

37. Otsuka, T.; Okimoto, N.; Taiji, M. Assessment and Acceleration of Binding Energy Calculations for Protein-Ligand Complexes by the Fragment Molecular Orbital Method. *J. Comput. Chem.* **2015**, *36*, 2209–2218, doi:10.1002/jcc.24055. [CrossRef] [PubMed]

38. Tognolini, M.; Lodola, A. Targeting the Eph-ephrin System with Protein-Protein Interaction (PPI) Inhibitors. *Curr. Drug Targets* **2015**, *16*, 1048–1056. [CrossRef] [PubMed]

39. Sokalski, W.A.; Roszak, S.; Pecul, K. An efficient procedure for decomposition of the SCF interaction energy into components with reduced basis set dependence. *Chem. Phys. Lett.* **1988**, *153*, 153–159. [CrossRef]

40. Góra, R.W.; Sokalski, W.A.; Leszczyński, J.; Pett, V. The nature of interactions in the ionic crystal of 3-pentenenitrile, 2-nitro-5-oxo, ion(-1) sodium. *J. Phys. Chem. B* **2005**, *109*, 2027–2033. [CrossRef] [PubMed]

41. Beker, W.; Langner, K.M.; Dyguda-Kazimierowicz, E.; Feliks, M.; Sokalski, W.A. Low-Cost Prediction of Relative Stabilities of Hydrogen-Bonded Complexes from Atomic Multipole Moments for Overly Short Intermolecular Distances. *J. Comput. Chem.* **2013**, *34*, 1797–1799. [CrossRef] [PubMed]

42. Wells, J.A.; McClendon, C.L. Reaching for high-hanging fruit in drug discovery at protein–protein interfaces. *Nature* **2007**, *450*, 1001–1009, doi:10.1038/nature06526. [CrossRef] [PubMed]

43. Pyrkov, T.V.; Chugunov, A.O.; Krylov, N.A.; Nolde, D.E.; Efremov, R.G. PLATINUM: A web tool for analysis of hydrophobic/hydrophilic organization of biomolecular complexes. *Bioinformatics* **2009**, *25*, 1201–1202, doi:10.1093/bioinformatics/btp111. [CrossRef] [PubMed]

44. Krammer, A.; Kirchhoff, P.D.; Jiang, X.; Venkatachalam, C.M.; Waldman, M. LigScore: A novel scoring function for predicting binding affinities. *J. Mol. Graph. Model.* **2005**, *23*, 395–407. [CrossRef] [PubMed]

45. Gehlhaar, D.K.; Verkhivker, G.M.; Rejto, P.A.; Sherman, C.J.; Fogel, D.B.; Fogel, L.J.; Freer, S.T. Molecular Recognition of the Inhibitor AG-1343 by HIV-1 Protease: Conformationally Flexible Docking by Evolutionary Programming. *Chem. Biol.* **1995**, *2*, 317–324. [CrossRef]

46. Gehlhaar, D.K.; Bouzida, D.; Rejto, P.A. *Rational Drug Design: Novel Methodology and Practical Applications*; American Chemical Society: Washington, DC, USA, 1999.

47. Jain, A.N. Scoring noncovalent protein–ligand interactions: A continuous differentiable function tuned to compute binding affinities. *J. Comput. Aided Mol. Des.* **1996**, *10*, 427–440. [CrossRef] [PubMed]

48. Muegge, I.; Martin, Y.C. A General and Fast Scoring Function for Protein-Ligand Interactions: A Simplified Potential Approach. *J. Med. Chem.* **1999**, *42*, 791–804. [CrossRef] [PubMed]

49. Muegge, I. PMF Scoring Revisited. *J. Med. Chem.* **2006**, *49*, 5895–5902. [CrossRef] [PubMed]

50. Böhm, H.J. The development of a simple empirical scoring function to estimate the binding constant for a protein–ligand complex of known three-dimensional structure. *J. Comput. Aided Mol. Des.* **1994**, *8*, 243–256. [CrossRef] [PubMed]

51. Böhm, H.J. Prediction of binding constants of protein ligands: A fast method for the prioritization of hits obtained from the de novo design or 3D database search programs. *J. Comput. Aided Mol. Des.* **1998**, *12*, 309–323. [CrossRef] [PubMed]

52. Dassault Systèmes BIOVIA. *Discovery Studio Modeling Environment*; Release 2017; Dassault Systèmes: San Diego, CA, USA, 2016.

53. Jones, G.; Willett, P.; Glen, R.C.; Leach, A.R.; Taylor, R. Development and Validation of a Genetic Algorithm for Flexible Docking. *J. Mol. Biol.* **1997**, *267*, 727–748. [CrossRef] [PubMed]

54. Trott, O.; Olson, A.J. AutoDock Vina: Improving the speed and accuracy of docking with a new scoring function, efficient optimization and multithreading. *J. Comput. Chem.* **2010**, *31*, 455–461. [CrossRef] [PubMed]

55. Korb, O.; Stutzle, T.; Exner, T.E. Empirical Scoring Functions for Advanced Protein-Ligand Docking with PLANTS. *J. Chem. Inf. Model.* **2009**, *49*, 84–96. [CrossRef] [PubMed]

56. Friesner, R.A.; Banks, J.L.; Murphy, R.B.; Halgren, T.A.; Klicic, J.J.; Mainz, D.T.; Repasky, M.P.; Knoll, E.H.; Shelley, M.; Perry, J.K.; et al. Glide: A new approach for rapid, accurate docking and scoring. 1. Method and assessment of docking accuracy. *J. Med. Chem.* **2004**, *47*, 1739–1749, doi:10.1021/jm0306430. [CrossRef] [PubMed]

57. Li, Y.; Liu, Z.; Li, J.; Han, L.; Liu, J.; Zhao, Z.; Wang, R. Comparative Assessment of Scoring Functions on an Updated Benchmark: 1. Compilation of the Test Set. *J. Chem. Inf. Model.* **2014**, *54*, 1700–1716, doi:10.1021/ci500080q. [CrossRef] [PubMed]

58. Li, Y.; Han, L.; Liu, Z.; Wang, R. Comparative Assessment of Scoring Functions on an Updated Benchmark: 2. Evaluation Methods and General Results. *J. Chem. Inf. Model.* **2014**, *54*, 1717–1736, doi:10.1021/ci500081m. [CrossRef] [PubMed]

59. Mahoney, M.W.; Jorgensen, W.L. A five-site model for liquid water and the reproduction of the density anomaly by rigid, nonpolarizable potential functions. *J. Chem. Phys.* **2000**, *112*, 8910–8922, doi:10.1063/1.481505. [CrossRef]

60. Brooks, B.R.; Bruccoleri, R.E.; Olafson, B.D.; States, D.J.; Swaminathan, S.; Karplus, M. CHARMM: A program for macromolecular energy, minimization, and dynamics calculations. *J. Comput. Chem.* **1983**, *4*, 187–217, doi:10.1002/jcc.540040211. [CrossRef]

61. Vanommeslaeghe, K.; Hatcher, E.; Acharya, C.; Kundu, S.; Zhong, S.; Shim, J.; Darian, E.; Guvench, O.; Lopes, P.; Vorobyov, I.; et al. CHARMM General Force Field: A Force Field for Drug-Like Molecules Compatible with the CHARMM All-Atom Additive Biological Force Fields. *J. Comput. Chem.* **2010**, *31*, 671–690. [CrossRef] [PubMed]

62. MacKerell, A.D.; Bashford, D.; Bellott, M.; Dunbrack, R.L.; Evanseck, J.D.; Field, M.J.; Fischer, S.; Gao, J.; Guo, H.; Ha, S.; et al. All-atom empirical potential for molecular modeling and dynamics studies of proteins. *J. Phys. Chem. B* **1998**, *102*, 3586–3616. [CrossRef] [PubMed]

63. Mackerell, A.D.; Feig, M.; Brooks, C.L. Extending the treatment of backbone energetics in protein force fields: Limitations of gas-phase quantum mechanics in reproducing protein conformational distributions in molecular dynamics simulations. *J. Comput. Chem.* **2004**, *25*, 1400–1415. [CrossRef] [PubMed]

64. Stote, R.H.; Karplus, M. Zinc binding in proteins and solution: A simple but accurate nonbonded representation. *Proteins* **1995**, *23*, 12–31. [CrossRef] [PubMed]

65. Vanommeslaeghe, K.; MacKerell, A.D., Jr. Automation of the CHARMM General Force Field (CGenFF) I: Bond Perception and Atom Typing. *J. Chem. Inf. Model.* **2012**, *52*, 3144–3154, doi:10.1021/ci300363c. [CrossRef] [PubMed]

66. Vanommeslaeghe, K.; Raman, E.P.; MacKerell, A.D., Jr. Automation of the CHARMM General Force Field (CGenFF) II: Assignment of Bonded Parameters and Partial Atomic Charges. *J. Chem. Inf. Model.* **2012**, *52*, 3155–3168, doi:10.1021/ci3003649. [CrossRef] [PubMed]

67. Yu, W.; He, X.; Vanommeslaeghe, K.; MacKerell, A.D., Jr. Extension of the CHARMM general force field to sulfonyl-containing compounds and its utility in biomolecular simulations. *J. Comput. Chem.* **2012**, *33*, 2451–2468, doi:10.1002/jcc.23067. [CrossRef] [PubMed]

68. *Maestro Version 9.3*; Schrödinger, LLC: New York, NY, USA, 2012.

69. Banks, J.L.; Beard, H.S.; Cao, Y.; Cho, A.E.; Damm, W.; Farid, R.; Felts, A.K.; Halgren, T.A.; Mainz, D.T.; Maple, J.R.; et al. Integrated Modeling Program, Applied Chemical Theory (IMPACT). *J. Comput. Chem.* **2005**, *26*, 1752–1780. [CrossRef] [PubMed]

70. Boys, S.F.; Bernardi, F. The calculation of small molecular interactions by the differences of separate total energies. Some procedures with reduced errors. *Mol. Phys.* **2002**, *100*, 65–73, doi:10.1080/00268970110088901. [CrossRef]

71. Schmidt, M.W.; Baldridge, K.K.; Boatz, J.A.; Elbert, S.T.; Gordon, M.S.; Jensen, J.H.; Koseki, S.; Matsunaga, N.; Nguyen, K.A.; Su, S.J.; et al. General Atomic and Molecular Electronic Structure System. *J. Comput. Chem.* **1993**, *14*, 1347–1363. [CrossRef]

72. Krishnan, R.; Binkley, J.S.; Seeger, R.; Pople, J.A. Selfconsistent molecular orbital methods. XX. A basis set for correlated wave functions. *J. Chem. Phys.* **1980**, *72*, 650–654, doi:10.1063/1.438955. [CrossRef]

73. McLean, A.D.; Chandler, G.S. Contracted Gaussian basis sets for molecular calculations. I. Second row atoms, Z = 11–18. *J. Chem. Phys.* **1980**, *72*, 5639–5648. [CrossRef]

74. Francl, M.M.; Pietro, W.J.; Hehre, W.J.; Binkley, J.S.; Gordon, M.S.; Defrees, D.J.; Pople, J.A. Self-consistent molecular orbital methods. XXIII. A polarization-type basis set for second-row elements. *J. Chem. Phys.* **1982**, *77*, 3654–3665, doi:10.1063/1.444267. [CrossRef]

75. Frisch, M.J.; Trucks, G.W.; Schlegel, H.B.; Scuseria, G.E.; Robb, M.A.; Cheeseman, J.R.; Scalmani, G.; Barone, V.; Mennucci, B.; Petersson, G.A.; et al. *Gaussian 09 Revision D.01*; Gaussian Inc.: Pittsburgh, PA, USA, 2009.

76. Tomasi, J.; Mennucci, B.; Cances, E. The IEF version of the PCM solvation method: An overview of a new method addressed to study molecular solutes at the QM ab initio level. *J. Mol. Struct.-THEOCHEM* **1999**, *464*, 211–226. [CrossRef]

77. Pascualahuir, J.L.; Silla, E.; Tunon, I. GEPOL: An improved description of molecular surfaces. III. A new algorithm for the computation of a solvent-excluding surface. *J. Comput. Chem.* **1994**, *15*, 1127–1138. [CrossRef]

78. Tomasi, J.; Mennucci, B.; Cammi, R. Quantum mechanical continuum solvation models. *Chem. Rev.* **2005**, *105*, 2999–3093. [CrossRef] [PubMed]

79. Improta, R.; Scalmani, G.; Frisch, M.J.; Barone, V. Toward effective and reliable fluorescence energies in solution by a new state specific polarizable continuum model time dependent density functional theory approach. *J. Chem. Phys.* **2007**, *127*, 074504. [CrossRef] [PubMed]

80. Improta, R.; Barone, V.; Scalmani, G.; Frisch, M.J. A state-specific polarizable continuum model time dependent density functional theory method for excited state calculations in solution. *J. Chem. Phys.* **2006**, *125*, 054103. [CrossRef] [PubMed]

81. Marenich, A.V.; Cramer, C.J.; Truhlar, D.G. Universal Solvation Model Based on Solute Electron Density and on a Continuum Model of the Solvent Defined by the Bulk Dielectric Constant and Atomic Surface Tensions. *J. Phys. Chem. B* **2009**, *113*, 6378–6396. [CrossRef] [PubMed]

82. Englebienne, P.; Moitessier, N. Docking Ligands into Flexible and Solvated Macromolecules. 4. Are Popular Scoring Functions Accurate for this Class of Proteins? *J. Chem. Inf. Model.* **2009**, *49*, 1568–1580, doi:10.1021/ci8004308. [CrossRef] [PubMed]

83. Shrake, A.; Rupley, J.A. Environment and exposure to solvent of protein atoms. Lysozyme and insulin. *J. Mol. Biol.* **1973**, *79*, 351–371, doi:10.1016/0022-2836(73)90011-9. [CrossRef]

84. Lee, B.; Richards, F.M. The interpretation of protein structures: Estimation of static accessibility. *J. Mol. Biol.* **1971**, *55*, 379, doi:10.1016/0022-2836(71)90324-x. [CrossRef]

85. Humphrey, W.; Dalke, A.; Schulten, K. VMD—Visual Molecular Dynamics. *J. Mol. Graph.* **1996**, *14*, 33–38. [CrossRef]

86. Stone, J. An Efficient Library for Parallel Ray Tracing and Animation. Master's Thesis, Computer Science Department, University of Missouri-Rolla, Rolla, MO, USA, 1998.

87. Falsafi, S.; Karimi, Z. SASA.tcl. Available online: http://www.ks.uiuc.edu/Research/vmd/mailing_list/vmd-l/att-18670/sasa.tcl (accessed on 30 May 2018).

88. *PyMOL(TM) Molecular Graphics System, Version 1.7.0.0.*; Schrödinger, LLC: New York, NY, USA, 2013.

89. Seeliger, D.; de Groot, B.L. Ligand docking and binding site analysis with PyMOL and Autodock/Vina. *J. Comput.-Aided Mol. Des.* **2010**, *24*, 417–422. [CrossRef] [PubMed]

90. Schrödinger LLC. *Schrödinger Release 2018-1, Glide*; Schrödinger, LLC: New York, NY, USA, 2018.

91. Langner, K.M.; Beker, W.; Sokalski, W.A. Robust Predictive Power of the Electrostatic Term at Shortened Intermolecular Distances. *J. Phys. Chem. Lett.* **2012**, *3*, 2785–2789. [CrossRef]

7

Role of Extracellular Loops and Membrane Lipids for Ligand Recognition in the Neuronal Adenosine Receptor Type 2A: An Enhanced Sampling Simulation Study

Ruyin Cao [1], **Alejandro Giorgetti** [1,2], **Andreas Bauer** [3], **Bernd Neumaier** [4]🆔,
Giulia Rossetti [1,5,6,*]🆔 and **Paolo Carloni** [1,7,8,9,*]

[1] Institute of Neuroscience and Medicine (INM-9) and Institute for Advanced Simulation (IAS-5),
Forschungszentrum Jülich, Wilhelm-Johnen-Strasse, 52425 Jülich, Germany;
caobb0214@gmail.com (R.C.); alejandro.giorgetti@univr.it (A.G.)

[2] Department of Biotechnology, University of Verona, Strada Le Grazie 15, 37134 Verona, Italy

[3] Institute for Neuroscience and Medicine (INM)-2, Forschungszentrum Jülich, 52428 Jülich, Germany;
an.bauer@fz-juelich.de

[4] Institute for Neuroscience and Medicine (INM)-5, Forschungszentrum Jülich, 52428 Jülich, Germany;
b.neumaier@fz-juelich.de

[5] Jülich Supercomputing Center (JSC), Forschungszentrum Jülich, 52428 Jülich, Germany

[6] Department of Oncology, Hematology and Stem Cell Transplantation, University Hospital Aachen,
52078 Aachen, Germany

[7] Department of Physics, RWTH Aachen University, 52078 Aachen, Germany

[8] Institute for Neuroscience and Medicine (INM)-11, Forschungszentrum Jülich, 52428 Jülich, Germany

[9] Department of Neurology, University Hospital Aachen, 52078 Aachen, Germany

* Correspondence: g.rossetti@fz-juelich.de (G.R.); p.carloni@fz-juelich.de (P.C.)

Abstract: Human G-protein coupled receptors (GPCRs) are important targets for pharmaceutical intervention against neurological diseases. Here, we use molecular simulation to investigate the key step in ligand recognition governed by the extracellular domains in the neuronal adenosine receptor type 2A (hA$_{2A}$R), a target for neuroprotective compounds. The ligand is the high-affinity antagonist (4-(2-(7-amino-2-(furan-2-yl)-[1,2,4]triazolo[1,5-a][1,3,5]triazin-5-ylamino)ethyl)phenol), embedded in a neuronal membrane mimic environment. Free energy calculations, based on well-tempered metadynamics, reproduce the experimentally measured binding affinity. The results are consistent with the available mutagenesis studies. The calculations identify a vestibular binding site, where lipids molecules can actively participate to stabilize ligand binding. Bioinformatic analyses suggest that such vestibular binding site and, in particular, the second extracellular loop, might drive the ligand toward the orthosteric binding pocket, possibly by allosteric modulation. Taken together, these findings point to a fundamental role of the interaction between extracellular loops and membrane lipids for ligands' molecular recognition and ligand design in hA$_{2A}$R.

Keywords: adenosine receptor; metadynamics; extracellular loops; allosterism

1. Introduction

The human adenosine receptor type 2A (hA$_{2A}$R, Figure 1) belongs to the human G protein-coupled receptors (GPCRs) [1], the largest membrane receptor family [2], essential for cell trafficking [3]. A$_{2A}$R, highly localized in the striatum of the brain [4], is considered a promising drug target for combating Parkinson's disease [5]. As in the other GPCRs, A$_{2A}$R folds in seven transmembrane helices (H1 to H7), connected by three extracellular loops (ECL1 to ECL3) and three intracellular loops (ICL1 to ICL3). The N-terminus is extracellular, while the C-terminus is intracellular (Figure 1). Agonists and antagonists bind to the receptors' orthosteric binding site (OBS), mostly from the extracellular space. The OBS is well extended into the hydrophobic core of the transmembrane bundles [6]. Agonist binding causes conformational changes of the receptor that ultimately lead to a variety of downstream processes.

A key role for ECLs in the early stages of molecular recognition of a variety of GPCRs is currently emerging [7]. They may influence ligand binding kinetics [8], serve as flexible gatekeepers along the ligand binding pathway [7,9], and act as selectivity filters against ligand subtypes [10]. ECLs may also contribute to the formation of an additional "vestibular" binding site (VBS) located well above the OBS [11–15].

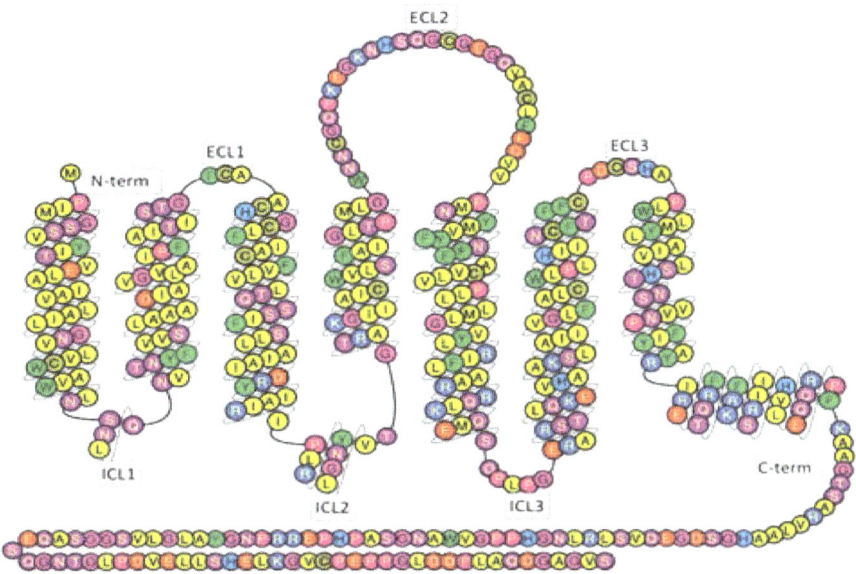

Figure 1. Snake view of hA$_{2A}$R sequence, generated by GPCRDB [16]. Residues are colored differently depending on their polarity.

Hence, a detailed understanding of ECLs' role for A$_{2A}$R/ligand interactions may provide new opportunities for designing novel ligands targeting neurodegenerative diseases [5]. Here we explore that role through well-tempered metadynamics [17,18]. This is a simulation method that accelerates the sampling of specific degrees of freedom by adding a history-dependent potential term that acts on a small number of collective variables (CVs) [18,19]. Not only can metadynamics accurately predict the absolute ligand binding free energy [17], but it also reconstructs a multi-dimensional, CV-dependent free energy surface, from which receptor interaction sites and ligand binding poses, corresponding to local free energy minima, can be identified. We focus on the human adenosine receptor type 2A, in complex with its high-affinity antagonist ZMA ((4-(2-(7-amino-2-(furan-2-yl)-[1,2,4]triazolo[1,5-a][1,3,5]triazin-5-ylamino)ethyl)phenol) or ZM241385, Figure 2) [20]. The system appears to be suitable for this research for several reasons. First, the structural determinants of the complex are well known [21–25]. Next, a comparison with biophysical and computational studies [26] allowed us to establish the accuracy of our predictions. Finally, our computational setup—in particular the modeling of the membrane and the choice of the force field—was shown to be able to correctly reproduce ligand/receptor interactions [27]. In particular, the inclusion of a realistic membrane environment turned out to impact on the description of the molecular recognition events [27–29], here and in other GPCRs [30–32].

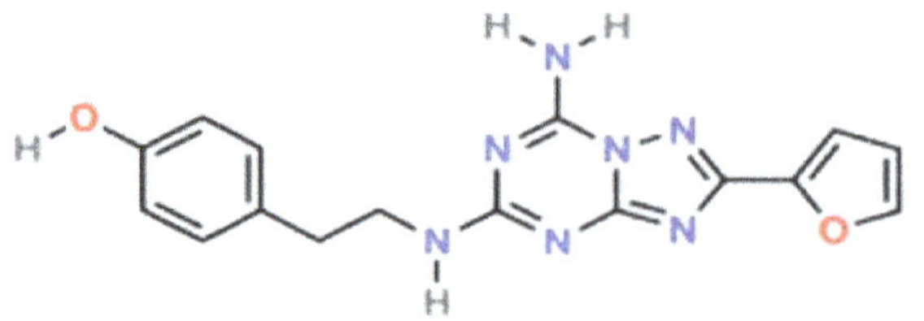

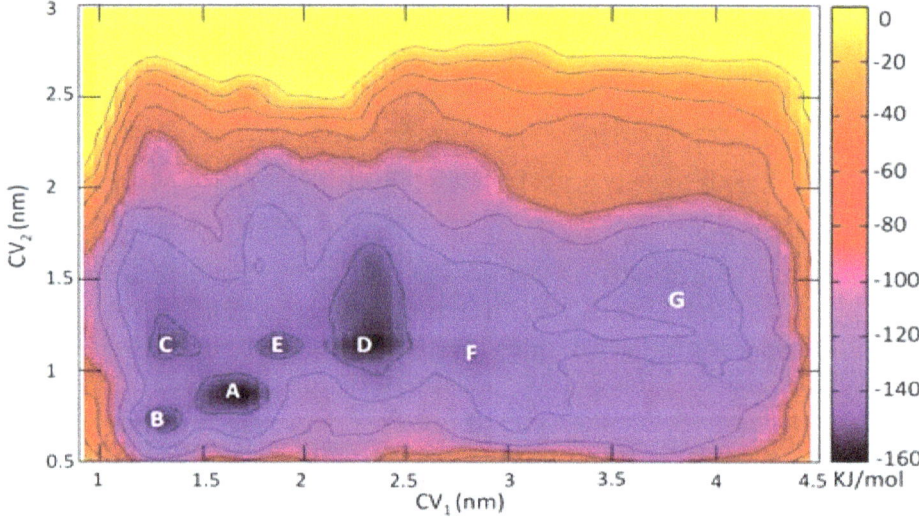

Figure 2. ZMA chemical structure, drawn with Maestro [33].

2. Results

We performed well-tempered metadynamics simulations [17,18] to investigate the role of ECLs in ligand binding by reconstructing the free energy landscape of ZMA from the extracellular space to its fully bound form to the receptor (see Section 4 and Section S1 of the Supplementary Materials for further details). The free energy is calculated as a function of two apt collective variables (Figure 3). The first (CV_1) has already been used to describe ligand binding/unbinding processes in GPCRs [19]. It is the distance between the centers of mass (COMs) of ZMA and the $C\alpha$ atoms of the transmembrane helical bundles of $hA_{2A}R$ along the membrane's normal axis. The second (CV_2) takes into account the distance between $H264^{7.29}$ and $E169^{ECL2}$ at the entrance of the orthosteric binding site (OBS) of $hA_{2A}R$. It is the distance between the $C\alpha$ atom of $E169^{ECL2}$ and the $C\alpha$ atom of and $H264^{7.29}$. These two residues can indeed form a salt bridge (see Section S2 and Table S1), which acts as a "gate" regulating the entrance of the ligand into the binding cavity [25,26]. The formation of this salt bridge is important for the ligand binding process [25,26]. Consistently, mutations of the residues in Ala and Gln impact the kinetics of unbinding [26]. During the 350-ns simulation, one cholesterol molecule binds to the hydrophobic cleft between helices H1 and H2, as previously observed [27].

Figure 3. Free-energy surface associated with ZMA/$hA_{2A}R$ interactions, as a function of collective variables, CV_1—a measure of ligand-OBS distance and CV_2—a measure of the $E169^{ECL2}$–$H264^{7.29}$ distance. The figure shows the minima associated with the ligand located in the OBS **A–C**, in the vestibular binding site **D** in the salt bridge **E** and in a solvent-exposed moiety of the ECL2 **F**. In the OBS, the free energy in **B** and **C** are higher than that in **A** by 10.0 and 14.6 kJ/mol, respectively. **G** indicates the unbound state.

The ligand bound to OBS in minimum **A** (Figure 3) represents the substate with the largest Boltzmann population, followed by minima **B** and **C**. However, the ligand turns out to also bind to an external or "vestibular" binding site (VBS), in a significant populated minimum (**D**). **D** is formed not only by helices' residues but also by ECL1 and ECL2 residues along with lipid molecules. ECL2 might play an additional role in retrieving the ligand (**F** in Figure 3).

2.1. The Orthosteric Binding Site

The ensemble of conformations forming **A** correspond to the OBS in the X-ray structures (Figure 4). The free energy difference between **A** and the unbound state **G** is -79.5 kJ (see Section 4 and Appendix A for a definition of **G**). The standard state free energy (ΔG^0) is calculated by taking into account: (i) The residual binding free energy on passing from the unbound state **G** to isolated ligand and receptor (ΔG^{Elec}), this is estimated by solving the nonlinear Poisson-Boltzmann equation [34] and (ii) The concentration of the protein in our simulation box (see Section 4). The calculated binding free energy without ΔG^{Elec} is 62.3 kJ/mol, with the correction is -58.2 ± 3.3 kJ/mol. This compares well with the experimental values found in the literature ($K_d = 1.9$ nM, $\Delta G^0 = -54.4$ kJ/mol) [21] and that measured here ($K_d = 0.8$ nM, $\Delta G^0 = -54.0$ kJ/mol, see lower panel of Figure 4). The ligand assumes an extended conformation similar to the ones present in other X-ray structures (Figure S1) with a Root Main Square Deviation (RMSD) lower than 0.24 nm of the Cα residues in the binding site (Figure S2). However, (i) the ligand's bicyclic ring moiety flips by around 60 degrees relative to the initial binding pose; (ii) the ZMA's furan ring moiety stretches towards H1 and H2, while it interacts with N253$^{6.55}$ in this and other X-ray structures of the complex (see Figure S1). E169^{ECL2}–H264$^{7.29}$ salt bridge is present for 90% of the structures belonging to **A**. Consistently, this salt bridge is present in most PDB structures of hA$_2$R/ZMA complex (3EML [22], 4EIY [23], 3VG9 [24], 3VGA [24], 5UI7 [25], 5K2A/B/C/D [35], 5UVI [36], 5JTB [37], 5VRA [38], 6AQF [39] among others) except two (3PWH [21], 5NM2 [40]; see Section S2 and Table S1 for a complete list).

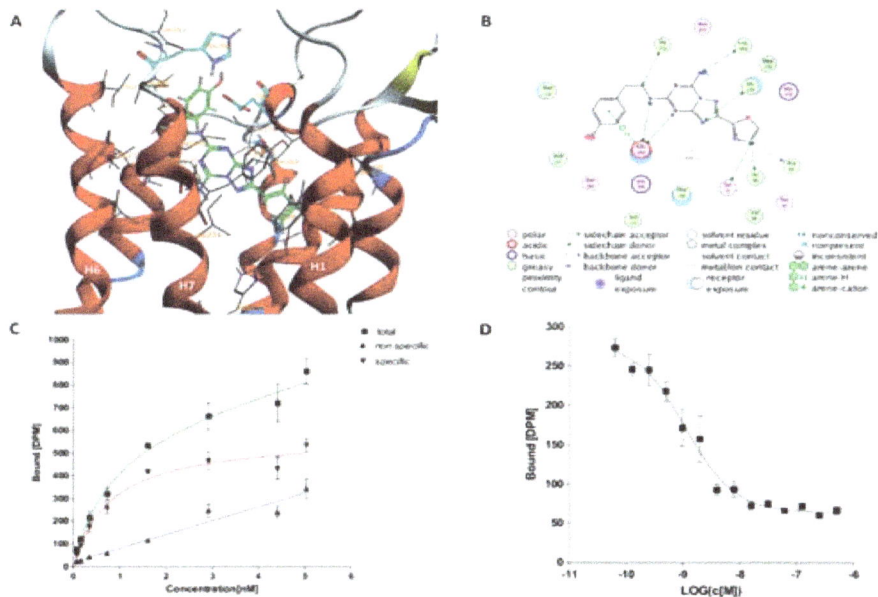

Figure 4. Lowest energy binding pose of ZMA in the orthosteric binding site (OBS, minimum **A** in Figure 3) in 3D (**A**) and 2D (**B**) representation. In (**A**) the protein backbone is render as cartoon, ZMA is shown as a green licorice, residues interacting with ZMA are shown as gray lines. The E169^{ECL2}-H264$^{7.29}$ salt bridge is shown in cyan licorice. Hydrogen, oxygen, and nitrogen atoms are specifically colored in white, red, and light blue, respectively. (**B**) 2D scheme of these binding pose in (**A**). Saturation binding assay result (**C**) and competition binding assay result (**D**) of ZMA/hA$_2$AR complex as performed in this work. The other two binding poses of ZMA in **B** and **C** minima are shown in Figure S3.

Minima **B** and **C** are higher in free energy by 10.0 kJ/mol and 14.6 kJ/mol. Here, the protein residues are less packed around the ligand: The volume of the OBS cavity increases from **A** to **B** and from **B** to **C** (0.38 nm^3, 0.42 nm^3, 0.45 nm^3, for **A**, **B**, and **C**, respectively; see Table S2). The bicyclic core of ZMA in binding poses in **B** and **C** is more deeply extended into the OBS of hA$_2$AR than in state **A** (see Figures S3 and S4). Interestingly, the E169^{ECL2}–H264$^{7.29}$ salt bridge interaction is formed in **B** but absent in **C** (99% and 2% of occurrence, respectively), as the $C\alpha$-$C\alpha$ distance between E169^{ECL2} and H264$^{7.29}$ increases from 0.6 nm to 1.3 nm. This indicates that the E169^{ECL2}–H264$^{7.29}$ salt bridge interaction is affected by ligand binding in the OBS, as previously noted by Guo et al. [26].

2.2. Role of ECLs in Molecular Recognition

Minimum **D** is higher in free energy than **A** by approximately two times $k_B T$ ($\approx$4 kJ/mol). It is associated with two "vestibular" binding sites (VBS and VBS' hereafter) located on the extracellular surface of the receptor at opposite sides of ECL2. Only the minimum associated with VBS is significantly populated (90% of the structures in **D**) and hence discussed here. Loops ECL1 and ECL2 form the VBS along with the extracellular ends of helices H1, H2, H7, and one lipid molecule (Figure 5). Lipids periodically find their way to that area and when the ligand is in the vestibular binding pocket, they establish water-mediated interactions. Two of the residues involved in ZMA binding, S67$^{2.65}$ and L267$^{7.32}$, in the VBS, if mutated, increase the residence time of ZMA for hA$_{2A}$R by 1.5–2.3 folds, while showing negligible influence on the ligand binding affinity [26].

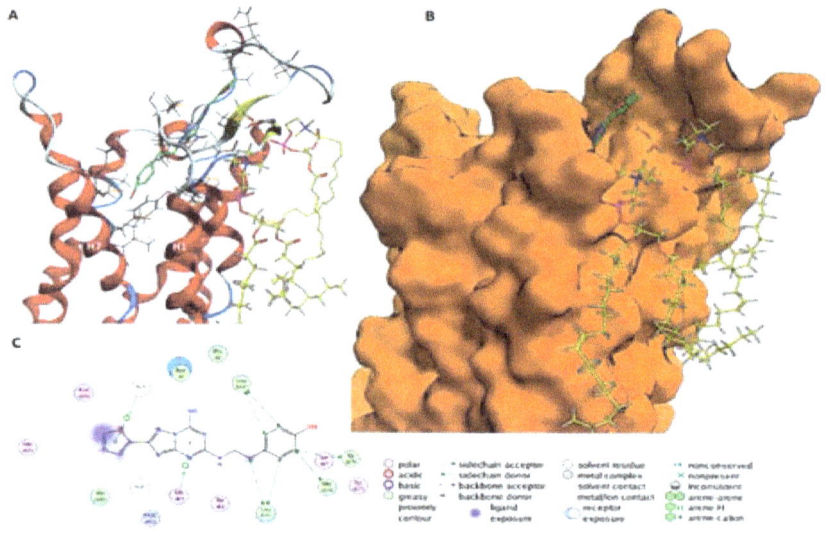

Figure 5. ZMA binding poses in the minimum **D** of Figure 3 is shown in the (**A–C**) panels as 3D, surface, and 2D representation, respectively. In (**A**) the protein backbone is rendered as a cartoon, ZMA and POPC molecules are shown as a green and yellow licorice, respectively, residues interacting with ZMA are shown as gray lines. Hydrogen, oxygen and nitrogen atoms are specifically colored in white, red and light blue, respectively. In (**B**) the solid protein surface, based on Van der Waal atom radii, is shown in orange.

Among the residues that comprise the VBS, those located on ECL2, e.g., N154^{ECL2}, H155^{ECL2} and A165^{ECL2}, and H7, e.g., L267$^{7.32}$, are not conserved across the human adenosine receptor subfamilies (Table S3). On the other hand, most of the residues located on the head of the remaining helices are better conserved, including Y9$^{1.35}$ (100% conservation), E13$^{1.39}$ (100% conservation), S67$^{2.65}$ (75% conservation), M270$^{7.35}$ (50% conservation), Y271$^{7.36}$ (75% conservation) across the human adenosine receptor subtypes. Similar trend of conservation of these residues in A$_{2A}$R across species is found (Table S3).

In the minimum **F**, ZMA interacts with a solvent-exposed motif of ECL2 (Figure 6): its 4-hydroxyphenyl moiety forms a hydrogen bond with E161^{ECL2}, a water-mediated hydrogen bond with K150^{ECL2} and hydrophobic interactions with G152^{ECL2}, K153^{ECL2}, N154^{ECL2}, H155^{ECL2} alkyl groups.

Although this minimum is not significantly populated (**F** is -20.92 kJ/mol higher in free energy than **A**), we suggest here it might play a role for ZMA's binding to the receptor. Mutagenesis experiments found K153^{ECL2}A mutation significantly decreased the dissociation rate of ZMA for hA$_{2A}$R [26]. Mutations of two glutamic residues (E151^{ECL2} and E161^{ECL2}) which are also located on the same solvent-exposed region of ECL2 have been shown to exert strong effects on ligand binding affinity [41]. The residues composing this solvent-exposed motif (K150–E161) are overall non-conserved (Figure S5) across the four human adenosine subfamilies. However, the conservation of the two glutamic residues is significant in A$_{2A}$R across species (28% for E151^{ECL2} and 50% for E161^{ECL2}, Figure S6).

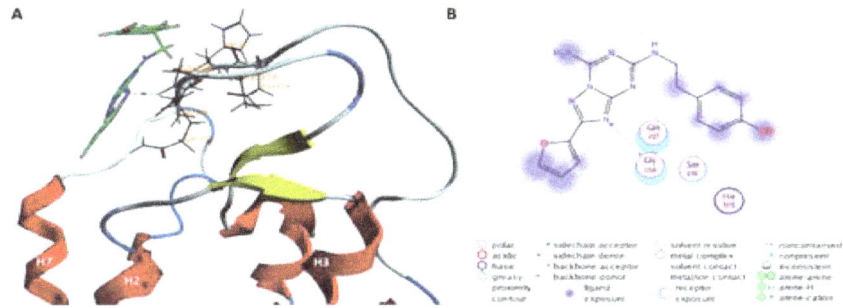

Figure 6. ZMA binding poses in the minimum **F** of Figure 1 are shown in the (**A,B**) panels, as 3D and 2D representation, respectively. In (**A**) the protein backbone is render as cartoon, ZMA is shown as a green licorice, residues interacting with ZMA are shown as gray lines. Hydrogen, oxygen, and nitrogen atoms are specifically colored in white, red, and light blue, respectively.

2.3. An Access Control Binding Site

In **E**, ZMA interferes with the E169^{ECL2}–H264$^{7.29}$ salt bridge (Figure 7) by H-bonding E169^{ECL2}. The ligand additionally forms hydrophobic interactions with I66$^{2.64}$ and water-mediated hydrogen bonding interaction with S67$^{2.65}$, as in [26]. Consistently, H264^{ECL2}A and E169^{ECL2}Q variants [26] impact on a ligand's dissociation rate, as do I66$^{2.64}$A and S67$^{2.65}$A variants on a ligand's residence time in A$_{2A}$R [26]. Interestingly, most of the residues involved in this binding site, specifically I66$^{2.64}$, S67$^{2.65}$, Y9$^{1.35}$, M270$^{7.35}$, and Y271$^{7.36}$, correspond to a recently identified cryptic allosteric pocket [42]. The latter was suggested to be responsible for the selective binding of a novel bitopic antagonist against other adenosine receptor subtypes [42].

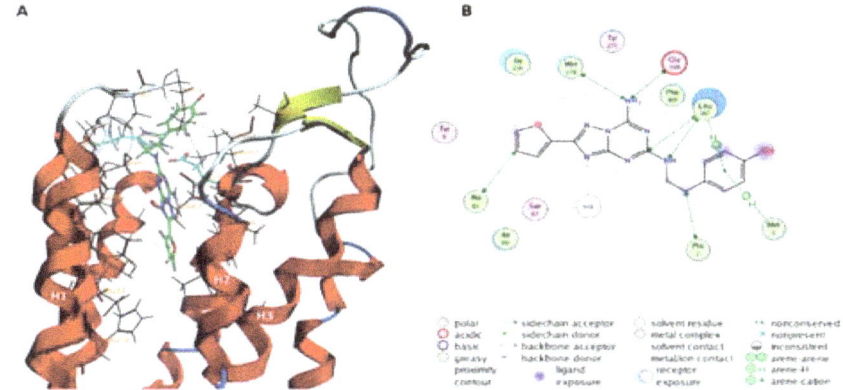

Figure 7. ZMA binding poses in the minimum **E** of Figure 3 is shown in (**A,B**) panels, as 3D and 2D representation, respectively. In (**A**) the protein backbone is render as cartoon, ZMA is shown as a green licorice, residues interacting with ZMA are shown as gray lines. The E169^{ECL2} and H264$^{7.29}$ residues are shown in cyan licorice. Hydrogen, oxygen and nitrogen atoms are specifically colored in white, red and light blue, respectively.

The E169^{ECL2}–H264$^{7.29}$ salt bridge is moderately conserved across human adenosine receptor subfamily members (see Section S3). Residues located in the lower region of the OBS are generally more conserved than residues located in the upper part of the binding pocket [43]. The first have been suggested to play a role for ligand affinity, the second for ligand specificity [43]. Here we demonstrate that the E169^{ECL2}–H264$^{7.29}$ salt bridge is better conserved in A$_1$Rs and A$_{2A}$Rs than A$_{2B}$Rs and A$_3$Rs (see Section S3). Granier et al. [10] has suggested that diversity in amino acid composition in the outer part the binding pocket may contribute to selection filter for larger ligands. Given these, we speculate that the gated access control site present at the binding pocket entrance of hA$_{2A}$R, may be one important structural property that can selectively modulate ligands' entering the OBS of adenosine receptor subtypes.

In conclusion of this section, let us analyze some common trends across the identified binding sites. As ZMA moves from the solvent-exposed minimum **F** to the membrane-facing vestibular minimum **D**, the number of water molecules decreases and the volume of OBS is the smallest (Table S2 number of water molecules around the ligand decreases from 33 (**F**) to 21 (**D**) while the volume of OBS increases from 0.33 nm^3 to 0.34 nm^3 (see Table S2). In minimum **E** and, even more in OBS, the number of water molecules decreases and OBS volume increases (Table S2). These trends are consistent with those uncovered in a microsecond-scale MD study [14] of the family A GPCR sphingosine-1-phosphate receptor [44].

2.4. Allosterism

We next asked ourselves whether the VBS and ECLs might act as allosteric sites for OBS. To address this issue, we focused on coevolving residues (residue pairs that are mutated in concert more frequently than random genetic events) between receptor binding sites and other protein regions [45–47]. Indeed, the latter are likely to play a role in the allosteric modulation of ligand binding [48].

The presence of allosterism is then identified by the so-called residue-paired coevolution score (PCS) [46]. The score ranges from 0 (no covariation) to 0.5 (moderate covariation) and 1 (complete covariation) [46]. In VBS, E13$^{1.39}$ and Y9$^{1.35}$ show moderate coevolutionary relation (PCS > 0.4) with residues located in the OBS, including V84$^{3.32}$, L85$^{3.33}$, N181$^{5.42}$ and I274$^{7.39}$ and H278$^{7.43}$ (Figure 8 and Table S4). Also, G69^{ECL1} show moderate coevolutionary relation (PCS > 0.4) with H278$^{7.43}$. Moreover, we find that M177$^{5.38}$, I274$^{7.39}$ and H278$^{7.43}$ in the OBS [49] coevolve with G142^{ECL2} and W143^{ECL2} belonging to VBS' (PCS 0.4–0.6). Interestingly, these two residues also coevolve with E13$^{1.39}$ in VBS. Inspection of the structure allowed identifying a possible network of interactions connecting VBS and ECLs with OBS. E13 was shown to play a role in the stabilization of the UK432097 agonist and has been suggested to play a role in the on-rate ligand binding [50].

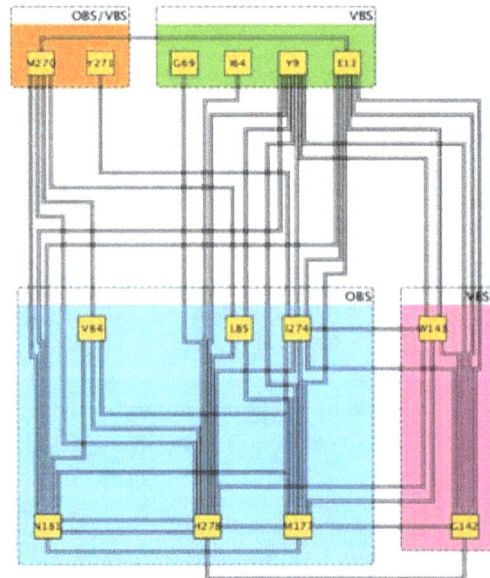

Figure 8. Coevolution relationships between amino acids of the relevant regions studied in this article (Table S4) based on Coeviz web server analyses [46].

Hence, we conclude that ECL2 and VBS residues might be allosterically coupled with OBS. PCS analysis of the available crystallized 27 human GPCRs with OBS-bound ligand complex structures (subclasses A, B, C, and F) shows that, in the majority of them (22 structures), extracellular loops residues coevolve with OBS ones (Table S5). A section is offered in the Supplementary Materials on the sodium allosteric binding site (Section S4).

3. Discussion

Our metadynamics simulations have provided the free energy landscape of ZMA binding to hA$_{2A}$R, embedded in a solvated neuronal-like membrane environment. Our calculations are consistent with the available experimental (i) and simulated (ii) data: (i) the predicted K$_d$ is in agreement with that derived by measurements available in the literature [21] and performed in this work. (ii) The free-energy profile features a 'multi-minima' landscape, consistent with a multi-step dissociation process suggested by previous temperature accelerated molecular dynamics simulation [26]. In particular, the residues interacting with the ligand in [26] are the same ones in our minima.

The ligand is located in the OBS as in X-ray structure [21] (minimum **A** in Figure 4), but it slightly differs in the orientation of its bicyclic ring. This might be ascribed, at least in part, to the dramatic differences in the protein environment. Indeed, the environment changes from a detergent micelle of the X-ray structure [21], to a membrane-mimicking environment, rich in cholesterol, in the MD [27,51]. Notably, cholesterol binds to a pocket located between helices H1 and H2 (see Figure 5 in [27]). This may affect ligand binding poses (via an allosteric mechanism [27]) and affinity [30] in GPCRs.

Our metadynamics simulations further reveal the existence of significantly populated states (minima **B**, **C** and **E**), where ZMA interferes with the E169^{ECL2}–H264$^{7.29}$ salt bridge located between ECL2 and ECL3 of hA2AR. In minima **B** and **C,** the ligand is still in the OBS but in **B**, the phenol moiety of ZMA form a hydrogen bond with E169^{ECL2}, possibly weakening the electrostatic strength of the salt bridge, while in **C** the phenol moiety is exposed toward the solvent and the salt bridge is broken. Notably, despite the fact that the geometrical position of ZMA is very similar (see Figure S4), the free energy increases on passing from **B** to **C** with respect to **A**, pointing toward a key role of such salt bridge in controlling the dissociation kinetics of ligands, as also suggested in [8]. Accordingly, H264$^{7.29}$A and E169^{ECL2}Q mutations impact on the ligand's dissociation rate [26]. In **E**, ZMA is located between the OBS and the VBS and, although the E169^{ECL2}–H264$^{7.29}$ is formed, its electrostatic strength is possibly decreased by a hydrogen bond of the ligand's triazin moiety with E169^{ECL2}. In this binding site, ZMA also interacts directly with I66$^{2.64}$ and forms a water-mediated hydrogen bonding interaction with S67$^{2.65}$. These residues, if mutated in alanine, I66$^{2.64}$A and S67$^{2.65}$A, impact on the ligand's residence time in A$_2$AR [26].

The E169^{ECL2}–H264$^{7.29}$ salt bridge therefore seems to act as a "narrowing gate", similar to what was observed in human GPCR β$_2$ adrenergic receptor. Here, the D192^{ECL2}–K305$^{7.32}$ salt bridge acts as a gate and the salt bridge is located at the entrance of the OBS of this receptor, deeply buried in the transmembrane region [11,52].

In addition to these minima, the extracellular loops ECL1, ECL2, as well as the heads of helices H1, H2, H7, contribute to the formation of a previously unnoticed, significantly populated vestibular binding site (VBS) accommodating the ligand (minimum **D**). Consistently, mutating S67$^{2.65}$ and L267$^{7.32}$, two VBS residues, increase the residence time of ZMA for hA$_{2A}$R by 1.5–2.3-fold, while showing negligible influence on the ligand binding affinity [26]. The role of some residues in ECL1 ECL2 of hA$_2$A in ligand binding affinity was already shown elsewhere [53], and the discovery that ECL2 forms the VBS is in line with several studies on other GPCRs [11–13,15]. However, we suggest here that such VBS is not transient but is actually significantly populated. Interestingly, the recently resolved structure of hA$_{2A}$R in complex with 5-amino-N-[(2-methoxyphenyl)methyl]-2-(3-methylphenyl)-2H-1,2,3-triazole-4-carboximidamide (Cmpd-1, hereafter) identified one potential allosteric pocket [42] that is located in the helical part of the VBS, as the one identified from our metadynamics simulation. Specifically, the site accommodating the methoxyphenyl group of Cmpd-1 consists of $Y9^{1.35}$, $A63^{2.61}$, $I66^{2.64}$, $S67^{2.65}$, $L267^{7.32}$, $M270^{7.35}$, $Y271^{7.36}$ and $I274^{7.39}$, five of which are located in the helical part of the VBS (see Figure 5).

A further interesting feature of our identified VBS, is that a lipid molecule contributes to the stabilization of the ligand binding. A key role of lipids for ligand binding has been already pointed out in other studies (see, for instance, [54,55]). The lipid bilayer was found to form the determinant entry pathway along which the ligand gains access to GPCRs [56] and even form a "membrane vestibule" that controls ligand binding kinetics [14]. Lipid composition in membrane could also modulate stability of specific ligand binding pose [27,28]. Therefore, altered lipid composition in the neuronal membrane could affect ligand binding. This, in turn, could alter the function of the receptor [57,58].

ECL2 forms a third binding site on the solvent-exposed region of the receptor, topographically distinct from the VBS (minimum F). The existence of the site might be consistent with mutagenesis experiments [26,41], since mutations of residues $E151^{ECL2}$, $K153^{ECL2}$, and $E161^{ECL2}$, which are found to directly interact with ZMA in our simulations, significantly influence ligand binding affinity or dissociation speed. At the speculative level, we suggest that ECL2 might function as a 'fishing' moiety for the ligand in the extracellular compartment, redirecting it toward the VBS, consistent with the fact that the volume of the OBS increases upon ligand binding from ECL2 to the OBS.

Specific residues belonging to VBS or located in the ECLs, turn out to co-evolve with residues in the OBS, suggesting an allosteric pathway connecting the extracellular domains of the receptor to OBS (Figure 8). The pathway might impact on ligand binding. A similar conclusion was reached for another GPCR, the dopamine D_2 receptor. For the latter, coevolved residues pairs show functional coupling in controlling responses to dopamine [59].

We close this section by analyzing major limitations of this work. First, experimental evidence indicates that $hA_{2A}R$ can form homo- and/or hetero-assemblies of two or more monomers [60,61]. It is expected that oligomeric order and architecture of the supramolecular assembly, not considered here, may affect ligand binding [62]. Second, the prediction of energetics and binding poses is determined by a priori choice of a set of CVs [19,63]. In this case, this issue might be alleviated by the fact that a wide range of optimal CVs are available to describe ligand binding to a GPCR [19]. Third, the calculated absolute ligand binding free energy might contain a significant source of inaccuracy from the use of necessarily approximate force fields [64] as well as nonlinear Poisson–Boltzmann calculations [34]. Fourth, the level of theory we employed inherently neglect the electronic degrees of freedom, that might be relevant for ligand binding. However, in this case no covalent binding occurs and therefore the polarization effects are negligible. Fifth, other components of cellular membrane, such as polyunsaturated chains and sphingolipids, have not been included in our membrane model [27]. The content of these is far less than cholesterol, however, also these biomolecules might impact on GPCRs function [65]. Finally, the sodium ion, recently discovered in the high-resolution structure of $hA_{2A}R$ [66], was not considered here. The consistency with experiments makes us suggest that these issues do not affect substantially the main predictions of the paper, namely the contribution of ECL2 to two significantly populated binding sites other than the OBS, along with the key role of lipids for the molecular recognition process.

4. Materials and Methods

4.1. System Preparation and MD Simulations

We have shown elsewhere that the conformation of the $hA_{2A}R$ is affected by membrane composition [27]. One of the main players in membrane-driven modulation of $hA_{2A}R$ is cholesterol, that specifically binds in a cleft between H1 and H2 and can allosterically affect the shape of the orthosteric binding site (OBS). Despite the fact that in cellular membranes, cholesterol content varies from 33% to 50% [67], unfortunately, cholesterol-driven allosteric effects are not captured in X-ray structures since artificial detergent-based environment or solubilizing antibodies are used [68]. In an effort to model $hA_{2A}R$ in a membrane environment mimicking the real cellular membrane, we embedded the receptor in a membrane of 42% POPC, 34% POPE and 25% of cholesterol molecules, mimicking the ratio among the three components in human cellular plasma membranes [69]). In our previous study, we showed that cholesterol affects the receptor structure, in the equilibrated part of the simulation. Therefore, to include the cholesterol-driven allosteric effects in the model, we used as starting conformation for $hA_{2A}R$ the last snapshot of the our previous 800-ns MD simulation of cholesterol-bound $hA_{2A}R$ with caffeine, embedded in a membrane with the same composition [27].

An educated guess of the ZMA binding pose in the cholesterol-bound $hA_{2A}R$ was obtained by superimposing ZMA via Pymol [70] software in the binding cavity, using as a template the configuration that ZMA has in the 3PWH X-ray structure [21]. Structural comparison (Figure S1) and Root Mean Square Deviation (RMSD) (Figure S2) across most of the X-ray structures of ZMA available so far in complex with $hA_{2A}R$, is offered in the Supplementary Materials.

The protonation state of histidine residues, and in particular of H264 involved in forming the salt bridge with E169, was evaluated by PROPKA [71] and cross-checked within available $hA_{2A}R$ crystal structures (see Section S1). The AMBER99SB-ILDN force fields [72], the Slipids [73,74] and the TIP3P [75] force fields were used for the protein and ions, the lipids, and the water molecules respectively. The General Amber force field (GAFF) parameters [76] were used for ZMA, along with the RESP atomic charge using Gaussian 09 [77] with the HF-6-31G* basis set [78,79]. MD simulations were performed using Gromacs v4.5.5 package [80]. The total system is a 14.3 nm $\times$ 10.8 nm $\times$ 9.6 nm box, including 248 POPC lipids, 204 POPE lipids, and 141 cholesterol molecules. The total number of atoms in the system is 151,850. The computational protocols utilized in the previous study [27] was applied here for the MD simulation of ZMA/$hA_{2A}R$ complex. Specifically, MD simulation was conducted in the NPT ensemble (constant pressure and temperature) under periodic boundary conditions. Constant temperature and pressure conditions were achieved via independently coupling protein, lipids, solvent and ions to Nosè-Hoover thermostat [81] at 310 K and Andersen-Parrinello-Rahman Barostat [82] at 1 atm. The Particle Mesh Ewald method [83] was used to treat the long-range electrostatic interaction with a real space cutoff of 1.2 nm. A 1.2-nm cutoff was used for the short-range non-bonded interaction. A time step of 2 fs was set. The LINCS algorithm [84] was applied to constrain all bonds involving hydrogen atoms. The final system was equilibrated for 20 ns under constant pressure and temperature (NPT ensemble) before metadynamics simulation.

4.2. Metadynamics Simulations

The well-tempered metadynamics approach [17,18], an enhanced sampling algorithm within the framework of classical MD, was applied together with the computational protocol above to delineate the free energy profile for the binding of ZMA to $hA_{2A}R$ within the solvated neuronal-like membrane model (see Section S1 for further details on the methods). The deposition rate of the Gaussian bias terms was set to 1 ps and the initial height to 1.0 kJ/mol, with a bias factor of 15. To obtain the free energy profile of ZMA binding to $hA_{2A}R$, we used two different collective variables, termed here CV_1 and CV_2 [19]. Specifically, CV_1 was defined as the distance between the center of mass (COM) of ZMA and COM of $C\alpha$ atoms of the transmembrane helical bundles of $hA_{2A}R$ along the membrane normal (Z-axis). CV_2 corresponded to the distance between $C\alpha$ atoms of H264 and E169. Gaussian widths of 0.05 nm were selected for CV_1 and CV_2, respectively, based on inspection of the initial dynamics of the system during equilibration. To restrict the sampling of conformational states in which the ligand was in contact with the protein, lower and upper limits of 1.5 nm and 3.8 nm, respectively, for the values of CV_1 were enforced using steep harmonic potentials with an elastic constant of 250 kJ/nm^2. Besides, one unbiased CV_3 representing the XY component of the distance between the COM of ligand and the COM of $C\alpha$ atoms of the transmembrane helical bundles of $hA_{2A}R$ was enforced below 1.2 nm so that the ligand would not diffuse to solvent regions that are far away from the receptor. All calculations used the Gromacs 4.5.5 program with the Plumed 1.3 plugin [85]. The unbinding free-energy was calculated as in [86–88]. The contribution to the free energy of binding from the metadynamics ΔG^{MetaD} was calculated as the free energy difference between the local minimum A (CV_1 = [1.58 nm, 1.75 nm], CV_2 = [0.82 nm, 0.91 nm]) and the unbound state G (CV_1 = [4.40 nm, 4.50 nm], CV_2 = [1.10 nm, 1.30 nm]); see Appendix A for further details of G definition. The contribution for CV_1 > 4.5 nm (ΔG^{Elec}) was estimated through the nonlinear Poisson-Boltzmann equation by using APBS 1.4 program [34]. The setup for ΔG^{Elec} calculation is the following: The interior dielectric constant of the $hA_{2A}R$ was set to 4 and that of the solvents to 80. The concentration of sodium and chloride ions are set to 0.15 M. The total calculated value for the free energy of binding was obtained as $\Delta G = \Delta G^{MetaD} + \Delta G^{Elec}$. The standard-state free energy of binding was calculated by $\Delta G^0 = \Delta G - RT \ln \Delta \left(\frac{[P]}{[P]^0} \right)$. R is the molar constant, $[P]$ is the concentration of the protein in our simulation box, and $[P]^0$ = 1 M is the standard-state concentration [88]. ΔG^0 was compared with the experimental binding free energies through the relationship $\Delta G^0 = RTlnk_{eq}$, where k_{eq} is the experimental equilibrium constant.

Volume analysis of the OBS of $hA_{2A}R$ was performed with trj_cavity 2.1 [89]. The residues comprising the OBS of $hA_{2A}R$ are defined as those within 0.6 nm of ZMA in the X-ray structure (PDBid:3PWH) [21]. Calculation of number of water molecules was performed with VMD [90]. The number of water molecules within 4 Å of ZMA is averaged over frames collected for each state. Coevolution analysis was performed with the web-based tool CoeViz [46] integrated in the web server POLYVIEW-2D [91].

4.3. Experimental Affinity Testing

4.3.1. Cell Culture

The cells were grown at 37 °C in 5% CO_2/95% air adherently and kept in Ham's F12 Nutrient Mixture, containing penicillin (100 U/mL), 10% fetal bovine serum, Geneticin (G418, 0.2 mg/mL), streptomycin (100 µg/mL), and L-glutamine (2 mM). Cells were split two or three times per week at a ratio between 1:5 and 1:20. The culture medium was removed and the cells were washed with PBS buffer (pH 7.4), scraped off, suspended in 1 mL PBS per dish, and stored at −80 °C, to prepare them for binding assays.

Membrane preparation for radioligand binding experiments. The cell were prepared as in [92]. The frozen cell suspension was thawed and homogenized on ice (Ultra-Turrax, 1 × 30 s at full speed). The homogenate was next centrifuged for 10 min (4 °C) at 600× g. The supernatant was then centrifuged for 60 min at 50,000× g after that, the membrane pellet was suspended again in 50 mM Tris/HCl buffer (pH 7.4) and frozen in liquid nitrogen at a protein concentration of 6 mg/mL. Finally, it was stored at −80 °C. Protein estimation used a naphthol blue black photometric assay [93] after solubilization in 15% NH_4OH containing 2% SDS (w/v); human serum albumin served as a standard.

4.3.2. Experimental Binding Affinity

Binding experiments used membranes from CHO K1 cells stably expressing the human A2A adenosine receptor. [3H]ZM 241385 (0.8 nM in competition experiments) as radioligand was used to obtain dissociation constant of [3H]ZM 241385 and the inhibition constant of not tritiated ZM 241385. Membrane homogenates with a protein content of 15 µg immobilized in a gel matrix were incubated with the radioligands in a total volume of 1500 µL 50 mM Tris/HCl buffer pH 7.4. This method produces the same results as conventional separation techniques and will be published in detail elsewhere. After an incubation time of 70 min the immobilized membrane homogenates were washed with water and transferred into scintillation cocktail (5 mL each, Ultima Gold, Perkin Elmer, Waltham, MA, USA). Liquid scintillation counter (Beckman Coulter, Brea, CA, USA) was used to measure the radioactivity of the samples (bound radioactivity). All binding data were calculated by non-linear curve fitting with a computer-aided curve-fitting program (Prism version 4.0, GraphPad Software, Inc., La Jolla, CA, USA).

5. Conclusions

Neuronal $hA_{2A}Rs$, like other human GPCRs, are important pharmaceutical targets [94]. Here, we have presented a metadynamics study of the interaction between the high-affinity ligand ZMA and $hA_{2A}R$, embedded in a solvated neuronal-like membrane environment. The calculations are consistent with the available experimental data and point to a clear and important role of lipids and of the second extracellular loop for ZMA's molecular recognition process.

Supplementary Materials: The following are available online, Section S1: Well-tempered Metadynamics, Section S2: $H264^{7.29}$–$E169^{ECL2}$ salt bridge: intramolecular interactions, Section S3: $H264^{7.29}$ and $E169^{ECL2}$ salt bridge: conservation across A2Ars, Section S4: Sodium allosteric binding site, Table S1: $H264^{7.29}$ protonation state across $hA_{2A}R$ X-ray structures, Table S2: Ligand hydration (defined here as the number of water molecules within 4 Å of ZMA) and OBS volume for free energy minima A, B, C, D, E and F, Table S3. Conservation of residues composing the VBS of $hA_{2A}R$, as emerging from our calculations, Table S4. Amino acid coevolution profile, Table S5. Presence of residue coevolution between orthosteric binding site (OBS) and extracellular loops (ECLs) of human receptors in class A, B, C and F, Figure S1: Receptor ligand interaction 2D scheme obtained by MOE (Molecular Operating Environment), Figure S2. Pairwise Root Main Square Deviation (RMSD) matrix, Figure S3. ZMA binding poses in the orthosteric binding site corresponding to minima B and C, Figure S4: Superimposition of hA_2AR representative structure in the minima **B** (yellow tube) and **C** (cyan tube), Figure S5: Conservation of solvent-exposed motif of ECL2 in human Adenosine receptor subfamily, Figure S6: Conservation of solvent-exposed motif of ECL2 in Adenosine receptor A_2R across different species, Figure S7: Conservation of $H264^{7.29}$ and $E169^{ECL2}$ in Adenosine receptor $A_{2A}R$ across different species, Figure S8: Conservation of $H264^{7.29}$ and $E169^{ECL2}$ in human Adenosine receptor subtypes hA_1R, $hA_{2A}R$, $hA_{2B}R$, hA_3R.

Author Contributions: Conceptualization, G.R., P.C., A.B., and B.N.; methodology and formal analysis R.C. and A.G.; investigation, all data curation, R.C; writing—original draft preparation, R.C., P.C. and G.R.; writing—review and editing, all; supervision, B.N., A.B., A.G., G.R. and P.C.; Resources P.C. and B.N.

Acknowledgments: We have to express our appreciation to Dirk Bier of the Institute of Neuroscience and Medicine at Forschungszentrum Jülich for conducting the binding assay experiments and sharing affinity data with us. The authors also gratefully acknowledge the computing time provided by John von Neumann Institute for Computing on the supercomputer JURECA at Jülich Supercomputer center.

Appendix A

The binding free energy depends on the difference between the free energy of fully bound state G_B and that of the unbound state (G_U) in which ZMA is located at infinite distance from the receptor:

$$\Delta G = G_U - G_B$$

However, this calculation of G_U is not possible given the necessarily finite size of the simulation box. To circumvent this problem, let us rewrite ΔG as the sum of two contributions:

$$\Delta G = G_U - G_G + G_G - G_B = G_{WTM} - \Delta G_{residual}$$

G is the relative minimum represented in Figure 3. G_{WTM} is by far the largest contribution, and it calculated by well-tempered metadynamics. $G_{residual}$ is well approximated by the free energy associated with long-range electrostatic interactions with the protein ($\Delta G_{residual} \approx \Delta G_{electr}$). Indeed, the ligand in **G** does not form direct H-bonds and/or hydrophobic contacts with the protein. It is at about 0.8 nm from the protein atoms, separated by water molecules. Hence, it forms only long-range electrostatic interactions with the membrane and the protein. However, the latter are vanishingly small because the ligand is at least 2.5 nm from the membrane. In addition, we notice that in **G** the conformational degrees of freedom of the ligand are not partially restricted. Hence, the entropic contribution associated with these degrees of freedom is expected to be also vanishingly small.

ΔG_{electr} is expected to be much smaller than ΔG_{WTM}, as ΔG_{electr} values lower than $2k_BT$ have been calculated for other ligand/protein interactions [88,95]. Indeed, a posteriori, ΔG_{electr}, as calculated using the APBS 1.4 program [34], turns out to be only 0.95 Kcal/mol. We conclude that the corrections due to the finite size of the simulation box are small. In other words, the ligand in **G** interacts very weakly with the protein. Hence, the errors in the calculations of this term are not expected to dramatically affect the binding free energy.

References

1. Fredholm, B.B.; IJzerman, A.P.; Jacobson, K.A.; Klotz, K.N.; Linden, J. International Union of Pharmacology. XXV. Nomenclature and classification of adenosine receptors. *Pharmacol. Rev.* **2001**, *53*, 527–552. [PubMed]

2. Kroeze, W.K.; Sheffler, D.J.; Roth, B.L. G-protein-coupled receptors at a glance. *J. Cell Sci.* **2003**, *116*, 4867–4869. [CrossRef] [PubMed]

3. Schöneberg, T.; Schulz, A.; Biebermann, H.; Hermsdorf, T.; Römpler, H.; Sangkuhl, K. Mutant G-protein-coupled receptors as a cause of human diseases. *Pharmacol. Ther.* **2004**, *104*, 173–206. [CrossRef] [PubMed]

4. Fink, J.S.; Weaver, D.R.; Rivkees, S.A.; Peterfreund, R.A.; Pollack, A.E.; Adler, E.M.; Reppert, S.M. Molecular cloning of the rat A2 adenosine receptor: Selective co-expression with D2 dopamine receptors in rat striatum. *Brain Res. Mol. Brain Res.* **1992**, *14*, 186–195. [CrossRef]

5. Xu, K.; Bastia, E.; Schwarzschild, M. Therapeutic potential of adenosine A(2A) receptor antagonists in Parkinson's disease. *Pharmacol. Ther.* **2005**, *105*, 267–310. [CrossRef] [PubMed]

6. Gimpl, G. Interaction of G protein coupled receptors and cholesterol. *Chem. Phys. Lipids* **2016**, *199*, 61–73. [CrossRef] [PubMed]

7. Peeters, M.C.; van Westen, G.J.P.; Li, Q.; IJzerman, A.P. Importance of the extracellular loops in G protein-coupled receptors for ligand recognition and receptor activation. *Trends Pharmacol. Sci.* **2011**, *32*, 35–42. [CrossRef] [PubMed]

8. Katritch, V.; Cherezov, V.; Stevens, R.C. Structure-function of the G protein-coupled receptor superfamily. *Annu. Rev. Pharmacol.* **2013**, *53*, 531–556. [CrossRef] [PubMed]

9. Avlani, V.A.; Gregory, K.J.; Morton, C.J.; Parker, M.W.; Sexton, P.M.; Christopoulos, A. Critical role for the second extracellular loop in the binding of both orthosteric and allosteric G protein-coupled receptor ligands. *J. Biol. Chem.* **2007**, *282*, 25677–25686. [CrossRef] [PubMed]

10. Granier, S.; Kobilka, B. A new era of GPCR structural and chemical biology. *Nat. Chem. Biol.* **2012**, *8*, 670–673. [CrossRef] [PubMed]

11. Dror, R.O.; Pan, A.C.; Arlow, D.H.; Borhani, D.W.; Maragakis, P.; Shan, Y.; Xu, H.; Shaw, D.E. Pathway and mechanism of drug binding to G-protein-coupled receptors. *Proc. Natl. Acad. Sci. USA* **2011**, *108*, 13118–13123. [CrossRef] [PubMed]

12. Sandal, M.; Behrens, M.; Brockhoff, A.; Musiani, F.; Giorgetti, A.; Carloni, P.; Meyerhof, W. Evidence for a transient additional ligand binding site in the TAS2R46 bitter taste receptor. *J. Chem. Theory Comput.* **2015**, *11*, 4439–4449. [CrossRef] [PubMed]

13. Kruse, A.C.; Hu, J.; Pan, A.C.; Arlow, D.H.; Rosenbaum, D.M.; Rosemond, E.; Green, H.F.; Liu, T.; Chae, P.S.; Dror, R.O.; et al. Structure and dynamics of the M3 muscarinic acetylcholine receptor. *Nature* **2012**, *482*, 552–556. [CrossRef] [PubMed]

14. Stanley, N.; Pardo, L.; Fabritiis, G.D. The pathway of ligand entry from the membrane bilayer to a lipid G protein-coupled receptor. *Sci. Rep.* **2016**, *6*, 22639. [CrossRef] [PubMed]

15. Provasi, D.; Bortolato, A.; Filizola, M. Exploring molecular mechanisms of ligand recognition by opioid receptors with metadynamics. *Biochemistry* **2009**, *48*, 10020–10029. [CrossRef] [PubMed]

16. Horn, F.; Weare, J.; Beukers, M.W.; Horsch, S.; Bairoch, A.; Chen, W.; Edvardsen, O.; Campagne, F.; Vriend, G. GPCRDB: An information system for G protein-coupled receptors. *Nucleic Acids Res.* **1998**, *26*, 275–279. [CrossRef] [PubMed]

17. Barducci, A.; Bussi, G.; Parrinello, M. Well-tempered metadynamics: A smoothly converging and tunable free-energy method. *Phys. Rev. Lett.* **2008**, *100*, 020603. [CrossRef] [PubMed]

18. Laio, A.; Parrinello, M. Escaping free-energy minima. *Proc. Natl. Acad Sci. USA* **2002**, *99*, 12562–12566. [CrossRef] [PubMed]

19. Schneider, S.; Provasi, D.; Filizola, M. The dynamic process of drug-GPCR binding at either orthosteric or allosteric sites evaluated by metadynamics. *Methods Mol. Biol. (Clifton, N.J.)* **2015**, *1335*, 277–294. [CrossRef]

20. Poucher, S.M.; Keddie, J.R.; Singh, P.; Stoggall, S.M.; Caulkett, P.W.; Jones, G.; Coll, M.G. The in vitro pharmacology of ZM 241385, a potent, non-xanthine A2a selective adenosine receptor antagonist. *Br. J. Pharmacol.* **1995**, *115*, 1096–1102. [CrossRef] [PubMed]

21. Dore, A.S.; Robertson, N.; Errey, J.C.; Ng, I.; Hollenstein, K.; Tehan, B.; Hurrell, E.; Bennett, K.; Congreve, M.; Magnani, F.; et al. Structure of the adenosine A(2A) receptor in complex with ZM241385 and the xanthines XAC and caffeine. *Structure* **2011**, *19*, 1283–1293. [CrossRef] [PubMed]

22. Jaakola, V.P.; Griffith, M.T.; Hanson, M.A.; Cherezov, V.; Chien, E.Y.; Lane, J.R.; Ijzerman, A.P.; Stevens, R.C. The 2.6 angstrom crystal structure of a human A2A adenosine receptor bound to an antagonist. *Science* **2008**, *322*, 1211–1217. [CrossRef] [PubMed]

23. Liu, W.; Chun, E.; Thompson, A.A.; Chubukov, P.; Xu, F.; Katritch, V.; Han, G.W.; Roth, C.B.; Heitman, L.H.; IJzerman, A.P.; et al. Structural basis for allosteric regulation of GPCRs by sodium ions. *Science* **2012**, *337*, 232–236. [CrossRef] [PubMed]

24. Hino, T.; Arakawa, T.; Iwanari, H.; Yurugi-Kobayashi, T.; Ikeda-Suno, C.; Nakada-Nakura, Y.; Kusano-Arai, O.; Weyand, S.; Shimamura, T.; Nomura, N.; et al. G-protein-coupled receptor inactivation by an allosteric inverse-agonist antibody. *Nature* **2012**, *482*, 237–240. [CrossRef] [PubMed]

25. Segala, E.; Guo, D.; Cheng, R.K.; Bortolato, A.; Deflorian, F.; Doré, A.S.; Errey, J.C.; Heitman, L.H.; IJzerman, A.P.; Marshall, F.H. Controlling the dissociation of ligands from the adenosine A2A receptor through modulation of salt bridge strength. *J. Med. Chem.* **2016**, *59*, 6470–6479. [CrossRef] [PubMed]

26. Guo, D.; Pan, A.C.; Dror, R.O.; Mocking, T.; Liu, R.; Heitman, L.H.; Shaw, D.E.; IJzerman, A.P. Molecular basis of ligand dissociation from the adenosine A2A receptor. *Mol. Pharmacol.* **2016**, *89*, 485–491. [CrossRef] [PubMed]

27. Cao, R.Y.; Rossetti, G.; Bauer, A.; Carloni, P. Binding of the antagonist caffeine to the human adenosine receptor hA(2A)R in nearly physiological conditions. *PLoS ONE* **2015**, *10*, e0126833. [CrossRef]

28. Grouleff, J.; Irudayam, S.J.; Skeby, K.K.; Schiott, B. The influence of cholesterol on membrane protein structure, function, and dynamics studied by molecular dynamics simulations. *Biochim. Biophys. Acta* **2015**, *1848*, 1783–1795. [CrossRef] [PubMed]

29. Guixà-González, R.; Albasanz, J.L.; Rodriguez-Espigares, I.; Pastor, M.; Sanz, F.; Martí-Solano, M.; Manna, M.; Martinez-Seara, H.; Hildebrand, P.W.; Martín, M. Membrane cholesterol access into a G-protein-coupled receptor. *Nat. Commun.* **2017**, *8*, 14505. [CrossRef] [PubMed]

30. Pucadyil, T.J.; Chattopadhyay, A. Cholesterol modulates ligand binding and G-protein coupling to serotonin(1A) receptors from bovine hippocampus. *Biochim. Biophys. Acta* **2004**, *1663*, 188–200. [CrossRef] [PubMed]

31. Klein, U.; Gimpl, G.; Fahrenholz, F. Alteration of the myometrial plasma membrane cholesterol content with beta.-cyclodextrin modulates the binding affinity of the oxytocin receptor. *Biochemistry* **1995**, *34*, 13784–13793. [CrossRef] [PubMed]

32. Nguyen, D.H.; Taub, D. CXCR4 function requires membrane cholesterol: Implications for HIV infection. *J. Immunol.* **2002**, *168*, 4121–4126. [CrossRef] [PubMed]

33. Schrödinger, M. LLC New York, NY: 2009. Available online: https://www.schrodinger.com (accessed on 8 October 2018).

34. Baker, N.A.; Sept, D.; Joseph, S.; Holst, M.J.; McCammon, J.A. Electrostatics of nanosystems: Application to microtubules and the ribosome. *Proc. Natl. Acad. Sci. USA* **2001**, *98*, 10037–10041. [CrossRef] [PubMed]

35. Batyuk, A.; Galli, L.; Ishchenko, A.; Han, G.W.; Gati, C.; Popov, P.A.; Lee, M.Y.; Stauch, B.; White, T.A.; Barty, A.; et al. Native phasing of X-ray free-electron laser data for a G protein-coupled receptor. *Sci. Adv.* **2016**, *2*, e1600292. [CrossRef] [PubMed]

36. Martin-Garcia, J.M.; Conrad, C.E.; Nelson, G.; Stander, N.; Zatsepin, N.A.; Zook, J.; Zhu, L.; Geiger, J.; Chun, E.; Kissick, D.; et al. Serial millisecond crystallography of membrane and soluble protein microcrystals using synchrotron radiation. *IUCrJ* **2017**, *4*, 439–454. [CrossRef] [PubMed]

37. Melnikov, I.; Polovinkin, V.; Kovalev, K.; Gushchin, I.; Shevtsov, M.; Shevchenko, V.; Mishin, A.; Alekseev, A.; Rodriguez-Valera, F.; Borshchevskiy, V.; et al. Fast iodide-SAD phasing for high-throughput membrane protein structure determination. *Sci. Adv.* **2017**, *3*, e1602952. [CrossRef] [PubMed]

38. Broecker, J.; Morizumi, T.; Ou, W.L.; Klingel, V.; Kuo, A.; Kissick, D.J.; Ishchenko, A.; Lee, M.Y.; Xu, S.; Makarov, O.; et al. High-throughput in situ X-ray screening of and data collection from protein crystals at room temperature and under cryogenic conditions. *Nat. Protoc.* **2018**, *13*, 260–292. [CrossRef] [PubMed]

39. Eddy, M.T.; Lee, M.-Y.; Gao, Z.-G.; White, K.L.; Didenko, T.; Horst, R.; Audet, M.; Stanczak, P.; McClary, K.M.; Han, G.W.; et al. Allosteric coupling of drug binding and intracellular signaling in the A2A adenosine receptor. *Cell* **2018**, *172*, 68–80.e12. [CrossRef] [PubMed]

40. Weinert, T.; Olieric, N.; Cheng, R.; Brunle, S.; James, D.; Ozerov, D.; Gashi, D.; Vera, L.; Marsh, M.; Jaeger, K.; et al. Serial millisecond crystallography for routine room-temperature structure determination at synchrotrons. *Nat. Commun.* **2017**, *8*, 542. [CrossRef] [PubMed]

41. Kim, J.; Jiang, Q.; Glashofer, M.; Yehle, S.; Wess, J.; Jacobson, K.A. Glutamate residues in the second extracellular loop of the human A2a adenosine receptor are required for ligand recognition. *Mol. Pharmacol.* **1996**, *49*, 683–691. [PubMed]

42. Sun, B.; Bachhawat, P.; Chu, M.L.; Wood, M.; Ceska, T.; Sands, Z.A.; Mercier, J.; Lebon, F.; Kobilka, T.S.; Kobilka, B.K. Crystal structure of the adenosine A2A receptor bound to an antagonist reveals a potential allosteric pocket. *Proc. Natl. Acad. Sci. USA* **2017**, *114*, 2066–2071. [CrossRef] [PubMed]

43. Jaakola, V.-P.; Lane, J.R.; Lin, J.Y.; Katritch, V.; IJzerman, A.P.; Stevens, R.C. Identification and characterization of amino acid residues essential for human A2A adenosine receptor: ZM241385 binding and subtype selectivity. *J. Biol. Chem.* **2010**, *285*, 13032–13044. [CrossRef] [PubMed]

44. Allende, M.L.; Dreier, J.L.; Mandala, S.; Proia, R.L. Expression of the sphingosine 1-phosphate receptor, S1P1, on T-cells controls thymic emigration. *J. Biol. Chem.* **2004**, *279*, 15396–15401. [CrossRef] [PubMed]

45. Wagner, J.R.; Lee, C.T.; Durrant, J.D.; Malmstrom, R.D.; Feher, V.A.; Amaro, R.E. Emerging computational methods for the rational discovery of allosteric drugs. *Chem. Rev.* **2016**, *116*, 6370–6390. [CrossRef] [PubMed]

46. Baker, F.N.; Porollo, A. CoeViz: A web-based tool for coevolution analysis of protein residues. *BMC Bioinform.* **2016**, *17*, 119. [CrossRef] [PubMed]

47. Burger, L.; van Nimwegen, E. Disentangling direct from indirect co-evolution of residues in protein alignments. *PLoS Comput. Biol.* **2010**, *6*, e1000633. [CrossRef] [PubMed]

48. De Juan, D.; Pazos, F.; Valencia, A. Emerging methods in protein co-evolution. *Nat. Rev. Genet.* **2013**, *14*, 249–261. [CrossRef] [PubMed]

49. Pang, X.; Yang, M.; Han, K. Antagonist binding and induced conformational dynamics of GPCR A2A adenosine receptor. *Proteins* **2013**, *81*, 1399–1410. [CrossRef] [PubMed]

50. Lee, J.Y.; Lyman, E. Agonist dynamics and conformational selection during microsecond simulations of the A(2A) adenosine receptor. *Biophys. J.* **2012**, *102*, 2114–2120. [CrossRef] [PubMed]

51. Seddon, A.M.; Curnow, P.; Booth, P.J. Membrane proteins, lipids and detergents: Not just a soap opera. *Biochim. Biophys. Acta Biomembr.* **2004**, *1666*, 105–117. [CrossRef] [PubMed]

52. Wacker, D.; Fenalti, G.; Brown, M.A.; Katritch, V.; Abagyan, R.; Cherezov, V.; Stevens, R.C. Conserved binding mode of human beta2 adrenergic receptor inverse agonists and antagonist revealed by X-ray crystallography. *J. Am. Chem. Soc.* **2010**, *132*, 11443–11445. [CrossRef] [PubMed]

53. Naranjo, A.N.; Chevalier, A.; Cousins, G.D.; Ayettey, E.; McCusker, E.C.; Wenk, C.; Robinson, A.S. Conserved disulfide bond is not essential for the adenosine A2A receptor: Extracellular cysteines influence receptor distribution within the cell and ligand-binding recognition. *Biochim. Biophys. Acta* **2015**, *1848*, 603–614. [CrossRef] [PubMed]

54. Chattopadhyay, A. GPCRs: Lipid-dependent membrane receptors that act as drug targets. *Adv. Biol.* **2014**, *2014*, 1–12. [CrossRef]

55. Dijkman, P.M.; Watts, A. Lipid modulation of early G protein-coupled receptor signalling events. *Biochim. Biophys. Acta* **2015**, *1848*, 2889–2897. [CrossRef] [PubMed]

56. Hurst, D.P.; Grossfield, A.; Lynch, D.L.; Feller, S.; Romo, T.D.; Gawrisch, K.; Pitman, M.C.; Reggio, P.H. A Lipid pathway for ligand binding is necessary for a cannabinoid G protein-coupled receptor. *J. Biol. Chem.* **2010**, *285*, 17954–17964. [CrossRef] [PubMed]

57. Olanow, C.W. Oxidation reactions in Parkinson's disease. *Neurology* **1990**, *40*, 37–39.

58. Ikeda, K.; Kurokawa, M.; Aoyama, S.; Kuwana, Y. Neuroprotection by adenosine A2A receptor blockade in experimental models of Parkinson's disease. *J. Neurochem.* **2002**, *80*, 262–270. [CrossRef] [PubMed]

59. Sung, Y.M.; Wilkins, A.D.; Rodriguez, G.J.; Wensel, T.G.; Lichtarge, O. Intramolecular allosteric communication in dopamine D2 receptor revealed by evolutionary amino acid covariation. *Proc. Natl. Acad. Sci. USA* **2016**, *113*, 3539–3544. [CrossRef] [PubMed]

60. Canals, M.; Burgueno, J.; Marcellino, D.; Cabello, N.; Canela, E.I.; Mallol, J.; Agnati, L.; Ferre, S.; Bouvier, M.; Fuxe, K.; et al. Homodimerization of adenosine A2A receptors: Qualitative and quantitative assessment by fluorescence and bioluminescence energy transfer. *J. Neurochem.* **2004**, *88*, 726–734. [CrossRef] [PubMed]

61. Franco, R.; Martinez-Pinilla, E.; Lanciego, J.L.; Navarro, G. Basic pharmacological and structural evidence for class A G-protein-coupled receptor heteromerization. *Front. Pharmacol.* **2016**, *7*, 76. [CrossRef] [PubMed]

62. Fanelli, F.; Felline, A. Dimerization and ligand binding affect the structure network of A2A adenosine receptor. *Biochim. Biophys. Acta Biomembr.* **2011**, *1808*, 1256–1266. [CrossRef] [PubMed]

63. Matsunaga, Y.; Komuro, Y.; Kobayashi, C.; Jung, J.; Mori, T.; Sugita, Y. Dimensionality of collective variables for describing conformational changes of a multi-domain protein. *J. Phys. Chem. Lett.* **2016**, *7*, 1446–1451. [CrossRef] [PubMed]

64. Gilson, M.K.; Zhou, H.X. Calculation of protein-ligand binding affinities. *Annu. Rev. Biophys. Biomol. Struct.* **2007**, *36*, 21–42. [CrossRef] [PubMed]

65. Jafurulla, M.; Chattopadhyay, A. Sphingolipids in the function of G protein-coupled receptors. *Eur. J. Pharmacol.* **2015**, *763*, 241–246. [CrossRef] [PubMed]

66. Massink, A.; Gutierrez-de-Teran, H.; Lenselink, E.B.; Ortiz Zacarias, N.V.; Xia, L.; Heitman, L.H.; Katritch, V.; Stevens, R.C.; AP, I.J. Sodium ion binding pocket mutations and adenosine A2A receptor function. *Mol. Pharmacol.* **2015**, *87*, 305–313. [CrossRef] [PubMed]

67. Pfrieger, F.W. Role of cholesterol in synapse formation and function. *Biochim. Biophys. Acta* **2003**, *1610*, 271–280. [CrossRef]

68. Serebryany, E.; Zhu, G.A.; Yan, E.C. Artificial membrane-like environments for in vitro studies of purified G-protein coupled receptors. *Biochim. Biophys. Acta* **2012**, *1818*, 225–233. [CrossRef] [PubMed]

69. Andreoli, T.E.; Hoffman, J.F.; Fanestil, D.D. *Membrane Physiology*; Springer: Boston, MA, USA, 1980.

70. DeLano, W.L. *The PyMol Molecular Graphics System*; DeLano Scientific LLC: San Carlos, CA, USA, 2002.

71. Dolinsky, T.J.; Czodrowski, P.; Li, H.; Nielsen, J.E.; Jensen, J.H.; Klebe, G.; Baker, N.A. PDB2PQR: Expanding and upgrading automated preparation of biomolecular structures for molecular simulations. *Nucleic Acids Res.* **2007**, *35*, W522–W525. [CrossRef] [PubMed]

72. Best, R.B.; Hummer, G. Optimized molecular dynamics force fields applied to the helix-coil transition of polypeptides. *J. Phys. Chem. B* **2009**, *113*, 9004–9015. [CrossRef] [PubMed]

73. Jambeck, J.P.; Lyubartsev, A.P. Derivation and systematic validation of a refined all-atom force field for phosphatidylcholine lipids. *J. Phys. Chem. B* **2012**, *116*, 3164–3179. [CrossRef] [PubMed]

74. Jambeck, J.P.; Lyubartsev, A.P. An extension and further validation of an all-atomistic force field for biological membranes. *J. Chem. Theory Comput.* **2012**, *8*, 2938–2948. [CrossRef] [PubMed]

75. Jorgensen, W.L.; Chandrasekhar, J.; Madura, J.D.; Impey, R.W.; Klein, M.L. Comparison of simple potential functions for simulating liquid water. *J. Chem. Phys.* **1983**, *79*, 926–935. [CrossRef]

76. Wang, J.; Wolf, R.M.; Caldwell, J.W.; Kollman, P.A.; Case, D.A. Development and testing of a general amber force field. *J. Comput. Chem.* **2004**, *25*, 1157–1174. [CrossRef] [PubMed]

77. Frisch, M.J.; Trucks, G.W.; Schlegel, H.B.; Scuseria, G.E.; Robb, M.A.; Cheeseman, J.R.; Scalmani, G.; Barone, V.; Mennucci, B.; Petersson, G.A.; et al. *Gaussian 09, Revision A.02*; Gaussian Inc.: Wallingford, CT, USA, 2009.

78. Wang, J.; Cieplak, P.; Kollman, P.A. How well does a restrained electrostatic potential (RESP) model perform in calculating conformational energies of organic and biological molecules? *J. Comput. Chem.* **2000**, *21*, 1049–1074. [CrossRef]

79. Case, D.A.; Cheatham, T.E., III; Darden, T.; Gohlke, H.; Luo, R.; Merz, K.M., Jr.; Onufriev, A.; Simmerling, C.; Wang, B.; Woods, R.J. The amber biomolecular simulation programs. *J. Comput. Chem.* **2005**, *26*, 1668–1688. [CrossRef] [PubMed]

80. Van Der Spoel, D.; Lindahl, E.; Hess, B.; Groenhof, G.; Mark, A.E.; Berendsen, H.J. GROMACS: Fast, flexible, and free. *J. Comput. Chem.* **2005**, *26*, 1701–1718. [CrossRef] [PubMed]

81. Hünenberger, P. Thermostat algorithms for molecular dynamics simulations. *Adv. Comput. Simul.* **2005**, 105–149. [CrossRef]

82. Parrinello, M.; Rahman, A. Polymorphic transitions in single-crystals—A new molecular-dynamics method. *J. Appl. Phys.* **1981**, *52*, 7182–7190. [CrossRef]

83.   Darden, T.; York, D.; Pedersen, L. Particle mesh ewald—An N.Log(N) method for ewald sums in large systems. *J. Chem. Phys.* **1993**, *98*, 10089–10092. [CrossRef]

84.   Hess, B.; Bekker, H.; Berendsen, H.J.C.; Fraaije, J.G.E.M. LINCS: A linear constraint solver for molecular simulations. *J. Comput. Chem.* **1997**, *18*, 1463–1472. [CrossRef]

85.   Bonomi, M.; Branduardi, D.; Bussi, G.; Camilloni, C.; Provasi, D.; Raiteri, P.; Donadio, D.; Marinelli, F.; Pietrucci, F.; Broglia, R.A.; et al. PLUMED: A portable plugin for free-energy calculations with molecular dynamics. *Comput. Phys. Commun.* **2009**, *180*, 1961–1972. [CrossRef]

86.   Bochicchio, A.; Rossetti, G.; Tabarrini, O.; Kraubeta, S.; Carloni, P. Molecular view of ligands specificity for CAG repeats in anti-Huntington therapy. *J. Chem. Theory Comput.* **2015**, *11*, 4911–4922. [CrossRef] [PubMed]

87.   Nguyen, T.H.; Rossetti, G.; Arnesano, F.; Ippoliti, E.; Natile, G.; Carloni, P. Molecular recognition of platinated DNA from chromosomal HMGB1. *J. Chem. Theory Comput.* **2014**, *10*, 3578–3584. [CrossRef] [PubMed]

88.   Kranjc, A.; Bongarzone, S.; Rossetti, G.; Biarnes, X.; Cavalli, A.; Bolognesi, M.L.; Roberti, M.; Legname, G.; Carloni, P. Docking ligands on protein surfaces: The case study of prion protein. *J. Chem. Theory Comput.* **2009**, *5*, 2565–2573. [CrossRef] [PubMed]

89.   Paramo, T.; East, A.; Garzon, D.; Ulmschneider, M.B.; Bond, P.J. Efficient characterization of protein cavities within molecular simulation trajectories: Trj_cavity. *J. Chem. Theory Comput.* **2014**, *10*, 2151–2164. [CrossRef] [PubMed]

90.   Humphrey, W.; Dalke, A.; Schulten, K. VMD: Visual molecular dynamics. *J. Mol. Graph.* **1996**, *14*, 33–38. [CrossRef]

91.   Porollo, A.A.; Adamczak, R.; Meller, J. POLYVIEW: A flexible visualization tool for structural and functional annotations of proteins. *Bioinformatics* **2004**, *20*, 2460–2462. [CrossRef] [PubMed]

92.   Klotz, K.-N.; Hessling, J.; Hegler, J.; Owman, C.; Kull, B.; Fredholm, B.; Lohse, M. Comparative pharmacology of human adenosine receptor subtypes–characterization of stably transfected receptors in CHO cells. *Naunyn-Schmiedeberg's Arch. Pharmacol.* **1997**, *357*, 1–9. [CrossRef]

93.   Neuhoff, V.; Philipp, K.; Zimmer, H.G.; Mesecke, S. A Simple, Versatile, Sensitive and Volume-Independent Method for Quantitative Protein Determination which is Independent of Other External Influences. *Biol. Chem.* **1979**, *360*, 1657–1670. [CrossRef]

94.   Stevens, R.C.; Cherezov, V.; Katritch, V.; Abagyan, R.; Kuhn, P.; Rosen, H.; Wuthrich, K. The GPCR network: A large-scale collaboration to determine human GPCR structure and function. *Nat. Rev. Drug Discov.* **2013**, *12*, 25–34. [CrossRef] [PubMed]

95.   Pietrucci, F.; Marinelli, F.; Carloni, P.; Laio, A. Substrate binding mechanism of HIV-1 protease from explicit-solvent atomistic simulations. *J. Am. Chem. Soc.* **2009**, *131*, 11811–11818. [CrossRef] [PubMed]

8

Truly Target-Focused Pharmacophore Modeling: A Novel Tool for Mapping Intermolecular Surfaces

Jérémie Mortier [ID], Pratik Dhakal and Andrea Volkamer * [ID]

In-Silico Toxicology Group, Institute of Physiology, Charité—Universitätsmedizin Berlin, Virchowweg 6, 10117 Berlin, Germany; jeremie.mortier@gmail.com (J.M.); dhakal.pratik@gmail.com (P.D.)
* Correspondence: andrea.volkamer@charite.de

Abstract: Pharmacophore models are an accurate and minimal tridimensional abstraction of intermolecular interactions between chemical structures, usually derived from a group of molecules or from a ligand-target complex. Only a limited amount of solutions exists to model comprehensive pharmacophores using the information of a particular target structure without knowledge of any binding ligand. In this work, an automated and customable tool for truly target-focused (T^2F) pharmacophore modeling is introduced. Key molecular interaction fields of a macromolecular structure are calculated using the AutoGRID energy functions. The most relevant points are selected by a newly developed filtering cascade and clustered to pharmacophore features with a density-based algorithm. Using five different protein classes, the ability of this method to identify essential pharmacophore features was compared to structure-based pharmacophores derived from ligand-target interactions. This method represents an extremely valuable instrument for drug design in a situation of scarce ligand information available, but also in the case of underexplored therapeutic targets, as well as to investigate protein allosteric pockets and protein-protein interactions.

Keywords: target-focused pharmacophore modeling; density-based clustering; structure-based drug design; AutoGrid; grid maps; probe energies; method development

1. Introduction

Events of intermolecular recognition are mediated through forces of attraction and repulsion between interacting chemical molecules. From the transduction of extracellular signals to DNA recognition by transcription factors, all combinations of interaction partners result in unique molecular complexes with a particular biological significance for a given cellular time and space. Therefore, studying the specificity of particular interactions at the atomic level is essential for comprehending biochemical mechanisms and predicting molecular behaviors. Depicting the ensemble of key interactions required for a specific intermolecular recognition event is the working principle of the pharmacophore approach. A pharmacophore model is made of a set of interactions, typically consisting of hydrogen bonds, electrostatic interactions, and π-stacking, as well as hydrophobic contacts, and may also include steric information, such as exclusion volumes. Recording the 3D-arrangement of all interaction features included in a pharmacophore model has been highly facilitated with the advent of computational chemistry [1].

Pharmacophore modeling is a computationally efficient and pragmatic strategy for the discovery and optimization of biologically active compounds [2–5], as well as the analysis of intermolecular interactions in silico [6,7]. Because of their simplicity, these intuitively understandable models can support binding event prediction for a selected group of molecules or to conduct high-throughput virtual screening of large compound libraries [8]. For a set of molecules sharing a similar biological response, a ligand-based pharmacophore model can be derived by superposing them and determining the maximum number of overlapping chemical features [2]. This method is called ligand-based approach. When structural information is available for an intermolecular surface involving multiple partners, a pharmacophore model can be derived from the interactions detected in this complex. This method is called structure-based approach, and is typically conducted by geometrically deriving features from a ligand-target complex [9]. While the label, structure-based approach, primarily evokes an investigation focused on a protein's structure, only a limited amount of solutions exists to model comprehensive pharmacophores using information of a particular target structure without considering the binding mode of a ligand [10]. The consequences of this situation are that (i) a restricted number of options are available to build pharmacophore models for targets when no binding ligand is known, and (ii) when a small amount of ligands that bind to a target are known, the resulting structure-based pharmacophores are limited to the interactions these particular molecules are forming, poorly representing the range of pharmacomodulations available for rational drug design (or pharmacophoric space). With the rising number of protein structures available, target-focused techniques will play an increasingly important role to elucidate biological functions and to support target-oriented drug design. Nowadays, protein structures are solved sometimes before anything is known about their biological function, e.g., by proteomics approaches. It appears clearly then that reliable methods capable of deriving pharmacophore models from ligand-free protein structures (and any other relevant biomolecular surfaces) are needed.

Target-focused pharmacophore models can be derived based on evolutionary pocket residue conservation (alignment), by minimizing probes (molecular dynamics), or by identifying favorable energetic properties (grid-based). The former two will only be shortly covered, while the latter one is the focus of this work.

Sequence alignment-based methods rely on the detection of key residues of a binding site for subsequent pharmacophore features assignment [11,12]. As an example, this approach was followed and developed for the construction of a G-protein coupled receptor (GPCR) pharmacophore model [13]. In this work, Kratochwil et al. aligned over 1000 sequences of GPCRs to identify patterns of residues conserved in the transmembrane domain across this protein family. Once aligned to the target structure, important positions for ligand binding and key residues for selectivity were selected to assign pharmacophore features. Clearly, this strategy can be effective in a scenario with sufficient data available, such as in the case of GPCRs, but has little chance of success with a less explored protein class.

An alternative approach is to simulate the dynamic behavior of chemical probes (water and organic solvents) on a flexible molecular surface [14,15]. The minimized probe molecules can unveil favorable interaction sites on the protein surface, which can be converted into pharmacophore features. A first attempt by Miranker and Karplus [14,15] in 1991 to include protein flexibility in a pharmacophore model led to a number of promising developments of computer-based investigations of macromolecules using molecular dynamics (MD). Although more time consuming, this approach can be successful when water, the natural solvent of proteins, is used as a probe. However, organic probes used to detect the hydrophobic regions of a macromolecule create a non-natural environment for most proteins, inducing conformations in silico that are unlikely to be observed in vivo. The relevance of the resulting pharmacophore model is therefore unclear. Current challenges and perspectives of this approach are discussed in this recent review [16].

To date, the most established method to derive pharmacophore models from an empty molecular surface or cavity is to identify regions with the most favorable energetic properties for ligand binding. In 1985, Goodford developed an efficient and straightforward method for sampling protein cavities [17]. This method is based on the determination of the chemical nature of a molecular surface by calculating interaction energies in the presence of chemical probes with different electronic properties, such as hydrogen bond donating or accepting groups, as well as hydrophobic ones. A three-dimensional (3D) Cartesian grid box subjected to energy calculation is spanned around the area of interest [17,18], and grid points with the most favorable energetics identify regions of attraction between a potential ligand and the studied macromolecule. Among the known grid-based approaches developed for scanning a target surface to identify hot spots (Table 1), *Pocket V2* [19],

GBPM [20], Tintori et al. [21], *Hydro-Pharm* [22], *FLAP* [23], and *PharmDock* [24] opted for a solely energy-based 3D grid approach, while *Ph4Dock* [18] builds upon a sphere-based cavity detection method. Some additional methods are available in modeling suites, such as the geometry-based GridMap approach in LigandScout, to derive a pharmacophore model from an empty cavity, but the methodology has not been published and therefore cannot be discussed in detail in this overview. Most of these methods were developed for docking applications. Ph4Dock [24] compares the derived protein pharmacophores to those of ligands to rank docking conformations. *FLAP* [23] converts the derived molecular interaction fields into fingerprints and is applied for high-throughput virtual screening. Hydro-Pharm [22] and PharmDock [24] use ChemScore-based energies [25] calculated on grid points to identify pharmacophore features for subsequent molecular docking. In contrast, PocketV2 [19], GBPM [20] (focused on protein-protein interactions), and the method by Titori et al. [21] were primarily developed for pharmacophore detection, all using an energy grid for hot spot detection based on point clustering (former) or minima extraction (latter two).

Table 1. List of the discussed energy-based methods relying on geometry or a grid for target-focused pharmacophore modeling.

Method	Cavity Definition	Approach	Clustering Method	Evaluation	Year	Refs.
Ph4Dock	cavity detection (Delaunay triangulation/ α spheres)	electrostatic interactions (MMFF94 [26]) of charged dummy atoms	single-linkage	CCDC/Astex valida-tion set [27] [d]	2004	[18]
Pocket V2	grid box around ligand (or user-defined pocket residues)	grid (Score) [28]	unclear clustering method [a]	CDK2, HIV1-PR, ER, 17b-HSD	2006	[19]
FLAP + BioGPS	grid box around ligand or FLAPsite detection	grid (GRID software) [17]	region-based energy minima	Patel set [29], DUD [30] [d]	2007	[23,31]
Tintori et al.	grid box around binding site	grid (GRID software) [17]	no clustering [b] (GRID minima + interpolation)	TrxR (MTB), HIV1 IN, HIV-1 RT dimer	2008	[21]
Hydro-Pharm	grid box around ligand (3 Å)	grid (ChemScore [25]) + MD-based hydration site feature reduction [c]	k-means	HIV1-PR, DHFR, FXa	2012	[22]
PharmDock	grid box around bound ligand (3 Å)	grid (ChemScore [25])	k-means	PDB bind, DUD [30] [d]	2014	[24,32]
T^2F-*Pharm*	grid box around ligand or user-defined center (& cavity point reduction)	grid (AutoDock) [33]	CNN [34]	Patel set [29] + A_{2A} receptor	2018	This paper

[a] Exact clustering method not specified in publication. [b] No clustering method used, points are reduced using the Minim and Filmap programs implemented in the GRID package, collecting all points within a certain energy threshold value and interpolating the closest ones. [c] Hydro-Pharms adds a second step to further restrict the pharmacophore points by calculating an overlap between the grid-based pharmacophores and molecular dynamics (MD)-derived hydration sites. [d] Evaluation focused on enrichment in docking/virtual screening, rather than evaluation of pharmacophoric features.

Pocket definition: For applying target-based methods to rather unexplored—ideally, to apo—structures, the location of the (potential) ligand binding site needs to be determined. However, some methods, such as *Hydro-Pharm* [22] or *PharmDock* [24], require a ligand to define the cavity volume (centered on the ligand coordinates). Some others, such as *Pocket V2*, additionally allow the user to input selected binding site residues to define the pocket. In contrast, *Ph4Dock* [18] is a ligand docking program that provides an inbuilt pocket detection routine by scanning the surface with a collection of spheres using a modified Delaunay triangulation [35]. *BioGPS* [31], which builds upon *FLAP* [23], includes the *FLAPsite* algorithm for pocket detection, using the GRID H-probe for sampling the complete protein surface together with a distance and buriedness filter, combined with an erosion (removing small anomalies) and dilation (filling holes) procedure. An alternative to an inbuilt pocket detection step would be the use of external binding site detection methods. Several such approaches exist, e.g., *SiteMap* [36], *Fpocket* [37], or *DoGSiteScorer* [38], that use geometric or energetic features of the protein surface to identify points in protrusions and cluster them to cavities [39]. Note that, in this case, the binding site detection method needs to return e.g., the center of the cavity and the size, and the pharmacophore method must be able to process this input.

Filter & clustering procedures: Once the binding site is defined, energy levels can be calculated using different probes sampled along the 3D grid and the most favorable regions for binding can be identified and translated into pharmacophore features. For each interaction type, grid points with the best energetics must be assembled accurately to derive the corresponding pharmacophore feature. To achieve this non-trivial task, a clustering method must be implemented, and the choice of the clustering algorithm will have a major impact on the resulting model. *Ph4Dock* [18] opted for the single-linkage clustering algorithm, a hierarchical clustering method grouping clusters from bottom to top (agglomerative clustering). This method allows fast processing of large data sets by grouping points separated by the smallest distances. As a consequence, the single-linkage method creates long clusters in which two points at the opposite ends of the same cluster can be more distant to each other than other points from neighboring clusters [40]. K-means clustering, used in *Hydro-Pharm* [22] and *PharmDock* [24], belongs to the class of partitioning cluster algorithms. The approach of k-means is to divide the dataset into k clusters. The algorithm starts by randomly selecting k seeds (cluster centroids) and annotates the points to the closest centroid. Iteratively, the cluster centroids are recalculated and the points are annotated to the updated centroids. This is repeated until the assignments remain constant and the system converges. A disadvantage of the k-means algorithm is that it requires the user to specify the amount of clusters, k, as an input. When the dataset is a group of points distributed on a 3D grid, defining a strict number of clusters significantly affects the resulting pharmacophore model, e.g., by placing the center of a cluster between two hot spots if they end up in the same cluster. A third clustering approach, named common nearest neighbor (CNN) algorithm and developed by Keller et al. [41], reproduces what human intelligence intuitively distinguishes as clusters in a group of points [42]. In contrast to most geometric cluster algorithms, which are founded on the notion that members of a cluster are closer to each other than to all other points in the data set (such as k-means), the cluster definition in this approach is based on a measure for the local data point density. The CNN algorithm displayed the best aptitude to cluster five two-dimensional test cases in a comparison that included a simple hierarchical clustering algorithm (related to single-linkage) and a partitional algorithm (related to k-means) [42,43]. Initially developed and tested for clustering molecular dynamic trajectories [34,42], the CNN algorithm can be applied to any set of 3D grid points and has, to the best of our knowledge, not yet been used for pharmacophore perception.

Application & novelty: Since methods, like *Ph4Dock* [18] or *PharmDock* [24], were primarily developed for improving docking algorithm performances or for efficient high-throughput virtual screening (*FLAP* [44]), these methods mostly calculate a high amount of pharmacophore features in the cavity to improve scoring. However, deriving a large number of features is not suitable for the representation of a pharmacophoric space in a simple, straightforward, and readable manner for a medicinal chemist (typically, structure-based and ligand-based models are made of three to eight pharmacophore features). For the obvious reason of efficiency, a reduced amount of features is also required for improving the calculation time in the frame of large library screening. However, most importantly, an ability to detect the most critical interactions for optimal binding clearly emerges as a key feature for the development of a new target-focused method. Even though pharmacophore modeling is historically bound to medicinal chemistry, the spectrum of applications related with the study of intermolecular recognitions is broader than simply the docking of potential new drugs. All researchers investigating multivalent systems can benefit from a fast and reliable method for molecular surface analysis and target-focused pharmacophore modeling. As it appears from this literature overview, a simple and robust method to automatically derive reliable pharmacophore models from empty macromolecule structures is urgently needed. In this paper, we report the development and evaluation of a novel computational tool combining the AutoGRID energy function, an advanced cavity annotation, and the CNN clustering algorithm for the design of truly target-focused pharmacophore (T^2F-Pharm) models.

2. Materials and Methods

2.1. Pharmacophore Generation

The algorithm developed in the T^2F approach relies on the following steps, as illustrated in Figure 1.

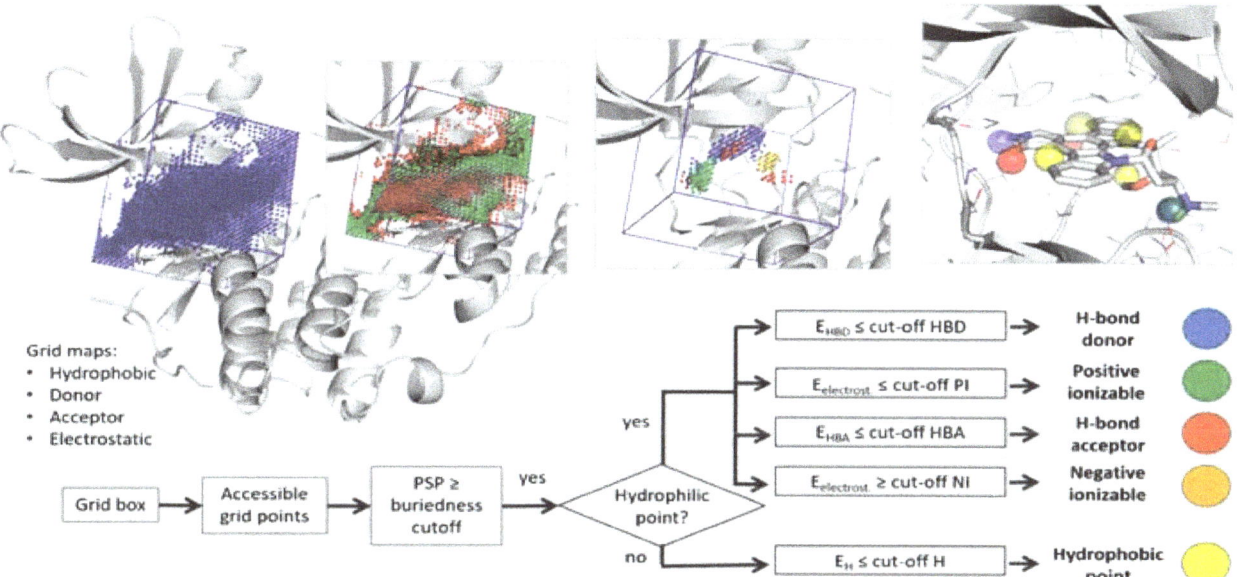

Figure 1. Graphical representation of the main steps of the decision tree invoked by the truly target-focused (T^2F) pharmacophore method.

Grid box: In the first step of the T^2F method, the domain of interest for the hot spot calculation is defined, with or without a co-crystallized ligand. If a ligand is present in the crystal structure, the center of mass of all ligand heavy atoms is automatically calculated and chosen as the grid center. If no ligand information is given, the grid center coordinates can be defined manually. In this case, either user knowledge about the location of the active site is necessary or a pocket detection method, such as DoGSiteScorer [38] (freely available on the ProteinsPlus server [45]), can be evoked. The size of the 3D grid is determined by the edge length of a cubic volume, and the density of the grid by the space between two points. A box edge size of 16 Å and a grid spacing of 0.6 Å are used by default.

Energy calculation: To sample the macromolecular surface of interest, a 3D grid is spanned around the specified targeted surface [46]. The freely available AutoGrid functionality within AutoDock [33] is then used to sample the energies on the grid points. Energy grid maps are calculated by positioning four different probes on each point of the 3D grid to determine the chemical nature of the macromolecular surface. The chosen probes, included in the AutoDock package [46], are (i) an aliphatic carbon for hydrophobic contacts (H-probe), (ii) a hydrogen that donates a hydrogen bond (HBD-probe),

(iii) an oxygen accepting hydrogen bonds (HBA-probe), and (iv) a charged group to calculate electrostatic interactions (i.e., to describe positive (PI) and negative ionizable (NI) groups).

Cavity points filtering: Accessible points are distinguished from those occupied by protein atoms using the energy values obtained from the hydrophilic hydrogen acceptor probe. Sterical clashes with the protein surface result in high energy values. Thus, grid points with an energy value below a given cut-off are considered as accessible, and as occupied otherwise. To further discard points that are too far from the surface, and thus outside of the cavity boundary, the buriedness of each grid point is calculated. Therefore, so called *protein-solvent-protein* (PSP) events, as described by Hendlich et al. [47], are determined. Each grid point is scanned in seven directions (x-, y-, z-axis, and the four diagonals) and the number of PSP events per grid point is calculated. Only grid points above a certain PSP buriedness cut-off are kept for the pharmacophore perception step. Per default, a PSP value of 4 is used.

Pharmacophore perception: To assign pharmacophore features, the remaining (accessible and buried) grid points are filtered and grouped based on their interaction type. First, hydrophilic and hydrophobic points are separated. To determine whether the region around a point is hydrophilic, its surrounding protein residues within a defined radius are scanned for hydrophilic atom types. PyMol [48] is used to span a sphere around each grid point and to collect all amino acid atoms inside this sphere. Points that are surrounded by at least one atom with the potential to function as a hydrogen bond donor or acceptor are considered hydrophilic. The assignment as a potential donor or acceptor is based on the annotation of the respective function encoded in the atom names in the PDB file using the IUPAC-IUB rules. If no hydrophilic atom is detected in this environment, the points are considered hydrophobic. Second, after separating the hydrophilic from the hydrophobic points, they are filtered by energy level using a type-specific cut-off. This minimum energy value is different for each probe. All points with an energy level above the cut-off are discarded. Third, the remaining points are clustered into low energy hot spots for each interaction type using the CNN clustering algorithm [41,42]. To sort low energy hot spots into clusters, two parameters are needed: Distance and similarity cut-offs. The distance cut-off defines the area of a neighborhood for each point, and the similarity cut-off assigns the minimum number of neighbors that two points need to have in common to belong to the same cluster [42]. Larger clusters (more than 80 grid points) are re-clustered with a higher similarity value to be fractioned into multiple moderate sized clusters. This additional step is useful to better describe two neighboring hotspots of the same interaction type, which would otherwise be represented as one dense cluster (often observed for hydrophobic pockets). Subsequently, clusters with less relevance, i.e., those including less than 15 points, are discarded. Finally, a T^2F pharmacophore model is derived by assigning the center of each remaining cluster as one pharmacophore feature. Importantly, the volume of a feature is relative to the amount of points included in the cluster. For this, a sigmoid function variation is used where the feature size responds to the cluster volume change within a radius range of 0.75–2.75 Å.

Pharmacophore processing: Output files generated by the T^2F software are readable (i) in PyMol for visual analysis (pseudo pdb and cgo files), and (ii) in LigandScout (pml file) for further pharmacophore alignment, evaluation, and subsequent virtual screening, as well as (iii) in LeadIT (phm files) for pharmacophore-based docking.

2.2. Evaluation Data Sets

Target data set: The Patel set [29] contains a collection of five well-described protein families, for which a large amount of structural data is available and has been used to evaluate other tools in the context of pharmacophore modelling. In our set, four groups of structurally diverse proteins were collected (the zinc-containing protein, thermolysin, had to be discarded in this first evaluation as this version of the T^2F method is not suitable for the detection of metal coordination). The set contains ligand-enzyme complexes for (a) cyclin-dependent kinase 2 (CDK2, 6 entries), (b) dihydrofolate reductase (DHFR, 6 entries), (c) thrombin (7 entries), and (d) HIV-reverse transcriptase (RT, 10 entries). In addition, the adenosine A_{2A} receptor was included in our set (3 entries) to represent the GPCR class of proteins. The target data set is summarized in Table 2.

Table 2. Evaluation data set.

Group	Reference Structure (Ligand)	Water	Others Structures
Cyclin-dependent kinase 2 (CDK2)	1AQ1 (STU)	-	1E1X, 1FVV, 1DI8, 1E1V, 1FIN
Dihydrofolate reductase (DHFR)	1DRF (FOL)	-	1BOZ, 1DLR, 2DHF, 1OHK, 1HFP
Thrombin	1C4V (IH2)	HOH 404, 408, 410 and 477	1D4P, 1D6W, 1D9I, 1DWD, 1TOM, 1FPC
HIV-reverse transcriptase (RT)	1TVR (TB9)	-	1DTT, 1EP4, 1FK9, 1RT1, 1RT3, 1VRU, 1RT5, 1KLM, 1BMQ
A_{2A} receptor	2DYO (ADN)	-	2YDV, 3EML

Structure preparation: All structures were downloaded from the Protein Databank (PDB) [49] and ProToss (included in the *ProteinsPlus* server [45]) was used to calculate optimal hydrogen bonding networks. All structures were pre-aligned per protein group to allow comparison of pharmacophores derived from different structures (using the *super* function in PyMOL [48]). If a cofactor or water molecule is important for ligand binding, it can be conserved in the protein file.

Experiment set-up: A T^2F pharmacophore model was generated for one representative ligand-protein complex of the five protein families of the target set after removing the co-crystallized ligand. The name and detailed information of the selected reference structure are reported in Table 2.

In order to compare T^2F pharmacophores to structure-based (SB) models, LigandScout [50] was used to derive an SB pharmacophore from each ligand-protein complex of the target set [29]. Except for thrombin (PDB entry 1C4V), water molecules were removed, SB pharmacophore models were built using default parameters, and exported in a pml format for subsequent analysis. If known key interactions were not detected with the default parameters, the small molecule was minimized using MMFF94 [26] and a new SB pharmacophore was built. If this key interaction was detected then the model with the minimized molecule was kept. Otherwise, the SB model derived from the non-minimized conformation was chosen.

Match calculation: A match between a T^2F and an SB pharmacophore feature is reported if two features of the same type are found within a maximum distance of 2 Å from one another. For T^2F features, the center of the feature is taken as the reference point for all types. For SB features, the same holds true for the hydrophobic features (represented as spheres), whereas, for H-bond features (represented by arrows in LigandScout), the position of the H-bond donating or accepting heavy atom is taken as the reference. Note that the HBD feature location is different to the T^2F method, which samples favorable positions for the H-bond donating H-atom. This difference results in a slight feature shift observed in the following evaluation. Finally, the root means square deviation (RMSD) between all matching features is calculated for each pair of T^2F—SB pharmacophore models.

3. Results and Discussion

In the following section, the selected parameters will be discussed, then the generated T^2F-Pharm models for the five proteins of our target dataset will be presented and analyzed.

3.1. Parameter Selection

Grid box: The typical volume of a druggable cavity is around 900 $Å^3$ [51,52]. To cover the complete volume of typical drug binding sites (including larger ligands) with a cubic box, the grid is spanned over a box with a 16 Å edge length. For grid spacing, values between 0.4 Å [32] and 1.0 Å [44] have been reported in the literature for grid representations of cavities. While the usage of smaller grid spacing can compensate for discretization effects, it comes with an exponential increase in grid points and, thus, in calculation time. A grid delta (space between two grid points) of 0.6 Å was an ideal compromise between accuracy and efficiency.

Cavity points: To determine the actual ligand accessible pocket volume in the grid box, first, points clashing with the protein are rejected. Then, points that are too far from the protein surface are filtered out.

Since energies are calculated on each grid point, those occupied by protein atoms will receive unfavorable energy values independently of the probe type. In this work, the energy of the hydrogen bond acceptor probe (HBA) was chosen for discriminating free from occupied grid points, using an energy cut-off of 0.6 kcal/mol as a good compromise of excluding points in small voids while keeping those close to the protein surface. The chosen occupancy value reduces the grid points to be considered to, on average, 30% of the total amount of grid points (about 20.000 points in a box) in the target set discussed hereafter. To exclude points distant from the protein surface and outside the typical interaction radius of a ligand, a buriedness filter is applied. PSP values calculated as described in LIGSITE [47] range from 0 (the most solvent-exposed) to 7 (the most buried). A PSP value ≥ 4 delivered meaningful results in previous pocket-related work [38] and was therefore chosen for this study. This buriedness cut-off further restricts the number of considered grid points to, on average, 14% of all grid points in the original grid box.

Hydrophilic points: To separate hydrophilic from hydrophobic points, the surrounding protein atoms of each grid point are scanned. A point is considered hydrophilic if at least one hydrogen-bond donating or accepting protein heavy atom is detected within a radius of 3 Å. This value corresponds to the typical distance of 2.8–3.2 Å found in protein-ligand complexes between two heavy atoms forming a hydrogen bond [53]. As the distance increases, hydrophilic interactions tend to become weaker and, hence, less reliable (note that the optimal angle for forming a hydrogen bond is not taken into consideration given that the probe is a single atom).

Energy cut-offs per probe: Default parameters per grid probe were selected based on an analysis of the individual probes energy value distributions. The cut-offs were chosen in a way that only the most energetically favorable grid points are retained for each interaction type, as illustrated by the energy distributions in the four histograms shown in Figure 2. Points with a calculated energy below the individual cut-offs are selected for the clustering procedure. A summary of all energy cut-offs can be found in Table 3. While the default energy cut-offs delivered good results for most of the reported cases, the active site of HIV-reverse transcriptase (PDB entry 1TVR) was found to be highly hydrophobic. Thus, cut-offs were adapted to avoid large hydrophobic clusters and to derive the most meaningful pharmacophore features.

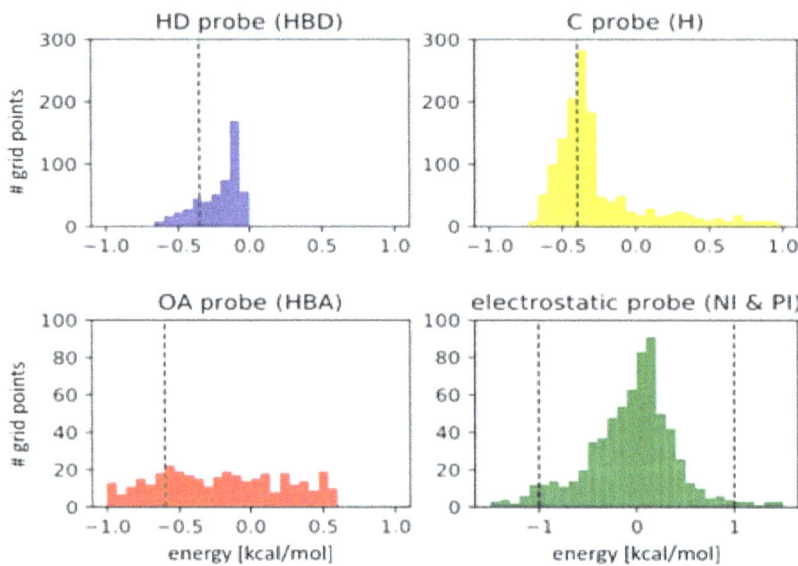

Figure 2. Energy distribution of the free and buried cavity points from the cavity of cyclin-dependent kinase 2 (CDK2) (Protein Databank (PDB) entry 1AQ1) for the four different probes: H-bond donor (HBD in blue), hydrophobic (H in yellow), H-bond acceptor (HBA in red), and electrostatics (in green). The respective energy cut-offs are represented by a dotted vertical line (points with an energy value above the cut-off are discarded). Note that for electrostatics both extrema are kept, describing positive and negative ionizable (PI and NI) areas, respectively.

CNN clustering: The CNN clustering algorithm is a novel local density-based method [34] and the script was kindly shared by their authors for implementation in the T^2F method [41]. For the neighbor distance cut-off, a value was chosen that spans a sphere with a radius of 1.21 Å around each point, thus, including the two surrounding shells of points on a 0.6 Å-spaced grid. Two points need to have at least six neighboring points in common to be allocated to the same cluster. Since hydrophobic areas tend to be more bulky, we introduced a hierarchical re-clustering if the resulting clusters are too large (>80 points), as suggested by the authors of CNN [43]. Two more CNN clustering rounds are introduced to split bulky clusters by increasing the required number of common neighbors (12 and 16). Finally, all clusters with less than 15 points are discarded, retaining only point clouds that span a volume of at least 3.24 Å^3.

Table 3. Default parameters used for T^2F pharmacophore elucidation.

	Center	Ligand CoM * or center coordinates
Grid box	Size of the edge of the cubic box	16 Å
	Distance between two grid points	0.6 Å
Cavity	Occupancy	0.6 kcal/mol
	Buriedness (PSP)	4
Feature type	Hydrophilic radius	3 Å
Type specific energy cut-off **	Hydrophobic (H)	−0.4 kcal/mol (−0.6)
	H-bond donor (HBD)	−0.35 kcal/mol (−0.3)
	H-bond acceptor (HBA)	−0.6 kcal/mol (−0.5)
	Negative/Positive ionizable (NI/PI)	±1.0 kcal/mol
Clustering	Neighbor distance cut-off	1.21 Å
	Number of common neighbors	6 (12, 16) ***
	Min. number of points per cluster	15

* CoM = center of mass. ** Numbers in parentheses are parameter values used for the hydrophobic pocket of reverse transcriptase (PDB entry 1TVR). *** Numbers in parentheses are parameter values used for splitting larger clusters in second and third rounds.

3.2. Evaluation

To evaluate the quality of the developed method, the following key questions will be addressed. First, is the T^2F method able to identify energy hot spots in a cavity where a classical structure-based (SB) approach derives pharmacophore features from the geometry of ligand-target interactions? Second, can additional features (not detected with the SB approach) be highlighted with the T^2F method and, if yes, how relevant are they for protein binding? To answer these questions, co-crystallized ligands were extracted from PDB files and the T^2F method was applied on an artificial apo-form of a crystal structure. For each of the five protein classes, one ligand-target complex was randomly picked and a T^2F model was built after extraction of the ligand and the water molecules (see method section for detailed protocol). Then, this model was aligned and compared to all SB pharmacophore models built for each ligand-protein complex in the same protein family of the set.

3.2.1. Cyclin-Dependent Kinase

A T^2F model was built for Cyclin-dependent kinase 2 (CDK2) using PDB entry 1AQ1 after removing its co-crystalized ligand (STU, an analog of the pan kinase inhibitor, staurosporine). The resulting pharmacophore model includes nine features (Table 4): Four hydrophobic contacts (H), two H-bond donors (HBD), two H-bond acceptors (HBA), and one positive ionizable (PI) feature, as shown in Figure 3. This 3D model was then compared to the SB pharmacophore models derived from the six ligand-kinase complexes of this protein group, including 1AQ1.

Comparing the T^2F and SB models derived from the same CDK2 structure, 1AQ1, allows a first evaluation of the T^2F approach (Figure 3A). Seven pharmacophore features are derived from the inhibitor-kinase complex with the SB approach (Figure 3B), of which four are matching T^2F features with a RMSD of 0.94 Å. The H-bond network (one HBD and one HBA) characteristic of a kinase ATP-binding site (Gln81 and Leu83) in the so-called "hinge region" are identified in both models.

Given the importance of this interaction for kinase inhibition [54–56], the presence of these two features in the T^2F model is an essential first validation. In the region of the cavity accommodating the positively charged amine of the ligand (interaction with Gln131 and Asp86), the two models share a second HBD and a PI feature. Interestingly, the hydrophobic features in both models are detected in the same region, but are not perfectly matching (distance of feature center > 2.0 Å). One match is in the backpocket around the aromatic ring (2.1 Å) and another one is at the aromatic ring in the front pocket (2.7 Å). This shift can be explained by the basic principle of SB pharmacophore design, which centers pharmacophore features on chemical groups of the ligand. On the other hand, a method focused on the target identifies the center of the most energetically favorable area for creating hydrophobic contacts, sometimes shifted slightly compared to the ligand coordinates, and sometimes broader than the very position of one particular chemical group of a bound molecule. Furthermore, a large hydrophobic core was identified in the cavity of CDK2, which results in a bulky area of hydrophobic points split up into nearby clusters. While grid points detected as hydrophobic characterize in detail the geometry of a particular hot spot, deriving this complex 3D-volume in a spherical pharmacophore feature is a simplification where some geometrical information can be lost (Figure S1).

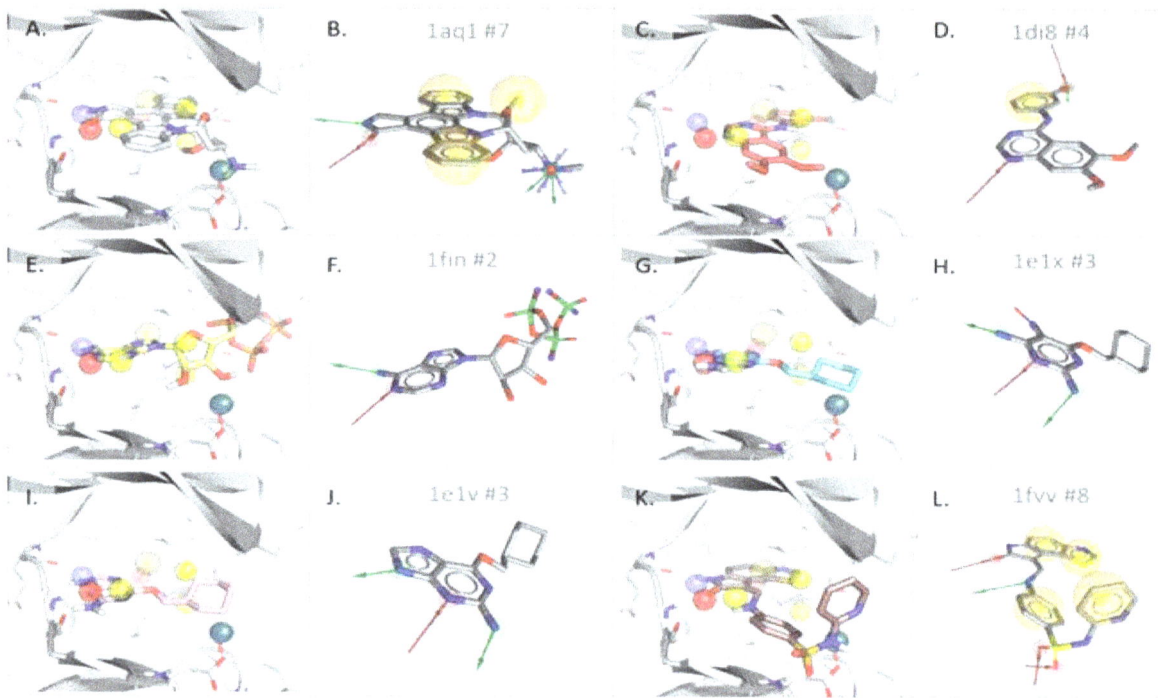

Figure 3. T^2F model derived from the empty CDK2 cavity (PDB entry 1AQ1) superposed to all CDK2-ligands structures used for the evaluation ((**A**) 1AQ1, (**C**) 1DI8, (**E**) 1FIN, (**G**) 1E1X, (**I**) 1E1V, and (**K**) 1FVV) compared to their corresponding structure-based (SB) pharmacophore models (LigandScout: (**B**) 1AQ1, (**D**) 1DI8, (**F**) 1FIN, (**H**) 1E1X, (**J**) 1E1V, and (**L**) 1FVV). The number of features comprised in the SB models is indicated in the respective subfigures. Color coding in the T^2F models (first and third column, drawn with PyMol): HBD = blue, HBA = red, H = yellow, and PI = green. Color coding in the SB models (second and fourth column, drawn with LigandScout): HBD = green, HBA = red, H = yellow, and PI = blue.

Additionally, the T^2F model of CDK2 identified one HBA that is absent in the SB model of 1AQ1. This feature is identified in a subpocket of CDK2 that is not reached by the co-crystallized ligand. This hot spot highlights a possible interaction with the backbone NH of Asp145, which lies at the entry of a small tunnel in the backpocket filled by several water molecules (e.g., HOH 391 and 392). One H feature not matched by the SB model is in a lipophilic region of the CDK2 pocket, spanned by two aliphatic side chains (Leu134 and Ala144). In total, four hydrophilic features of the SB model are matched in the

T^2F model, the same hydrophobic regions are identified, and one extra HBA feature could be derived, providing additional information on the targeted cavity.

To further evaluate our method, the T^2F model derived from an emptied CDK2 structure (1AQ1) was compared to SB models derived from other structures of CDK2 co-crystallized with different ligands (Table 4). In the case of the SB pharmacophore of the PDB entry, 1DI8, three H-bond (HB) and one hydrophobic contact (H) were derived from the ligand-kinase complex. Among the three HB features in the SB model (Figure 3D), only the HBA of the hinge region is matched in the T^2F model (note that the ligand, DTQ500, only forms one hinge HB). Also, the hydrophobic back pocket feature matches one H feature of the T^2F model (distance 1.84 Å). The two HBs involving the hydroxyl group of DTQ500 in the back pocket were not recorded as a hot spot by the T^2F approach. This can be explained (i) by the flexibility of this region of the pocket, closed in 1AQ1 and more open in 1DI8 due to ligand binding, and (ii) by the presence of a water molecule (HOH604) interacting with the ligand in the 1DI8 structure. For the PDB entries, 1FIN (Figure 3E,F), 1E1X (Figure 3G,H), and 1E1V (Figure 3I,J), respectively two, three, and three features are detected using an SB approach. These features are HBs formed with the hinge region of CDK2, of which two are matched in the T^2F model. Finally, the SB model derived from structure 1FVV (holding eight SB features) only matches two features of the T^2F model: One hydrophobic contact (matched by two SB features) and one HBA (in the hinge region). The binding conformation of the 1FVV ligand (Figure 3K,L) highlights interactions that are different to those of the five other inhibitors of the set. These interactions are located in a region that is very solvent-exposed and not buried enough to be detected using the default settings of the T^2F method. This explains why two HBA (with the sulfonamide) and one hydrophobic feature (pyridine) were not detected with the T^2F approach.

In summary, all SB-models shared with the T^2F model one HB feature in the hinge region, and four SB models shared the two key HB anchor features. However, in other regions of the cavity, four features derived with a T^2F approach were either too far apart to match or simply absent in the SB models. Therefore, this first step of the evaluation using CDK2 structures demonstrates the ability of the T^2F tool to not only identify key features for ligand binding, but also new regions that are invisible in an SB approach. Starting from one single structure, the T^2F model not only detects most features presented by the ensemble of six SB models, but also unveiled novel hot spots that are not covered by any of the six co-crystallized CDK2 inhibitors.

Table 4. Feature overlap for the T^2F model derived from the empty CDK2 cavity (empty 1AQ1).

Type	Dist **	Freq ***	1AQ1	1DI8	1FIN	1E1V	1E1X	1FVV
#match *	-	-	4/7	2/4	2/2	2/3	2/3	3/8
rmsd ****			0.94	1.59	0.50	0.69	0.96	1.18
HBD	0.56	4	X		X	X	X	
HBA	0.45	6	X	X	X	X	X	X
HBD	1.00	1	X					
PI	1.17	1	X					
H	1.39	2	X (2.1 Å)	X				2 * X
H			X (2.7 Å)	Slightly shifted front pocket H feature				
H			Surrounding of Leu134 and Ala144					
H			Not detected in SB models					
HBA			HBA towards back pocket water channel (ASP 145, backbone NH)					

* Match: Number of matches between T^2F and SB pharmacophore features in relation to number of SB pharmacophores features in the respective SB model. ** dist: Minimum distance (in Å) of the respective matching features from the different SB models. *** freq: Number of protein structures that exhibit this T^2F -SB feature match. Notes in light grey are comments referring to features close to a match but more distant than 2 Å. **** RMSD describes the root-mean-square deviation (RMSD) in Å of the matching T^2F and structure-based (SB) features.

3.2.2. Dihydrofolate Reductase

Dihydrofolate Reductase (DHFR) reduces dihydrofolic acid using cofactor NADPH, which binds in a neighboring pocket to the one targeted with small molecule inhibitors. To derive a target-focused pharmacophore model of this enzyme, the ligand was extracted from PDB entry 1DRF and the T^2F

method was applied. The 13 resulting pharmacophore features are three H, two PI, one NI, four HBA, and three HBD features (Figure 4A and Table S1). The comparison to the SB model derived from 1DRF with the co-crystallized ligand folic acid shows for each of the five SB features a perfect match with one of the T^2F features (RMSD = 1.17 Å). This result indicates that the T^2F approach fully covers the pharmacophoric space detectable with an SB approach for this particular structure, and also identifies additional hot spots in neighboring regions. Interestingly, among these non-matched T^2F features, two are found on positions very close to a water molecule, bridging an interaction between the ligand and the protein in the co-crystal (HBD with Ser59 and HBA Thr136). Two PI features were detected in the surroundings of the negatively charged side chains of Glu30 and Asp21, respectively. These two features also remained unmatched in the SB model due to the absence of ligand-target interactions with these residues (the water molecule, HOH648, was found to have exactly the same coordinates as the PI feature close to Asp21). Nevertheless, these positions can become important features to consider for DHFR binding. Finally, two more HB (one donor and one acceptor) T^2F features corresponding to potential interactions with the backbone CO of Val115 and the OH group of Tyr121 remained unmatched. Interestingly, these residues are in the pocket accommodating the dihydropteridin bicycle of the folic acid, in PDB entry 1DRF. Despite the proximity of a nitrogen of the dihydropteridin in the surroundings of these two residues, no interaction was detected with the SB approach. However, we surmise that these interactions can be formed if the dynamics of the system were to be considered. This assumption is also confirmed by the non-matched, but nearby HBD, features detected in other DHFR-ligand complexes of the set (PDB entries 1HFP, IDLR, 1OHK, and 1BOZ). This result indicates that a T^2F approach to derive a pharmacophore model from a single and static ligand-free structure can detect this type of potential interaction, while the SB approach would require molecular dynamics to access this information [57–59].

Comparing the unique T^2F model (derived from PDB entry 1DRF without a ligand) to the six SB models derived from ligand-DHFR complexes shows that some interactions are shared among this inhibitor class (Table S1). These interactions are represented by the following features, all matched in the T^2F model: (a) An HBD with Val8 and Glu30 (shared by four SB models); (b) H with Phe31 and 34, Leu67 and Ile60 (shared by four SB models); and (c) an HBA and NI with Arg70 (shared by four SB models). The six remaining features of the T^2F model are invisible to the SB approach, highlighting novel anchoring points for further pharmacomodulations. An overlay of the T^2F model with all ligand-DHFR complexes analyzed in this study is provided in the supporting information (Figure S2).

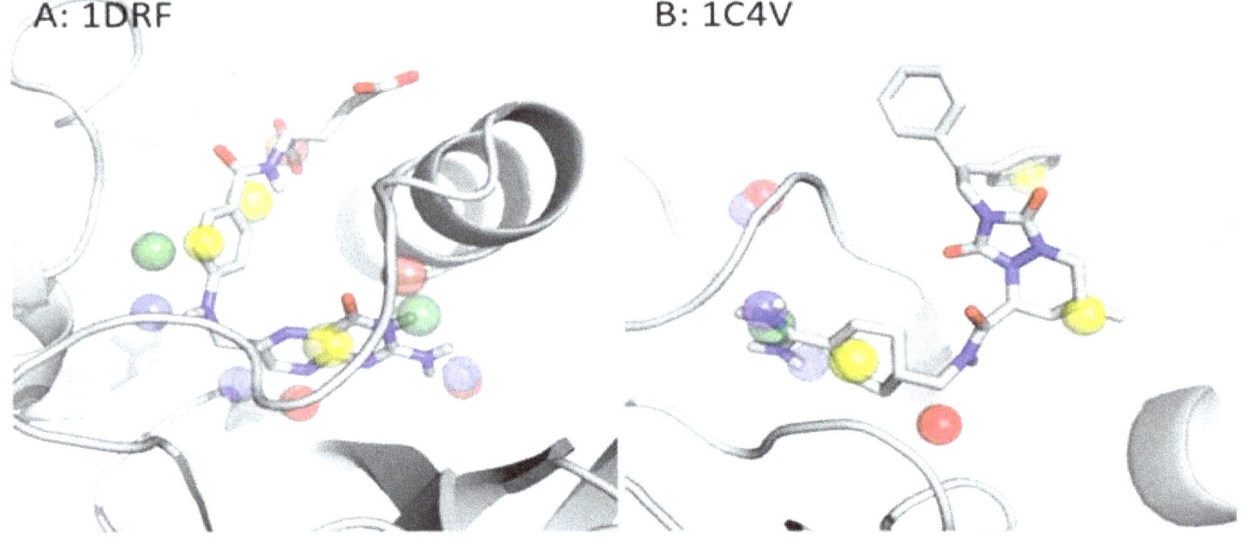

Figure 4. *Cont.*

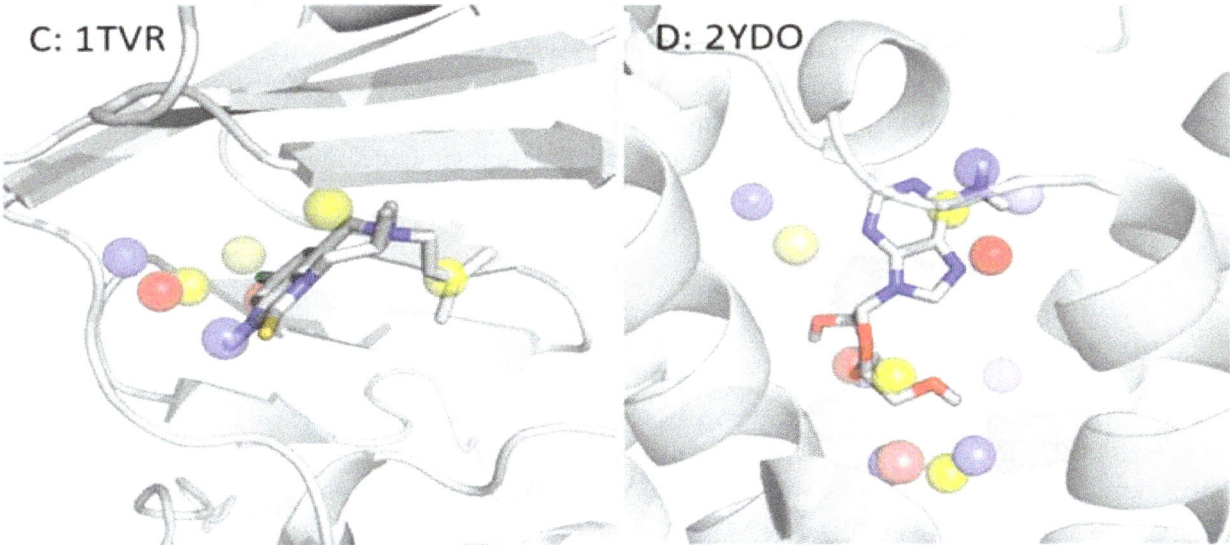

Figure 4. T^2F models overlaid with the ligand previously extracted from its cavity for the proteins, (**A**) dihydrofolate reductase, (**B**) thrombin, (**C**) reverse transcriptase, and (**D**) adenosine A_{2A} receptor. Color coding in the T^2F models: HBD = blue, HBA = red, H = yellow, PI = green, and NI = orange.

3.2.3. Thrombin

The third protein chosen to evaluate our method is thrombin, a serine protease for which seven ligand-protein complexes were assembled in the Patel evaluation set [29]. The particularity of the thrombin active site is the geometry of the P1, P2, and P3 pockets, which can recognize and hydrolyze specific peptide sequences. Competitive thrombin inhibitors bind to this region of the enzyme, in particular to the P1 pocket, which is negatively charged due to the presence of Asp189. The PDB entry selected for deriving a T^2F pharmacophore model is 1C4V. The ligand, IH2370, was extracted from the structure and a target-focused model was built. This model comprises nine features (Figure 4B): Five H-bond features, three H, and one PI features. Two HBD and the PI features are located at the bottom of the P1 subpocket and one H feature is at the entry of this subcavity. One HBA is found in the neighborhood of the catalytic residue, Ser195. Two more H features are detected: One in the subpocket, P2, and one in P3. The last two HB (one donor and one acceptor) detected by the T^2F approach are in a subpocket hosting two water molecules in the analyzed crystal structure (HOH405 and 408). None of these two HB features are matched by any of the SB models derived from the seven thrombin-ligand complexes in the set, as discussed hereafter.

Out of the nine features in the T^2F model, six are matched with an RMSD of 0.72 Å in the SB model derived from the same structure, 1C4V with ligand IH2370 (also comprising nine features, see Table S2). Features in pockets P1, P2, and P3 are all identified with an SB and T^2F approach. However, the T^2F method could not detect two H-bond features derived from the ligand-thrombin complex involving Gly216 found in the SB models. The reason for this is that this residue is located at the edge between the subpockets, P2 and P3, in a non-buried region that is solvent exposed in absence of a ligand. One of the four H features found in the SB model was also absent from the T^2F model for the same reason. Besides the two HB identified in an allosteric pocket, the singularity of the T^2F model is that one HBA was identified in the region of the backbone NHs of Ser195 and Gly193.

A very similar result comes out of the detailed analysis of the six other SB models derived from the remaining thrombin-ligand complexes in the set (the T^2F model superposed to all ligand-thrombin complexes is shown in Figure S3). On the one hand, all crystallized inhibitors also form between one and two H-bonds at the edge between the P2-P3 subpockets (Gly216) that are not detected with the T^2F approach due to the buriedness criterion (apart from PDB entry 1D4P, where no SB hydrogen bond features are found in this region for this particular ligand-thrombin complex). On the other hand, the pharmacophore features, PI and HBD, of the electrophilic P1 subpocket are detected in all structures by both methods (except for PDB entry 1D6W for which only two HBDs are derived due to the neutral charged assigned to the bound aminoimidazol group). Similarly, the H feature of the lipophilic P3 subpocket is found in all SB models, while the H feature of the P2 pocket is only detected for PDB entries, IC4V and 1FPC, leaving this key feature invisible to an SB approach in other crystal structures.

3.2.4. Reverse Transcriptase

HIV Reverse Transcriptase (RT) is a DNA polymerase extensively studied for the development of anti-retrovirus therapy. The binding site of this protein is very lipophilic, which is why some parameters were adjusted to derive the T^2F-Pharm model of this protein (see methods section). Scanning empty PDB entry 1TVR with the T^2F tool highlighted eight hot spots (Figure 4C): Four H features, two HBD, and two HBA features. Among them, four are matched by the SB model derived from the same structure with a ligand (three H and one HBD features), with an RMSD of 0.76 Å. Two SB features are not detected as hot spots in the T^2F approach. The first one is an HBA with the NH backbone of Lys101, which is detected in the SB model only if the ligand is minimized in the cavity. The second non-matched feature of the SB model is one of two H features derived by LigandScout from the short aliphatic chain flanking the core of the inhibitor. In this area, the T^2F-Pharm clustering procedure returns only one H hot spot.

Among the eight features detected with a T^2F approach, seven are matched by at least two SB-models from all 10 RT complexes (Table S3). Interestingly, not a single T^2F feature is matched by all SB models, indicating that no model fully covers the spectrum of possible interactions to interact with RT. In fact, no SB-derived feature is shared by all SB models, illustrating the diversity of the pharmacophoric space for this cavity. Finally, one T^2F-derived feature that is absent in all SB models is an HBA hotspot identified between the OH group of Tyr318 and the backbone NH of His 235. These results illustrate the quality of this model derived from a unique empty cavity in contrast with the ten SB-model (the T^2F model superposed to all ligand-RT complexes is shown in Figure S4).

3.2.5. Adenosine A_{2A} Receptor

G-protein coupled receptors (GPCRs) are an important group of proteins for which small molecule binders represent a large proportion of the drugs on the market [60]. Among them, adenosine receptors have become central therapeutic targets for the treatment of various pathologies (cardiovascular, renal, and nervous systems, as well as endocrine and pulmonary disorders) for more than a decade [61]. To further evaluate our method, we derived a T^2F pharmacophore model from the A_{2A} receptor structure after extracting the co-crystalized adenosine (PDB entry 2YDO) and compared it to SB models derived from crystal structures with (a) adenosine, (b) the synthetic agonist, NECA, and (c) the inverse agonist, ZM241385 [62] (Table 2).

With parameters adapted to the large and hydrophilic cavity of A_{2A}, a T^2F model, comprising 15 features, was derived, including seven HBD, four HBA, and four H features (Table S4). Out of the seven features detected with an SB approach with the bound adenosine, six are matched by the T^2F model (RMSD = 1.1 Å), as shown in Figure 4D. The non-matched feature in the SB model is a second HBD between a hydroxyl group of the ribose of the ligand and His278. The T^2F algorithm derives in this region one single hotspot and centers this cluster on the neighboring HBD of the SB model (adjacent to the OH group interacting with the same His278 residue as well as with Ser277). The comparison of the T^2F model extracted from the empty PDB entry 2YDO with the structure-based model derived from the A_{2A} receptor in the complex with the agonist NECA (PDB entry 2YDV) shows similarly good results. Here, eight out of nine features are matched (RMSD 1.2 Å) and the last HBD is not matched by the T^2F model for exactly the same reason as discussed in the previous example. Finally, an SB model derived from a slightly different receptor conformation (PDB entry 3EML) hosting an A_{2A}-inverse agonist was compared to the T^2F model. In this case, three out of five features of the SB model derived from the interactions detected with the ligand, ZM241385 (PDB entry 3EML), are matched by the T^2F model. The two non-matched SB features are hydrophobic contacts (H). The first one is 2.36 Å apart from the closest H features of the T^2F model, thus, almost matching the 2.00 Å distance cut-off. The second unmet SB feature is detected on the phenol ring at the pocket entrance, which is outside of the grid built from the 2YDO pocket in the reference T^2F model. Interestingly, eight features detected by T^2F-Pharm in the large GPCR binding site are invisible to the SB approach. Note that water molecules were not included in the T^2F calculation, while the adenine ring in the x-ray structure is stabilized through a water network. Three of the unmet T^2F features are overlapping with the position of these water molecules. Furthermore, one unfulfilled hydrophobic feature partially overlaps with the aromatic adenine ring moiety of all three ligand structures. The remaining features are located in sub-pockets not reached by these three ligands, delivering information of additional potential hotspots for ligand binding.

3.2.6. Sensitivity to Conformational Changes

The presented evaluation was conducted for five different protein classes by choosing randomly one structure to derive a reference T^2F pharmacophore. Thus, it appears important to analyze the sensitivity of the method to conformational changes of the protein structure selected for T^2F modeling. Therefore, one T^2F-Pharm model was derived from each of the six CDK2 kinase structures considered in this study (after removing their co-crystallized ligands) and the number of matching features was measured for each pair of T^2F-Pharm models using the 2.00 Å distance cut-off. While some pairs exhibit high similarity (e.g., pairs: 1E1V-1E1X: 100%, 1FIN-1FVV: 80%, or 1AQ1-1FVV: 67% features matched), some show low similarity (e.g., pairs: 1E1V-1FIN: 20%), indicating a sensitivity of the method to the protein conformation under investigation. After a superposition on the Cα atoms of all structures, an analysis of the RMSD of all atoms (including side chains) of the CDK2 binding site (residues within 6 Å around the co-crystalized ligand STU) shows deviations between 2.1 Å for the 1E1V-1E1X pair to 14.9 Å for the same residues for the 1E1V-1FIN pair. Thus, on the one hand, the ability of the method to distinguish between enzyme conformations can be considered a strength, while on the other hand, a too high sensitivity regarding minor side chain movements is not desirable.

4. Conclusions

The landscape of available tools for translating protein surfaces into hot spots for optimal binding in the absence of ligand information is extremely limited. For that reason, we developed the T^2F method for target-focused pharmacophore modeling, an innovative approach based on an elaborated cavity annotation method, combined with hot spot filtering and the advanced common nearest neighbor (CNN) clustering method [41]. The targeted biomolecular surface can be defined manually by the user, derived from the center of mass of a co-crystallized ligand, or by using a cavity detection tool.

An evaluation of our method was conducted with five structurally different enzymes, demonstrating its ability to identify key hot spots for binding to a biomolecular cavity. The presented work shows that most key features derived from the geometry of multiple ligand-target complexes with an SB approach can also be detected by scanning the energy landscape of a single empty protein structure. The five presented cases show how a T^2F model derived from one unique structure can highlight a set of hot spots that an SB approach only accesses if multiple ligand-protein structures are available. Moreover, we showed that the T^2F tool can detect hot spots that remained invisible to a classical SB approach.

In some cases, particular features identified with an SB method were not observed using default parameters. To overcome this issue, an acute knowledge of the studied structure is required to fine tune the parameters, in particular the buriedness cut-off PSP (e.g., thrombin case) or hydrophobicity cut-off (e.g., reverse transcriptase case). Also, opting for spherical pharmacophore features results in a loss of geometrical details on the studied cavity. However, this simplification is indispensable as most 3D modeling and pharmacophore screening software can only process pharmacophore models with features represented as spheres. Nevertheless, the size of the sphere representing a T^2F feature is proportional to the amount of grid points contained in the derived cluster, conserving key spatial information about the cavity that is absent in an SB approach.

Finally, an analysis of the variability of the derived pharmacophore features with respect to the selected reference structure showed that the method is sensitive to changes in side chain orientation observed in closely related crystal structures. To have a better control of the impact of side chain flexibility, we are currently working on incorporating pocket flexibility, e.g., using structural ensembles or molecular dynamic trajectories, to derive a dynamic T^2F pharmacophore.

In conclusion, this preliminary evaluation demonstrates a promising robustness of the T^2F method to derive important hot spots from empty biomolecules. This information can be particularly useful to analyze underexplored protein cavities or targets with no known inhibitors. Additionally, T^2F-Pharm outputs pharmacophore models that can be used for docking and virtual screening, as well as in the investigation of protein allosteric pockets or protein-protein interactions. Therefore, the presented approach not only represents a novel tool for drug discovery, but also a valuable instrument to investigate protein surfaces in the absence of known binding partners. We believe that this simple and straightforward tool can deliver meaningful pharmacophore models for unexplored proteins, but is also complementary to the SB approach, e.g., to identify potential interactions to a target in the context of a lead expansion program.

Author Contributions: J.M.: formal analysis, investigation, methodology, visualization, writing, review & editing; P.D.: programming, investigation & review; A.V.: conceptualization, management, resources, investigation, methodology, supervision, visualization, writing, review & editing.

Acknowledgments: We thank G. Wolber for providing us with a LigandScout license to compare the T^2F pharmacophores with structure-based pharmacophores. We thank the group of B. Keller, especially O. Lemke, for the freely available CNN code and the discussions on its usage.

References

1. Gund, P. Three-dimensional pharmacophoric pattern searching. In *Progress in Molecular and Subcellular Biology*; Springer: Berlin, Germany, 1977; pp. 117–143.

2. Wolber, G.; Seidel, T.; Bendix, F.; Langer, T. Molecule-pharmacophore superpositioning and pattern matching in computational drug design. *Drug Discov. Today* **2008**, *13*, 23–29. [CrossRef] [PubMed]

3. Al-Asri, J.; Gyemant, G.; Fazekas, E.; Lehoczki, G.; Melzig, M.F.; Wolber, G.; Mortier, J. α-Amylase Modulation: Discovery of Inhibitors Using a Multi-Pharmacophore Approach for Virtual Screening. *ChemMedChem* **2016**, *21*, 2372–2377. [CrossRef] [PubMed]

4. Al-Asri, J.; Fazekas, E.; Lehoczki, G.; Perdih, A.; Gorick, C.; Melzig, M.F.; Gyemant, G.; Wolber, G.; Mortier, J. From carbohydrates to drug-like fragments: Rational development of novel α-amylase inhibitors. *Bioorg. Med. Chem.* **2015**, *23*, 6725–6732. [CrossRef] [PubMed]

5. El-Houri, R.B.; Mortier, J.; Murgueitio, M.S.; Wolber, G.; Christensen, L.P. Identification of PPARgamma Agonists from Natural Sources Using Different In Silico Approaches. *Planta Med.* **2015**, *81*, 488–494. [PubMed]

6. Bock, A.; Bermudez, M.; Krebs, F.; Matera, C.; Chirinda, B.; Sydow, D.; Dallanoce, C.; Holzgrabe, U.; De Amici, M.; Lohse, M.J.; et al. Ligand Binding Ensembles Determine Graded Agonist Efficacies at a G Protein-coupled Receptor. *J. Biol. Chem.* **2016**, *291*, 16375–16389. [CrossRef] [PubMed]

7. Mortier, J.; Prevost, J.R.C.; Sydow, D.; Teuchert, S.; Omieczynski, C.; Bermudez, M.; Frederick, R.; Wolber, G. Arginase Structure and Inhibition: Catalytic Site Plasticity Reveals New Modulation Possibilities. *Sci. Rep.* **2017**, *7*, 13616. [CrossRef] [PubMed]

8. Murgueitio, M.S.; Bermudez, M.; Mortier, J.; Wolber, G. In silico virtual screening approaches for anti-viral drug discovery. *Drug Discov. Today Technol.* **2012**, *9*, e219–e225. [CrossRef] [PubMed]

9. Spitzer, G.M.; Heiss, M.; Mangold, M.; Markt, P.; Kirchmair, J.; Wolber, G.; Liedl, K.R. One concept, three implementations of 3D pharmacophore-based virtual screening: Distinct coverage of chemical search space. *J. Chem. Inf. Model.* **2010**, *50*, 1241–1247. [CrossRef] [PubMed]

10. Sanders, M.P.A.; McGuire, R.; Roumen, L.; de Esch, I.J.P.; de Vlieg, J.; Klomp, J.P.G.; de Graaf, C. From the protein's perspective: The benefits and challenges of protein structure-based pharmacophore modeling. *MedChemComm* **2012**, *3*, 28–38. [CrossRef]

11. Sanders, M.P.; Verhoeven, S.; de Graaf, C.; Roumen, L.; Vroling, B.; Nabuurs, S.B.; de Vlieg, J.; Klomp, J.P. Snooker: A structure-based pharmacophore generation tool applied to class A GPCRs. *J. Chem. Inf. Model.* **2011**, *51*, 2277–2292. [CrossRef] [PubMed]

12. Klabunde, T.; Giegerich, C.; Evers, A. Sequence-derived three-dimensional pharmacophore models for G-protein-coupled receptors and their application in virtual screening. *J. Med. Chem.* **2009**, *52*, 2923–2932. [CrossRef] [PubMed]

13. Kratochwil, N.A.; Malherbe, P.; Lindemann, L.; Ebeling, M.; Hoener, M.C.; Muhlemann, A.; Porter, R.H.; Stahl, M.; Gerber, P.R. An automated system for the analysis of G protein-coupled receptor transmembrane binding pockets: Alignment, receptor-based pharmacophores, and their application. *J. Chem. Inf. Model.* **2005**, *45*, 1324–1336. [CrossRef] [PubMed]

14. Miranker, A.; Karplus, M. Functionality maps of binding sites: A multiple copy simultaneous search method. *Proteins* **1991**, *11*, 29–34. [CrossRef] [PubMed]

15. Meagher, K.L.; Carlson, H.A. Incorporating protein flexibility in structure-based drug discovery: Using HIV-1 protease as a test case. *J. Am. Chem. Soc.* **2004**, *126*, 13276–13281. [CrossRef] [PubMed]

16. Ghanakota, P.; Carlson, H.A. Driving Structure-Based Drug Discovery through Cosolvent Molecular Dynamics. *J. Med. Chem.* **2016**, *59*, 10383–10399. [CrossRef] [PubMed]

17. Goodford, P.J. A computational procedure for determining energetically favorable binding sites on biologically important macromolecules. *J. Med. Chem.* **1985**, *28*, 849–857. [CrossRef] [PubMed]

18. Goto, J.; Kataoka, R.; Hirayama, N. Ph4Dock: Pharmacophore-based protein-ligand docking. *J. Med. Chem.* **2004**, *47*, 6804–6811. [CrossRef] [PubMed]

19. Chen, J.; Lai, L. Pocket v.2: Further developments on receptor-based pharmacophore modeling. *J. Chem. Inf. Model.* **2006**, *46*, 2684–2691. [CrossRef] [PubMed]

20. Ortuso, F.; Langer, T.; Alcaro, S. GBPM: GRID-based pharmacophore model: Concept and application studies to protein-protein recognition. *Bioinformatics* **2006**, *22*, 1449–1455. [CrossRef] [PubMed]

21. Tintori, C.; Corradi, V.; Magnani, M.; Manetti, F.; Botta, M. Targets looking for drugs: A multistep computational protocol for the development of structure-based pharmacophores and their applications for hit discovery. *J. Chem. Inf. Model.* **2008**, *48*, 2166–2179. [CrossRef] [PubMed]

22. Hu, B.; Lill, M.A. Protein pharmacophore selection using hydration-site analysis. *J. Chem. Inf. Model.* **2012**, *52*, 1046–1060. [CrossRef] [PubMed]

23. Baroni, M.; Cruciani, G.; Sciabola, S.; Perruccio, F.; Mason, J.S. A common reference framework for analyzing/comparing proteins and ligands. Fingerprints for Ligands and Proteins (FLAP): Theory and application. *J. Chem. Inf. Model.* **2007**, *47*, 279–294. [CrossRef] [PubMed]

24. Hu, B.; Lill, M.A. Exploring the potential of protein-based pharmacophore models in ligand pose prediction and ranking. *J. Chem. Inf. Model.* **2013**, *53*, 1179–1190. [CrossRef] [PubMed]

25. Eldridge, M.D.; Murray, C.W.; Auton, T.R.; Paolini, G.V.; Mee, R.P. Empirical scoring functions: I. The development of a fast empirical scoring function to estimate the binding affinity of ligands in receptor complexes. *J. Comput. Aided Mol. Des.* **1997**, *11*, 425–445. [CrossRef] [PubMed]

26. Halgren, T.A. Merck molecular force field. I. Basis, form, scope, parameterization, and performance of MMFF94. *J. Comput. Chem.* **1996**, *17*, 490–519. [CrossRef]

27. Hartshorn, M.J.; Verdonk, M.L.; Chessari, G.; Brewerton, S.C.; Mooij, W.T.; Mortenson, P.N.; Murray, C.W. Diverse, high-quality test set for the validation of protein-ligand docking performance. *J. Med. Chem.* **2007**, *50*, 726–741. [CrossRef] [PubMed]

28. Wang, R.; Liu, L.; Lai, L.; Tang, Y. SCORE: A new empirical method for estimating the binding affinity of a protein-ligand complex. *J. Mol. Model.* **1998**, *4*, 379–394. [CrossRef]

29. Patel, Y.; Gillet, V.J.; Bravi, G.; Leach, A.R. A comparison of the pharmacophore identification programs: Catalyst, DISCO and GASP. *J. Comput. Aided Mol. Des.* **2002**, *16*, 653–681. [CrossRef] [PubMed]

30. Huang, N.; Shoichet, B.K.; Irwin, J.J. Benchmarking sets for molecular docking. *J. Med. Chem.* **2006**, *49*, 6789–6801. [CrossRef] [PubMed]

31. Siragusa, L.; Cross, S.; Baroni, M.; Goracci, L.; Cruciani, G. BioGPS: Navigating biological space to predict polypharmacology, off-targeting, and selectivity. *Proteins* **2015**, *83*, 517–532. [CrossRef] [PubMed]

32. Hu, B.; Lill, M.A. PharmDock: A pharmacophore-based docking program. *J. Cheminform.* **2014**, *6*, 14. [CrossRef] [PubMed]

33. Morris, G.M.; Huey, R.; Lindstrom, W.; Sanner, M.F.; Belew, R.K.; Goodsell, D.S.; Olson, A.J. AutoDock4 and AutoDockTools4: Automated docking with selective receptor flexibility. *J. Comput. Chem.* **2009**, *30*, 2785–2791. [CrossRef] [PubMed]

34. Lemke, O.; Keller, B.G. Density-based cluster algorithms for the identification of core sets. *J. Chem. Phys.* **2016**, *145*, 164104. [CrossRef] [PubMed]

35. Edelsbrunner, H.; Facello, M.; Fu, P.; Liang, J. Measuring proteins and voids in proteins. In Proceedings of the Twenty-Eighth Hawaii International Conference on System Sciences, Wailea, HI, USA, 3–6 January 1995; IEEE: Piscataway, NJ, USA, 1995; pp. 256–264.

36. Halgren, T. New method for fast and accurate binding-site identification and analysis. *Chem. Biol. Drug Des.* **2007**, *69*, 146–148. [CrossRef] [PubMed]

37. Le Guilloux, V.; Schmidtke, P.; Tuffery, P. Fpocket: An open source platform for ligand pocket detection. *BMC Bioinform.* **2009**, *10*, 168. [CrossRef] [PubMed]

38. Volkamer, A.; Kuhn, D.; Rippmann, F.; Rarey, M. DoGSiteScorer: A web server for automatic binding site prediction, analysis and druggability assessment. *Bioinformatics* **2012**, *28*, 2074–2075. [CrossRef] [PubMed]

39. Volkamer, A.; Rarey, M. Exploiting structural information for drug-target assessment. *Future Med. Chem.* **2014**, *6*, 319–331. [CrossRef] [PubMed]

40. Everitt, B. *Cluster Analysis*, 5th ed.; Wiley: Chichester, UK, 2011.

41. Lemke, O.; Keller, B.G. Common Nearest Neighbor Clustering. Available online: https://github.com/ BDGSoftware/CNNClustering (accessed on 21 November 2017).

42. Keller, B.; Daura, X.; van Gunsteren, W.F. Comparing geometric and kinetic cluster algorithms for molecular simulation data. *J. Chem. Phys.* **2010**, *132*, 074110. [CrossRef] [PubMed]

43. Lemke, O.; Keller, B. Common Nearest Neighbor Clustering—A Benchmark. *Algorithms* **2018**, *11*, 19. [CrossRef]

44. Cross, S.; Baroni, M.; Goracci, L.; Cruciani, G. GRID-based three-dimensional pharmacophores I: FLAPpharm, a novel approach for pharmacophore elucidation. *J. Chem. Inf. Model.* **2012**, *52*, 2587–2598. [CrossRef] [PubMed]

45. Bietz, S.; Urbaczek, S.; Schulz, B.; Rarey, M. Protoss: A holistic approach to predict tautomers and protonation states in protein-ligand complexes. *J. Cheminform.* **2014**, *6*, 12. [CrossRef] [PubMed]

46. Trott, O.; Olson, A.J. AutoDock Vina: Improving the speed and accuracy of docking with a new scoring function, efficient optimization, and multithreading. *J. Comput. Chem.* **2010**, *31*, 455–461. [CrossRef] [PubMed]

47. Hendlich, M.; Rippmann, F.; Barnickel, G. LIGSITE: Automatic and efficient detection of potential small molecule-binding sites in proteins. *J. Mol. Graph. Model.* **1997**, *15*, 359–363. [CrossRef]

48. Schrodinger, L.L.C. *The PyMOL Molecular Graphics System, Version 1.8*; DeLano Scientific: San Carlos, CA, USA, 2015.

49. Berman, H.M.; Westbrook, J.; Feng, Z.; Gilliland, G.; Bhat, T.N.; Weissig, H.; Shindyalov, I.N.; Bourne, P.E. The Protein Data Bank. *Nucleic Acids Res.* **2000**, *28*, 235–242. [CrossRef] [PubMed]

50. Wolber, G.; Langer, T. LigandScout: 3-D pharmacophores derived from protein-bound ligands and their use as virtual screening filters. *J. Chem. Inf. Model.* **2005**, *45*, 160–169. [CrossRef] [PubMed]

51. Volkamer, A.; Kuhn, D.; Grombacher, T.; Rippmann, F.; Rarey, M. Combining global and local measures for structure-based druggability predictions. *J. Chem. Inf. Model.* **2012**, *52*, 360–372. [CrossRef] [PubMed]

52. Egner, U.; Hillig, R.C. A structural biology view of target drugability. *Expert Opin. Drug Discov.* **2008**, *3*, 391–401. [CrossRef] [PubMed]

53. Klebe, G. *Wirkstoffdesign: Entwurf und Wirkung von Arzneistoffen*; Springer: Berlin, Germany, 2009.

54. Mortier, J.; Masereel, B.; Remouchamps, C.; Ganeff, C.; Piette, J.; Frederick, R. NF-kappaB inducing kinase (NIK) inhibitors: Identification of new scaffolds using virtual screening. *Bioorg. Med. Chem. Lett.* **2010**, *20*, 4515–4520. [CrossRef] [PubMed]

55. Mortier, J.; Frederick, R.; Ganeff, C.; Remouchamps, C.; Talaga, P.; Pochet, L.; Wouters, J.; Piette, J.; Dejardin, E.; Masereel, B. Pyrazolo[4,3-*c*]isoquinolines as potential inhibitors of NF-kappaB activation. *Biochem. Pharmacol.* **2010**, *79*, 1462–1472. [CrossRef] [PubMed]

56. Kooistra, A.J.; Volkamer, A. Chapter Six—Kinase-Centric Computational Drug Development. In *Annual Reports in Medicinal Chemistry*; Goodnow, R.A., Ed.; Academic Press: Cambridge, MA, USA, 2017; Volume 50, pp. 197–236.

57. Bermudez, M.; Mortier, J.; Rakers, C.; Sydow, D.; Wolber, G. More than a look into a crystal ball: Protein structure elucidation guided by molecular dynamics simulations. *Drug Discov. Today* **2016**, *21*, 1799–1805. [CrossRef] [PubMed]

58. Christin, R.; Marcel, B.; Keller, B.G.; Jérémie, M.; Gerhard, W. Computational close up on protein–protein interactions: How to unravel the invisible using molecular dynamics simulations? *Wiley Interdiscip. Rev. Comput. Mol. Sci.* **2015**, *5*, 345–359.

59. Mortier, J.; Rakers, C.; Bermudez, M.; Murgueitio, M.S.; Riniker, S.; Wolber, G. The impact of molecular dynamics on drug design: Applications for the characterization of ligand-macromolecule complexes. *Drug Discov. Today* **2015**, *20*, 686–702. [CrossRef] [PubMed]

60. Tautermann, C.S. GPCR structures in drug design, emerging opportunities with new structures. *Bioorg. Med. Chem. Lett.* **2014**, *24*, 4073–4079. [CrossRef] [PubMed]

61. Jacobson, K.A.; Gao, Z.-G. Adenosine receptors as therapeutic targets. *Nat. Rev. Drug Discov.* **2006**, *5*, 247–264. [CrossRef] [PubMed]

62. Lebon, G.; Warne, T.; Edwards, P.C.; Bennett, K.; Langmead, C.J.; Leslie, A.G.; Tate, C.G. Agonist-bound adenosine A2A receptor structures reveal common features of GPCR activation. *Nature* **2011**, *474*, 521–525. [CrossRef] [PubMed]

9

Discovery of Potential Inhibitors of Squalene Synthase from Traditional Chinese Medicine Based on Virtual Screening and In Vitro Evaluation of Lipid-Lowering Effect

Yankun Chen [1] (ID), Xi Chen [1], Ganggang Luo [1], Xu Zhang [1] (ID), Fang Lu [1], Liansheng Qiao [1], Wenjing He [2], Gongyu Li [1] and Yanling Zhang [1,*]

[1] School of Chinese Material Medica, Beijing University of Chinese Medicine, Beijing 100102, China; 18811791975@163.com (Y.C.); chenxi_cx95@163.com (X.C.); 17801080765@163.com (G.L.); 18003381008@163.com (X.Z.); lufang1017@163.com (F.L.); b20100222012@163.com (L.Q.); lidoc2727@163.com (G.L.)

[2] College of Traditional Chinese Medicine Xinjiang Medical University, Urumqi 830054, China; wenjhe@163.com

* Correspondence: zhangyanling@bucm.edu.cn

Abstract: Squalene synthase (SQS), a key downstream enzyme involved in the cholesterol biosynthetic pathway, plays an important role in treating hyperlipidemia. Compared to statins, SQS inhibitors have shown a very significant lipid-lowering effect and do not cause myotoxicity. Thus, the paper aims to discover potential SQS inhibitors from Traditional Chinese Medicine (TCM) by the combination of molecular modeling methods and biological assays. In this study, cynarin was selected as a potential SQS inhibitor candidate compound based on its pharmacophoric properties, molecular docking studies and molecular dynamics (MD) simulations. Cynarin could form hydrophobic interactions with PHE54, LEU211, LEU183 and PRO292, which are regarded as important interactions for the SQS inhibitors. In addition, the lipid-lowering effect of cynarin was tested in sodium oleate-induced HepG2 cells by decreasing the lipidemic parameter triglyceride (TG) level by 22.50%. Finally. cynarin was reversely screened against other anti-hyperlipidemia targets which existed in HepG2 cells and cynarin was unable to map with the pharmacophore of these targets, which indicated that the lipid-lowering effects of cynarin might be due to the inhibition of SQS. This study discovered cynarin is a potential SQS inhibitor from TCM, which could be further clinically explored for the treatment of hyperlipidemia.

Keywords: hyperlipidemia; squalene synthase (SQS); molecular modeling; drug discovery; Traditional Chinese Medicine

1. Introduction

Hyperlipidemia, characterized by abnormally-elevated levels of cholesterol in the blood, is one of the main risk factors for atherosclerosis and visceral obesity [1]. Reduction of cholesterol can be achieved by inhibiting cholesterol biosynthesis [2]. To date, human HMG-CoA reductase (hHMGR) inhibitors such as statins are the most effective medicines for reducing cholesterol levels. However, these statins have potential adverse effects, such as myotoxicity, hepatotoxicity and even rhabdomyolysis [3]. The major cause of these side effects is the inhibition of HMG-CoA reductase that will interfere with the synthesis of many nonsteroidal isoprenoid molecules, which plays a major role in diverse cellular functions [4]. Compared to HMG-CoA reductase, squalene synthase (SQS), a key downstream enzyme involved in the cholesterol biosynthetic pathway, is regarded as an attractive target for anti-hyperlipidemia [5]. SQS is the first step of the steroid synthesis pathway, which means the inhibition of SQS can prevent the cholesterol biosynthesis without interrupting isoprenoid production [6]. Due to its strategic location in the pathway, inhibitors of SQS are promising drugs for the treatment of hyperlipidemia.

At present, chemical synthesis [7] and genetic engineering methods [8] are utilized to discover SQS inhibitors, which requires much time and money. Traditional Chinese Medicine (TCM) has been widely used in the treatment of hyperlipidemia with low cost and minimal adverse effects. For example, *Fructus Crataegi* and *Salviae Miltiorrhizae* are the most well-known used Chinese herbs for treating hyperlipidemia [9,10]. Although TCM has played an important role in drug discovery for treating hyperlipidemia for a long time due to its rich natural resources, there are few studies at present on the discovery of SQS inhibitors from TCM. Thus, it is of great importance to discover potential SQS inhibitors from TCM. In [11] the authors researched SQS inhibitors by using molecular docking and virtual screening methods but the shortcoming of the study was the lack of biological assays to verify the accuracy of the results.

In our study, we provide a reliable strategy to discover potential SQS inhibitors from TCM by the combination of molecular modeling methods and biological assays. First, ten HipHop pharmacophore models were generated based on known SQS inhibitors. The optimal pharmacophore model was selected by four validation indices and used as a query to screen potential SQS inhibitors from the Traditional Chinese Medicine Database (TCMD, Version 2009). Molecular docking was employed to refine the pharmacophore model hits and analyze the protein-ligand binding modes. Then, MD simulations were performed to validate the binding stability between the compounds and the protein. The potential SQS inhibitors were selected based on the fitvalue, docking score, and interactions formed between the ligands and SQS. In addition, the compounds were evaluated for the lipid-lowering effect in sodium oleate-induced HepG2 cells. Finally, the active compounds were utilized to reversely identify the other anti-hyperlipidemia targets existed in HepG2 cells to further evaluate the lipid-lowering effect was due to the inhibition of SQS. This study aims to discover potential SQS inhibitors from TCM, which also provide the candidate compounds for the clinical treatment of hyperlipidemia.

2. Results

2.1. Pharmacophore Model Studies

Ten pharmacophore models were generated based on twenty-two SQS inhibitors by the HipHop method within the Discovery Studio 4.0 (DS) from Accelrys (San Diego, CA, USA). All of the models had high rank scores (154.43–157.40, Table 1), which indicated that compounds in the training set mapped well with generated pharmacophore models. The test set was applied for evaluating the generated ten pharmacophore models based on the three evaluation indices as follows: hit rate of active compounds (*HRA*), identify effective index (*IEI*) and comprehensive appraisal index (*CAI*). *HRA*, *IEI* and *CAI* are defined by Equations (1)–(3), where D represents the total number of compounds in the test set and A represents the number of active compounds in the test set. Ht is the total number of hit compounds from the test set and Ha represents the number of active hit compounds from the test set. *HRA* represents the ability to identify active compounds from the test set. *IEI*, the index of effective identification, is used to evaluate the ability of the models to identify active compounds from the inactive compounds. *CAI* is the comprehensive evaluation of pharmacophore model [12]:

$$HRA = \left(\frac{Ha}{A} \right) \times 100 \tag{1}$$

$$IEI = \frac{\left(\frac{Ha}{Ht} \right)}{\frac{A}{D}} \tag{2}$$

$$CAI = HRA \times IEI \tag{3}$$

The evaluation results of the 10 pharmacophore models are shown in Table 1. The calculation of the *HRA* index returned values greater than 80% for nine of 10 models, revealing the high accuracy of the generated pharmacophore models. The rank score represents the total score of how the training set fits the pharmacophore, and the best model has the highest rank [13]. Hypo1 had the highest rank score of 157.40. Therefore, Hypo1 was selected as the optimal pharmacophore model. In general, scores of *HRA*, *IEI* and *CAI* above the values of 80%, 2, and 2 are considered excellent. *HRA*, *IEI* and *CAI* of Hypo1 were 94.16%, 2.26, and 2.12, respectively. As shown in Figure 1a, Hypo1 contained one hydrogen bond acceptor (A), two hydrophobic features (H), one aromatic ring (R), and five excluded volumes (Ev). In order to validate the veracity of the best pharmacophore model, the crystallographic ligand of D99 and the positive SQS inhibitor of TAK-475 [14] were mapped with the optimal pharmacophore model. Both compounds mapped well with all the features of Hypo 1, which are shown in Figure 1b,c.

Table 1. The Validation Results of the Pharmacophore Models.

Hypo	Feature	Rank	D	A	Ha	Ht	*HRA*	*IEI*	*CAI*
1	RHHAEv5	157.40	616	154	145	256	94.16%	2.26	2.12
2	RHHAEv5	156.97	616	154	147	290	95.45%	2.03	1.93
3	RHHAEv5	156.45	616	154	138	271	89.61%	2.04	1.83
4	RHHAEv5	155.73	616	154	138	278	89.61%	1.99	1.78
5	RHHAEv5	155.62	616	154	147	265	95.45%	2.22	2.12
6	RHHAEv5	155.54	616	154	151	268	98.05%	2.25	2.21
7	RHHAEv5	154.89	616	154	106	247	68.83%	1.72	1.18
8	RHHAEv5	154.67	616	154	126	219	81.81%	2.30	1.88
9	RHHAEv5	154.43	616	154	144	267	93.50%	2.16	2.02
10	RHHAEv5	154.43	616	154	143	254	92.86%	2.25	2.09

Note: D is the total number of compounds in test set; A is the number of active compounds in the test set; Ha is the hits number of active molecules mapped pharmacophores; Ht is the total hits number of molecules mapped pharmacophores; *HRA* (hit rate of active compounds); *IEI* (identify effective index); *CAI* (comprehensive appraisal index).

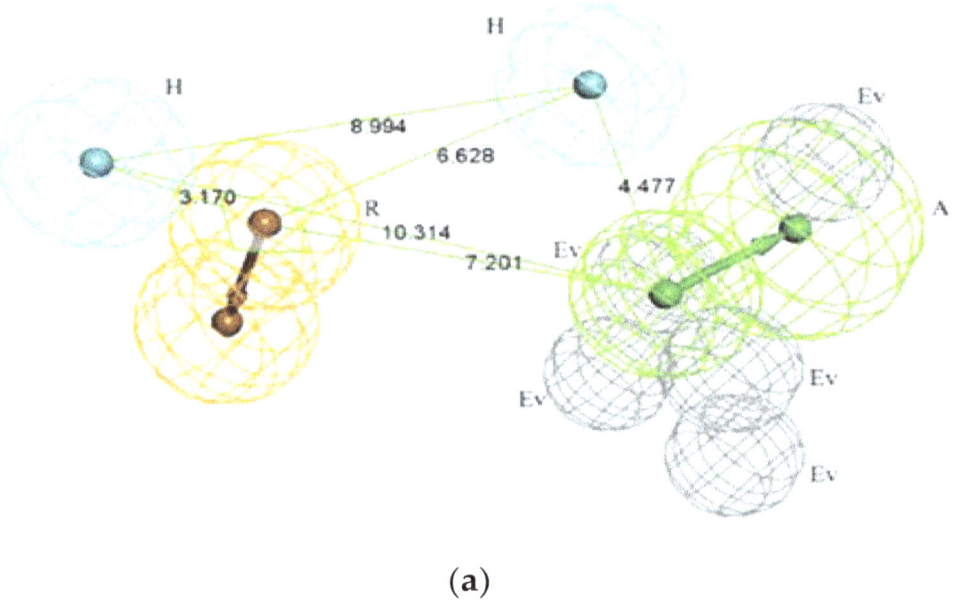

(a)

Figure 1. *Cont.*

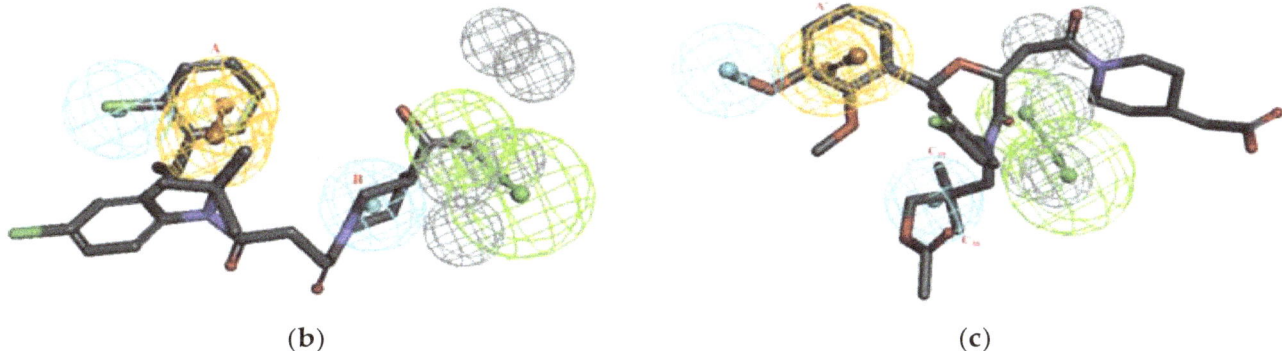

Figure 1. (**a**) The optimal pharmacophore model Hypo1; Wherein, green features represent hydrogen bond acceptor (A), light blue features represent hydrophobic features (H), orange features represent ring aromatic (R) and gray features represent excluded volumes (Ev); (**b**) The mapping of the crystallographic ligand with the optimal pharmacophore model Hypo1; (**c**) mapping of TAK-475 with the Hypo1.

According to the literature, researchers have constructed pharmacophore models of SQS [15,16]. We further compared our pharmacophore model to those of these researchers. First, the method used for constructing the pharmacophore model was different. The pharmacophore models in the literature were constructed by using the three-dimensional quantitative structure-activity relationship (3D-QSAR) method, which belongs to the quantitative hypothesis models, while we built the pharmacophore models by using HipHop method, which belongs to the qualitative hypothesis models. Second, the structure of the training sets was different. The structures of the training set in the articles were relatively simple, aimed at directing the structural modification of the potential compounds. Our training set with structural diversity was used to screen active compounds with novel structures from the database. Third, the purposes of the papers were different. The researchers used a training set of ligands with activity values to derive 3D-QSAR pharmacophore models for prediction. Our HipHop pharmacophore was built by using a training set of some active ligands to derive common feature pharmacophores for lead identification. Fourth, the similarity analysis. The features of the 3D-QSAR pharmacophore model and the HipHop pharmacophore such as hydrogen bond acceptor, hydrophobic features, aromatic ring, and excluded volumes, were consistent, which indicated that our HipHop pharmacophore was reliable and could be applicable to screen potential SQS inhibitors.

What is more, to further evaluate the reliability of the pharmacophore model, a 2D similarity search was used to compare the similarity between the TAK-475 and the 22 ligands used in the construction of the pharmacophore model based on 2D fingerprints [17]. During this process, the positive SQS inhibitor of TAK-475 as the template molecule was chosen to search for similar molecules in the 22 ligands, as the top-ranked molecules are likely to exhibit similar biological activity [18]. The Tanimoto coefficient [19] was used to measure the similarity to find ligands that are similar to TAK-475. In general, the range of Tanimoto coefficient values is from zero to one. A value closer to one indicates a greater similarity between the ligand and TAK-475. There is no specific standard for the threshold of Tanimoto coefficient to identify ligands, the Tanimoto coefficient value of 0.3 was also set as threshold in some references to identify ligands [20]. From the results (Table 2), the 22 ligands had Tanimoto coefficient values all higher than 0.45. In addition, the ligands with Tanimoto coefficient values higher than 0.7 account for more than 60% of the 22 ligands. The result indicated that these 22 ligands had similar structures compared to TAK-475, with similar biological activity and could be used to construct the pharmacophore model.

Table 2. Similarity search results of 22 ligands.

Tanimoto Coefficient [a]	Number [b]	Percent [c]
$0 < T \leq 0.4$	0	0
$0.4 < T \leq 0.5$	4	0.18%
$0.5 < T \leq 0.7$	4	0.18%
$0.7 < T \leq 0.8$	7	0.32%
$0.8 < T < 0.9$	5	0.23%
$0.8 < T < 1.0$	2	0.09%

[a] The Tanimoto coefficient is a similarity index. [b] Number is the number of ligands of the training set within in the corresponding threshold value of the Tanimoto coefficient. [c] Percent is percentage of the number of ligands.

Then the Hypo 1 program was used to screen potential SQS inhibitors from the Traditional Chinese Medicine Database (TCMD, Version 2009), before which the TCMD database was filtered based upon the Lipinski's rules, leaving 13,905 compounds. Then, a hit list of 1775 TCM compounds were obtained for further docking studies.

2.2. Molecular Docking Studies

The binding pocket was defined with a default parameter of sphere radius of 9.16 Å around D99 of SQS. The D99 was re-docked into the active pocket by using two docking algorithms, LibDock and CDOCKER, respectively. The RMSD values of D99 were 7.98 Å and 0.69 Å for the corresponding two docking algorithms. The reason for such a high RMSD returned by LibDock, in comparison to CDOCKER, may be ascribed to the differences between the two docking algorithms. LibDock is a kind of semi-flexible docking method and CDOCKER is a flexible one. In addition, the LibDOCK algorithm is a high-throughput algorithm for docking ligands into receptor binding sites [21]. The CDOCKER algorithm uses a CHARMm-based molecular dynamics (MD) method to dock ligands into an active receptor site [22]. The ligand can generate random conformations to form a favorable interaction with the protein, which may cause a lower RMSD compared to LibDock. In general, an RMSD less than 2.00 Å shows that the docking algorithm is fit for this protein-ligand binding mode. The closer the RMSD is to zero, the better is the docking result [23]. Therefore, the CDOCKER algorithm is appropriate and employed to perform molecular docking studies. The CDOCKER energy (kcal/mol) and CDOCKER interaction energy (kcal/mol) of D99 were 51.30 and 61.78, respectively, which were the scoring function of the CDOCKER algorithm. The CDOCKER energy indicated the energy of the ligand-protein complexes, and the CDOCKER interaction energy represented the energy of the ligands [24]. The interaction between the D99 and the protein was analyzed in detail, which is shown in Figure 2a. D99 could form hydrogen bond interactions with PHE54, SER51, ARG52, SER53, and generated hydrophobic interactions with PHE54, TYR73, VAL179, LEU183, LEU211, and PRO292.

TAK-475 was then successfully docked into the active pocket, which further indicated the docking model was reasonable. The CDOCKER energy and CDOCKER interaction energy of TAK-475 were 55.34 and 74.39, which were both higher than the scores of D99. The interaction between TAK-475 and the active site was further analyzed. TAK-475 formed the hydrogen bond interactions with GLN212, and formed the hydrophobic interactions with PHE54, ALA176, VAL179, LEU183, LEU211, and PRO292 (shown in Figure 2b). D99 and TAK-475 both formed hydrophobic interactions with PHE54, VAL179, LEU183, LEU211, and PRO292. Thus, these amino acids were regarded as key residues, which is consistent with the literature [25,26].

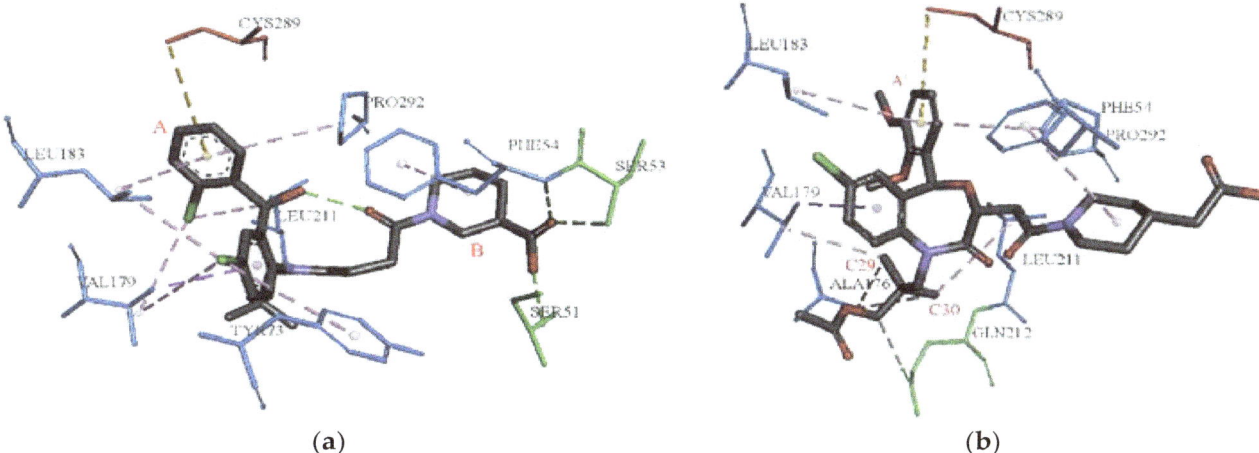

(a) (b)

Figure 2. (**a**) the docking result of the crystallographic ligand with the crystal structure of SQS; (**b**) the docking result of TAK-475; the pink dash line represented hydrophobic effect; the green dash line represented hydrogen bond donor; the green amino acids represent hydrogen bond interactions; blue amino acids represent hydrophobic interactions.

After that, the 22 ligands used in the construction of the pharmacophore model were docked into the binding site of SQS for further demonstrating the key amino acids in receptor-ligand interaction. By counting the frequency of hydrophobic amino acids formed by 22 compounds, the receptor-ligand hydrophobic interactions column diagram shown in Figure 3 was generated. From the result, most of the active compounds could form the hydrophobic interactions with LEU211, VAL179, LEU183, ALA176, PHE54, PRO292, and MET207. This indicated that D99, TAK-475 and the 22 active compounds all could form hydrophobic interactions with PHE54, VAL179, LEU183, LEU211, and PRO292, which were considered to be important key amino acids and used as the reference for selecting potential inhibitors.

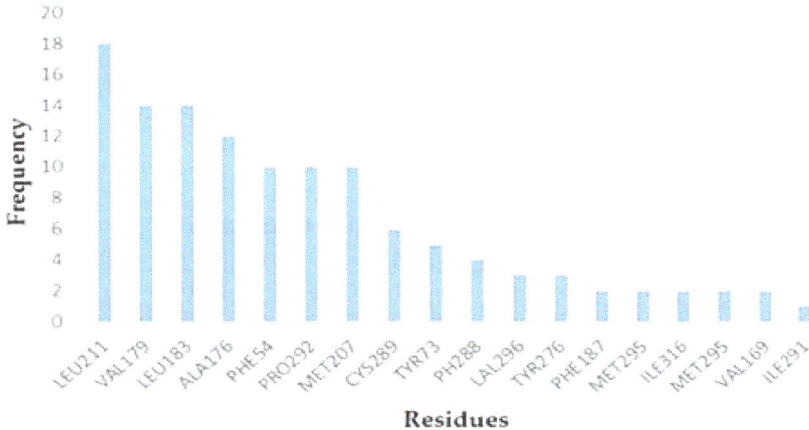

Figure 3. The frequency of hydrophobic amino acids formed by 22 compounds.

Then 1775 drug-like characteristic compounds which were filtered by the optimal pharmacophore model and Lipinski's rules were docked into the binding pocket of SQS. The threshold of the docking score, which is mentioned in the material section of molecular docking, was used to select the potential compounds, and then a hit list of 37 compounds was obtained. Among the 37 potential compounds, cynarin, which got the high docking score and formed an important binding mode with SQS was considered as the most promising candidate.

More specifically, cynarin obtained a CDOCKER energy of 42.08 and CDOCKER interaction energy of 52.92, and formed hydrogen bond interactions with PHE288, GLN212, ARG77, and CYS289, and hydrophobic interactions with PHE54, LEU183, LEU211, and PRO292. The details are shown in Figure 4. In addition, the docked pose of cynarin was screened with the pharmacophore model to further ensure the docked pose fit the pharmacophore model. The result indicated that cynarin was mapped with three features of the optimal pharmacophore model and the fitvalue was 0.66. Moreover, one benzene ring A" of cynarin could form hydrophobic interactions with PHE54, LEU211, and PRO292, which mapped with one H feature in the pharmacophore model. Another benzene ring B" of cynarin formed hydrophobic interactions LEU183 and PRO292, and also mapped with another H feature in the pharmacophore. Compared with D99 and TAK-475, cynarin formed similar hydrophobic interactions with PHE54, LEU183, LEU211, and PRO292. Moreover, the features contained in the pharmacophore model of Hypo1 and the specific hydrophobic interactions formed between cynarin and SQS were consistent. The rationality of our pharmacophore model and molecular model were also confirmed.

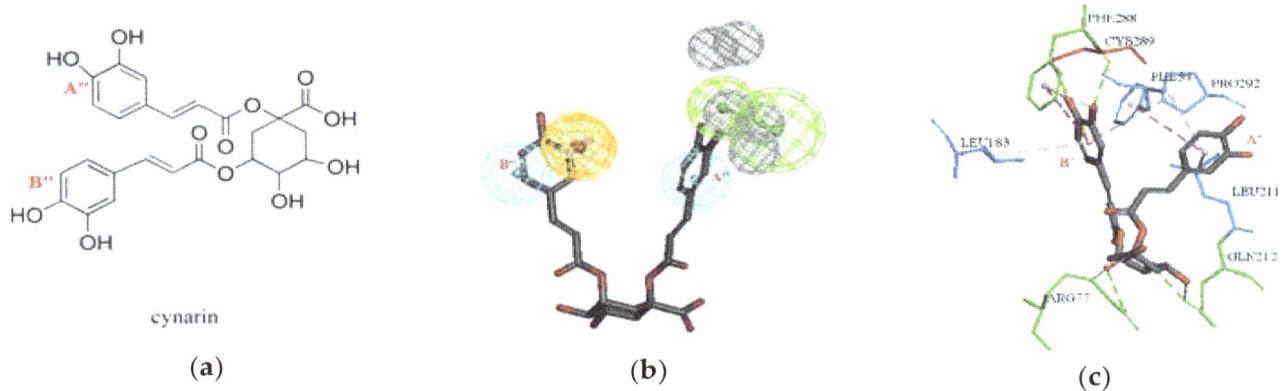

(a)   (b)   (c)

Figure 4. (a) The 2D structures of cynarin; (b) The mapping results of cynarin with Hypo1; (c) the docking result of cynarin with the crystal structure of SQS; the green amino acids represent hydrogen bond interactions; blue amino acids represent hydrophobic interactions.

2.3. MD Simulations

MD simulations were implemented to analyze the binding stability of SQS-cynarin, SQS-D99, and SQS-TAK-475 under dynamic conditions. The RMSD of the protein backbone of each protein-ligand complex were calculated to evaluate the stability of the system [27]. The RMSD trajectories of the SQS-cynarin, SQS-D99 and SQS-TAK-475 complexes were equilibrated after 15 ns (shown in Figure 5a). The root mean square fluctuation (RMSF) was further calculated to evaluate the flexibility of the residues. The results were plotted using residue numbers at the simulation trajectory, which is shown in Figure 5b. It can be observed that the SQS-cynarin complex exhibited a similar RMSF value in comparison to the SQS-D99 and SQS-TAK475 complexes. The protein residues with lower RMSF value are regarded as more stable [28]. Then, by analyzing the flexibility of the important hydrophobic residues, including PHE54, LEU183, LEU211, and PRO292, these amino acids in the cynarin complex had similar RMSF values as in the D99 and TAK-475 complexes (shown in Figure 5c), which were regarded as important and stable hydrophobic interactions between cynarin and SQS.

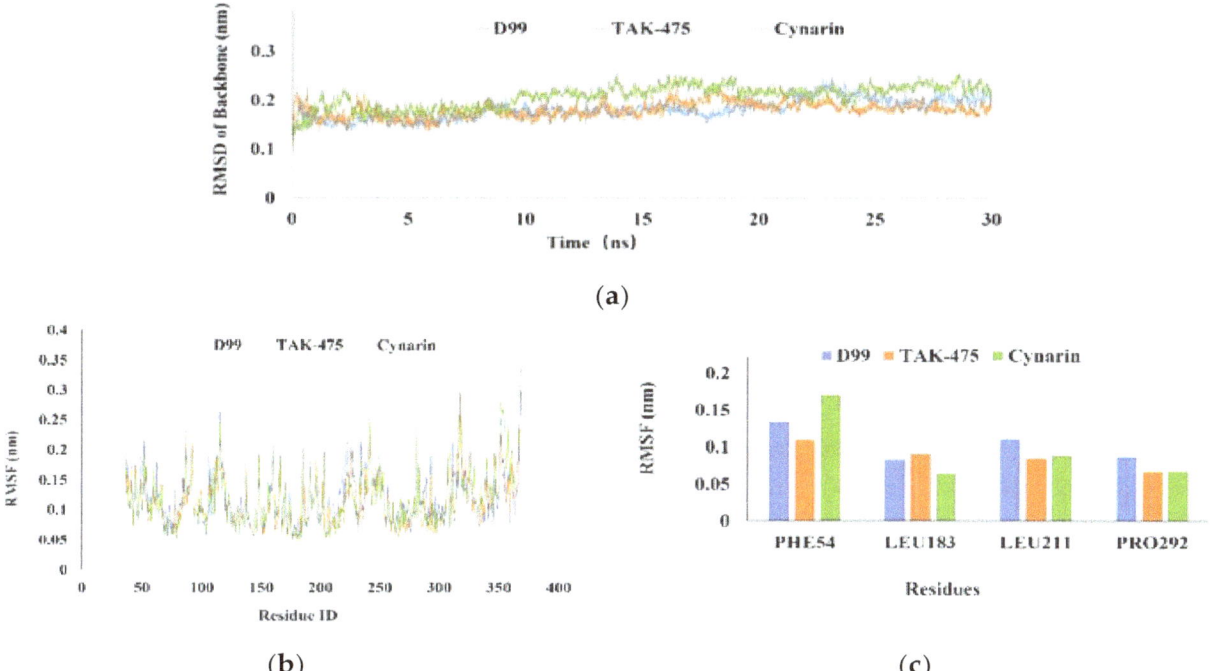

Figure 5. (a) The trajectory of MD simulations of three complexs: average protein RMSD; Blue, red and green bars represent for the data of D99, TAK-475 and cynarin, respectively; (b) Root mean square fluctuation (RMSF) corresponds to MD trajectory; (c) the analysis of hydrophobic residues implicated in docking.

Then the binding free energy of the SQS-cynarin, SQS-D99, and SQS-TAK-475 complexes was calculated by the Molecular Mechanic-Poisson Boltzmann Surface Area (MM-PBSA) with GROMACS v5.0.2 [29], with the results listed in Table 3. The results indicated that SQS-cynarin, SQS-D99, and SQS-TAK-475 complexes possessed a negative binding free energy of -210.39, -253.03 and -285.36 kJ/mol. Moreover, van der Waals, electrostatic interactions and non-polar solvation energy negatively contributed to the total interaction energy, while only polar solvation energy positively contributed to total free binding energy. Thus, the relative binding free energies of the SQS-cynarin, SQS-D99, and SQS-TAK-475 complexes indicated the strong binding in the dynamic system. To obtain a more detailed thermodynamic description of the residue contributions to the binding free energy, we decomposed the binding energy $\Delta G_{\text{MM-GBSA}}$ on a per-residue level depicted in Table 4. The contribution of residue PHE54, LEU183, LEU211, and PRO292 to binding varies from -2.32 to -11.56 kJ/mol, which could be identified as the key residues of SQS. Based on the consensus results among the pharmacophore based virtual screening and the docking/MD simulations, cynarin exhibited a key and stable interaction profile with SQS, being regarded as a potential SQS inhibitor.

Table 3. The binding free energy (kJ/mol) of the three complexes.

Complex	Binding Energy	Van der Waal Energy	Electrostattic Energy	Polar Solvation Energy	SASA Energy
SQS-cynarin	-210.39 ± 11.00	-291.56 ± 10.01	-39.10 ± 1.36	144.01 ± 0.25	-23.83 ± 0.61
SQS-D99	-253.03 ± 4.59	-310.59 ± 13.49	-36.47 ± 1.89	118.63 ± 6.81	-24.60 ± 0.20
SQS-TAK-475	-285.36 ± 6.50	-374.76 ± 7.76	-32.18 ± 0.97	149.79 ± 1.23	-28.20 ± 0.95

Table 4. The contribution of residues to binding free energy (kJ/mol).

Complex	PHE54	LEU183	LEU211	PRO292
SQS-cynarin	-10.56 ± 0.90	-2.32 ± 1.04	-11.55 ± 0.32	-5.40 ± 0.26
SQS-D99	-11.56 ± 0.53	-5.04 ± 0.53	-10.15 ± 0.95	-9.44 ± 1.13
SQS-TAK-475	-7.68 ± 0.49	-6.15 ± 0.10	-11.42 ± 1.39	-8.69 ± 0.96

2.4. Experimental Result

To test the lipid-lowering effect of cynarin (CAS number: 19870-46-3), sodium oleate-induced HepG2 cells were treated with various doses of cynarin (5, 10, 20, 40, and 80 $\mu mol \cdot L^{-1}$), and the positive compound pravastatin, respectively. The control group cells were cultured with only HepG2 cells. The model control group cells were the hyperlipidemia cell model. The positive control group cells were cultured with pravastatin. Firstly, the MTT assay was utilized for the detection of cell viability, with the result shown in Figure 6a. From the result, the five different concentrations of cynarin were not cytotoxic to HepG2 cells compared to the control group ($p > 0.05$).

Then, the lipid-lowering effect of cynarin was evaluated in sodium oleate-induced HepG2 cells, which is shown in Figure 6b. Compared to the control group, the plasma triglyceride (TG) level of the model control group shows a significant difference with the control group ($p < 0.001$), which indicates that the hyperlipidemia cell model could be used for evaluating the lipid-lowering activity of cynarin. In addition, the pravastatin could decrease the TG level compared to the model group ($p < 0.001$), which demonstrated the hyperlipidemia cell model was reliable. From the result, 20 $\mu mol \cdot L^{-1}$ cynarin and 40 $\mu mol \cdot L^{-1}$ cynarin could both decrease the TG level, and there was no difference between these two groups in statistics ($p > 0.05$). However, the result of 20 $\mu mol \cdot L^{-1}$ cynarin for reducing the TG level was more reliable with a higher confidence interval ($p < 0.01$) compared to 40 $\mu mol \cdot L^{-1}$ cynarin ($p < 0.05$). Thus, the optimum concentration of cynarin was 20 $\mu mol \cdot L^{-1}$, which could decrease the TG level by 22.50%. Cynarin was mildly cytotoxic to the sodium oleate-induced HepG2 cells at 80 $\mu mol \cdot L^{-1}$, so it may be speculated that the sodium oleate-induced HepG2 cells were more sensitive compared to normal HepG2 cells. On the basis of the above analysis, cynarin could be a potential SQS inhibitor for the treatment of hyperlipidemia.

Cynarin, also called 1,3-dicaffeoylquinic acid, was identified as a potential SQS inhibitor. Cynarin is a common component of various TCM herbs such as *Cynara scolymus*, *Cynara cardunculus*, and *Senecio nemorensis*. It was proved to have positive pharmacological choleretic, hepatoprotective, anti-atherosclerotic, anti-oxidant, anti-cholinergic, antioxidative, anticarcinogenic effects and so on. To be specific, for the anti-atherosclerotic effects, the researchers demonstrated that cynarin could reduce the nitric oxide synthase (iNOS) activity and cynarin was the most effective with 3 μM [30]. For the hepatoprotective effects, the study with the rat hepatocytes indicated that 3 μM cynarin could reduce *tert*-butylhydroperoxide (t-BPH)-induced malondialdehyde (MDA) production and EC50 value of cynarin was 15.2 $\mu g/mL$ [31]. For the anti-diabetic effects, the study demonstrated the potential antiglycative effects of cynarin in the bovine serum albumin-glycose system, and cynarin could inhibit the ability of advanced glycation end products (AGE) in a dose dependent manner (3 μM–40 μM) [32]. Meanwhile, consulting the literature, there are no reports about drug interactionz between cynarin and other SQS inhibitors. Combining these results with our research, cynarin was proved to be a potential SQS inhibitor, and in view of the extremely low toxicity of the cynarin, which provided a new perspective for the treatment of hyperlipidemia.

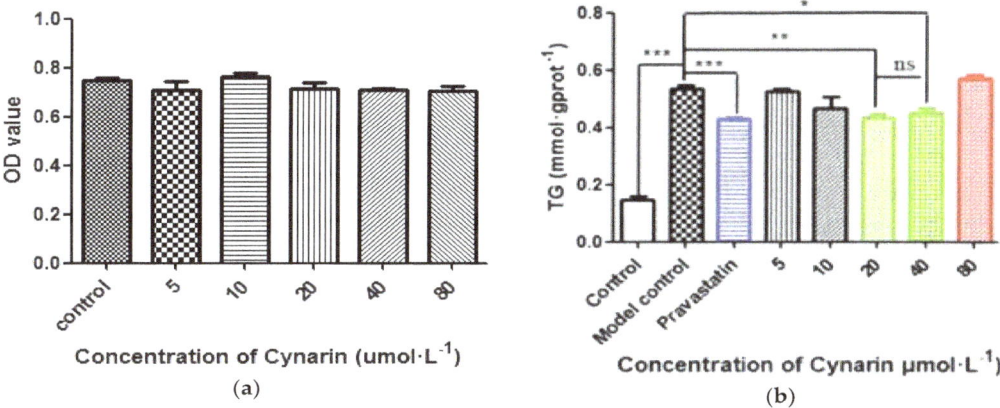

Figure 6. (**a**) Cell-viability of different concentration of cynarin on HepG2 cells by the MTT assay; (**b**) Effect of different concentration of cynarin on the TG content in sodium oleate-induced HepG2 cells (* means $p < 0.05$, ** means $p < 0.01$ and *** means $p < 0.001$ compared with the model control group).

2.5. Anti-Hyperlipidemia Target Identification by Pharmacophore

To provide more evidence for the lipid-lowering effect of cynarin on SQS activity at the molecular level, cynarin was utilized to reversely screen it against the pharmacophore models of other anti-hyperlipidemia targets that exist in HepG2 cells. The fitvalue was used as an important judgment index to represent the overlap degree between the compound and pharmacophore model [33]. According to the screening results, cynarin was unable to map with the pharmacophore models of these commonly used targets, including 3-hydroxy-3-methylglutaryl coenzyme A (HMG-CoA) [34], peroxisome proliferator-activated receptor-α (PPAR-α) [35], liver X receptor β (LXRβ) [36], cholesteryl ester transfer protein (CETP) [37], and microsomal triglyceride transfer protein (MTP) [37], which is shown in Figure S1. The result indicated that the lipid-lowering effects in HepG2 cells of cynarin might due to the inhibition of SQS. In addition, based on the above results, cynarin is regarded as a promising SQS inhibitor candidate and could be explored for the treatment of hyperlipidemia. The biological activity of cynarin against other targets should also be studied in the future research.

3. Materials and Methods

3.1. HipHop Pharmacophore Hypotheses Generation

Among the library compounds 22 active compounds were selected as the training set and were used to generate HipHop pharmacophore models by using DS 4.0 from Accelrys (San Diego, CA, USA). The structure, ID numbers, and biological activity (IC$_{50}$) values of the compounds are shown in Figure 7. Then, 154 active compounds and 462 inactive compounds [38], which selected randomly from the Binding Database, were regarded as the test set in order to validate the pharmacophore model. The 3D structures of all the compounds were generated using the 'Prepare Ligands' module and minimized in CHARMm force field [39]. The conformations of these compounds were created within an energy threshold of 20 kcal/mol by using the BEST method. The maximum ligand conformations were set to 255.

The HipHop pharmacophore models were constructed by extracting the common pharmacological features from the 3D structure features of each compound in the training set [40]. The Principal and MaxOmitFeat values are used to describe the activity of the compounds. The range of "Principle" and "MaxOmitFeat" values are 0, 1 and 2. The "Principal" value is set to 2, representing the superior activity of the compounds. The corresponding "MaxOmitFeat value is set to 0, which indicates that no features that are allowed to be missed for each compound. Then, the "Principal" value is set to 0, indicating the lower activity of the compounds.

The corresponding "MaxOmitFeat value" is set to 2 to suggest that all features can be ignored for these compounds [38]. The maximum excluded volumes (Ev) value was set to 5, and all the other parameters were set at default values. The optimal pharmacophore model was selected based on rank score, *HRA*, *IEI*, and *CAI*. Then, the crystallographic ligand and the positive SQS inhibitor TAK-475 were used to map the optimal model to further evaluate the accuracy of the pharmacophore model. In addition, in order to validate the reliability of the best pharmacophore model, based on 2D fingerprints, similarity search method was utilized to compare the similarity between the 22 ligands and the TAK-475.

The selected optimal pharmacophore model was then utilized to screen potential SQS inhibitors from TCMD [41], before which the TCMD database was filtered based upon the Lipinski's rules for drug–likeness prediction [42]. The list of compounds with drug-like characteristics was regarded as potential SQS inhibitors and were retained for molecular docking study.

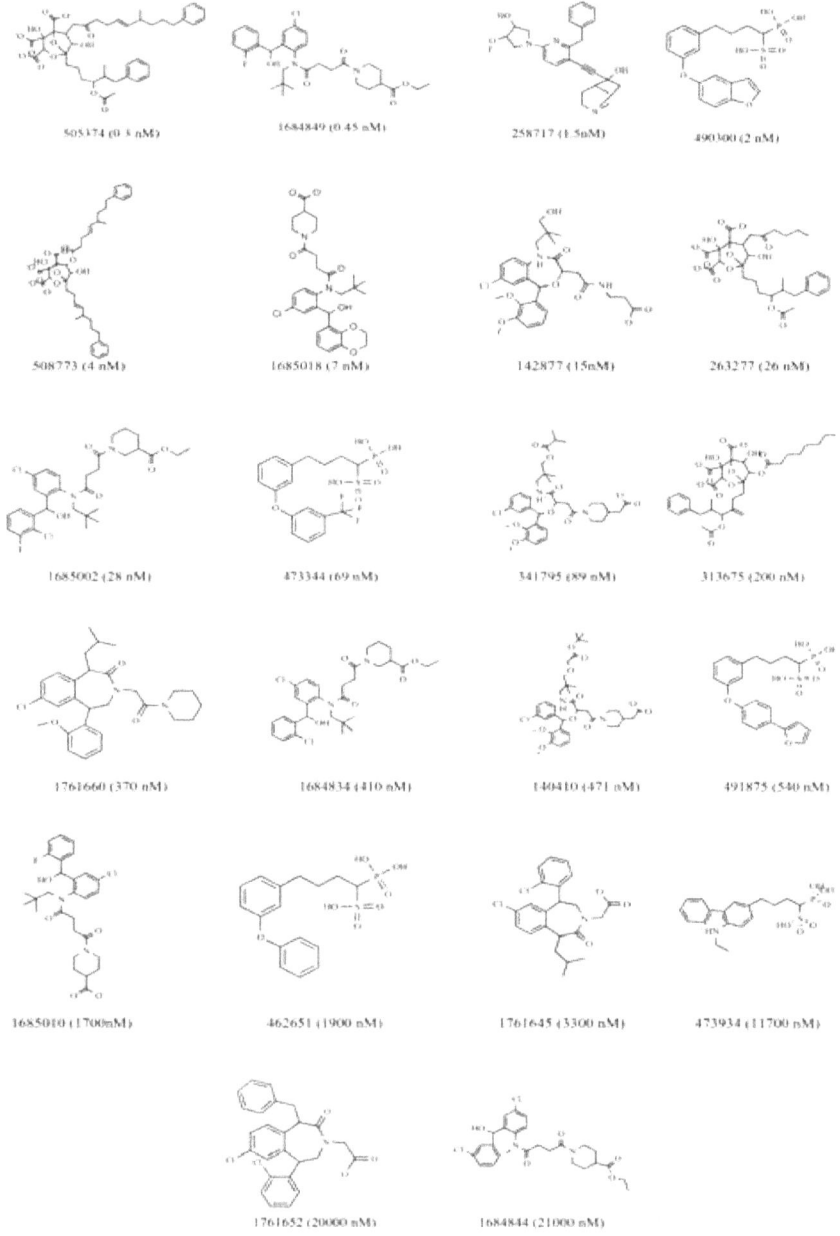

Figure 7. Structures, ID and the value of IC$_{50}$ of 22 compounds in the training set for pharmacophore model generation of SQS.

3.2. Molecular Docking Studies

The crystal structure of the human SQS (PDB entry 3ASX, resolution 2.0 Å) was obtained from the RCSB Protein Data Bank (PDB), which is complexed with an inhibitor, (3R)-1-{4-[{4-chloro-2-[(S)-(2-chlorophenyl)(hydroxy)methyl]phenyl}(2,2-dimethylpropyl)amino]-4-oxobutanoyl}piperidine- 3-carboxylic acid (D99) [43]. The protein was automatically cleaned up by the Prepare Protein protocol for some common problems, such as incomplete residues, the lack of hydrogens, the existence of crystallographic water and ligands [44]. The binding active pocket of 3ASX was determined around the crystallographic ligand using the Define and Edit Binding Site tools. LibDock and CDOCKER, two common docking algorithms, were utilized to evaluate the applicability for the docking studies. The crystallographic ligand D99 was extracted from the active pocket and was then re-docked into the crystal structure by these two docking methods. The docking algorithm with the smallest RMSD was used for the study. In addition, the positive SQS inhibitor TAK-475 was docked into the active pocket of SQS, which further evaluated the rationality of the docking model. Then the interactions between D99, TAK-475 and the active pocket of SQS were analyzed.

After that, the 22 ligands used in the construction of the pharmacophore model were docked into the active binding pocket of SQS to further analyze the key amino acids. Then, the hit compounds screened by the optimal pharmacophore model were docked into the binding site. Eighty percent of the docking scores of D99 were regarded as the threshold value for identifying potential SQS inhibitors from TCMD [45]. Finally, the compounds which got a high docking scores and formed similar interactions to D99 and TAK-475 were obtained to evaluate the stability of the complex.

3.3. MD Simulations

A 30 ns MD simulation was employed to investigate the dynamic binding stability of the complexes with GROMACS v5.0.2 using GROMOS96 43a1 force field [46]. Initially, the topology parameters of SQS were obtained using the GROMACS program and the force field parameters for the three ligands were derived from PRODRG server [47]. In each simulation, the complex was solvated using simple point charge (SPC) water molecules [48] and five sodium ions were added by replacing solvent molecules in order to neutralize the system. Each system consisted of ~22,800 waters molecules and the solvent and ions around the protein were first equilibrated before collecting frames for analysis. The energy minimizations were carried out using the steepest descent method with 5000 steps. The system was then subjected to two phases of equilibration for a period of 1500 ps at 300 K with position restraints on the protein and ligands (fc = 1000). A first 500 ps NVT equilibration was performed using V-rescale thermostat coupling method [49] for temperature control in order to relieve any bad contacts at the residues solvent interface [50]. Then a 1000 ps NPT equilibration was conducted at 1.0 bar using Parrinello-Rahman barostat method [50] for pressure control. Upon the two equilibration phases, the position restraints were released and MD simulations were produced.

By consulting the related literatures, for example, the researchers performed a relatively short time (such as 10–30 ns) MD simulation to evaluate the binding stability during a dynamic environment and analyze the key amino acids by a series of MD analysis tools such as RMSF, RMSD and the total energy [51,52]. It makes sense and contributes to the whole paper for the discovery of the potential compounds. Actually, 30 ns might still be a little short, but literatures have showed it could also give key information for molecular modeling [53,54].

In addition, the MM-PBSA method has been widely utilized to study the receptor-ligand interaction. For the three complexes including SQS-cynarin, SQS-D99, and SQS-TAK-475 system, free energy calculations were performed for 10 snapshots extracted from the last 1 ns stable MD trajectory using g_mmpbsa tool [55]. The MM-PBSA method can be summarized by the following equations.

For each snapshot, the free energy was calculated for each molecular species (complex, protein and ligand) and the binding free energy was computed by Equation (4). The free energy of each component Gx in Equation (4) could be calculated taking in account three terms (Equations (5)–(8)):

$$\Delta G_{binding} = G_{complex} - (G_{protein} + G_{ligand}) \tag{4}$$

$$G_x = E_{MM} + G_{solv} - T\Delta S \tag{5}$$

$$E_{MM} = E_{vdW} + E_{ele} \tag{6}$$

$$G_{solv} = G_{polar} + G_{nonpolar} \tag{7}$$

$$G_{nonpolar} = \gamma SASA + \beta \tag{8}$$

G_{MM}, the molecular mechanics energy, was calculated by the electrostatic and van der waals interactions. G_{solv}, the solvation free energy, was composed of the polar and the nonpolar contributions. Polar solvation free energy could be obtained by solving the Poisson-Boltzmann equation for MM/PBSA method, whereas nonpolar solvation free energy was determined using Solvent Accessible Surface Area (SASA) model. $T\Delta S$ represents the entropy term.

3.4. Experimental Validation

The lipid-lowering activity of the potential compound was evaluated by examining the inhibition of the formation of lipid droplets in HepG2 cells in vitro. Bligh et al. [56] have reported an efficient and rapid method of total lipid extraction and purification. Compared with this method, we used sodium oleate-induced HepG2 cells to generate the lipid droplets [57]. The cells were grown at 37 °C with 5% CO_2 in DMEM solution containing 10% FBS and 1% penicillin/streptomycin.

Then cell viabilities were determined by 3-(4,5-dimethylthiazol-2-yl)-2,5-diphenyltetrazolium bromide (MTT) method [58]. In 96-well plates, HepG2 cells were seeded for 24 h at a density 2×10^4 cells/well, and then incubated at various concentrations of the compounds for another 24 h. Then, each well was treated with 200 uL MTT working solution (5 mg·mL^{-1}) and cultured for a further 4 h. After removing the MTT, 150 uL dimethylsulphoxide (DMSO) was added to each well for terminating response, and the plate was set to the table shaker for 5 min at a low speed. Then the absorbance of cells was measured at 570 nm using microplate reader. The maximum concentration of the compound that can be used for the assay was determined by the MTT cytotoxicity assay in HepG2 cells.

To evaluate the lipid-lowering effect of the potential compounds, HepG2 cells were induced by sodium oleate for establishing a model of hyperlipidemia [59]. The HepG2 cells were seeded in 6-well plates at 20×10^4 cells/well for 24 h. Then, sodium oleate was added into the each well for producing fat accumulation as model cells at 60 μg/mL concentration and incubated at another 24 h. The control group cells were cultured without sodium oleate. It has been proved that SQS inhibitors could reduce TG level through an LDL receptor-independent mechanism [60]. Tavridou et al. [61] demonstrated that SQS inhibitors could significantly reduce the TG level in HepG2 cells. Moreover, other related literature has indicated that SQS inhibitors can decrease the TG level in in vivo experiments [26,62]. To measure the lipidemic parameter triglyceride (TG) level, appropriate kits were utilized to analyze the TG content in HepG2 cells.

3.5. Anti-Hyperlipidemia Target Profiling

Ligand profiler module is an important method to reversely identify the action targets for candidate compound, and it is widely used for drug poly-pharmacology prediction of TCM [63]. In order to further illustrate the lipid-lowering effects of the active compound was caused by the inhibition of SQS at the molecular level, a pharmacophore database of other anti-hyperlipidemia targets, which exist in HepG2 cells, was built to assess the activity of the candidate. This anti-hyperlipidemia database contained five commonly used targets, including HMG-CoA, PPAR-α, LXRβ, CETP and MTP. Initially, diverse conformations of the active compound were generated by BEST mode with 255 conformations, and the relative energy threshold was less than 20.0 kcal/mol. The generated conformations were regarded as query to map with the anti-hyperlipidemia pharmacophore database by flexible searching method.

4. Conclusions

The main purpose of this study was to screen potential SQS inhibitors from Chinese herbs using a series of methods, including molecular modeling methods including pharmacophore model, molecular docking, MD simulations, lipid-lowering experiments in HepG2 cells, and anti-hyperlipidemia target profiling. From the result, cynarin, with high fitvalue, docking scores and predicted to form similar and stable interactions with SQS (as suggested by the MD simulations) was selected as a potential SQS inhibitor. Then, cynarin was investigated for its lipid-lowering effect on sodium oleate-induced HepG2 cells, and it was shown to decrease the lipidemic parameter triglyceride (TG) level by 22.50% using appropriate kits. Finally, to provide more evidence for the lipid-lowering effect of cynarin on SQS activity, cynarin was utilized to reversely identify other anti-hyperlipidemia targets existing in HepG2 cells, where it was unable to map with pharmacophores of these targets, which indicated that lipid-lowering effect of cynarin was due to the inhibition of SQS to some extent.

By the combination of three different computational approaches and biological assays, cynarin was selected as a potential SQS inhibitor and could be explored for the treatment of hyperlipidemia. Furthermore, the established assay of sodium oleate-induced steatosis on HepG2 cells provided a rapid method for evaluating the lipid-lowering effect of other compounds. According to the related literature, it is very difficult and complex to obtain and purify the SQS protein. With the development of biological experimental technique, the follow-up study can be a further validation that cynarin actually targets SQS by western blotting. In conclusion, this study provided a promising SQS inhibitor candidate compound for the treatment of hyperlipidemia. The combination of computational approaches and biological assays contributed to the discovery of active compounds from TCM.

Author Contributions: Y.C. and Y.Z. have been involved in designing the experiment project and wrote the manuscript; X.C. and G.L. contributed to pharmacophore model studies and cultured HepG2 cell; F.L., X.Z. and L.Q. addressed molecular docking and revised the manuscript; W.H. and G.L. analyzed the data. All authors read and approved the final manuscript.

Acknowledgments: The authors gratefully acknowledge the support of this work by the National Natural Science Foundation of China (No. 81573831; No. 81430094; No. 81173522).

References

1. Klop, B.; Elte, J.W.F.; Cabezas, M.C. Dyslipidemia in obesity: Mechanisms and potential targets. *Nutrients* **2013**, *5*, 1218–1240. [CrossRef] [PubMed]
2. Maxfield, F.R.; Tabas, I. Role of cholesterol and lipid organization in disease. *Nature* **2005**, *438*, 612–621. [CrossRef] [PubMed]
3. Masters, B.A.; Palmoski, M.J.; Flint, O.P.; Gregg, R.E.; Wangiverson, D.; Durham, S.K. In vitro myotoxicity of the 3-hydroxy-3-methylglutaryl coenzyme a reductase inhibitors, pravastatin, lovastatin, and simvastatin, using neonatal rat skeletal myocytes. *Toxicol. Appl. Pharmacol.* **1995**, *131*, 163–174. [CrossRef] [PubMed]
4. Ichikawa, M.; Ohtsuka, M.; Ohki, H.; Ota, M.; Haginoya, N.; Itoh, M.; Shibata, Y.; Ishigai, Y.; Terayama, K.; Kanda, A. Discovery of DF-461, a potent squalene synthase inhibitor. *ACS Med. Chem. Lett.* **2013**, *4*, 932–936. [CrossRef] [PubMed]
5. Davidson, M.H. Squalene synthase inhibition: A novel target for the management of dyslipidemia. *Curr. Atheroscler. Rep.* **2007**, *9*, 78–80. [CrossRef] [PubMed]
6. Ichikawa, M.; Ohtsuka, M.; Ohki, H.; Haginoya, N.; Itoh, M.; Sugita, K.; Usui, H.; Suzuki, M.; Terayama, K.; Kanda, A. Discovery of novel tricyclic compounds as squalene synthase inhibitors. *Bioorg. Med. Chem.* **2012**, *20*, 3072–3093. [CrossRef] [PubMed]

7.   Brusselmans, K.; Timmermans, L.; Van, D.S.T.; Van Veldhoven, P.P.; Guan, G.; Shechter, I.; Claessens, F.; Verhoeven, G.; Swinnen, J.V. Squalene synthase, a determinant of raft-associated cholesterol and modulator of cancer cell proliferation. *J. Biol. Chem.* **2007**, *282*, 18777–18785. [CrossRef] [PubMed]

8.   Warchol, I.; Gora, M.; Wysocka-Kapcinska, M.; Komaszylo, J.; Swiezewska, E.; Sojka, M.; Danikiewicz, W.; Plochocka, D.; Maciejak, A.; Tulacz, D. Genetic engineering and molecular characterization of yeast strain expressing hybrid human-yeast squalene synthase as a tool for anti-cholesterol drug assessment. *J. Appl. Microbiol.* **2016**, *120*, 877. [CrossRef] [PubMed]

9.   Xie, W.; Zhao, Y.; Du, L. Emerging approaches of traditional chinese medicine formulas for the treatment of hyperlipidemia. *J. Ethnopharmacol.* **2012**, *140*, 345–367. [CrossRef] [PubMed]

10. Rong, Q.; Jiang, D.; Chen, Y.; Shen, Y.; Yuan, Q.; Lin, H.; Zha, L.; Zhang, Y.; Huang, L. Molecular cloning and functional analysis of squalene synthase 2 (SQS2) insalvia miltiorrhizabunge. *Front. Plant Sci.* **2016**, *7*, 1274. [CrossRef] [PubMed]

11. Zhan, D.; Zhang, Y.; Song, Y.; Sun, H.; Li, Z.; Han, W.; Liu, J. Computational studies of squalene synthase from panax ginseng: Homology modeling, docking study and virtual screening for a new inhibitor. *J. Theor. Comput. Chem.* **2012**, *11*, 1101–1120. [CrossRef]

12. Wang, X.; Ren, Z.; He, Y.; Xiang, Y.; Zhang, Y.; Qiao, Y. A combination of pharmacophore modeling, molecular docking and virtual screening for inos inhibitors from chinese herbs. *Bio. Med. Mater. Eng.* **2014**, *24*, 1315–1322.

13. Jiang, L.; He, Y.; Luo, G.; Yang, Y.; Li, G.; Zhang, Y. Discovery of potential novel microsomal triglyceride transfer protein inhibitors via virtual screening of pharmacophore modelling and molecular docking. *Mol. Simul.* **2016**, *42*, 1223–1323. [CrossRef]

14. Stein, E.A.; Bays, H.; O'Brien, D.; Pedicano, J.; Piper, E.; Spezzi, A. Lapaquistat acetate: Development of a squalene synthase inhibitor for the treatment of hypercholesterolemia. *Circulation* **2011**, *123*, 1974–1985. [CrossRef] [PubMed]

15. Hou, M.; Yan, G.; Ma, X.; Luo, J.; Hou, X.; Zhou, M.; Pu, C.; Han, X.; Zhang, W.; Zhang, M. Identification of hit compounds for squalene synthase: Three-dimensional quantitative structure-activity relationship pharmacophore modeling, virtual screening, molecular docking, binding free energy calculation, and molecular dynamic simulation. *J. Chemom.* **2017**, *31*, e2923. [CrossRef]

16. Fairlamb, I.J.; Dickinson, J.M.; O'Connor, R.; Higson, S.; Grieveson, L.; Marin, V. Identification of novel mammalian squalene synthase inhibitors using a three-dimensional pharmacophore. *Bioorg. Med. Chem.* **2002**, *10*, 2641–2656. [CrossRef]

17. Willett, P. Similarity methods in chemoinformatics. *Annu. Rev. Inf. Sci. Technol.* **2009**, *43*, 1–117. [CrossRef]

18. Johnson, M.A.; Maggiora, G.M. Similarity in chemistry. (book reviews: Concepts and applications of molecular similarity). *Science* **1991**, *252*, 1189.

19. Willett, P. Similarity-based virtual screening using 2d fingerprints. *Drug Discov. Today* **2006**, *11*, 1046–1053. [CrossRef] [PubMed]

20. Yao, S.; Lu, T.; Zhou, Z.; Liu, H.; Yuan, H.; Ran, T.; Lu, S.; Zhang, Y.; Ke, Z.; Xu, J. An efficient multistep ligand-based virtual screening approach for GPR40 agonists. *Mol. Divers.* **2014**, *18*, 183–193. [CrossRef] [PubMed]

21. Sarvagalla, S.; Singh, V.K.; Ke, Y.Y.; Shiao, H.Y.; Lin, W.H.; Hsieh, H.P.; Hsu, J.T.A.; Coumar, M.S. Identification of ligand efficient, fragment-like hits from an hts library: Structure-based virtual screening and docking investigations of 2 h- and 3 h-pyrazolo tautomers for aurora kinase a selectivity. *J. Comput. Aided Mol. Des.* **2015**, *29*, 89–100. [CrossRef] [PubMed]

22. Wu, G.; Robertson, D.H.; Brooks, C.L., 3rd; Vieth, M. Detailed analysis of grid-based molecular docking: A case study of cdocker-a charmm-based md docking algorithm. *J. Comput. Chem.* **2003**, *24*, 1549–1562. [CrossRef] [PubMed]

23. Qiao, L.; Li, B.; Chen, Y.; Li, L.; Chen, X.; Wang, L.; Lu, F.; Luo, G.; Li, G.; Zhang, Y. Discovery of anti-hypertensive oligopeptides from adlay based on in silico proteolysis and virtual screening. *Int. J. Mol. Sci.* **2016**, *17*, 2099. [CrossRef] [PubMed]

24. Duan, Y.T.; Yao, Y.F.; Huang, W.; Makawana, J.A.; Teraiya, S.B.; Thumar, N.J.; Tang, D.J.; Tao, X.X.; Wang, Z.C.; Jiang, A.Q. Synthesis, biological evaluation, and molecular docking studies of novel 2-styryl-5-nitroimidazole derivatives containing 1,4-benzodioxan moiety as Fak inhibitors with anticancer activity. *Bioorg. Med. Chem.* **2014**, *22*, 2947–2954. [CrossRef] [PubMed]

25. Pandit, J.; Danley, D.E.; Schulte, G.K.; Mazzalupo, S.; Pauly, T.A.; Hayward, C.M.; Hamanaka, E.S.; Thompson, J.F.; Jr, H.H. Crystal structure of human squalene synthase. A key enzyme in cholesterol biosynthesis. *J. Biol. Chem.* **2000**, *275*, 30610–30617. [CrossRef] [PubMed]

26. Ladopoulou, E.M.; Matralis, A.N.; Nikitakis, A.; Kourounakis, A.P. Antihyperlipidemic morpholine derivatives with antioxidant activity: An investigation of the aromatic substitution. *Bioorg. Med. Chem.* **2015**, *23*, 7015–7023. [CrossRef] [PubMed]

27. Szefler, B.; Diudea, M.; Putz, M.; Grudzinski, I. Molecular dynamic studies of the complex polyethylenimine and glucose oxidase. *Int. J. Mol. Sci.* **2016**, *17*, 1796. [CrossRef] [PubMed]

28. Junaid, M.; Muhseen, Z.T.; Ullah, A.; Wadood, A.; Liu, J.; Zhang, H. Molecular modeling and molecular dynamics simulation study of the human Rab9 and RhoBTB3 C-terminus complex. *Bioinformation* **2014**, *10*, 757. [CrossRef] [PubMed]

29. Ren, X.; Zeng, R.; Tortorella, M.; Wang, J.; Wang, C. Structural insight into inhibition of CsrA-RNA interaction revealed by docking, molecular dynamics and free energy calculations. *Sci. Rep.* **2017**, *7*. [CrossRef] [PubMed]

30. Xia, N.; Pautz, A.; Wollscheid, U.; Reifenberg, G.; Li, H. Artichoke, cynarin and cyanidin downregulate the expression of inducible nitric oxide synthase in human coronary smooth muscle cells. *Molecules* **2014**, *19*, 3654–3668. [CrossRef] [PubMed]

31. Gebhardt, R. Antioxidative and protective properties of extracts from leaves of the artichoke (*Cynara scolymus* L.) against hydroperoxide-induced oxidative stress in cultured rat hepatocytes. *Toxicol. Appl. Pharmacol.* **1997**, *144*, 279–286. [CrossRef] [PubMed]

32. Z, S.; J, C.; J, M.; Y, J.; M, W.; G, R.; F, C. Cynarin-rich sunflower (*Helianthus annuus*) sprouts possess both antiglycative and antioxidant activities. *J. Agric. Food Chem.* **2012**, *60*, 3260.

33. Kapil, J.; Dara, A.; Elizabeth, S.M. Targeting PKC-β II and PKB connection: Design of dual inhibitors. *Mol. Inf.* **2011**, *30*, 329–344.

34. Nakagawa, S.; Kojima, Y.; Sekino, K.; Yamato, S. Effect of polyphenols on 3-Hydroxy-3-methylglutaryl-Coenzyme a lyase activity in human hepatoma HepG2 cell extracts. *Biol. Pharm. Bull.* **2013**, *36*, 1902. [CrossRef] [PubMed]

35. Takahashi, N.; Kang, M.S.; Kuroyanagi, K.; Goto, T.; Hirai, S.; Ohyama, K.; Lee, J.Y.; Yu, R.; Yano, M.; Sasaki, T. Auraptene, a citrus fruit compound, regulates gene expression as a PPARα agonist in HepG2 hepatocytes. *Biofactors* **2008**, *33*, 25–32. [CrossRef] [PubMed]

36. Hoang, M.H.; Jia, Y.; Jun, H.J.; Lee, J.H.; Lee, D.H.; Hwang, B.Y.; Kim, W.J.; Lee, H.J.; Lee, S.J. Ethyl 2,4,6-trihydroxybenzoate is an agonistic ligand for liver X receptor that induces cholesterol efflux from macrophages without affecting lipid accumulation in HepG2 cells. *Bioorg. Med. Chem. Lett.* **2012**, *22*, 4094. [CrossRef] [PubMed]

37. Tchoua, U.; D'Souza, W.; Mukhamedova, N.; Blum, D.; Niesor, E.; Mizrahi, J.; Maugeais, C.; Sviridov, D. The effect of cholesteryl ester transfer protein overexpression and inhibition on reverse cholesterol transport. *Cardiovasc. Res.* **2008**, *77*, 732–739. [CrossRef] [PubMed]

38. Qiao, L.S.; Zhang, X.B.; Jiang, L.D.; Zhang, Y.L.; Li, G.Y. Identification of potential acat-2 selective inhibitors using pharmacophore, svm and svr from chinese herbs. *Mol. Divers.* **2016**, *20*, 933–944. [CrossRef] [PubMed]

39. Kusuma, S.S.; Tanneeru, K.; Didla, S.; Devendra, B.N.; Kiranmayi, P. Antineoplastic activity of monocrotaline against hepatocellular carcinoma. *Anti-Cancer Agents Med. Chem.* **2014**, *14*, 1237–1248. [CrossRef]

40. Vadivelan, S.; Sinha, B.N.; Rambabu, G.; Boppana, K.; Jagarlapudi, S.A.R.P. Pharmacophore modeling and virtual screening studies to design some potential histone deacetylase inhibitors as new leads. *J. Mol. Graph. Model.* **2008**, *26*, 935–946. [CrossRef] [PubMed]

41. Jiang, L.; Li, Y.; Qiao, L.; Chen, X.; He, Y.; Zhang, Y.; Li, G. Discovery of potential negative allosteric modulators of mGluR5 from natural products using pharmacophore modeling, molecular docking, and molecular dynamics simulation studies. *Can. J. Chem.* **2015**, *93*. [CrossRef]

42. Zhang, X.; Lu, F.; Chen, Y.K.; Luo, G.G.; Jiang, L.D.; Qiao, L.S.; Zhang, Y.L.; Xiang, Y.H. Discovery of potential orthosteric and allosteric antagonists of P2Y1R from chinese herbs by molecular simulation methods. *Evid.-Based Complement. Altern. Med.* **2016**, *2016*, 4320201. [CrossRef] [PubMed]

43. Ichikawa, M.; Yokomizo, A.; Itoh, M.; Usui, H.; Shimizu, H.; Suzuki, M.; Terayama, K.; Kanda, A.; Sugita, K. Discovery of a new 2-aminobenzhydrol template for highly potent squalene synthase inhibitors. *Bioorg. Med. Chem.* **2011**, *19*, 1930–1949. [CrossRef] [PubMed]

44. Jiang, L.; Zhang, X.; Chen, X.; He, Y.; Qiao, L.; Zhang, Y.; Li, G.; Xiang, Y. Virtual screening and molecular dynamics study of potential negative allosteric modulators of mGluR1 from chinese herbs. *Molecules* **2015**, *20*, 12769. [CrossRef] [PubMed]

45. Doman, T.N.; Mcgovern, S.L.; Witherbee, B.J.; Kasten, T.P.; Kurumbail, R.; Stallings, W.C.; Connolly, D.T.; Shoichet, B.K. Molecular docking and high-throughput screening for novel inhibitors of protein tyrosine Phosphatase-1B. *J. Med. Chem.* **2002**, *45*, 2213. [CrossRef] [PubMed]

46. Pol-Fachin, L.; Fernandes, C.L.; Verli, H. GROMOS96 43a1 performance on the characterization of glycoprotein conformational ensembles through molecular dynamics simulations. *Carbohydr. Res.* **2009**, *344*, 491–500. [CrossRef] [PubMed]

47. Schuttelkopf, A.W.; van Aalten, D.M.F. Prodrg: A tool for high-throughput crystallography of protein-ligand complexes. *Acta Crystallogr. D Biol. Crystallogr.* **2004**, *60*, 1355–1363. [CrossRef] [PubMed]

48. Berendsen, H.J.C.; Postma, J.P.M.; Gunsteren, W.F.V.; Hermans, J. Interaction models for water in relation to protein hydration. *Intermol. Forces* **1981**, *14*, 331–342.

49. Bussi, G.; Donadio, D.; Parrinello, M. Canonical sampling through velocity rescaling. *J. Chem. Phys.* **2007**, *126*, 014101. [CrossRef] [PubMed]

50. Nandy, S.K.; Bhuyan, R.; Seal, A. Modelling family 2 cystatins and their interaction with papain. *J. Biomol. Struct. Dyn.* **2013**, *31*, 649–664. [CrossRef] [PubMed]

51. Yang, S.C.; Chang, S.S.; Chen, H.Y.; Chen, Y.C. Identification of potent EGFR inhibitors from TCM database@taiwan. *PLoS Comput. Biol.* **2011**, *7*, e1002189. [CrossRef] [PubMed]

52. Ke, Y.Y.; Singh, V.K.; Coumar, M.S.; Hsu, Y.C.; Wang, W.C.; Song, J.S.; Chen, C.H.; Lin, W.H.; Wu, S.H.; Hsu, J.T. Homology modeling of DFG-in FMS-like tyrosine kinase 3 (FLT3) and structure-based virtual screening for inhibitor identification. *Sci. Rep.* **2015**, *5*. [CrossRef] [PubMed]

53. Rampogu, S.; Baek, A.; Zeb, A.; Lee, K.W. Exploration for novel inhibitors showing back-to-front approach against vegfr-2 kinase domain (4AG8) employing molecular docking mechanism and molecular dynamics simulations. *BMC Cancer* **2018**, *18*, 264. [CrossRef] [PubMed]

54. Fu, Y.; Sun, Y.N.; Yi, K.H.; Li, M.Q.; Cao, H.F.; Li, J.Z.; Ye, F. Combination of virtual screening protocol by *in silico* toward the discovery of novel 4-hydroxyphenylpyruvate dioxygenase inhibitors. *Front. Chem.* **2018**, *6*. [CrossRef] [PubMed]

55. Kumari, R.; Kumar, R.; Lynn, A. G_mmpbsa–a gromacs tool for high-throughput mm-pbsa calculations. *J. Chem. Inf. Model.* **2014**, *54*, 1951–1962. [CrossRef] [PubMed]

56. Bligh, E.L.G.; Dyer, W.J.A. A rapid method of total lipid extraction and purification, Can. J. Biochem. Physiol. 37, 911–917. *Can. J. Biochem. Physiol.* **1959**, *37*, 911–917. [CrossRef] [PubMed]

57. Yang, Y.; Piao, X.; Zhang, M.; Wang, X.; Bing, X.; Zhu, J.; Fang, Z.; Hou, Y.; Lu, Y.; Yang, B. Bioactivity-guided fractionation of the triglyceride-lowering component and *in vivo* and *in vitro* evaluation of hypolipidemic effects of *Calyx seu fructus physalis*. *Lipids Health Dis.* **2012**, *11*, 38. [CrossRef] [PubMed]

58. Lu, F.; Luo, G.; Qiao, L.; Jiang, L.; Li, G.; Zhang, Y. Virtual screening for potential allosteric inhibitors of cyclin-dependent kinase 2 from traditional chinese medicine. *Molecules* **2016**, *21*, 1259. [CrossRef] [PubMed]

59. Ou, T.T.; Hsu, M.J.; Chan, K.C.; Huang, C.N.; Ho, H.H.; Wang, C.J. Mulberry extract inhibits oleic acid-induced lipid accumulation via reduction of lipogenesis and promotion of hepatic lipid clearance. *J. Sci. Food Agric.* **2011**, *91*, 2740. [CrossRef] [PubMed]

60. Hiyoshi, H.; Yanagimachi, M.; Ito, M.; Saeki, T.; Yoshida, I.; Okada, T.; Ikuta, H.; Shinmyo, D.; Tanaka, K.; Kurusu, N. Squalene synthase inhibitors reduce plasma triglyceride through a low-density lipoprotein receptor-independent mechanism. *Eur. J. Pharmacol.* **2001**, *431*, 345–352. [CrossRef]

61. Tavridou, A.; Kaklamanis, L.; Megaritis, G.; Kourounakis, A.P.; Papalois, A.; Roukounas, D.; Rekka, E.A.; Kourounakis, P.N.; Charalambous, A.; Manolopoulos, V.G. Pharmacological characterization in vitro of EP2306 and EP2302, potent inhibitors of squalene synthase and lipid biosynthesis. *Eur. J. Pharmacol.* **2006**, *535*, 34–42. [CrossRef] [PubMed]

62. Kourounakis, A.P.; Matralis, A.N.; Nikitakis, A. Design of more potent squalene synthase inhibitors with multiple activities. *Bioorg. Med. Chem.* **2010**, *18*, 7402. [CrossRef] [PubMed]

63. Shao, Y.; Qiao, L.; Wu, L.; Sun, X.; Zhu, D.; Yang, G.; Zhang, X.; Mao, X.; Chen, W.; Liang, W. Structure identification and anti-cancer pharmacological prediction of triterpenes from *Ganoderma lucidum*. *Molecules* **2016**, *21*, 678. [CrossRef] [PubMed]

10

Be Aware of Aggregators in the Search for Potential Human *ecto*-5′-Nucleotidase Inhibitors

Lucas G. Viviani [1] (iD), **Erika Piccirillo** [1,2] (iD), **Arquimedes Cheffer** [2], **Leandro de Rezende** [1,2], **Henning Ulrich** [2], **Ana Maria Carmona-Ribeiro** [2] (iD) and **Antonia T.-do Amaral** [1,*] (iD)

[1] Departamento de Química Fundamental, Instituto de Química, Universidade de São Paulo, Av. Prof. Lineu Prestes, 748, São Paulo 05508-000, Brazil; lucas.viviani@usp.br (L.G.V.); erika@iq.usp.br (E.P.); lrezende@iq.usp.br (L.d.R.)

[2] Departamento de Bioquímica, Instituto de Química, Universidade de São Paulo, Av. Prof. Lineu Prestes, 748, São Paulo 05508-000, Brazil; arquiqbq@iq.usp.br (A.C.); henning@iq.usp.br (H.U.); mcribeir@iq.usp.br (A.M.C.-R.)

* Correspondence: atdamara@iq.usp.br

Academic Editor: Rebecca C. Wade

Abstract: Promiscuous inhibition due to aggregate formation has been recognized as a major concern in drug discovery campaigns. Here, we report some aggregators identified in a virtual screening (VS) protocol to search for inhibitors of human *ecto*-5′-nucleotidase (*ecto*-5′-NT/CD73), a promising target for several diseases and pathophysiological events, including cancer, inflammation and autoimmune diseases. Four compounds (**A**, **B**, **C** and **D**), selected from the ZINC-11 database, showed IC_{50} values in the micromolar range, being at the same time computationally predicted as potential aggregators. To confirm if they inhibit human *ecto*-5′-NT *via* promiscuous mechanism, forming aggregates, enzymatic assays were done in the presence of 0.01% (v/v) Triton X-100 and an increase in the enzyme concentration by 10-fold. Under both experimental conditions, these four compounds showed a significant decrease in their inhibitory activities. To corroborate these findings, turbidimetric assays were performed, confirming that they form aggregate species. Additionally, aggregation kinetic studies were done by dynamic light scattering (DLS) for compound **C**. None of the identified aggregators has been previously reported in the literature. For the first time, aggregation and promiscuous inhibition issues were systematically studied and evaluated for compounds selected by VS as potential inhibitors for human *ecto*-5′-NT. Together, our results reinforce the importance of accounting for potential false-positive hits acting by aggregation in drug discovery campaigns to avoid misleading assay results.

Keywords: aggregation; promiscuous mechanism; human *ecto*-5′-nucleotidase; virtual screening; enzymatic assays; turbidimetry; dynamic light scattering

1. Introduction

Virtual screening (VS) and high-throughput screening (HTS) approaches have been well established as the main techniques for identification of bioactive compounds as potential drug candidates from large chemical libraries [1–4], showing significant success rates. However, currently it is well recognized that many screened hits are further recognized as not truly actives against their specific biological targets [5–8]. These compounds, usually termed "false hits" or "false positives", act by a variety of mechanisms, including covalent protein reactivity, redox cycling, absorbance and/or fluorescence assay interference, membrane disruption, metal complexation, decomposition in assay buffers and formation of aggregates [8–10]. Thus, their activities do not depend on specific interactions with a binding site on the corresponding target protein. Accordingly, most of them do not show any structure-biological function relationship [10].

Small molecule aggregation, leading to promiscuous inhibition, in particular, seems to be the major source of false-positive results in drug discovery campaigns [5,8]. Molecular aggregates are formed in solution at micromolar or submicromolar concentrations, inhibiting or activating proteins nonspecifically in vitro, mainly by adsorption to protein surfaces [11]. Therefore, compounds classified in the literature as "aggregators" are usually not suitable as drug candidates and their early identification can contribute to save time and money in drug discovery projects [5,6,12].

In order to minimize the impact of this important issue in drug design, computational methods, based mainly on physical and structural properties, have been proposed to identify and predict potential aggregators [5,12–14]. Despite the relevance of these methods, they have had only limited applicability and success rates, since the formation of aggregates depends on many different factors, such as temperature, ionic strength and both inhibitor and target protein concentrations, being very difficult to predict [5,15]. For this reason, such computational models should not be used to filter out potential aggregators from screening libraries, but only to quickly identify compounds that are potentially able to aggregate [5].

Thus, it has been stressed in the literature that the use of experimental procedures is the best way to detect aggregate formation and promiscuous inhibition mechanism in drug discovery projects as early as possible, reducing the number of data reports based on these artifacts [5,6,8]. It has been established that a molecule can be classified as an aggregator when it meets two or more of the following experimental criteria [5,8,11]: (i) attenuated activity in the presence of small amounts of a nonionic detergent, such as 0.01% (v/v) Triton X-100 or 0.025% (v/v) Tween-80 [11]; (ii) formation of aggregate particles in dispersion as detected by DLS [16–18]; (iii) noncompetitive inhibition with high Hill slopes [19]; (iv) attenuated inhibition by increasing target concentration [7,20]; (v) detergent-dependent inhibition of a well-established "counter-screen enzyme" [21], such as AmpC β-lactamase, trypsin or malate dehydrogenase, which show high sensitivity to compound aggregation; (vi) for cell based-assays, decreased activity after centrifugation of the medium, since aggregate particles can be precipitated by centrifugation [22].

Despite the importance of using suitable experimental procedures for detecting aggregation in drug discovery campaigns, so far only a few studies have drawn attention to compounds that showed typical aggregation behavior [2,6,13,23–25]. In addition, in most examples, the promiscuous behavior of some designed inhibitors is investigated just after they have already been reported as promising hits by scientific journals [8].

Here, in order to address and stress the issues of false positives and promiscuous inhibition mechanism in drug discovery campaigns, we describe some promiscuous aggregator inhibitors identified in a VS search for potential inhibitors of human *ecto*-5′-nucleotidase (*ecto*-5′-NT, CD73). *Ecto*-5′-NT is a key-enzyme in purinergic signaling pathways [26], which catalyzes the hydrolysis of AMP into adenosine and phosphate, playing a major role in the control of extracellular adenosine concentrations. Human *ecto*-5′-NT has been recognized as a promising biological target for many diseases and pathophysiological events [27], including cancer [28–32], autoimmune diseases [33], infections [34–36], atherosclerosis [37,38], ischemia-reperfusion injury [39] and central nervous system disorders [40]. Additionally, human *ecto*-5′-NT expression and activity have been used as a prognostic factor for multiple cancer types [41]. Considering its importance for therapy, the screening for *ecto*-5′-NT inhibitors has become urgent. Although numerous studies describing *ecto*-5′-NT inhibitors have been published in the literature [42–50], the corresponding procedures and controls concerning compound aggregation have not been systematically described so far for this target enzyme.

In this study, we observed that four compounds, designed and selected by a VS procedure as specific inhibitors of human *ecto*-5′-NT, significantly lost their inhibitory activities in the presence of 0.01% (v/v) Triton X-100), as well as at a 10-fold enzyme concentration increase. To corroborate these enzymatic study results, turbidimetric assays were performed, strongly suggesting that all these compounds probably form aggregates. In addition, aggregation kinetic studies were done, for one of them, by dynamic light scattering (DLS). These observations suggest typical aggregate formation and reinforce the need to control artifactual inhibition in drug discovery campaigns.

2. Results and Discussion

To search for novel potential human *ecto*-5′-NT inhibitors, a VS consisting of two consecutive filters (pharmacophore and docking complemented by visual inspection) was performed. Initially, a pharmacophore model was built, using LigandScout (Inte:Ligand, Maria Enzersdorf, Austria) [51], based on the 3D crystallographic structure of human *ecto*-5′-NT (in an open conformation) complexed with a peptidonucleoside inhibitor, PSB11552 (PDB code: 4H1Y) [52]. The generated pharmacophore model consists of five chemical features: one aromatic ring, one hydrogen bond donor and three hydrogen bond acceptors (Figure 1). Exclusion volume spheres were also considered, mimicking the cavity environment.

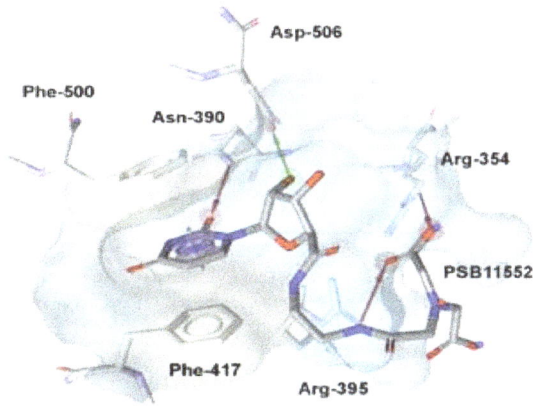

Figure 1. Pharmacophore model generated for PSB11552 complexed with human *ecto*-5′-NT, using LigandScout [51]. Green sphere: hydrogen-bond donor; red spheres: hydrogen-bond acceptors; blue circles: aromatic ring. The surface corresponding to PSB11552 binding site is colored according to lipophilic potential, ranging from white (highest lipophilic area surface) to cyan (highest hydrophilic area surface).

The pharmacophore model was applied to the ZINC-11 database (~23 × 10⁶ compounds) [53], from which 58 compounds matched all pharmacophore features. All of them were submitted to docking into the inhibitor binding site, using ChemPLP scoring function [54], available in GOLD. Subsequently, the best scored docking pose of each compound was submitted to visual inspection. In this last step, the following criteria were considered: (1) observation of mutual surface complementarity between ligand and protein; (2) presence of interactions with key-residues of the inhibitor binding site, specially π-stacking interactions with Phe-500 and Phe-417 side chains; hydrogen-bonds with backbone and/or side chain atoms from Asn-390, Asp-506, Arg-354 and Arg-395; hydrophobic interactions with Phe-500 and Phe-417; cation-π interactions with Arg-354 and Arg-395; (3) presence of additional interactions with residues located near the inhibitor binding site (e.g., hydrophobic interactions with Leu-415, Phe-421, Leu-389 and Thr-446 side chains); and (4) quality of the overall binding conformation to avoid clearly constrained conformations.

Finally, 12 compounds, which met these visual inspection criteria, were selected as potential human *ecto*-5′-NT inhibitors, from which six were purchased and tested by enzymatic inhibition assays for VS experimental validation. Among the tested compounds, four showed IC_{50} values in the micromolar range (compounds **A**, **B**, **C** and **D**; Table 1) and two showed no significant inhibitory activity until *c.a.* 100 μM (i.e., less than 25% inhibition). The corresponding concentration-inhibition/dose-response curves are shown in Figure 2.

Table 1. Chemical structures, physical-chemical properties (molecular weight and cLogP values) and IC_{50} values obtained for four *ecto*-5′-NT inhibitors (**A**, **B**, **C** and **D**) selected by VS.

Compound (ID)	Structure	Molecular Weight (g·mol^{-1})	cLogP [1]	IC_{50} (μM) [2]
A		331.28	2.4	82.9 ± 1.1
B		354.36	4.2	1.9 ± 1.0
C		361.45	3.6	16.3 ± 1.1
D		414.46	4.5	2.2 ± 1.2

[1] Values calculated with LigandScout 4.1 [51], using the topological cLogP estimation algorithm of Wildman and Crippen [55]. [2] Values obtained from a four-parameter logistic nonlinear model used to fit the experimental data from dose-response curves (Figure 2). All experiments were performed in a reaction mixture containing HEPES buffer (10 mM; pH = 7.4), $MgCl_2$ (2 mM), $CaCl_2$ (1 mM), human *ecto*-5′-NT (3.6 nM), AMP (500 μM) as substrate and each tested compound over a range of concentration values (0–500 μM for **A** and 0–100 μM for **B**, **C** and **D**). The concentration of DMSO in all samples was kept at 1.0% (*v/v*). Inorganic phosphate released in the reaction was quantified spectrophotometrically (at λ = 630 nm), using the malachite green method, as described in the literature [56].

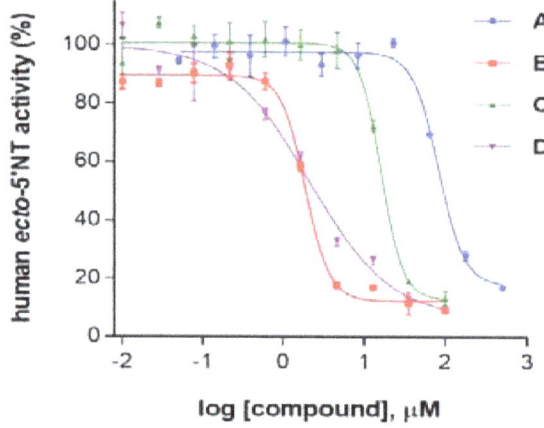

Figure 2. Dose-response curves for each tested compound (**A**, **B**, **C** and **D**). All assays were carried out in a reaction mixture containing HEPES buffer (10 mM; pH = 7.4), $MgCl_2$ (2 mM), $CaCl_2$ (1 mM), human *ecto*-5′-NT (3.6 nM), AMP (500 μM) as substrate, and each tested compound over a range of concentration values (0–500 μM for **A** and 0–100 μM for **B**, **C** and **D**). The concentration of DMSO in all samples was kept at 1.0% (*v/v*). After incubation for 10 min at 37.0 ± 0.2 °C, the reactions were stopped by heating the system for 5 min at 99.0 ± 0.2 °C. Inorganic phosphate released in the reaction was quantified spectrophotometrically (at λ = 630 nm), using the malachite green method, as described in the literature [56]. Data are expressed as the percentage of human *ecto*-5′-NT activity. Each experiment was done in triplicate. A four-parameter logistic non-linear regression model was used to fit the experimental data, using GraphPad Prism (GraphPad, San Diego, CA, USA).

Although four compounds have shown at least moderate inhibitory activities against human *ecto*-5'-NT, one should note that steep concentration-inhibition curves were obtained for **A**, **B** and **C** (Hill slope values of -2.75, -2.90 and -3.19, respectively). For these three compounds, it is observed a sharp transition to almost full inhibition over a narrow range of concentrations (Figure 2). It is described that one possible interpretation for concentration-inhibition curves steepness is inhibition due to aggregation [19]. Additionally, compounds **B**, **C** and **D** have fairly high cLogP values (>3.0; see Table 1), which has also been recognized to be a typical physical chemical feature of aggregate-forming compounds [5].

Thus, to initially verify if the identified inhibitors are prone to aggregate, we used Aggregator Advisor tool (online available at *http://advisor.bkslab.org/*; provided by Shoichet Laboratory, UCSF, San Francisco, CA, USA) [5], which helps to distinguish between true and artifactual screening hits, based on Tanimoto structural similarity index (compared to known aggregators) and on lipophilicity criteria (based on calculated LogP). According to Aggregation Advisor predictions, **A**, **B** and **D** show high structural similarity with aggregators previously reported in the literature, as can be confirmed by their calculated Tanimoto index values (Table 2). Using the same similarity index, compound **C** did not show any structural similarity with aggregators comprised in the Aggregator Advisor database, but was also flagged as a potential aggregator, probably due to its high calculated Log P value (~3.6).

Table 2. Chemical structures of compounds **A**, **B**, **C** and **D**, chemical structures of some previously reported aggregators, and the corresponding Tanimoto similarity index values (%), obtained using Aggregator Advisor tool [5].

Compound (ID)	Structure	Previously Reported Aggregator (Structure)	Tanimoto Similarity Index Value (%) [1]	Reference
A			72	[2]
B			72	[2]
C		n.s.[2]		
D			81	[2]

[1] Values calculated using Aggregator Advisor Tool (online available at *http://advisor.bkslab.org/*) [5]. [2] n.s. means not similar to any compound from Aggregator Advisor database.

These computational predictions findings led us to use experimental controls to further investigate if compounds **A**, **B**, **C** and **D** are truly specific human *ecto*-5'-NT inhibitors or if they in fact act via aggregation. With this purpose, two experiments were initially performed, as suggested in the literature [5,6,8,11]: (i) enzymatic inhibition assays using a nonionic detergent (0.01% (*v/v*) Triton X-100) and (ii) enzymatic inhibition assays with a 10-fold increase in enzyme concentration.

Inhibitory activities of compounds **A** and **C** were almost fully reversed by Triton X-100 addition (Figure 3a,c), as attested by the increase in their corresponding IC_{50} values (from 82.9 ± 1.1 µM to >500 µM for **A** and from 16.3 ± 1.1 µM for >100 µM for **C**). Compounds **B** and **D** had their inhibitory activities partially lost when Triton X-100 was added in the assays (Figure 3b,d), as also can be verified by the increase in their corresponding IC_{50} values (from $1.9 \pm 1.0\ \mu M$ to $2.3 \pm 1.2\ \mu M$ for **B** and from $2.2 \pm 1.2\ \mu M$ to $> c.a.$ 36 μM for **D**). Additionally, it should be emphasized that IC_{50} value calculated for **B** in the presence of Triton X-100 (0.01% (v/v)) is probably underestimated, since the minimum plateau value from its dose-response curve is far from zero, which means that full inhibition was not achieved for this compound (Figure 3b). It is important to report that adenosine diphosphate (ADP), known to be a specific, competitive and well-behaved inhibitor of mammalian *ecto*-5'-NT [57,58], was used as a negative control for aggregation studies. As expected, addition of detergent did not significantly affect ADP inhibitory activity against human *ecto*-5'-NT (Figure 3e), as attested by the IC_{50} values obtained in the absence ($29.7 \pm 1.2\ \mu M$) and in the presence ($31.7 \pm 1.2\ \mu M$) of 0.01% (v/v) Triton X-100).

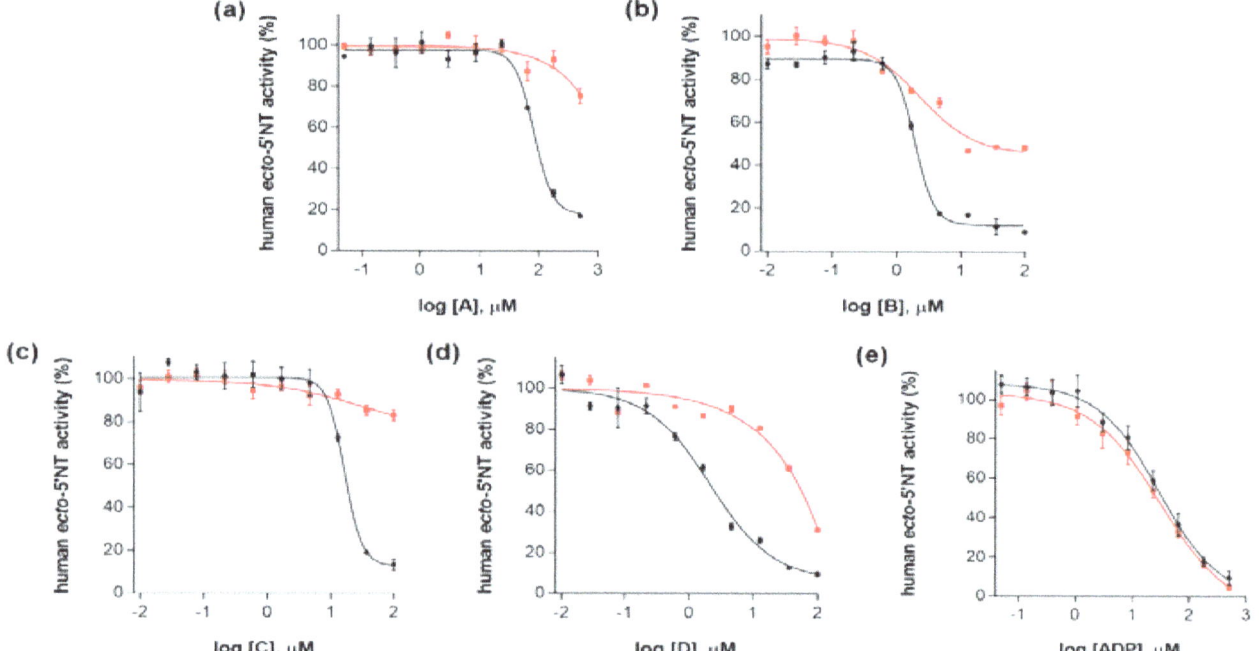

Figure 3. Dose-response curves for (**a**) compound **A**; (**b**) compound **B**; (**c**) compound **C**; (**d**) compound **D** and (**e**) ADP (negative control), without 0.01% (v/v) Triton X-100 (curves in black) and with 0.01% (v/v) Triton X-100 (curves in red). All assays were carried out in a reaction mixture containing HEPES buffer (10 mM; pH = 7.4), $MgCl_2$ (2 mM), $CaCl_2$ (1 mM), human *ecto*-5'-NT (3.6 nM), AMP (500 μM) as substrate, and tested compound over a range of concentration values (0–500 μM for **A** and ADP; and 0–100 μM for **B**, **C** and **D**), with or without 0.01% (v/v) Triton X-100. After incubation for 10 min at 37.0 ± 0.2 °C, the reactions were stopped by heating the system for 5 min at 99.0 ± 0.2 °C. Inorganic phosphate released in the reaction was quantified spectrophotometrically (at $\lambda = 630$ nm), using the malachite green method, as described in the literature [56]. For compounds **A**–**D**, the concentration of DMSO in all samples was kept at 1.0% (v/v). Data are expressed as the percentage of human *ecto*-5'-NT activity. Each experiment was done in triplicate. A four-parameter logistic non-linear regression model was used to fit the experimental data, using GraphPad Prism (GraphPad, San Diego, CA, USA).

These results suggest that the inhibitory activities of compounds **A**, **B**, **C** and **D** can be attributed, at least in part, to aggregate formation. According to the aggregation model proposed for protein inhibition, when an aggregate specie is formed in solution, proteins adsorb to its surface, being partially denatured, which leads to nonspecific inhibition [5,11]. Addition of a non-ionic detergent, such as Triton X-100, can disrupt the aggregates, leading to inhibitory activity loss [5,6,11].

In agreement with our results obtained using 0.01% (v/v) Triton X-100, inhibitory activities of compounds **A**, **B**, **C** and **D** were, at least, partially lost when human *ecto*-5'-NT concentration was increased by 10-fold (i.e., from 3.6 nM to 36 nM). For compound **B**, IC_{50} value has increased from

1.9 ± 1.0 μM (at 3.6 nM *ecto*-5'-NT) to $> c.a.$ 36 μM (at 36 nM *ecto*-5'-NT). Compound **D** had its IC_{50} value increased from 2.2 ± 1.2 μM (at 3.6 nM *ecto*-5'-NT) to $> c.a.$ 36 μM (at 36 nM *ecto*-5'-NT). The IC_{50} values for compounds **A** and **C** at 36 nM of human *ecto*-5'-NT could not be properly obtained, since the minimum plateau values from their corresponding dose-response curves are far from zero (Figure 4a,c, curves colored in red). Nevertheless, it is reasonable to consider that the inhibitory activity for these two compounds were also reduced, by comparing their corresponding dose-response curves obtained at 3.6 nM (colored in black) and at 36 nM of human *ecto*-5'-NT (colored in red) (Figure 4a,c).

The partial loss of inhibitory activity observed for compounds **A**–**D**, when enzyme concentration was increased from 3.6 nM to 36 nM, suggests inhibition due to aggregation. It is well known that enzyme concentration dependence is typically observed for aggregate-based inhibitors [11,20], since the molar ratio of aggregate particles to enzyme is much lower than the corresponding molar ratio of a well-behaved inhibitor to enzyme. Accordingly, a considerable increase ($\geq$10-fold) in enzyme concentration easily overwhelms the ability of aggregate particles to inhibit enzymatic activity [11,20].

Not surprisingly, for the negative control (ADP), the IC_{50} value obtained when the concentration of the enzyme was increased by 10-fold (27.2 ± 1.1 μM) was comparable to that obtained using 3.6 nM *ecto*-5'-NT (29.7 ± 1.2 μM) (Figure 4e). This observation agrees with the assumption that even a 10-fold increase in human *ecto*-5'-NT concentration was not enough to significantly affect the free concentration of ADP, a well behaved competitive inhibitor, which was present at micromolar concentrations.

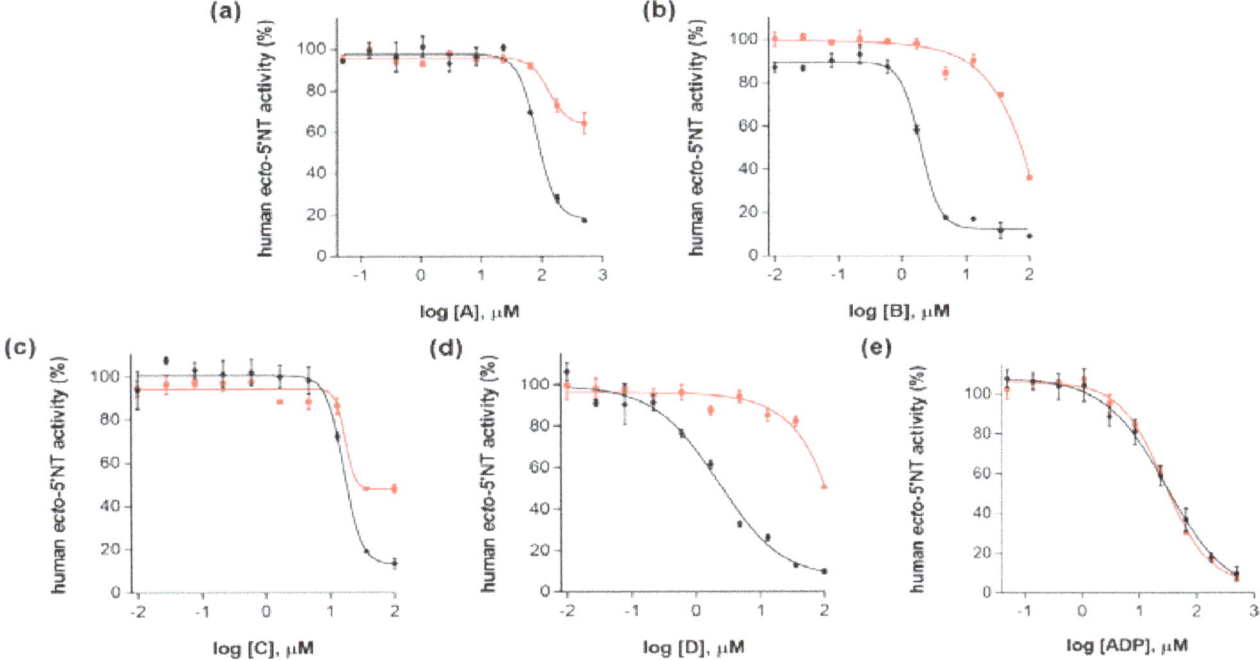

Figure 4. Dose-response curves for (**a**) compound **A**; (**b**) compound **B**; (**c**) compound **C**; (**d**) compound **D** and (**e**) ADP (negative control), at 3.6 nM (curves in black) and at 36 nM human *ecto*-5'-NT (curves in red). All assays were carried out in a reaction mixture containing HEPES buffer (10 mM; pH = 7.4), $MgCl_2$ (2 mM), $CaCl_2$ (1 mM), human *ecto*-5'-NT (3.6 nM or 36 nM), AMP (500 μM) as substrate, each tested compound over a range of concentration values (0–500 μM for **A** and ADP; and 0–100 μM for **B**, **C** and **D**), with or without 0.01% (*v/v*) Triton X-100. After incubation for 10 min at 37.0 ± 0.2 °C, the reactions were stopped by heating the system for 5 min at 99.0 ± 0.2 °C. Inorganic phosphate released in the reaction was quantified spectrophotometrically (at λ = 630 nm), using the malachite green method, as described in the literature [56]. For compounds **A**–**D**, the concentration of DMSO in all samples was kept at 1.0% (*v/v*). Data are expressed as the percentage of human *ecto*-5'-NT activity. Each experiment was done in triplicate. A four-parameter logistic non-linear regression model was used to fit the experimental data, using GraphPad Prism 7.0 (GraphPad, San Diego, CA, USA).

To support our findings based on enzymatic assays, turbidimetric assays were done. As shown in Figure 5, from a critical concentration value, turbidity measured at 400 nm starts increasing, suggesting aggregation. This value corresponds to the estimated compound solubility in the assay buffer (Table 3). Interestingly, a reasonable correlation was observed between compound solubility and the corresponding predicted cLogP value for **A** and **D**. Compound **A**, which has the lower cLogP value (2.4), has the highest estimated solubility (79.1 μM). Compound **D**, in contrast, has been predicted to be the most lipophilic one (cLogP = 4.5) and shows the lowest estimated solubility (lower than 0.5 μM).

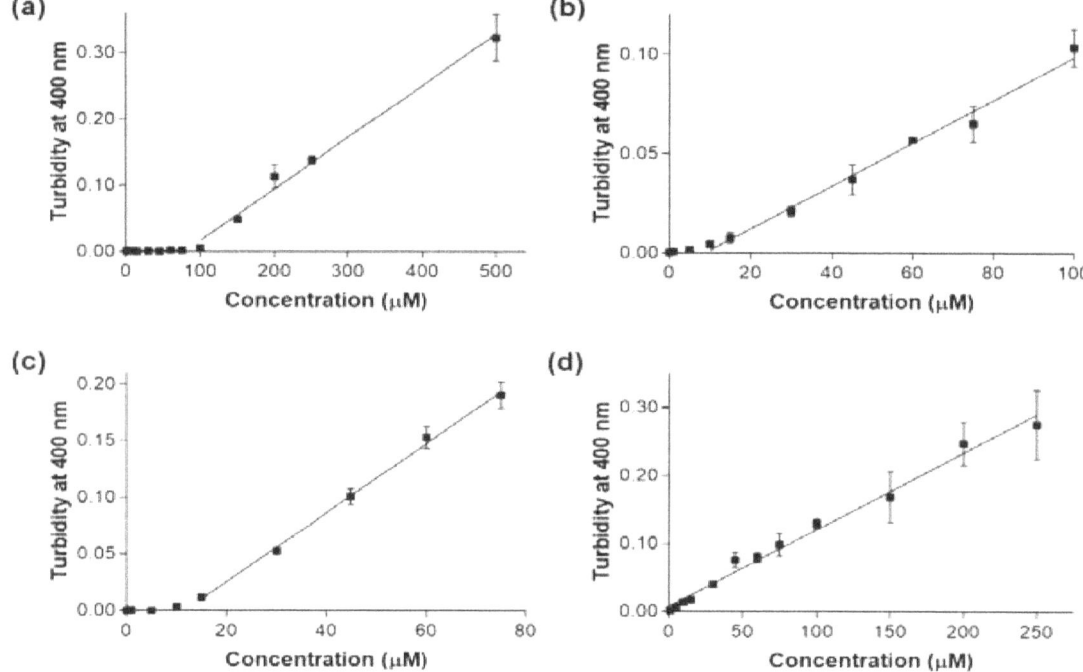

Figure 5. Turbidity at 400 nm as a function of concentration values measured for (**a**) compound **A**, (**b**) compound **B**, (**c**) compound **C** and (**d**) compound **D**. All solutions were prepared in HEPES buffer (10 mM, pH = 7.4) containing MgCl$_2$ (2 mM) and CaCl$_2$ (1 mM) salts. The final DMSO concentration in each sample was 1.0% (v/v). Each experiment was performed in triplicate.

Table 3. cLogP and estimated solubility values for each compound (**A**, **B**, **C** and **D**).

Compound (ID)	cLogP [1]	Estimated Solubility (μM) [2]
A	2.4	79.1
B	4.2	8.8
C	3.6	11.7
D	4.5	< 0.5 *

[1] Values calculated with LigandScout [51], using the topological cLogP estimation algorithm of Wildman and Crippen [55]. [2] Values calculated from turbidimetric solubility assays (Figure 5). * The estimated solubility could not be accurately calculated for compound **D**, due to method sensitivity limitations.

Additionally, turbidity at 400 nm as a function of time was followed for compounds **A–D** (Figure 6a). The concentration of each compound in these assays was near to the maximum that could be obtained, so that DMSO concentration was kept at 1.0% (v/v) in the assay buffer. For compounds **A**, **B** and **D**, a decrease in turbidity is observed as a function of time, in agreement with precipitation of these compounds verified in the assay buffer. In fact, after 60 min, precipitates at the bottom of the cuvettes were clearly observed by visual inspection (data not shown). Precipitation itself revealed rapid and massive aggregation with formation of heavy and large aggregates. For compound **C**, aggregate particle size slowly increased with time as shown by means of DLS (Figure 6b).

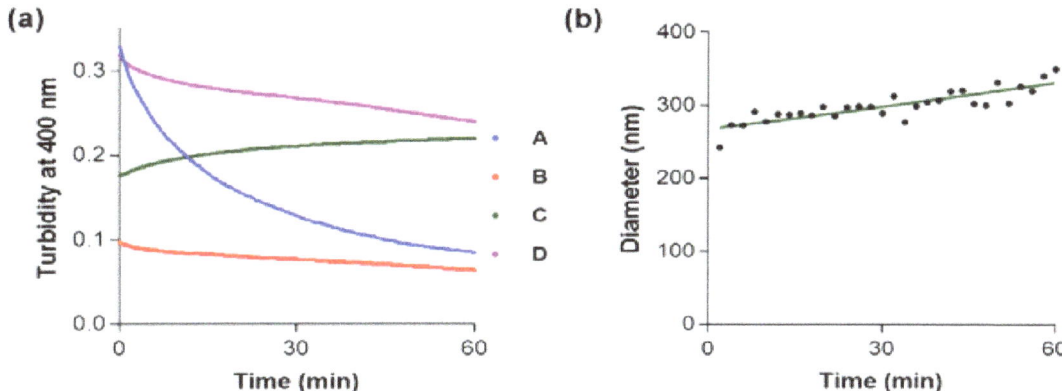

Figure 6. (**a**) Turbidity at 400 nm as a function of time measured for compounds **A**, **B**, **C** and **D** (at 500 μM, 100 μM, 80 μM and 250 μM, respectively). Each solution was prepared in HEPES buffer (10 mM), containing $MgCl_2$ (2 mM) and $CaCl_2$ (1 mM), pH = 7.4. Final concentration of DMSO in each sample was 1.0% (*v/v*); (**b**) Mean diameter (*D*) values as a function of time for compound **C** (80 μM) as determined by DLS. A solution of **C** was prepared in HEPES buffer (10 mM), containing $MgCl_2$ (2 mM) and $CaCl_2$ (1 mM), pH = 7.4. Final concentration of DMSO in each sample was 1.0% (*v/v*).

To expand our analysis concerning aggregation-based inhibition in the search for *ecto*-5'-NT inhibitors, we further analyzed 49 known *ecto*-5'-NT inhibitors described in the literature [42–44,46,52,57,58] to verify if they would be flagged as potential aggregators, using the Aggregator Advisor tool. These inhibitors were clustered considering: (i) structural similarity with compounds previously described as aggregators and (ii) calculated LogP values (Tables S1–S3, Supplementary Material). We observed that 12 of them (~25%), grouped as Cluster **1** (Table S1), were not flagged as potential aggregators since they are not structurally similar to any known aggregator and have calculated LogP values lower than 3.0. Cluster **2** (Table S2) includes 32 compounds (~65%), which are structurally similar to one aggregator from the database, but have calculated LogP values lower than 3. A critical analysis of the structures from this cluster reveals that the majority of them have a negatively charged or a polar group (compounds **LIT-13** to **LIT-43**, Table S2), which probably contributes to make them more hydrophilic. For this reason, they are likely not prone to aggregate. Alarmingly, however, one of the compounds from this cluster is quercetin, a well-known aggregator [2,13,21]. Cluster **3** (Table S3) comprises 5 compounds (~10%), which are not similar to previously described aggregators, but were appointed as possible aggregators due to their fairly high calculated LogP values (>3.0). Despite all these compounds contain a polar group in their structures, some of them have calculated LogP values up to 4.0. In summary, this preliminary analysis of known *ecto*-5'-NT inhibitors [42–44,46,52,57,58], using only a computational tool, warns the scientific community about the necessity to perform further experimental assays, in a systematic way, to discard the possibility of false-positive results among the human *ecto*-5'-NT inhibitors already described in the literature [42–44,46,52,57,58].

Taken together, the results obtained in our study suggest that the inhibitory activity of compounds **A**, **B**, **C** and **D**, selected by a VS protocol as potential human *ecto*-5'-NT inhibitors, can be explained, at least in part, by aggregation taking place over a range of micromolar concentrations. Thus, most likely these compounds are false-positive and promiscuous hits, which inhibit human *ecto*-5'-NT nonspecifically. To the best of our knowledge, they have not been previously reported as aggregators in the literature. One should notice that compound **C** was not shown to be significantly structurally similar to any other compound from the Aggregator Advisor tool database, despite its similarity with compounds **B** and **D** (Tanimoto similarity index values of 62% and 63%, respectively), which were recognized to be structurally similar to an aggregator from Aggregator Advisor (Table 2). These observations reinforce that computational methods to "advise" aggregation are constantly under development and should always be complemented by experimental procedures. Additionally, compounds **A**, **B** and **D** themselves are not reported as aggregators in Aggregator Advisor, despite their relatively high Tanimoto similarity index values in relation to previously reported aggregators (Table 2). In this respect, this study provides novel data and information to feed Aggregator Advisor tool as well as other knowledge-based devices, thus contributing to increase the prediction power of such computational methods, which have been continuously refined over time.

For the first time, we describe aggregators identified on a VS search for human *ecto*-5′-NT. Due to its key role in purinergic signaling pathways regulation, *ecto*-5′-NT has been recognized as a promising biological target for multiple diseases and pathophysiological events, including cancer, autoimmune diseases, inflammation, infections and ischemia-reperfusion injury. The remarkable efforts that have been made by scientific community towards discovery of novel *ecto*-5′-NT inhibitors can be attested by the numerous studies that account for potential bioactive compounds and/or drug candidates targeting this enzyme [42–47,49,59]. Despite the encouraging results obtained by most of them, controls for inhibitors aggregation and/or precipitation have not been systematically reported so far.

Finally, our study reinforces the importance of performing accurate experimental procedures to control for aggregation as a fundamental step in experimental validation of VS results. Although it has been well accepted in the drug discovery community that identifying artifactual inhibition due to aggregation as early as possible is essential to save time and money, just a few studies have directly addressed this issue.

3. Materials and Methods

Materials. Purified recombinant human *ecto*-5′-nucleotidase was obtained from OriGene Technologies, Inc (Rockville, MD, USA); adenosine monophosphate (≥99%), adenosine diphosphate (≥99%), calcium chloride dihydrate (≥99%) and Triton X-100 were obtained from Sigma Aldrich, Inc (St. Louis, MO, USA); compound **A** ([(2,6-difluorophenyl)carbamoyl]methyl 1*H*-indazole-3-carboxylate) was obtained from Enamine Ltd (Kiev, Ukraine); compound **B** (*N*-(6-fluoro-1,3-benzothiazol-2-yl)-3-(2-hydroxyphenyl)-1*H*-pyrazole-5-carboxamide) was obtained from Pharmex, Ltd (Moscow, Russia); compound **C** (3-(2-hydroxy-3,5-dimethylphenyl)-*N*-[5-(methylsulfanyl)-1,3,4-thiadiazol-2-yl]-1*H*-pyrazole-5-carboxamide) and compound **D** (5-(2-hydroxyphenyl)-*N*-(6-methanesulfonyl-1,3-benzothiazol-2-yl)-1*H*-pyrazole-3-carboxamide) were obtained from Vitas-M Laboratory, Ltd, (Champaign, IL, USA); HEPES (2-[4-(2-hydroxyethyl)piperazin-1-yl]ethanesulfonic acid) (high purity grade) was obtained from Amresco, Inc (Solon, OH, USA); magnesium chloride anhydrous (≥99.9%) was obtained from USBiological Life Sciences, Co (Salem, MA, USA); green malachite oxalate, ammonium molybdate tetrahydrate (99%) and polyvinyl alcohol 98–99% hydrolyzed, high molecular weight, were obtained from Alfa Aesar, Co (Tewksbury, MA, USA); dimethyl sulfoxide (DMSO) was obtained from Merck, KGaA (Darmstadt, Germany).

Turbidimetric assays were done using a Hitachi U-2010 spectrophotometer (Hitachi, Chiyoda, Tokyo, Japan). DLS analysis was done using a Zeta Plus Zeta-Potential Analyzer (Brookshaven Instruments Corporation, Hotsville, NY, USA) equipped with a 570 nm laser for dynamic light scattering at 90°. For enzymatic assays, absorbance measurements were done using a FlexStation 3 Multi-Mode Microplate Reader (Molecular Devices, LLC, San Jose, CA, USA).

Aggregator Advisor tool (available online on *http://advisor.bkslab.org/*; provided by Shoichet Laboratory, UCSF, San Francisco, CA, USA) [5] was used to predict potential aggregators.

ZINC-11 database (~23 × 10⁶ compounds) [53] was used for virtual screening.

Virtual screening. In a first step, a pharmacophore model (generated using the LigandScout 4.1 program, Inte:Ligand GmbH, Maria Enzersdorf, Austria; *www.inteligand.com*) [51], based on the available crystallographic 3D structure of human *ecto*-5′-NT complexed with a peptidonucleoside inhibitor (PSB11552) (PDB code: 4H1Y) [52], was generated and applied to the ZINC-11 database (conformers generated by OMEGA 2.4.3 program, OpenEye Scientific Software, Santa Fe, NM, USA) [60]). H-bond acceptor and donor features have 1.95 Å tolerance radius and the aromatic ring feature has 0.90 Å tolerance radius. Exclusion volume spheres were created based on the binding-site residues positions. Subsequently, compounds from ZINC-11 that matched all pharmacophore features were docked into human *ecto*-5′-NT adenosine binding site (in *ecto*-5′-NT open conformation), using GOLD 5.2 (CCDC, Cambridge, UK) [61], scoring function ChemPLP [54]. The binding site was defined as a sphere with 10 Å radius, centered at $X = 13.817$; $Y = 11.61$ and $Z = 37.81$. In all docking calculations, GOLD default settings were applied, using the maximum search efficiency. For each compound, 10 docking runs were performed. Finally, the best pose of each docked compound was subjected to a visual inspection and those that best fitted into adenosine binding site were selected as potential *ecto*-5′-NT inhibitors.

LogP values calculation. cLogP (n-octanol/water as partition model system) values were obtained with LigandScout 4.01 [51], using the topological cLogP estimation algorithm of Wildman and Crippen [55].

Tanimoto index values calculation. Instant JChem was used for calculating the Tanimoto values between compounds **C** and **B** and **C** and **D** applying the default Chemical Hashed Fingerprint, Instant JChem 18.13.0, ChemAxon (Budapest, Hungary) (https://www.chemaxon.com).

Enzyme inhibition assays (without Triton X-100). Following procedures described in the literature [62], with some modifications, all assays were carried out in a reaction mixture containing HEPES buffer (10 mM; pH = 7.4), $MgCl_2$ (2 mM), $CaCl_2$ (1 mM), human *ecto*-5′-NT (3.6 nM), AMP (500 μM) as substrate and variable concentration of each tested compound (from 0 to 500 μM for **A** and from 0 to 100 μM for **B**, **C** and **D**). Stock solutions of each compound were prepared in DMSO. The final concentration of DMSO in all samples/assays/experiments was 1.0% (v/v). Results were controlled for the effect of DMSO on enzymatic activity. After incubation for 10 min at 37.0 ± 0.2 °C, the reactions were stopped by heating the system for 5 min at 99.0 ± 0.2 °C. Inorganic phosphate concentrations were quantified spectrophotometrically (at $\lambda = 630$ nm), using the malachite green method, as described in the literature [56]. Each experiment was done in triplicate. A four-parameter logistic non-linear regression model was used to fit the experimental data, using GraphPad Prism 7.0 (GraphPad, San Diego, CA, USA). From the corresponding fitted curves, we obtained the IC_{50} values, except when the minimum plateau value from the dose-response curve was far from zero. For such curves, IC_{50} ranges were estimated based on the inhibition (%) achieved at the maximum tested concentration.

Promiscuous inhibition mechanism aggregation studies: As proposed in the literature [5,8,11], promiscuous inhibition mechanism was analyzed through the following experiments:

(*i*) Non-ionic detergent-sensitivity evaluation: For each compound (**A**, **B**, **C** and **D**), enzyme inhibition assays were done, similarly as described above, using however Triton X-100 (a non-ionic detergent) at a final concentration of 0.01% (v/v) in the reaction mixture.

(*ii*) Enzyme concentration sensitivity evaluation: For each compound (**A**, **B**, **C** and **D**), enzyme inhibition assays were done, similarly as described above, but using human *ecto*-5′-NT at 36 nM (increased by 10-fold).

(*iii*) Turbidimetric solubility assays: Solutions of each compound (**A**, **B**, **C** and **D**) were prepared at multiple concentrations by diluting concentrated DMSO stock solutions into HEPES buffer (10 mM, pH = 7.4) containing $MgCl_2$ (2 mM) and $CaCl_2$ (1 mM) salts. The final DMSO concentration in each sample was 1.0% (v/v). Increased turbidity (light scattering) was measured at 400 nm, since all compounds have absorbance peaks below this wavelength. Each sample was prepared and measured in triplicate. All measurements were done using a Hitachi U-2010 spectrophotometer.

(*iv*) Dynamic light scattering (DLS): Particle size (mean zeta-average diameter D) for compound **C** was determined using a Zeta Plus Zeta-Potential Analyzer (Brookshaven Instruments Corporation, Hotsville, NY, USA) equipped with a 570 nm laser for dynamic light scattering at 90° [63]. Solutions of Compound **C** (80 μM) were prepared in HEPES buffer (10 mM), pH = 7.4. The final concentration of DMSO in each sample was 1.0% (v/v).

4. Conclusions

This study reports the identification of four false positive hits selected on a VS search for human *ecto*-5′-NT inhibitors. These compounds inhibited human *ecto*-5′-NT nonspecifically, most likely acting by aggregate formation, as suggested by computational predictions and confirmed by experimental procedures, including non-ionic detergent-based assays, evaluation of enzyme concentration effect on inhibitory activity, turbidimetric assays and, eventually, DLS experiments. To the best of our knowledge, none of the identified compounds has previously been reported as an aggregator in the literature. For the first time, the aggregation and promiscuous inhibition issues were systematically studied and evaluated for compounds selected as potential inhibitors of human *ecto*-5′-NT (CD73), an enzyme that has increasingly attracted attention of scientific community due to its potential as a biological target for many diseases and pathophysiological conditions, especially inflammation, immune imbalance and cancer.

Together, the results and data reported here reinforce the importance of performing accurate experimental procedures to identify aggregators, which are recognized as a major source of false-positives in drug discovery campaigns. Early identification of aggregate-forming compounds, acting by promiscuous mechanism, contributes to avoid misleading results, saving time and money in drug discovery projects.

Author Contributions: L.G.V., A.C., E.P. and H.U. conceived and designed the enzymatic experiments with human *ecto*-5′-nucleotidase; L.G.V., E.P., A.M.C.-R. and A.T.-d.A. conceived and designed turbidimetric and DLS experiments; L.G.V. performed all experiments; L.G.V., E.P., L.d.R. and A.T.-d.A. analyzed the data; L.G.V. and A.T.-d.A. wrote the paper.

Acknowledgments: We would like to thank the Fundação de Amparo à Pesquisa do Estado de São Paulo (FAPESP), Brazil, and Conselho Nacional de Desenvolvimento Científico e Tecnológico (CNPq), Brazil, for the grants to support this study. A.T.-do A. is member of the CEPID Redoxoma (No. 2013/07937-8) and of the NAP Redoxoma (PRPUSP). L.G.V., E.P. and A.T.A. acknowledge their fellowships from FAPESP (2014/07248-0, 2012/06633-2 and 2016/12392-9). H.U. acknowledges a FAPESP grant (No. 2012/50880-4). A.M.C.-R. acknowledges a CNPq grant (No. 302352/2014-7). The authors also acknowledge Openeye Scientific Sofware, Inc., Santa Fe, for the Omega program and ChemAxon for the Instant JChem program.

References

1. Feng, B.Y.; Shoichet, B.K. A detergent-based assay for the detection of promiscuous inhibitors. *Nat. Protoc.* **2006**, *1*, 550–553. [CrossRef] [PubMed]
2. Ferreira, R.S.; Simeonov, A.; Jadhav, A.; Eidam, O.; Mott, B.T.; Keiser, M.J.; McKerrow, J.H.; Maloney, D.J.; Irwin, J.J.; Shoichet, B.K. Complementarity between a docking and a high-throughput screen in discovering new cruzain inhibitors. *J. Med. Chem.* **2010**, *53*, 4891–4905. [CrossRef] [PubMed]
3. Scior, T.; Bender, A.; Tresadern, G.; Medina-Franco, J.L.; Martínez-Mayorga, K.; Langer, T.; Cuanalo-Contreras, K.; Agrafiotis, D.K. Recognizing pitfalls in virtual screening: A critical review. *J. Chem. Inf. Model.* **2012**, *52*, 867–881. [CrossRef] [PubMed]
4. Malvezzi, A.; Queiroz, R.F.; De Rezende, L.; Augusto, O.; Amaral, A.T. Do MPO inhibitors selected by virtual screening. *Mol. Inform.* **2011**, *30*, 605–613. [CrossRef] [PubMed]
5. Irwin, J.J.; Duan, D.; Torosyan, H.; Doak, A.K.; Ziebart, K.T.; Sterling, T.; Tumanian, G.; Shoichet, B.K. An Aggregation Advisor for Ligand Discovery. *J. Med. Chem.* **2015**, *58*, 7076–7087. [CrossRef] [PubMed]
6. Malvezzi, A.; de Rezende, L.; Izidoro, M.A.; Cezari, M.H.S.; Juliano, L.; Amaral, A.T.d. Uncovering false positives on a virtual screening search for cruzain inhibitors. *Bioorgan. Med. Chem. Lett.* **2008**, *18*, 350–354. [CrossRef] [PubMed]
7. McGovern, S.L.; Shoichet, B.K. Kinase inhibitors: Not just for kinases anymore. *J. Med. Chem.* **2003**, *46*, 1478–1483. [CrossRef] [PubMed]
8. Aldrich, C.; Bertozzi, C.; Georg, G.I.; Kiessling, L.; Lindsley, C.; Liotta, D.; Merz, K.M.; Schepartz, A.; Wang, S. The Ecstasy and Agony of Assay Interference Compounds. *J. Med. Chem.* **2017**, *60*, 2165–2168. [CrossRef] [PubMed]
9. Baell, J.B.; Holloway, G.A. New substructure filters for removal of pan assay interference compounds (PAINS) from screening libraries and for their exclusion in bioassays. *J. Med. Chem.* **2010**, *53*, 2719–2740. [CrossRef] [PubMed]

10. Baell, J.; Walters, M.A. Chemistry: Chemical con artists foil drug discovery. *Nature* **2014**, *513*, 481–483. [CrossRef] [PubMed]

11. McGovern, S.L.; Helfand, B.T.; Feng, B.; Shoichet, B.K. A specific mechanism of nonspecific inhibition. *J. Med. Chem.* **2003**, *46*, 4265–4272. [CrossRef] [PubMed]

12. Yang, J.J.; Ursu, O.; Lipinski, C.A.; Sklar, L.A.; Oprea, T.I.; Bologa, C.G. Badapple: Promiscuity patterns from noisy evidence. *J. Cheminform.* **2016**, *8*, 1–14. [CrossRef] [PubMed]

13. Seidler, J.; McGovern, S.L.; Doman, T.N.; Shoichet, B.K. Identification and prediction of promiscuous aggregating inhibitors among known drugs. *J. Med. Chem.* **2003**, *46*, 4477–4486. [CrossRef] [PubMed]

14. Feng, B.Y.; Shelat, A.; Doman, T.N.; Guy, R.K.; Shoichet, B.K. High-throughput assays for promiscuous inhibitors. *Nat. Chem. Biol.* **2005**, *1*, 146–148. [CrossRef] [PubMed]

15. Mateen, R.; Ali, M.M.; Hoare, T. A printable hydrogel microarray for drug screening avoids false positives associated with promiscuous aggregating inhibitors. *Nat. Commun.* **2018**, *602*, 1–9. [CrossRef] [PubMed]

16. Pacheco, L.F.; Carmona-Ribeiro, A.M. Effects of synthetic lipids on solubilization and colloid stability of hydrophobic drugs. *J. Colloid Interface Sci.* **2003**, *258*, 146–154. [CrossRef]

17. Eliete, G.L.; Luciano, R.G.; Carmona-Ribeiro, A. Stable Indomethacin Dispersions in Water from Drug, Ethanol, Cationic Lipid and Carboxymethyl-Cellulose. *Pharm. Nanotechnol.* **2016**, *4*, 126–135. [CrossRef]

18. Coan, K.E.D.; Shoichet, B.K. Stoichiometry and Physical Chemistry of Promiscuous Aggregate-Based Inhibitors. *J. Am. Chem. Soc.* **2008**, *130*, 9606–9612. [CrossRef] [PubMed]

19. Shoichet, B.K. Interpreting Steep Dose-Response Curves in Early Inhibitor Discovery. *J. Med. Chem.* **2006**, *49*, 7274–7277. [CrossRef] [PubMed]

20. McGovern, S.L.; Caselli, E.; Grigorieff, N.; Shoichet, B.K. A common mechanism underlying promiscuous inhibitors from virtual and high-throughput screening. *J. Med. Chem.* **2002**, *45*, 1712–1722. [CrossRef] [PubMed]

21. Babaoglu, K.; Simeonov, A.; Irwin, J.J.; Nelson, M.E.; Feng, B.; Thomas, C.J.; Cancian, L.; Costi, M.P.; Maltby, D.A.; Jadhav, A.; et al. Comprehensive mechanistic analysis of hits from high-throughput and docking screens against β-lactamase. *J. Med. Chem.* **2008**, *51*, 2502–2511. [CrossRef] [PubMed]

22. Sassano, M.F.; Doak, A.K.; Roth, B.L.; Shoichet, B.K. Colloidal aggregation causes inhibition of G protein-coupled receptors. *J. Med. Chem.* **2013**, *56*, 2406–2414. [CrossRef] [PubMed]

23. Pohjala, L.; Tammela, P. Aggregating behavior of phenolic compounds—A source of false bioassay results? *Molecules* **2012**, *17*, 10774–10790. [CrossRef] [PubMed]

24. Alturki, M.S.; Fuanta, N.R.; Jarrard, M.A.; Hobrath, J.V.; Goodwin, D.C.; Rants'o, T.A.; Calderón, A.I. A multifaceted approach to identify non-specific enzyme inhibition: Application to Mycobacterium tuberculosis shikimate kinase. *Bioorg. Med. Chem. Lett.* **2018**, *28*, 802–808. [CrossRef] [PubMed]

25. Duan, D.; Doak, A.K.; Nedyalkova, L.; Shoichet, B.K. Colloidal Aggregation and the in Vitro Activity of Traditional Chinese Medicines. *ACS Chem. Biol.* **2015**, *10*, 978–988. [CrossRef] [PubMed]

26. Zimmermann, H.; Zebisch, M.; Sträter, N. Cellular function and molecular structure of *ecto*-nucleotidases. *Purinergic Signal.* **2012**, *8*, 437–502. [CrossRef] [PubMed]

27. Antonioli, L.; Pacher, P.; Vizi, E.S.; Haskó, G. CD39 and CD73 in immunity and inflammation. *Trends Mol. Med.* **2013**, *19*, 355–367. [CrossRef] [PubMed]

28. Stagg, J.; Beavis, P.A.; Divisekera, U.; Liu, M.C.P.; Möller, A.; Darcy, P.K.; Smyth, M.J. CD73-Deficient mice are resistant to carcinogenesis. *Cancer Res.* **2012**, *72*, 2190–2196. [CrossRef] [PubMed]

29. Stagg, J.; Divisekera, U.; McLaughlin, N.; Sharkey, J.; Pommey, S.; Denoyer, D.; Dwyer, K.M.; Smyth, M.J. Anti-CD73 antibody therapy inhibits breast tumor growth and metastasis. *Proc. Natl. Acad. Sci. USA* **2010**, *107*, 1547–1552. [CrossRef] [PubMed]

30. Loi, S.; Pommey, S.; Haibe-Kains, B.; Beavis, P.A.; Darcy, P.K.; Smyth, M.J.; Stagg, J. CD73 promotes anthracycline resistance and poor prognosis in triple negative breast cancer. *Proc. Natl. Acad. Sci. USA* **2013**, *110*, 11091–11096. [CrossRef] [PubMed]

31. Cappellari, A.R.; Rockenbach, L.; Dietrich, F.; Clarimundo, V.; Glaser, T.; Braganhol, E.; Abujamra, A.L.; Roesler, R.; Ulrich, H. Oliveira Battastini, A.M. Characterization of Ectonucleotidases in Human Medulloblastoma Cell Lines: *Ecto*-5'NT/CD73 in Metastasis as Potential Prognostic Factor. *PLoS ONE* **2012**, *7*, e47468. [CrossRef]

32. Cappellari, A.R.; Pillat, M.M.; Souza, H.D.N.; Dietrich, F.; Oliveira, F.H.; Figueiró, F.; Abujamra, A.L.; Roesler, R.; Lecka, J.; Sévigny, J.; et al. Ecto-5′-Nucleotidase Overexpression Reduces Tumor Growth in a Xenograph Medulloblastoma Model. *PLoS ONE* **2015**, *10*, e0140996. [CrossRef] [PubMed]

33. Flögel, U.; Burghoff, S.; Van Lent, P.L.E.M.; Temme, S.; Galbarz, L.; Ding, Z.; El-Tayeb, A.; Huels, S.; Bönner, F.; Borg, N.; et al. Selective activation of adenosine A2A receptors on immune cells by a CD73-dependent prodrug suppresses joint inflammation in experimental rheumatoid arthritis. *Sci. Transl. Med.* **2012**, *4*. [CrossRef] [PubMed]

34. Paletta-Silva, R.; Meyer-Fernandes, J.R. Adenosine and Immune Imbalance in Visceral Leishmaniasis: The Possible Role of Ectonucleotidases. *J. Trop. Med.* **2012**, *2012*, 650874. [CrossRef] [PubMed]

35. Russo-Abrahão, T.; Cosentino-Gomes, D.; Gomes, M.T.; Alviano, D.S.; Alviano, C.S.; Lopes, A.H.; Meyer-Fernandes, J.R. Biochemical properties of Candida parapsilosis ecto-5′-nucleotidase and the possible role of adenosine in macrophage interaction. *FEMS Microbiol. Lett.* **2011**, *317*, 34–42. [CrossRef] [PubMed]

36. Fan, J.; Zhang, Y.; Chuang-Smith, O.N.; Frank, K.L.; Guenther, B.D.; Kern, M.; Schlievert, P.M.; Herzberg, M.C. Ecto-5′-Nucleotidase: A Candidate Virulence Factor in Streptococcus sanguinis Experimental Endocarditis. *PLoS ONE* **2012**, *7*, e38059. [CrossRef] [PubMed]

37. Di Virgilio, F.; Solini, A. P2 receptors: New potential players in atherosclerosis. *Br. J. Pharmacol.* **2002**, *135*, 831–842. [CrossRef] [PubMed]

38. Reiss, A.B.; Cronstein, B.N. Regulation of foam cells by adenosine. *Arterioscler. Thromb. Vasc. Biol.* **2012**, *32*, 879–886. [CrossRef] [PubMed]

39. Hart, M.L.; Henn, M.; Köhler, D.; Kloor, D.; Mittelbronn, M.; Gorzolla, I.C.; Stahl, G.L.; Eltzschig, H.K. Role of extracellular nucleotide phosphohydrolysis in intestinal ischemia-reperfusion injury. *FASEB J.* **2008**, *22*, 2784–2797. [CrossRef] [PubMed]

40. Hernandez-Mir, G.; McGeachy, M.J. CD73 is expressed by inflammatory Th17 cells in experimental autoimmune encephalomyelitis but does not limit differentiation or pathogenesis. *PLoS ONE* **2017**, *12*, 1–13. [CrossRef] [PubMed]

41. Jiang, T.; Xu, X.; Qiao, M.; Li, X.; Zhao, C.; Zhou, F.; Gao, G.; Wu, F.; Chen, X.; Su, C.; et al. Comprehensive evaluation of NT5E/CD73 expression and its prognostic significance in distinct types of cancers. *BMC Cancer* **2018**, *18*, 267. [CrossRef] [PubMed]

42. Braganhol, E.; Tamajusuku, A.S.K.; Bernardi, A.; Wink, M.R.; Battastini, A.M.O. Ecto-5′-nucleotidase/CD73 inhibition by quercetin in the human U138MG glioma cell line. *Biochim. Biophys. Acta Gen. Subj.* **2007**, *1770*, 1352–1359. [CrossRef] [PubMed]

43. Ripphausen, P.; Freundlieb, M.; Brunschweiger, A.; Zimmermann, H.; Müller, C.E.; Bajorath, J. Virtual Screening Identifies Novel Sulfonamide Inhibitors of ecto-5′-Nucleotidase. *J. Med. Chem.* **2012**, *55*, 6576–6581. [CrossRef] [PubMed]

44. Baqi, Y.; Lee, S.; Iqbal, J.; Ripphausen, P.; Lehr, A.; Scheiff, A.B.; Zimmermann, H.; Bajorath, J.; Müller, C.E. Development of Potent and Selective Inhibitors of ecto-5′-Nucleotidase Based on an Anthraquinone Scaffold. *J. Med. Chem.* **2010**, *53*, 2076–2086. [CrossRef] [PubMed]

45. Iqbal, J.; Saeed, A.; Raza, R.; Matin, A.; Hameed, A.; Furtmann, N.; Lecka, J.; Sévigny, J.; Bajorath, J. Identification of sulfonic acids as efficient ecto-5′-nucleotidase inhibitors. *Eur. J. Med. Chem.* **2013**, *70*, 685–691. [CrossRef] [PubMed]

46. Bhattarai, S.; Freundlieb, M.; Pippel, J.; Meyer, A.; Abdelrahman, A.; Fiene, A.; Lee, S.Y.; Zimmermann, H.; Yegutkin, G.G.; Sträter, N.; et al. α,β-Methylene-ADP (AOPCP) Derivatives and Analogues: Development of Potent and Selective ecto-5′-Nucleotidase (CD73) Inhibitors. *J. Med. Chem.* **2015**, *58*, 6248–6263. [CrossRef] [PubMed]

47. Al-Rashida, M.; Batool, G.; Sattar, A.; Ejaz, S.A.; Khan, S.; Lecka, J.; Sévigny, J.; Hameed, A.; Iqbal, J. 2-Alkoxy-3-(sulfonylarylaminomethylene)-chroman-4-ones as potent and selective inhibitors of ectonucleotidases. *Eur. J. Med. Chem.* **2016**, *115*, 484–494. [CrossRef] [PubMed]

48. Saeed, A.; Ejaz, S.A.; Shehzad, M.; Hassan, S.; al-Rashida, M.; Lecka, J.; Sévigny, J.; Iqbal, J. 3-(5-(Benzylideneamino)thiazol-3-yl)-2H-chromen-2-ones: A new class of alkaline phosphatase and ecto-5′-nucleotidase inhibitors. *RSC Adv.* **2016**, *6*, 21026–21036. [CrossRef]

49. Rahimova, R.; Fontanel, S.; Lionne, C.; Jordheim, L.P.; Peyrottes, S.; Chaloin, L. Identification of allosteric inhibitors of the *ecto*-5'-nucleotidase (CD73) targeting the dimer interface. *PLoS Comput. Biol.* **2018**, *14*, 1–23. [CrossRef] [PubMed]

50. Figueiró, F.; Mendes, F.B.; Corbelini, P.F.; Janarelli, F.; Jandrey, E.H.F.; Russowsky, D.; Eifler-Lima, V.L.; Battastini, A.M.O. A monastrol-derived compound, LaSOM 63, inhibits *ecto*-5'-nucleotidase/CD73 activity and induces apoptotic cell death of glioma cell lines. *Anticancer Res.* **2014**, *34*, 1837–1842. [PubMed]

51. Wolber, G.; Langer, T. LigandScout: 3-D pharmacophores derived from protein-bound ligands and their use as virtual screening filters. *J. Chem. Inf. Model.* **2005**, *45*, 160–169. [CrossRef] [PubMed]

52. Knapp, K.; Zebisch, M.; Pippel, J.; El-Tayeb, A.; Müller, C.E.; Sträter, N. Crystal structure of the human *ecto*-5'-nucleotidase (CD73): Insights into the regulation of purinergic signaling. *Structure* **2012**, *20*, 2161–2173. [CrossRef] [PubMed]

53. Irwin, J.J.; Shoichet, B.K. ZINC—A Free Database of Commercially Available Compounds for Virtual Screening ZINC—A Free Database of Commercially Available Compounds for Virtual Screening. *J. Chem. Inf. Model* **2005**, *45*, 177–182. [CrossRef] [PubMed]

54. Korb, O.; Stützle, T.; Exner, T.E. Empirical scoring functions for advanced Protein-Ligand docking with PLANTS. *J. Chem. Inf. Model.* **2009**, *49*, 84–96. [CrossRef] [PubMed]

55. Wildman, S.A.; Crippen, G.M. Prediction of physicochemical parameters by atomic contributions. *J. Chem. Inf. Comput. Sci.* **1999**, *39*, 868–873. [CrossRef]

56. Chan, K.M.; Delfert, D.; Junger, K.D. A direct colorimetric assay for Ca^{2+}-stimulated ATPase activity. *Anal. Biochem.* **1986**, *157*, 375–380. [CrossRef]

57. Iqbal, J.; Jirovsky, D.; Lee, S.Y.; Zimmermann, H.; Müller, C.E. Capillary electrophoresis-based nanoscale assays for monitoring *ecto*-5'-nucleotidase activity and inhibition in preparations of recombinant enzyme and melanoma cell membranes. *Anal. Biochem.* **2008**, *373*, 129–140. [CrossRef] [PubMed]

58. Freundlieb, M.; Zimmermann, H.; Müller, C.E. A new, sensitive *ecto*-5'-nucleotidase assay for compound screening. *Anal. Biochem.* **2014**, *446*, 53–58. [CrossRef] [PubMed]

59. Channar, P.A.; Shah, S.J.A.; Hassan, S.; Nisa, Z.; Lecka, J.; Sévigny, J.; Bajorath, J.; Saeed, A.; Iqbal, J. Isonicotinohydrazones as inhibitors of alkaline phosphatase and *ecto*-5'-nucleotidase. *Chem. Biol. Drug Des.* **2017**, *89*, 365–370. [CrossRef] [PubMed]

60. Hawkins, P.C.D.; Skillman, A.G.; Warren, G.L.; Ellingson, B.A.; Stahl, M.T. Conformer generation with OMEGA: Algorithm and validation using high quality structures from the protein databank and cambridge structural database. *J. Chem. Inf. Model.* **2010**, *11*, 572–584. [CrossRef] [PubMed]

61. Verdonk, M.L.; Cole, J.C.; Hartshorn, M.J.; Murray, C.W.; Taylor, R.D. Improved Protein—Ligand Docking Using GOLD. *Proteins Struct. Funct. Bioinform.* **2003**, *623*, 609–623. [CrossRef] [PubMed]

62. Servos, J.; Reiländer, H.; Zimmermann, H. Catalytically active soluble *ecto*-5'-nucleotidase purified after heterologous expression as a tool for drug screening. *Drug Dev. Res.* **1998**, *276*, 269–276. [CrossRef]

63. Grabowski, E.; Morrison, I. Particle Size Distribution from Analysis of Quasieletric Light Scattering Data. In *Measurements of Suspended Particles by Quasielastic Light Scattering*; Dahneke, B.E., Ed.; Willey-Interscience: New York, NY, USA, 1983; pp. 199–236.

11

Comparative Study of Carborane- and Phenyl-Modified Adenosine Derivatives as Ligands for the A2A and A3 Adenosine Receptors Based on a Rigid in Silico Docking and Radioligand Replacement Assay

Marian Vincenzi [1,†,‡] 🆔, **Katarzyna Bednarska** [2,†] 🆔 **and Zbigniew J. Leśnikowski** [1,*] 🆔

[1] Laboratory of Molecular Virology and Biological Chemistry, Institute of Medical Biology of the Polish Academy of Sciences, 106 Lodowa St., 93-232 Lodz, Poland; marian.vincenzi@unina.it

[2] Laboratory of Experimental Immunology, Institute of Medical Biology, Polish Academy of Sciences, 106 Lodowa St., 93-232 Lodz, Poland; kbednarska@cbm.pan.pl

* Correspondence: zlesnikowski@cbm.pan.pl

† These authors contributed equally to this work.

‡ Current address: Institute of Biostructures and Bioimaging (IBB), National Research Council (CNR), Via Mezzocannone 16, I-80134 Naples, Italy.

Academic Editors: Rebecca C. Wade and Outi Salo-Ahen

Abstract: Adenosine receptors are involved in many physiological processes and pathological conditions and are therefore attractive therapeutic targets. To identify new types of effective ligands for these receptors, a library of adenosine derivatives bearing a boron cluster or phenyl group in the same position was designed. The ligands were screened in silico to determine their calculated affinities for the A2A and A3 adenosine receptors. An virtual screening protocol based on the PatchDock web server was developed. In the first screening phase, the effects of the functional group (organic or inorganic modulator) on the adenosine ligand affinity for the receptors were determined. Then, the lead compounds were identified for each receptor in the second virtual screening phase. Two pairs of the most promising ligands, compounds **3** and **4**, and two ligands with lower affinity scores (compounds **11** and **12**, one with a boron cluster and one with a phenyl group) were synthesized and tested in a radioligand replacement assay for affinity to the A2A and A3 receptors. A reasonable correlation of in silico and biological assay results was observed. In addition, the effects of a phenyl group and boron cluster, which is new adenosine modifiers, on the adenosine ligand binding were compared.

Keywords: in silico screening; adenosine; boron cluster; adenosine receptors; AR ligands

---

1. Introduction

Adenosine is a key endogenous molecule involved in the activation of the A1, A2A, A2B and A3 adenosine receptors (ARs), which belong to the P1 class of purinergic receptors. Each of these receptors promotes a different signaling pathways associated with specific, although with some overlap, effects. ARs are members of the G protein coupled receptors family, which also includes many well-known receptors, such as dopamine, adrenergic, histamine and serotonin receptors.

ARs in response to adenosine binding trigger essential signals into cells by activating one or more heterotrimeric G protein, located on the inner side of the cell membrane and subsequently influence multiple-effector systems (i.e., adenylate cyclase, ion channels, phospholipases). These receptors are important pharmacological and therapeutic targets [1].

Similar to other G protein-coupled receptors, ARs consist of seven transmembrane helices that contain a ligand binding site. Each helix is composed of approximately 21 to 28 amino acids. The transmembrane helices are connected by three extracellular and three cytoplasmic loops with different numbers of amino acids. The N-terminus and C-terminus are located on the extracellular and cytoplasmic sides, respectively, of the cell membrane (Figure 1) [2]. For more than three decades, medicinal chemistry research has focused on developing potent and selective synthetic AR agonists and antagonists as agents potentially useful in the treatment of inflammation, the central nervous system (CNS) disorders and pulmonary or cardiovascular diseases. In addition, several allosteric modulators of AR subtypes have also been synthesized [3].

AR action can be modulated directly by ligands or indirectly by availability of extracellular adenosine through its metabolism or cellular uptake [3]. The A2A and A3 adenosine receptors are considered to be among the attractive therapeutic targets for the inflammatory disorders and cancer treatment [3]. In the development of potential drugs targeting ARs, many adenosine derivatives and non-nucleoside molecules have been synthesized and tested [4–6]. The presence of the hydrophobic pharmacophore is considered as one of the essential features of the ligands in terms of the binding activity and their A2A AR selectivity. One new avenue of research in this field is the development of nucleoside-boron cluster conjugates, including adenosine derivative conjugates.

Medicinal chemists are increasingly utilizing boron clusters (polyhedral boron cages) as a new generation of 3-dimensional, abiotic privileged scaffolds, modifiers and pharmacophores in bioactive molecule design [7–10]. Many boron cluster conjugates with biologically important low molecular weight compounds, including amino acids, lipids, carbohydrates, porphyrins, nucleic acid bases, nucleosides and DNA groove binders, have been synthesized [11–14]. In addition, biopolymers bearing one or more boron cages (carboranes), including carboranyl peptides and proteins, carboranyl oligophosphates, and nucleic acids (RNA and DNA oligonucleotides), have been prepared [13,15,16].

The low molecular weight biomolecules that have been conjugated to a boron cluster include many receptor ligands, such as estrogen, androgen, retinoic acid, dihydrofolate, etc. [8]. Boron clusters bearing ligands to ARs have also been described [17,18].

We previously reported the chemical synthesis of various nucleoside-boron cluster conjugates [15], including those formed from adenosine and evaluated their activity as blood platelet aggregation inhibitors [18], reactive oxygen species (ROS) inhibitors [19], antivirals [20] and anti-tumor agents [21]. Some of these compounds are also potential ligands to purinergic receptors [19,22–24]. Herein, a series of adenosine derivatives modified with either phenyl or boron cluster (carborane group) were evaluated to compare the effects of organic and inorganic modification on the ligand affinity for the A2A and A3 receptors, in silico. The approach to computational ligand-adenosine receptor rigid docking was applied to screened virtually adenosine conjugates bearing such diverse structures as the phenyl group and the boron cluster. Two pairs of ligands with the highest and lowest affinity scores were selected based on in silico screening results, then were synthesized and tested in vitro in the radioligand replacement assay. Finally, the results of in silico and in vitro study were compared.

2. Results and Discussion

Due to the limitations of the available docking algorithms and the unique properties of boron clusters that prevent them from being defined in the same way as organic moieties [10,25], higher errors are obtained in in silico studies of boron clusters than in those of purely organic structures. Herein, a simple and versatile computational approach based on PatchDock web server was used providing preliminary information on ligand receptor interaction. The results of the in silico screening were verified by the synthesis of the real compound library followed by in vitro screening of the obtained ligands. This work provides insight into the effects of boron clusters on the adenosine affinity for the A2A and A3 receptors and the relationship between in silico and experimental results. Here, proof of concept of the presented in silico assay and its comparison with radioligand replacement test is described.

It is generally accepted that in silico target profiling methods for selecting lead compounds are efficient alternatives to expensive, time-consuming high-throughput in vitro target profiling of compound libraries. This type of approach was recently used to successfully identify molecules that selectively bind to the A1, A2A, A2B and A3 receptors [26]. The selected molecules were further evaluated in vitro under the same conditions. This work utilizes a similar approach though with other focuses. The main difference between the previous study and the study presented herein is that instead of screening large, diverse compound libraries, a small specialized library of adenosine derivatives was utilized, and the effects of specific, known phenyl group modification were compared to those of the corresponding inorganic modifications with boron cluster cage. This task oriented approach allowed to evaluate the effects of boron cluster modification on adenosine ligand properties [7,10,27] and to propose a practical protocol for comparative study in silico.

2.1. Protein Structure Selection and Modeling

2.1.1. A2A Adenosine Receptor

The choice of the protein structure is critical in virtual screening studies. In this work, we studied two different adenosine receptors i.e., A2AR and A3R. For the A2A adenosine receptor model, the crystal structure of the thermostabilized human A2A receptor with bound adenosine located in the binding pocket was selected (Protein Data Bank (PDB) code 2YDO, Figure 1a) [28]. The A2A receptor structure was extracted from the X-ray structure using the UCSF Chimera visualization system [29]. The receptor structure is fixed in the active state.

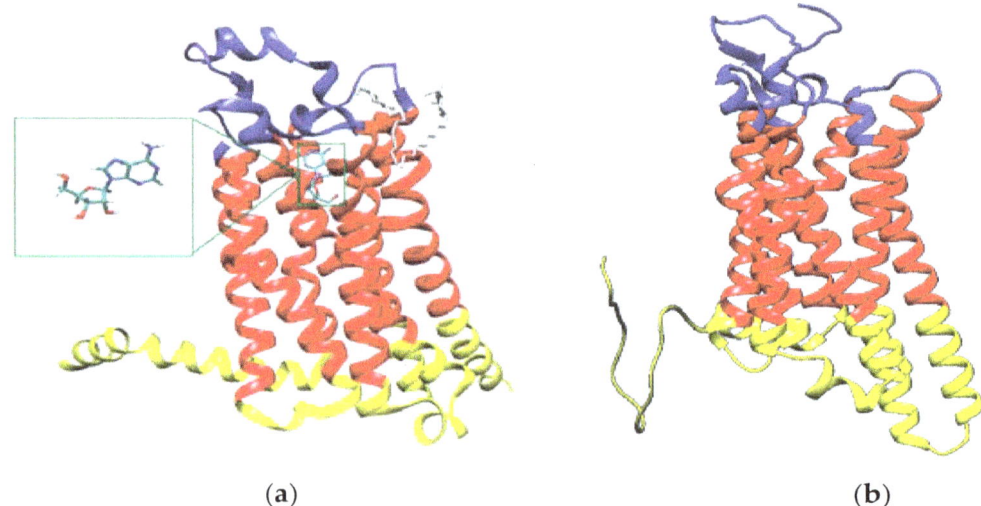

(a)   (b)

Figure 1. (**a**) X-ray structure of the thermostabilized human A2A receptor with bound adenosine (PDB code: 2YDO [28]; blue: extracellular region, red: transmembrane domain, yellow: cytoplasmic region); insert: adenosine; (**b**) A3 adenosine receptor structures obtained with LOMETS (cyan rectangle: A8.60-E318 region, blue: extracellular region, red: transmembrane domains, yellow: cytoplasmic region).

2.1.2. A3 Adenosine Receptor

Because the A3 adenosine receptor structure has not yet been deposited in the PDB, protein homology modeling [30–32] was performed using a procedure previously reported in the literature [30] and LOMETS (Local Meta-Threading-Server) [33], one of the many programs available for this purpose. The amino acid sequence of A3 AR was obtained from the UniProtKB/Swiss-Prot database (P0DMS8).

Using LOMETS, PDB 5IU4 structure of the complete A2 AR [34] was selected and verified as the best fit for the designing of A3 AR model with the highest score. According the validation of the

5IU4 model as the template for A3 AR, the conformation of 5IU4 structure gives similar fashion of adenosine binding like in 2YDO. The A3 adenosine receptor structure obtained by Lomets modeling is reported in Figure 1b.

2.2. Docking Method Validation and Optimization

2.2.1. A2A Adenosine Receptor

To validate the SwissDock docking methodology, a study was performed using adenosine as the reference ligand (Figure 1a), and the best docked structure (Figure 2b) was compared to the X-ray structure of the A2A receptor/adenosine complex (PDB code: 2YDO [28], Figures 1a and 2a. The superimposition of the two structures showed that the position and orientation of adenosine in the best docked pose are identical to those in the crystal 2YDO structure (Figure 2c,d), showing that the SwissDock web server method [35] is valid for this ligand-receptor system. In this context it is important to consider that SwissDock web server was chosen because it is a program that give the possibility to consider the flexibility of the side chains of residues into input protein, thus giving a chance to better describe the protein-ligand interaction.

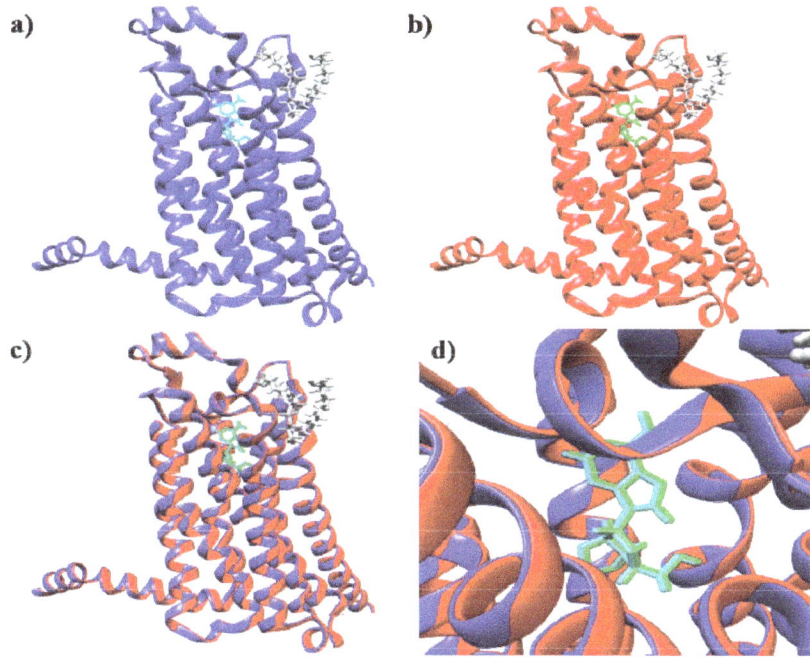

Figure 2. (**a**) X-ray structure of the thermostabilized human A2A receptor with bound adenosine (PDB code: 2YDO) [28]; (**b**) best docked pose for adenosine obtained with SwissDock; (**c**) superimposition of the reference complex (PDB code: 2YDO) and the best docked adenosine pose; (**d**) magnification of the binding pocket viewed from the top of the extracellular region (blue and cyan: A2A receptor and adenosine, respectively, in the X-ray reference structure; red and green: protein and adenosine, respectively, in the best docked structure; yellow: 1-S-octyl-β-D-thioglucoside molecules (thioglucoside), crystallization helper molecules).

The docking mode is "blind", i.e., without any specific box of analysis which results in consideration of the entire surface of the protein by the docking program. In our case the reference structure for A2A receptor (PDB code: 2YDO) is characterized by the presence of crystallization helper molecules that normally are not present and therefore were removed in docking experiments. One may judiciously expect that helper molecules would not influence the binding pose at the extracellular pocket attained by the docking, however to make sure that this custom change does not influence docking results docking experiments with and without the helper molecule were performed. In the selected A2A structure (PDB code: 2YDO), two sets of SwissDock control docking simulations, one with and one without the crystallization helper molecules [28], were performed with adenosine as the ligand (Figure S1a,b). As expected, the superimposition of the two best A2A receptor/adenosine structures obtained from these simulations revealed that the presence of the crystallization helper molecules does not influence the protein-ligand interactions (Figures S1 and S3).

2.2.2. A3 Adenosine Receptor

A3 AR structure obtained by homology modeling with LOMETS [33] was validated with the use of two different tools, VERIFY3D [36] and RAMPAGE (Ramachandran plot analysis) [37]. The VERIFY3D program compares the overall structure to the amino acid sequence using a 3D profile computed from the atomic coordinates of the given structure and it has been used to determine the accuracy of the A3R model obtained using LOMETS.

The VERIFY3D analysis revealed that the structure of A3 AR has high percentage of residues with an average 3D-1D score of <0.2, indicating that our 3D model is compatible with its sequence.

The RAMPAGE program provides tools for analyzing Ramachandran plots to assess the stereochemical quality of proteins and the distribution of residues between the different regions ("favored", "allowed" or "outlier"). It should be noted that in our study the number of residues classified as the favored regions is prevailing. Overall, the results obtained both by VERIFY3D and RAMPAGE homology modeling tools, confirmed the validity of our A3R model.

2.2.3. Ligand Modeling

Modeling of the set of ligands was the next stage of our docking study (Figure 3). Figure 3 shows the series of molecules **1–16** designed to test the effects of adenosine structural modifications on the interactions between the ligand and the A2A and A3 purinergic receptors. Four specific modifications were evaluated: (1) type of lipophilic group (phenyl ring vs. boron cluster); (2) sugar configuration (β-D-ribofuranose vs. β-D-arabinofuranose); (3) position of the adenosine modification (2, 8 or 2′) and (4) spacer flexibility (ethynyl vs. ethyl linker).

Here, the boron clusters present an interesting challenge. Various docking approaches for molecules with boron clusters have been reported in the literature. One method is to substitute a carborane cage ($C_2B_{10}H_{12}$) with one of its common bioisosteres, such as aryl, cycloalkyl and adamantyl groups [38–40]. The advantage of this method is that all the atom types are well described in the available docking programs; however, the steric properties of the boron cluster are not properly described. Therefore, others prefer to use protocols that can be directly applied to boron cluster structures.

It should be noted that most of available docking software such as AutoDock, FlexX, Glide, and Surflex do not have built-in parameters for hexacoordinated boron atoms, meaning calculations of molecules containing these atoms cannot be performed [40]. The most widely used approach to solve this problem is to change the boron atom type to the C.3 atom type [41–44]. In accordance with this approach, boron clusters are artificially treated as clusters of only carbon atoms. Using this protocol, some new information about the effect of the boron cluster structure on the protein-ligand interactions can be obtained. However, the effects of specific boron cluster properties [45], such as 3D aromaticity, hydridic character of the B-H hydrogens, dihydrogen bond formation or sigma-hole bonding, on ligand and receptors interaction cannot be determined in this simplified model. Therefore, herein the shape complementarity approach was applied to screen boron-bearing adenosines as the ligands for AR and the PatchDock software [46], as a tool for rigid ligand docking was used. The PatchDock provided a way to get preliminary information on ligand-protein interaction energies and contact surfaces without the change of the boron cage atom types, though although without consideration of the properties of boron cluster listed above.

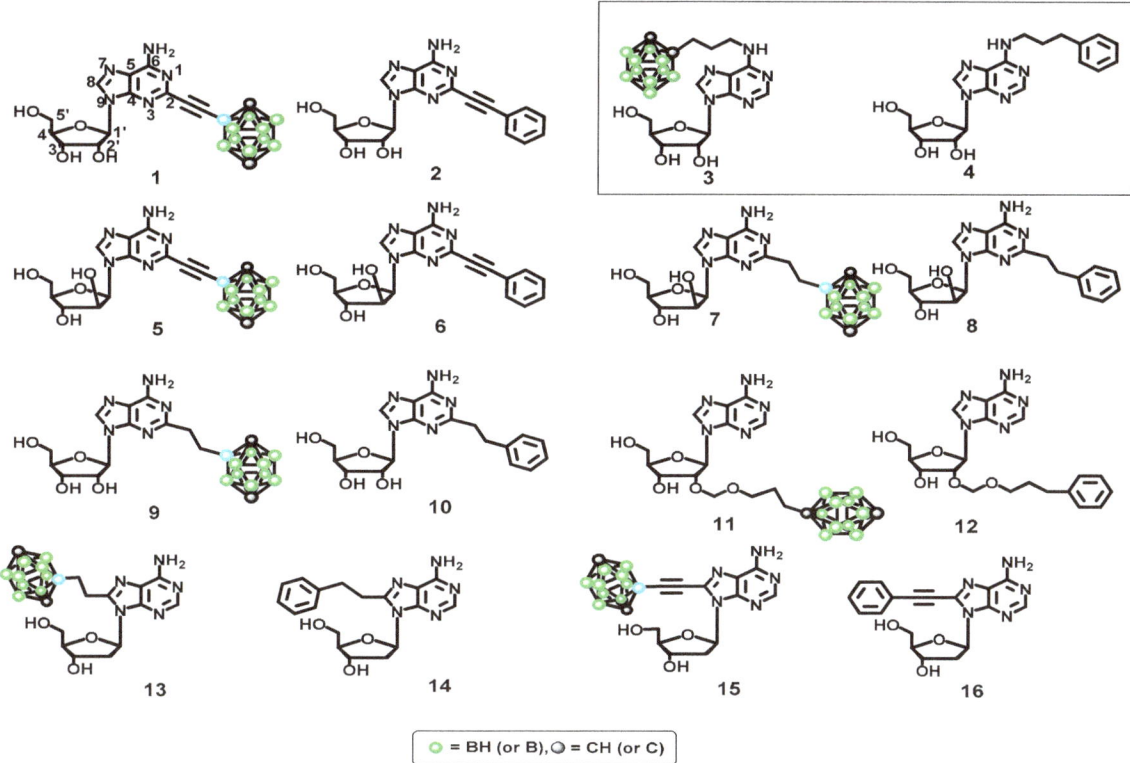

Figure 3. Modified adenosine derivatives designed as potential ligands for the A2A and A3 adenosine receptors screened in silico (dodecahedral structures correspond to the 1,12-dicarba-*closo*-dodecaboran-1-yl substituent, $C_2B_{10}H_{11}$).

PatchDock is a geometry-based molecular docking algorithm based on the object recognition and the image segmentation procedures used in Computer Vision that searches for docking configurations with good molecular shape complementarity. The PatchDock program assigns each molecule a geometric shape complementarity score (geometric score) that takes into account the interface area and desolvation energy associated with the protein-ligand interactions. Different candidate complexes are determined and ranked by a score that depends on the shape complementarity. The advantage of this docking program is the molecule input format. Whereas many methods require a ".mol2" input file format which cannot recognize the carborane boron and carbon atom types, PatchDock requires PDB input files. In this format the boron cluster structures can be more accurately described and thus the effect of this moiety on the space fitting of the entire molecule can be more precisely defined. Hence, all the docking studies in this work were performed using the PatchDock server. First, this program was validated using the same protocol as that used for the SwissDock web server. The obtained results revealed that as in the previous case, the method is reliable and provides correct information about the position of the molecule in the binding pocket of the A2A receptor (Figure S3). The slight shift of the adenosine position between the crystal structure and docking pose can be explained by the fact that PatchDock server does not consider flexibility of the residues side chains. However, this approximation did not affect the final result of the classification of our ligands.

2.2.4. Reference Ligands

For the docking simulations, selected molecules that were reported to be efficient ligands for the A2A and A3 receptors were used as references (Figure 4) The selective agonists of A2AR, i.e., apadenoson (also known as ATL-146e, $K_i = 0.68 \pm 0.1$ nM [47], phase IIb/III clinical trials [48]) and CGS 21680 ($K_i = 17.3 \pm 5.1$ nM) [47] were chosen as the reference ligands for docking study. Additionally, the antagonist A2A, SCH 58261 ($K_i = 1.3$ nM) [49], was considered in calculations.

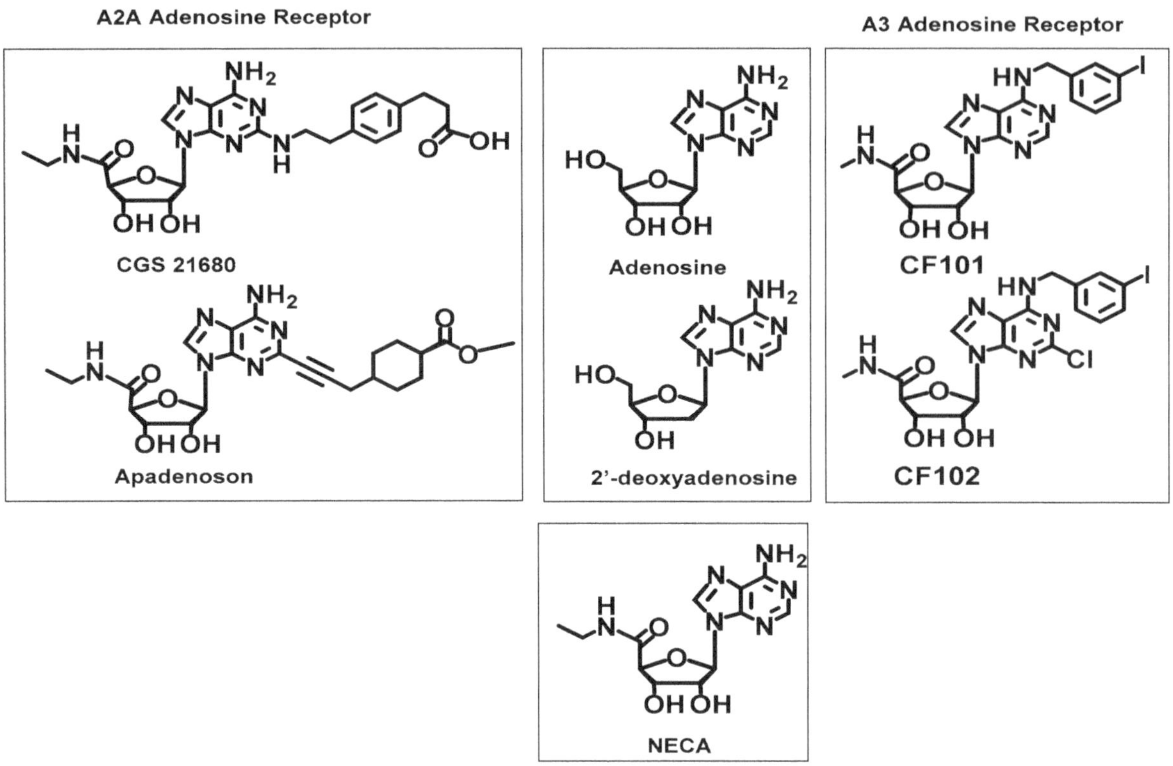

Figure 4. Reference ligands used to validate the PatchDock docking methods.

The selective agonists CF101 (known generically as IB-MECA, Ki = 1.8 nM, phase IIb/III clinical trials) [48] and CF102 (known generically as CI-IB-MECA, Ki = 1.4 nM, phase IIb/III clinical trials) [48] were chosen for the A3 receptor. Furthermore, NECA was also used for comparison (Figure 4) because it has a high affinity for adenosine receptors, although it does not exhibit receptor selectivity. Moreover, non-selective ligands, adenosine and 2′-deoxyadenosine, were used as the references.

2.3. Docking Results

As described in the "Docking method validation and optimization" section, the SwissDock and PatchDock docking methods were shown to give similar results. Indeed, the best docked adenosine pose was nearly identical in position and orientation to the reference crystal structure (PDB code: 2YDO) [28] (Figure 2d and Figure S2d). In the first screening phase, performed by PatchDock calculations, the effects of the functional group (organic or inorganic modulator) on the adenosine ligand affinity for the receptors were compared.

2.3.1. Phenyl Ring vs. Boron Cluster

Figure 5 shows the geometric scores for the docked molecules **1–16** in the A2A receptor (for the exact numerical values, see the Supplementary Information, Table S1). The calculations revealed that specific ligands were described by the score above 5000–5200 (apadenoson, CGS 21680 and SCH 58261), and non-specific ones by scores below this threshold (adenosine, 2′-deoxyadenosine, and NECA, Figure 5). A comparison of geometric scores for adenosine derivatives revealed that the molecules with phenyl groups (Figure 5) have similar (compounds **4**, **6**, **14**, and **16**) or higher (compounds **8**, **10**, and **12**) geometric scores than the corresponding molecules bearing boron clusters (compounds **1**, **3**, **5**, **7**, **9**, **11**, **13** and **15**). Interestingly, compound **1**, which has a boron cluster at position 2, has a higher geometric score than the corresponding molecule with a phenyl group (compound **2**) (Figure 5).

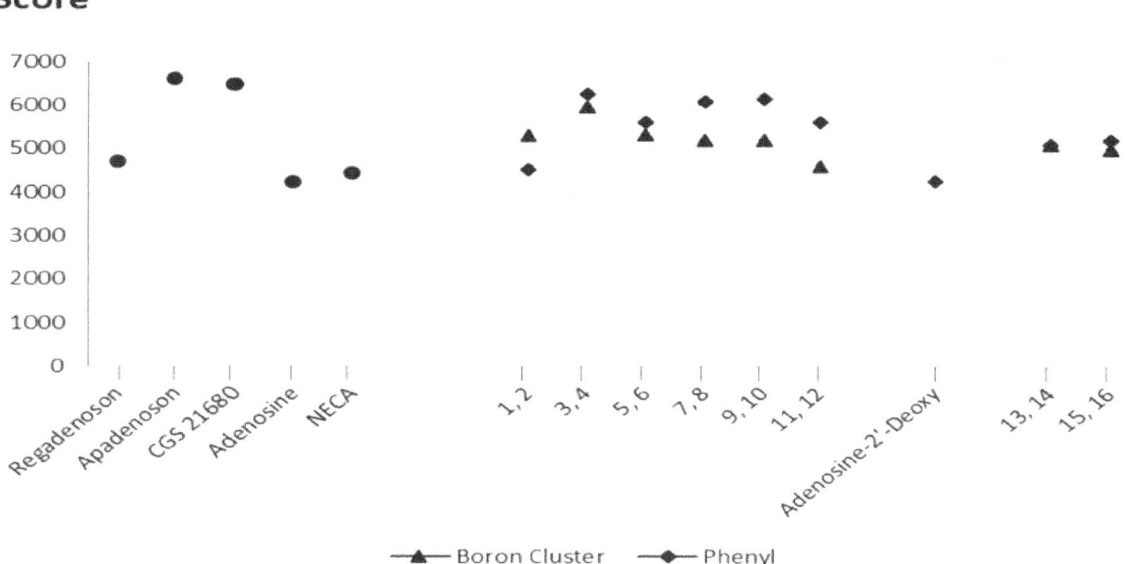

Figure 5. Geometric scores for the ligand shape complementarity to the A2A receptor. ● reference molecules (apadenoson, CGS 21680, adenosine, 2′-deoxyadenosine, NECA and SCH 58261; ▲ molecules bearing a boron cluster (**1, 3, 5, 7, 9, 11, 13, 15**); ◆ molecules with a phenyl group (**2, 4, 6, 8, 10, 12, 14, 16**).

For the compounds within the pairs **3–4** and **5–6**, the absolute differences in the affinity scores are small, but they vary slightly between the pairs. For the pairs **13–14** and **15–16**, the scores are nearly identical. However, the presence of a large, rigid group such as a boron cluster can hinder deep penetration of the ligand into the binding pocket of the receptor (compounds **7, 9,** and **11**).

As shown in Figure 6, the atomic contact energies (ACE), defined as the desolvation free energies required to transfer atoms from water to a protein's interior, of compounds **7** and **9** are consistent with this observation. It should be noted that the desolvation energy of compound **1** (boron cluster at position 2) is much lower than that of compound **2** (phenyl group at position 2) (Figure 6).

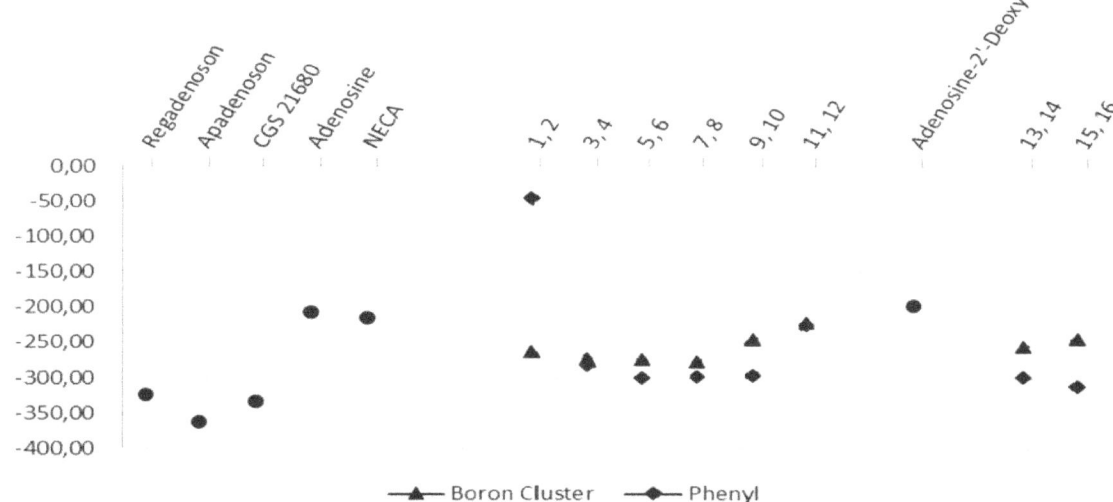

Figure 6. Desolvation energy (ACE) for the A2A protein-ligand interactions. ● reference molecules (apadenoson, CGS 21680, adenosine, 2′-deoxyadenosine, NECA and SCH 58261; ▲ molecules bearing a boron cluster (**1, 3, 5, 7, 9, 11, 13, 15**); ◆ ligands with a phenyl group (**2, 4, 6, 8, 10, 12, 14, 16**).

Similarly to results for A2AR, the score above 5000–5200 also described specific ligands for A3 receptor (CF101, CF102 and PSB 10 hydrochloride), and scores below this threshold non-specific ones (adenosine, 2′-deoxyadenosine and NECA, Figure 7). The affinity scores of compounds **1–16** for the A3 receptor are shown in Figure 7, and the exact values are listed in Table S2. In general, the affinity scores and desolvation energy profiles for binding to the A3 receptor are similar to those for binding to the A2A receptor (Figure 7 and Figure S4), although the differences between specific modifications are less pronounced. Furthermore, the A3 receptor appears to be less discriminative for the phenyl group than for the boron cluster. Similar to the A2A receptor results, both the phenyl and boron cluster modifications to the exo-amine group in position 6 are the most favorable.

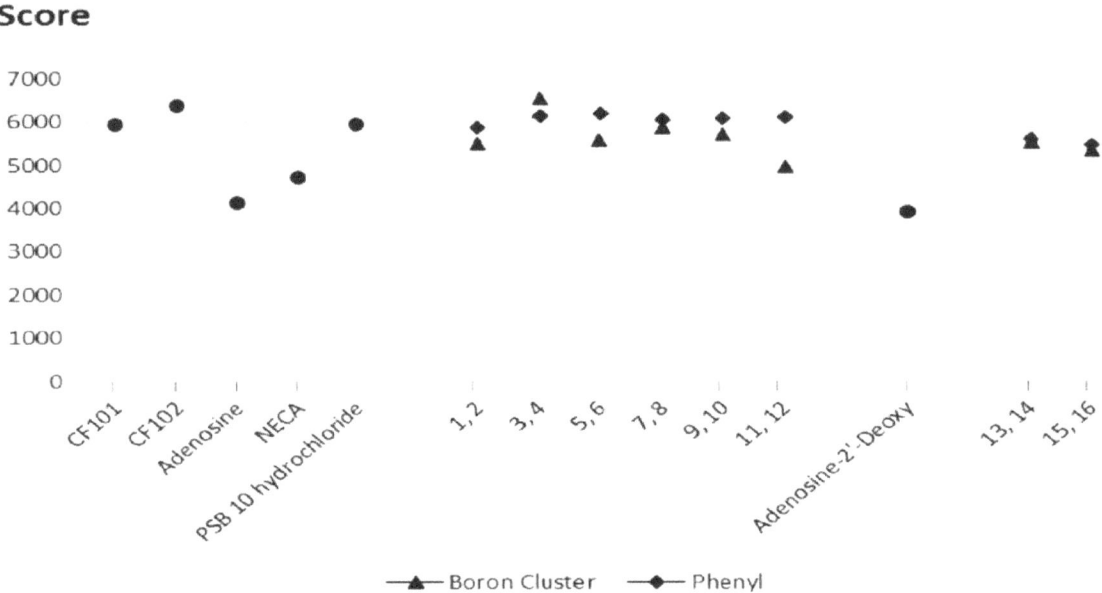

Figure 7. Geometric scores for the ligand shape complementarity to the A3 receptor. ● reference molecules (CF101, CF102, adenosine, 2′-deoxyadenosine, NECA and PSB 10 hydrochloride); ▲ molecules bearing a boron cluster (**1, 3, 5, 7, 9, 11, 13, 15**); ◆ ligands with a phenyl group (**2, 4, 6, 8, 10, 12, 14, 16**).

These results indicated that the use of phenyl groups or boron clusters in the design of selective ligands for these two receptors must be carefully considered. However, the fact that compound **4** (phenyl group tethered to the exo-amine group at position 6 by a propyl linker) has the highest docking score for the A2A receptor, whereas compound **3** (corresponding molecule with a boron cluster, Figure 3) has the highest docking score for the A3 receptor could be of interest.

2.3.2. Sugar Configuration

To determine the effects of the sugar stereochemistry on the ligand binding, β-D-ribofuranose containing compounds (**1–2** and **9–10**) were compared to β-D-arabinofuranose containing compounds (**5–6** and **7–8**, respectively), Figure 3, Table S1. It should be noted that the largest differences between the geometric scores and desolvation energies of the ribofuranose- and arabinofuranose-containing derivatives are observed when the phenyl ring is attached to the scaffold by a rigid ethynyl spacer (compounds **2** and **6**, Figures 5 and 7). This result shows that the protein-ligand interactions are influenced not only by the presence of a phenyl ring but also possibly by the spacer flexibility. If the sugar moiety is arabinofuranose, functionalizing the ligand with a phenyl group in the 2 position via a rigid ethynyl linker might increase its affinity for the A2A receptor.

For the A3 receptor, the nature of the linker and functional group has a smaller effect on the influence of the sugar configuration (Figure 7 and Table S2). Therefore, this moiety could impact the ligand selectivity for the A3 receptor.

2.3.3. Spacer Flexibility

The influence of spacer flexibility on A2A and A3 receptor-ligand interactions was already mentioned above. The geometric score profiles and desolvation energies reveal the significant effects of the spacer flexibility and length (score profiles shown in Figures 5 and 8, energy profiles shown in Figure 6 and Figure S4). Indeed, the molecules with the highest geometric scores and lowest desolvation energies have flexible (**7**, **8**, **9**, **10**, **13** and **14**, Figure 5) and/or long spacers (**3**, **4**, **11** and **12**, Figure 5).

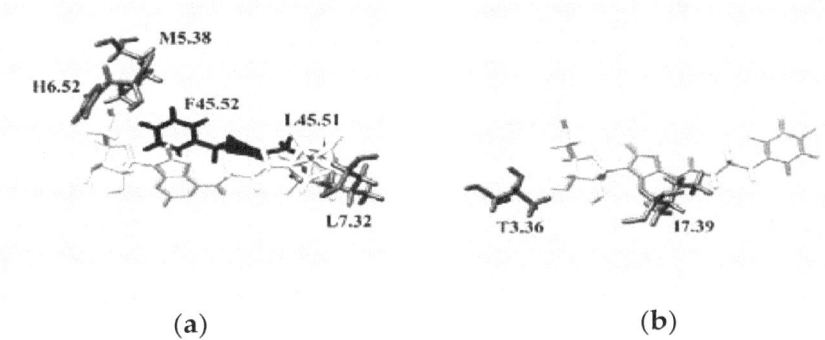

(a)   (b)

Figure 8. Views of the A2A receptor binding pocket showing the best docked poses of (**a**) compound **3** and (**b**) compound **4** (white: ligand, gray: transmembrane residues, black: extracellular amino acids).

2.3.4. Binding Pocket

Figure 8b shows the A2A binding pocket for the best docked pose of molecule **4** (Figure 3), which is best ligand for this receptor based on the in silico test results (Figures 5 and 7). Based on the number of unfavorable interactions (i.e., the number of clash contacts, or atom pairs separated by a distance of less than the sum of their van der Waals radii, Figure S5), substituting the phenyl ring with a carborane cage is unfavorable, although the geometric score is essentially unchanged after this modification (cf. the results for compounds **3** and **4**). This result might be due to the fact that new interactions between the ligand and the extracellular L45.51 and F45.52 residues arise when the phenyl group is substituted by a boron cluster (Figure 8b) (the reference for the residue distribution was the sequence of human A2 AR, UniProtKB code: P29274). These amino acids probably force the boron cluster to lie in a pocket that is too small, resulting in unfavorable protein-ligand interactions. In contrast, molecule **4** has a more complementary size and shape to the A2A into this adenosine receptor.

For the A3 receptor, replacing the phenyl ring with a boron cluster results in more favorable protein-ligand interactions (Figure 9). Indeed, of molecules **1–16**, molecule **3** was identified as the hit compound for A3 AR among the molecules containing carborane group. As shown in Figure 9b, when the phenyl group is substituted by a boron cluster to give molecule **3**, the ligand interacts with more transmembrane amino acids. Therefore, it was concluded that this region is fundamental for ligand binding to the A3 receptor. This hypothesis was also supported by the relative numbers of clash contacts for compounds **3** and **4** (Figure S6). Indeed, this parameter is reduced by 33% when the phenyl ring is replaced with a boron cluster, thus suggesting that, from a geometric shape complementarity point of view, boron cluster structure fits the A3 receptor channel in a better way if compared to phenyl group. A possible reason could be given by the reduced mobility of boron cluster which anchors the molecule more efficiently to inner surface of the channel. Anyway, the cause of this clash contacts reduction is unclear taking into account that the van der Waals volume of the carborane cage is ca. 50% higher than that of the phenyl group [38].

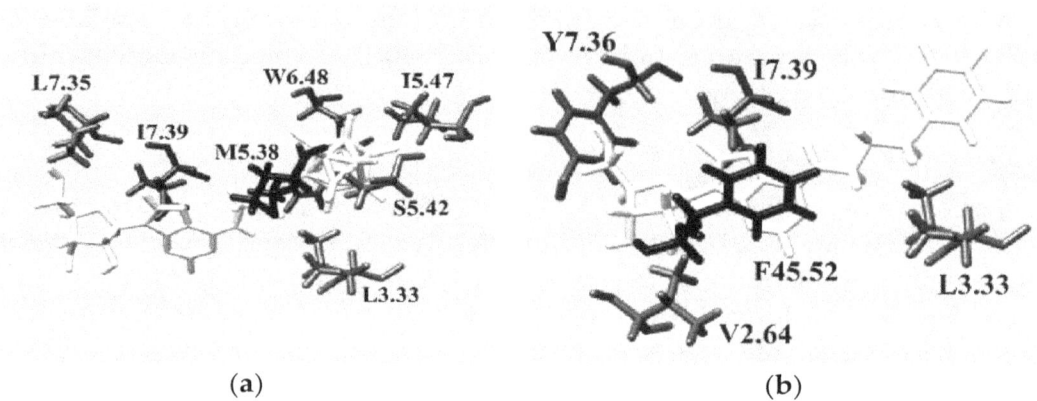

Figure 9. Views of the A3 receptor binding pocket for the best docked poses of (**a**) compound **3** and (**b**) compound **4** (white: ligand, gray: transmembrane residues, black: extracellular amino acids).

Nevertheless, the boron cluster in molecule **3** interacts with more intramembrane amino acids (L3.33, S5.42, I5.47 and W6.48) (the reference for the residue distribution was the sequence of human A3 AR, UniProtKB code: P0DMS8). These interactions are more favorable than those between the phenyl ring in molecule **4** and the surrounding protein residues (Figure 9b). Interestingly, when the phenyl ring is substituted by a boron cluster, the molecule can penetrate further into the receptor binding pocket towards the intramembrane region. Indeed, molecule **3** interacts with only one extracellular residue, which is located near the intramembrane region of the protein, whereas molecule **4** interacts with one residue located deeper in the extracellular region.

Overall, these results indicated that the distal region of extracellular loops of both the A2A and A3 receptors (blue in Figure 1) hinders ligand binding to them, whereas interactions between the ligand and the protein at extra/intramembrane interface, where binding pocket is located, are favorable. Indeed, the binding pockets of the best ligands (Figures 8b and 9b) are mostly localized in the transmembrane region, and these ligands exhibit the fewest clash contacts with the protein residues in rigid docking (Figures S5 and S6). Thus, the distal extracellular residues in A2A and A3 might not interact unfavorably with ligands; the A2A and A3 binding sites for ligands not fitting sterically (e.g., compound **11**) involve more distinct amino acids at the extracellular loop and have more clash contacts than those of the best ligands (Tables S1 and S3 for A2A and Tables S2 and S4 for A3). It is well-founded that extracellular domains of G protein-coupled receptors can be crucial for ligand binding and for activation/inhibition of the adenosine receptors. It would be of interest therefore to analyze the molecules described herein also against other members of this group of receptors to acquire information about their specificity for different members of this protein family. These study are however beyond the scope of the present communication.

2.4. Synthesis of Compounds **3**,**4** *and* **11**,**12**, *Which Contain a Boron Cluster or Phenyl Group*

Compounds **3** [19], **4** [50] and **11** [51] were synthesized as described. Compound **12** was analogously obtained as **11** using 3-phenyl-1-propanol instead of 3-(1,12-dicarba-*closo* dodecaboran-1-yl)-1-propanol (Figure 10). Thus, first, 6-*N*-benzoyl-3′,5′-*O*,*O*-(tetraisopropyldisiloxane-1,3-diyl)adenosine was prepared from adenosine according to the a previously reported procedure [52]; then, a reaction with DMSO in a mixture of acetic acid/acetic anhydride [53] provided a key intermediate 6-*N*-/benzoyl-3′,5′-*O*,*O*-(tetraisopropyldisiloxane-1,3-diyl)-2′-*O*-methylenethiomethyl-adenosine. The treatment of the intermediate with 3-phenyl-1-propanol and subsequent removal of the protecting groups produced compound **12**.

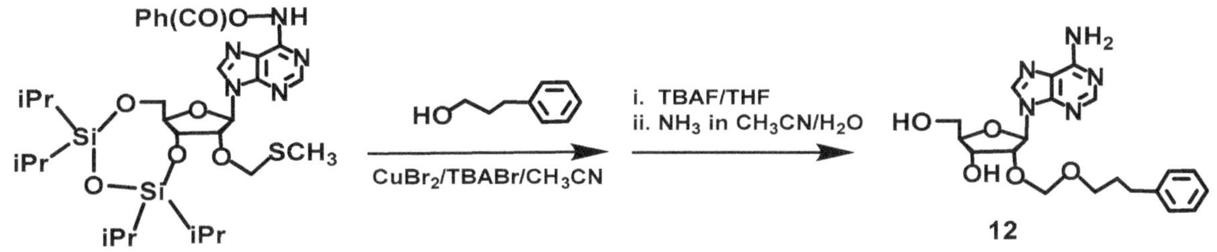

Figure 10. Synthesis of 2'-O-(3-phenylpropyleneoxymethyl]-adenosine (**12**) from a key intermediate 6-N-benzoyl-3',5'-O,O-(tetraisopropyldisiloxane-1,3-diyl)-2'-O-methylenethiomethyladenosine.

2.5. Radioligand Assay

The radioligand replacement assay based on the competition binding of the tested compound and a ligand with known affinity toward the receptor was performed under contractual service agreement with Plataforma de Screening de Farmacos (USEF), 15782 Santiago de Compostela, Spain. As the radioligand for the A2A receptor [³H]-ZM241385, a 2,8-substituted[1,2,4]triazolo[1,5-*a*][1,3,5]triazine, an adenine isoster and a high-affinity antagonist, which is selective for the adenosine A2A receptor, was used. As the standard control for the studied receptor-binding CGS15943, non-nucleoside agonist was applied. Non-specific binding was determined in the presence of NECA.

For the A3 receptor 10 nM [³H]-NECA, a 5'-N-ethylcarboxamide adenosine derivative, was used. Non-specific binding was determined in the presence of a high concentration of R-PIA, a N⁶-(2-phenylisopropyl)adenosine, a specific agonist for A1 receptor. The binding affinities were measured as a percent of the radioligand displacement by the tested compounds and are shown in Table 1. For both pairs of compounds, the binding of the adenosine ligands that were modified with the phenyl group is more efficient than that of the counterparts modified with the boron cluster. The high, in nM range, binding affinity of phenyl modified compound **4** (K_i = 7.5 nM) should be pointed out as a good starting point for the further improvements [54,55].

Table 1. Specific binding of the compounds **3**, **4**, and **11**, **12** to adenosine receptors A2A and A3 in radioligand competition binding assay.

Compound	% Inhib. 10 μM A2A	% Inhib. 10 μM A3	K_i (nM) A3
Adenosine 2'-deoxyadenosine	2 ± 1	10 ± 1	n.d.
	3 ± 1	5 ± 1	n.d.
3	2 ± 2	11 ± 4	n.d.
4	32 ± 1	98 ± 1	7.5
11	1 ± 1	23 ± 3	n.d.
12	2 ± 1	67 ± 1	2208

n.d. = not determined.

Interestingly the concentration-response curves of the compounds are qualitatively, if not quantitatively, consistent with the results of the in silico study. Thus, Figure 11a,b shows a high concentration-dependent binding of compounds **3** and **4** to the A2A receptor, where **4** > **3**, which is qualitatively consistent with the in silico calculation. In the case of compounds **11** and **12**, a substantial decrease in geometric score is observed for both compounds (Figure 5), which is reflected in the flat, non-binding, concentration-response curves (Figure 11c,d). However, the differences in binding affinities are much smaller (Figure 11c,d) than the expected values based on the geometric score disparity (Figure 5).

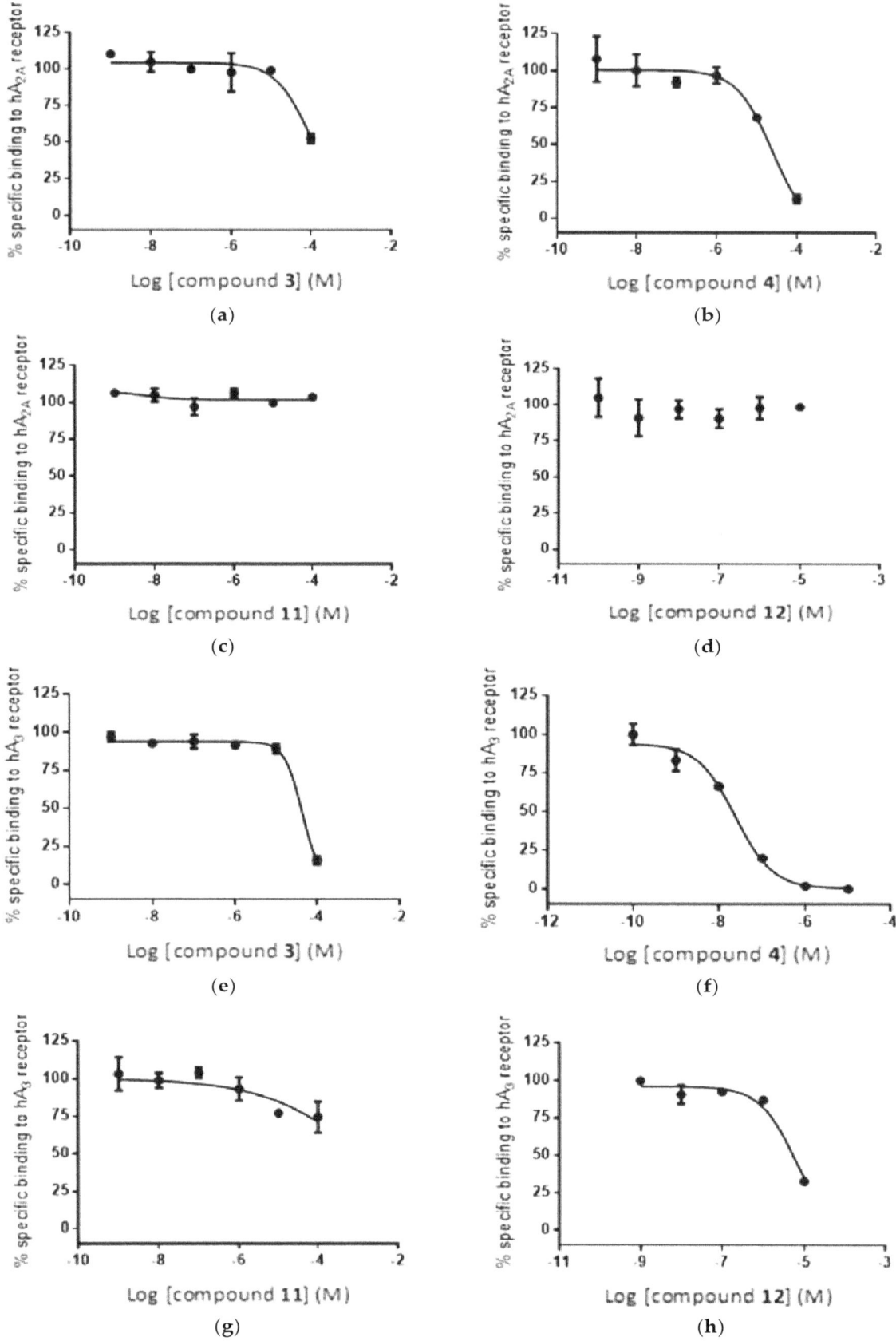

Figure 11. Radioligand competition binding assay: specific binding of compounds **3**, **4**, **11** and **12** to the adenosine receptors A2A (**a–d**) and A3 (**e–h**) (see Experimental Section).

For the binding of compounds **3**, **4** and **11**, **12** to receptor A3, a similar but less consistent relationship between in silico calculations and biological screening can be observed. Again, compounds **3** and **4** show reasonable binding to the receptor (Figure 11d,e) as predicted in the in silico calculations (Figure 7), although the order of affinities is reversed: **3** > **4** for the in silico assay and **4** > **3** in the biological test. In other words, the radioligand assay showed lower affinity of compound **3** (boron cluster modification) to A3 receptor than that of compound **4** (phenyl group modification), though the geometric scores (6530 vs. 6120) and clash contacts (11 vs. 18) do not reflect the respective differences in compounds **3** and **4** affinity for the receptor (Figure 7 and Table S2). Consequently, the observed difference in binding affinity between compounds **3** and **4**, in the radioligand replacement assay (Table 1, Figure 11d,e) is much higher than the predicted value from the geometric scores (Figure 7). For the compound pair of **11** and **12** (Figure 11f,g) the relationship goes back to the previous correlation in both the in silico assay and the biological test: the phenyl modification corresponds to higher affinity. These in vitro results are consistent with the observed trend in in silico study (Figure 12).

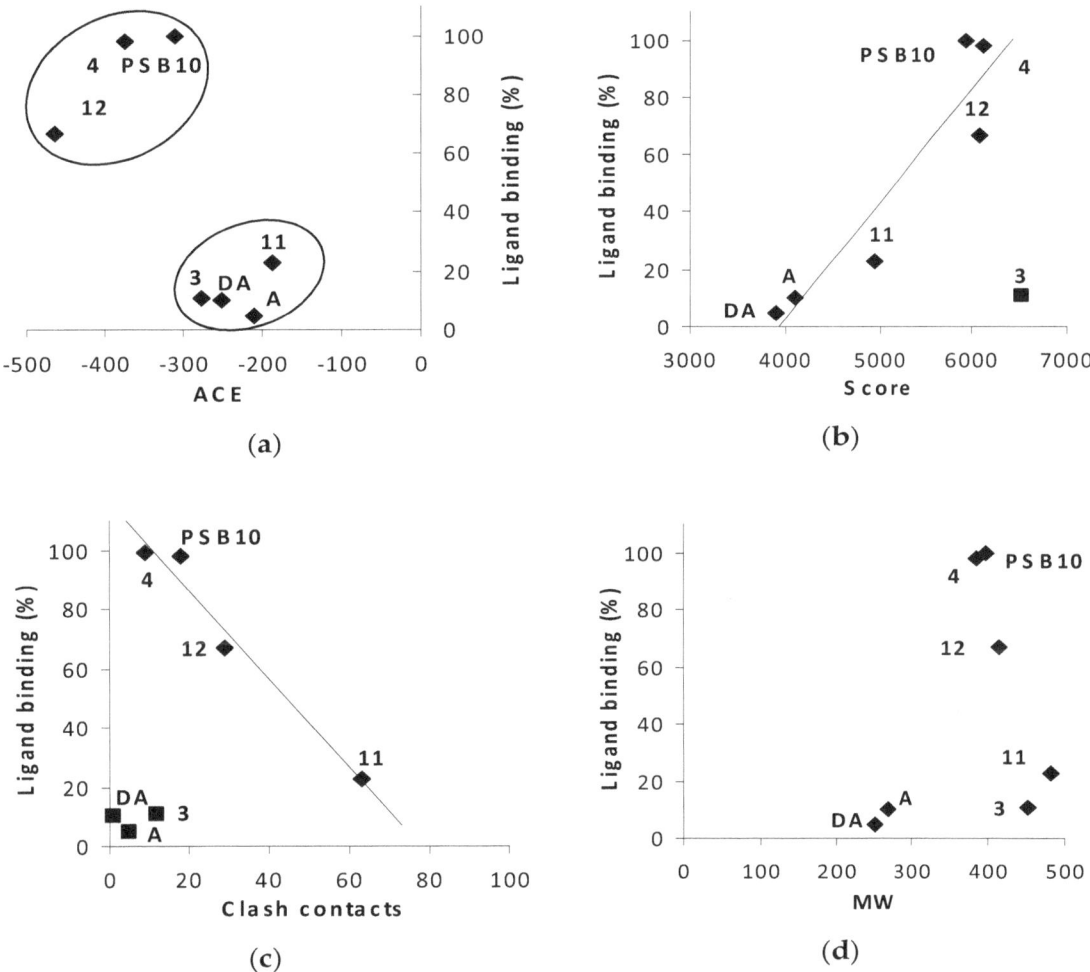

Figure 12. Relationships of desolvation energy, ACE (**a**), geometric scores (**b**) and clash contacts (**c**), calculated in silico, and ligand binding in vitro for A3 receptor, for the compounds **3**, **4**, **11**, **12**, PSB 10, adenosine (A) and 2′-deoxyadenosine (DA) as ligands; The relationship of ligand binding and molecular weight (MW) (**d**). Ligand binding was determined in radioligand competition binding assay as described in Materials and Methods, and expressed as % inhibition of a specific binding at 10 μM. Values of MW [D] for **3**, **4**, **11**, **12**, PSB 10, adenosine and 2′-deoxyadenosine are 451.53, 385.42, 481.56, 415.44, 398.67, 267.24 and 251.24, respectively.

The observed consistency in trends of the effect of the boron cluster or phenyl modification on the adenosine binding to the A2 and A3 receptors for both in silico and wet assays, although only approximate, is important as a base for further improvements.

In silico and in vitro results for A3 receptor and compounds **3**, **4**, **11**, **12** and adenosine, or $2'$-deoxyadenosine show similar trends (Figure 12a–c). The higher the score, the better the ligand binding to A3 receptor for **4**, **11**, **12**, PSB 10, adenosine and $2'$-deoxyadenosine (Figure 12b, correlation $R^2 = 0.884$). The score above 5000–5200 characterizes active compounds with an exception of compound **3** (outlier, Figure 12b). The reason for this discrepancy needs furthers study. The ACE parameter distinguish PSB 10 (highly specific ligand) and phenyl-modified compounds, **4** and **12**, with low ACE value and high activity form lower active adenosine or $2'$-deoxyadenosine (nonspecific ligands) and boron cluster-modified compounds, **3** and **11** (Figure 12b). Notably, the clash contacts *versus* radioligand inhibition relationship was also convergent for almost all compounds compared (Figure 12c). The higher the clash contacts value, the worse the ligand binding to A3 receptor of **4**, **11**, **12** and PSB 10 (Figure 12c, correlation $R^2 = 0.96$). Based on the correlation data we suggest that ACE can be the most specific parameter differentiating the compounds activity. We notice, that compounds with ACE < -300 were active to the receptor in in vitro study. Moreover, we can presume that compounds with activity equal to or greater than 40% (at 10 µM) can exhibit specific interactions with A3R. The theoretical and experimental properties of compounds **4** and **12** are in close proximity to selective A3R ligand, PSB 10 in all analyses (Figure 12a–c). On the other hand, with reference to low number of clash contacts and concomitant low binding to A3R, the compound **3** is similar to non-selective adenosine or $2'$-deoxyadenosine. In light of these observations, compound **11** with low score, high number of clash contacts and the highest molecular weight (whatever it means for the adenosine receptor), appears to be the worst candidate for the A3 receptor.

The plot analysis of the ligand binding *versus* molecular weight (MW) of compounds did not show any clear relationships (Figure 12d). Additionally, ligand efficiency instead of ligand binding plotted against in silico parameters (Figure 7) yielded similar results like for ligand binding factor (Figure 12a–c).

The consistency between in silico and wet assay results is still a difficult challenge. There are several possible reasons for not perfect overlapping of the calculation and experimental results described herein. First, despite the use of a new in boron cluster field calculation protocol based on the PatchDock server, the effects of the unique boron cluster properties on the action of molecules that target the A2A and A3 adenosine receptors could not be included into calculations. Second, for the docking experiments, the A2A receptor based on the structure that bound the endogenous adenosine, which is an agonist and fixes the receptor in the active conformation, was selected. Though, based on their structural similarity to adenosine, compounds **1**–**16** are likely agonists too, their agonist properties are currently not proven. Furthermore, the structure of the receptor A3 was not based on the crystal structure (as in the case of A2A), which is not yet available but is based on the homology modeling which introduces additional uncertainty. For receptor A3, the agonist radioligand was consistently used. Finally, the access to commercial radioligands with specific agonist or antagonist properties is limited. Thus, fitting the AR receptor structure, which is fixed in the agonist or antagonist conformation, with a tested ligand with agonist or antagonist properties, which are often unknown, in the docking experiment is difficult. Moreover, for the best result comparison of the in silico study and radioligand replacement assay, the radioligand with the identical agonist or antagonist properties to that in the docking experiment should be used in the biological test. Lack of conformity of all factors may affect the accuracy of prediction of the ligand behaviour in the biological environment based on the docking experiments. Therefore it may be of interest that compound **3** tested for affinity to A2A receptor using [³H]-ZM241385 antagonist radioligand described herein, displayed only moderate affinity, but it showed considerably higher inhibitory property when [³H]-CGS 21680, an agonist radioligand was used [18]. The reasonable, albeit not perfect overlapping of in silico and wet assay results described herein, while based on small set of compounds is encouraging and prompt study of larger library of adenosine derivatives using reported methodology.

3. Materials and Methods

3.1. General Information

Commercially available chemicals were of reagent grade and used as received. Adenosine and 2'-O-deoxyadenosine were purchased from Pharma-Waldhof GmbH (Düsseldorf, Germany), 3-phenylpropane-1-ol was from Sigma-Aldrich (Steinheim, Germany). Solvents were purchased in the highest available quality. Column chromatography was performed on silica gel 230-400 mesh and TLC was performed on silica gel F254 plates, both purchased from Sigma-Aldrich (Steinheim, Germany).

^{1}H-NMR spectra were recorded on a Bruker Avance III 600 MHz spectrometer. The spectra for ^{1}H nuclei were recorded at 600.26 MHz using a deuterated solvent as a standard All chemical shifts are reported in ppm relative to the internal standards. UV measurements were performed with a GBC Cintra10e UV-VIS spectrometer (Dandenong, Australia). Samples for UV experiments, ca. 0.5 A_{260} ODU for each compound, were dissolved in 96% C_2H_5OH or CH_3OH. The measurement was performed at ambient temperature.

3.2. Synthesis of 2'-O-(3-Phenylpropyleneoxymethyl)-adenosine (12)

6-N-Benzoyl-3',5'-O,O-(tetraisopropyldisiloxane-1,3-diyl)-2'-O-methylenethiomethyladenosine (1 eq, 40 mg, 0.059 mmol) was dissolved in acetonitrile (0.5 mL) then was treated with 3-phenylpropane-1-ol (2.7 eq., 22 mg, 0.16 mmol) at the presence of copper(II) bromide (1.1 eq, 14 mg, 0.065 mmol) and tetrabutylammonium bromide (1.1 eq, 21 mg, 0.065 mmol) as activators. The reaction progress was monitored by TLC (CH_2Cl_2/CH_3OH, 9:1). After reaction completion (ca. 48 at room temperature) solvent was evaporated under reduced pressure, then the oily residue was dissolved in dichloromethane (2 mL). The resultant solution was washed with water (3 × 1 mL) then the organic fraction was dried over anhydrous magnesium sulfate and evaporated to dryness under vacuum. Next, the crude product without purification was dissolved in THF (1.2 mL) then TBAF (1 mL, 3.5 mmol) was added. After 15 min to the reaction mixture a pyridine/methyl alcohol/water (3:1:1, 2.5 mL) followed by ion exchange resin Dowex 50Wx8 in pyridinium form, was added. After 30 min the ion exchange resin was filtered off and washed with pyridine/methyl alcohol/water (3:1:1, 3 × 5 mL). The filtrate and washings were combined together then whole was evaporated to dryness under vacuum yielding crude product. The crude product was purified by silica gel column chromatography (5 g, 230–400 mesh) using a linear gradient of CH_3OH in CH_2Cl_2 (0–5%) as a eluting solvent system. The obtained 6-N-benzoyl-2'-O-(3-phenylpropyleneoxymethyl)adenosine (ca. 30 mg) was dissolved in CH_3CN (0.2 mL) then concentrated aq. ammonia solution was added (2 M, 2 mL). After 2 h at room temperature (TLC control, CH_2Cl_2/CH_3OH, 9:1) solvents were evaporated under vacuum then final product was purified by silica gel column chromatography (5 g, 230–400 mesh) using a linear gradient of CH_3OH in CH_2Cl_2 (0–10%) as a eluting solvent system. Yield: 9 mg. TLC (CH_2Cl_2/CH_3OH, 9:1 v/v): R_f = 0.64; UV-Vis (95% C_2H_5OH) λ, nm: 249 (min), 280 (max); ^{1}H-NMR (600 MHz, CDCl$_3$) δ (ppm): 2.34–2.37 (m, 2H, CH$_2$), 3.33–3.35 (m, 2H, CH$_2$), 3.81–3.95 (m, 2H, H-5', 5''), 4.075 (d, 2H, CH$_2$-O, J = 6.6 Hz), 4.225 (d, 1H, H-4', J = 3.0 Hz), 4.51–4.52 (m, 1H, H-3'), 4.735 (d, 1H, H-2', J = 6.6 Hz), 6.305 (d, 1H, H-1', J = 6 Hz), 7.04–7.09 (m, 2H, phenyl ring), 7.18 (t, 2H, NH$_2$, J = 7.8 Hz), 7.58 (t, 2H, phenyl ring, J = 7.8 Hz), 7.67–7.68 (m, 1H, phenyl ring), 8.065 (d, 1H, H-8, J = 1.8 Hz), 8.725 (d, 1H, H-2, J = 7.8 Hz).

3.3. Docking Analysis

Computational docking simulations were conducted using two different web services: SwissDock (http://www.swissdock.ch) and PatchDock v1.3 (http://bioinfo3d.cs.tau.ac.il/PatchDock). The former web service is based on the EADock DSS (Evolutionary Algorithm for Docking) software [35]. Evolutionary algorithms are iterative stochastic optimization procedures in which an initial population of solutions is generated and evaluated with respect to a set of constraints described by the fitness function.

The PatchDock algorithm consists of three main phases [46]. In the first phase, the surface of the molecule is computed, and a segmentation process is subsequently performed to determine the geometric patches (concave, convex and flat surface sections). Then, only the patches with "hot spot" residues are retained and matched via a hybrid of the Geometric Hashing and Pose-Clustering matching techniques. During this step, the concave patches are matched to convex ones, and flat patches are matched with any type of patches. The resulting candidate complexes are examined to discard all complexes with unacceptable overlap between the receptor and ligand atoms. Finally, the remaining candidates are ranked based on their geometric shape complementarity scores. For the preliminary A2A receptor docking studies, the input target consisted of the A2A structure extracted from the A2A protein/adenosine complex X-ray structure (PDB code: 2YDO) [28]. All the molecules in Figures 3 and 4 were considered to be ligands, and their input files for the SwissDock server were generated with Chem3DPro 16.0 and then converted into the ".mol2" format with UCSF Chimera [29]. For the PatchDock simulations, all the compounds were drawn in Chem3DPro 16.0 and saved as ".pdb" files. The docking studies were performed using a clustering RMSD (root mean square deviation, parameter used to discard redundant solutions) of 4.0 Å. All the simulations were conducted without specifying a region of interest (ROI) to ensure that the chosen docking methods could locate the correct binding pocket.

3.4. Radioligand Replacement Assay for Human Adenosine Receptors

The 10 mM stock solutions were prepared by dissolving 0.5–1 mg of the tested compounds in suitable volume of DMSO. Next, dilutions in binding buffer were prepared to obtain concentration ranges 10^{-10}–10^{-5} or 10^{-9}–10^{-4} depending upon the compound potency. The concentration-response curves of the standard controls were made. For this purpose A3R antagonist, MRS 1220, and A2AR antagonist, CGS 15943 were used for the binding study in the same buffer/vehicle system as for tested compounds.

Adenosine A2A receptor competition binding experiments were carried out in a multiscreen GF/C 96-well plate (Millipore, Madrid, Spain) pretreated with binding buffer (Tris-HCl 50 mM, EDTA 1 mM, $MgCl_2$ 10 mM, 2 U/mL adenosine deaminase, pH = 7.4). In each well was incubated 5 µg of membranes from Hela-A2A cell line (Lot: A001/18-10-2010, protein concentration = 4288 µg/mL), 3 nM [3H]-ZM241385 (50 Ci/mmol, 1 mCi/mL, ARC-ITISA 0884) and compounds studied and standard. Non-specific binding was determined in the presence of NECA 50 µM (Sigma E2387, St. Louis, MO, USA). The reaction mixture (Vt: 200 µL/well) was incubated at 25 °C for 30 min, after was filtered and washed four times with 250 µl wash buffer (Tris-HCl 50 mM, EDTA 1 mM, $MgCl_2$ 10 mM, pH = 7.4), before measuring in a microplate beta scintillation counter (Microbeta Trilux, PerkinElmer, Madrid, Spain).

Adenosine A3 receptor competition binding experiments were carried out in a multiscreen GF/B 96-well plate (Millipore, Madrid, Spain) pretreated with binding buffer (Tris-HCl 50 mM, EDTA 1 mM, $MgCl_2$ 5 mM, 2 U/mL adenosine deaminase, pH = 7.4). In each well was incubated 70 µg of membranes from Hela-A3 cell line (Lot: A003/13-04-2016, protein concentration = 2449 µg/mL), 10 nM [3H]-NECA (27.6 Ci/mmol, 1 mCi/mL, Perkin Elmer NET811250UC) and compounds studied and standard. Non-specific binding was determined in the presence of R-PIA 100 µM (Sigma P4532, St. Louis, MO, USA). The reaction mixture (Vt: 200 µL/well) was incubated at 25 °C for 180 min, after was filtered and washed six times with 250 µL wash buffer (Tris-HCl 50 mM pH = 7.4), before measuring in a microplate beta scintillation counter (Microbeta Trilux, PerkinElmer, Madrid, Spain).

4. Conclusions

This work describes a practical protocol for evaluating adenosine derivatives as potential ARs ligands in silico. In this approach, the structural features of the ligand and their effect on interactions with receptor proteins can be assessed. Here, the binding of adenosine ligands modified with phenyl groups was compared to that of ligands modified with inorganic modifier, a boron cluster. The results can be used to guide future biological tests and identify lead molecules within this class of compounds for follow-up studies. In this work, compounds **3** and **4** (one with a boron cluster and one with a phenyl group) were found to be the most promising ligands. However, the observation made that a boron cluster can hinder the ligand from entering the receptor binding pocket calls for caution in the cases where the interaction between a cluster and spatially limited protein cavity can be expected. It shows also that the 50% higher van der Waals volume of the carborane cage than that of the rotating phenyl group can make a difference. This observation can be of importance not only for adenosine ligands/AR system but for medicinal chemistry of boron cluster in general. On the other hand, it should be stressed that the performed in silico screening was based on the effect of this moiety on the rigid docking and the space fitting approach, and that other unique, potentially beneficial, properties of boron clusters were not included due to the modeling softwares limitations.

In silico as well as in biological assay results show the noteworthy compatibility in trends although not quantitative consistency. Current our work is focused on improvements of the described modeling methodology to make it better applicable to A3 and other adenosine receptors. Further works both in silico and experimentally are also ongoing to provide more insight into adenosine modifications for more selective binding to different adenosine receptors and wet assays to validate this in silico approach.

Supplementary Materials: The following are available online, Figure S1, (a) Best docking pose of adenosine obtained with SwissDock web server and considering human A2A receptor (PDB code: 2YDO) with thioglucoside, crystallization helper molecules and without (b); Figure S2, (a) X-ray structure thermostabilized human A2A receptor with adenosine bound (PDB code: 2YDO), (b) PatchDock best-docked pose obtained with adenosine as a ligand and considering human A2A receptor (PDB code: 2YDO); Figure S3, (a) Best docking pose of adenosine obtained with PatchDock web server and considering human A2A receptor (PDB code: 2YDO), with crystallization helper molecule and without (b); Figure S4, Desolvation energy associated to the A3 protein-ligand interaction. ● for reference molecules (CF101, CF102, adenosine, 2′-deoxyadenosine, and NECA); ▲ for molecules bearing boron clusters (**1, 3, 5, 7, 9, 11, 13, 15**); ◆ for ligands with phenyl ring (**2, 4, 6, 8, 10, 12, 14, 16**); Figure S5, Clash contact involved in A2A-ligand interaction ● for reference molecules (regadenoson, apadenoson, CGS 21680, adenosine, 2′-deoxyadenosine and NECA); ▲ for molecules bearing boron clusters (**1, 3, 5, 7, 9, 11, 13, 15**); ◆ for ligands with phenyl ring (**2, 4, 6, 8, 10, 12, 14, 16**); Figure S6, Clash contact involved in A3-ligand interaction ● for reference molecules (CF101, CF102, adenosine, 2′-deoxyadenosine, and NECA); ▲ for molecules bearing boron clusters (**1, 3, 5, 7, 9, 11, 13, 15**); ◆ for ligands with phenyl ring (**2, 4, 6, 8, 10, 12, 14, 16**); Figure S7, Relationships of desolvation energy, ACE (a), geometric scores (b) and clash contacts (c) *versus* ligand efficiency for A3 receptor and for the compounds **3, 4, 11, 12**, PSB 10, adenosine and 2′-deoxyadenosine as ligands. Table S1, Docking results of molecules **1–16** towards adenosine receptor A2A; Table S2, Docking results of molecules **1–16** towards adenosine receptor A3; Table S3, Binding pockets of molecule **1–16** for adenosine receptor A2; Table S4, Binding pockets of molecule **1–16** for adenosine receptor A3.

Author Contributions: Conceptualization, M.V. and Z.L.; Data curation, K.B. and Z.L.; Formal analysis, K.B.; Funding acquisition, Z.L.; Investigation, M.V., K.B. and Z.L.; Methodology, M.V. and K.B.; Project administration, Z.L.; Resources, Z.L.; Software, M.V.; Supervision, Z.L.; Validation, K.B. and Z.L.; Visualization, M.V.; Writing—original draft, M.V.; Writing—review & editing, K.B. and Z.L., K.B. and M.V. contributed to this work equally.

Acknowledgments: The authors thank Aleksandra Kierozalska for the synthesis of compound **4**.

References

1. Haskó, G.; Linden, J.; Cronstein, B.; Pacher, P. Adenosine receptors: Therapeutic aspects for inflammatory and immune diseases. *Nat. Rev. Drug Discov.* **2008**, *7*, 759–770. [CrossRef] [PubMed]
2. Trincavelli, M.L.; Daniele, S.; Martini, C. Adenosine receptors what we know and what we are learning. *Curr. Top. Med. Chem.* **2010**, *10*, 860–877. [CrossRef] [PubMed]

3. Jacobson, K.A.; Gao, Z.G.; Ijzerman, A.P. Allosteric modulation of purine and pyrimidine receptors. *Adv. Pharmacol.* **2011**, *61*, 187–220. [CrossRef] [PubMed]

4. Jacobson, K.A.; Daly, J.W.; Manganiello, V.F. *Purines in Cellular Signaling—Targets for New Drugs*; Springer-Verlag: New York, NY, USA, 1990; ISBN 978-0-387-97244-2.

5. Tosh, D.K.; Paoletta, S.; Chen, Z.; Moss, S.M.; Gao, Z.G.; Salvemini, D.; Jacobson, K.A. Extended N^6 substitution of rigid C2-arylethynyl nucleosides for exploring the role of extracellular loops in ligand recognition at the A_3 adenosine receptor. *Bioorg. Med. Chem. Lett.* **2014**, *15*, 3302–3306. [CrossRef] [PubMed]

6. Tosh, D.K.; Phan, K.; Gao, Z.G.; Gakh, A.A.; Xu, F.; Deflorian, F.; Abagyan, R.; Stevens, R.C.; Jacobson, K.A.; Katritch, V. Optimization of adenosine 5′-carboxamide derivatives as adenosine receptor agonists using structure-based ligand design and fragment screening. *J. Med. Chem.* **2012**, *55*, 4297–4308. [CrossRef] [PubMed]

7. Scholz, M.; Hey-Hawkins, E. Carbaboranes as pharmacophores—Properties, synthesis, and application strategies. *Chem. Rev.* **2011**, *111*, 7035–7062. [CrossRef] [PubMed]

8. Issa, F.; Kassiou, M.; Rendina, L.M. Boron in drug discovery Carboranes as unique pharmacophores in biologically active compounds. *Chem. Rev.* **2011**, *111*, 5701–5722. [CrossRef] [PubMed]

9. Leśnikowski, Z.J. Recent developments with boron as a platform for novel drug design. *Exprt Opin. Drug Discov.* **2016**, *11*, 569–578. [CrossRef] [PubMed]

10. Leśnikowski, Z.J. Challenges and opportunities for the application of boron clusters in drug design. *J. Med. Chem.* **2016**, *59*, 7738–7758. [CrossRef] [PubMed]

11. Armstrong, A.F.; Valiant, J.F. The bioinorganic and medicinal chemistry of carboranes from new drug discovery to molecular imaging and therapy. *Dalton Trans.* **2007**, *38*, 4240–4251. [CrossRef] [PubMed]

12. Lesnikowski, Z.J. New opportunities in boron chemistry for medical applications. In *Boron Sciences. New Technologies and Applications*, 1st ed.; Hosmane, N.S., Ed.; CRC Press: Boca Raton, FL, USA, 2011; pp. 3–19. ISBN 9781439826621.

13. Lesnikowski, Z.J.; Shi, J.; Schinazi, R. Nucleic acids and nucleosides containing carboranes. *J. Organomet. Chem.* **1999**, *581*, 156–169. [CrossRef]

14. Druzina, A.A.; Bregadze, V.I.; Mironov, A.F.; Semioshkin, A.A. Synthesis of conjugates of polyhedral boron hydrides with nucleosides. *Russ. Chem. Rev.* **2016**, *85*, 1229–1254. [CrossRef]

15. Lesnikowski, Z.J.; Olejniczak, A.B.; Schinazi, R.F. *Frontiers in Nucleic Acids*, 1st ed.; Schinazi, R.F., Liotta, D.C., Eds.; IHL Press: Tucker, GA, USA, 2004; ISBN 0975418807.

16. Lesnikowski, Z.J. Boron Clusters—A New Entity for DNA-Oligonucleotide Modification. *Eur. J. Org. Chem.* **2003**, *23*, 4489–4500. [CrossRef]

17. Baxter, A.; Bent, J.; Bowers, K.; Braddock, M.; Brough, S.; Fagura, M.; Lawson, M.; McInally, T.; Mortimore, M.; Robertson, M.; et al. Hit-to-Lead studies: The discovery of potent adamantane amide P2X7 receptor antagonists. *Bioorg. Med. Chem. Lett.* **2003**, *13*, 4047–4050. [CrossRef] [PubMed]

18. Bednarska, K.; Olejniczak, A.B.; Wojtczak, B.; Sułowska, Z.; Leśnikowski, Z.J. Adenosine and 2′-deoxyadenosine modified with boron cluster pharmacophores as new classes of human blood platelet function modulators. *Chem. Med. Chem.* **2010**, *5*, 749–756. [CrossRef] [PubMed]

19. Bednarska, K.; Olejniczak, A.B.; Piskała, A.; Klink, M.; Sułowska, Z.; Leśnikowski, Z.J. Effect of adenosine modified with a boron cluster pharmacophore on reactive oxygen species production by human neutrophils. *Bioorg. Med. Chem.* **2012**, *20*, 6621–6629. [CrossRef] [PubMed]

20. Białek-Pietras, M.; Olejniczak, A.B.; Paradowska, E.; Studzińska, M.; Suski, P.; Jabłońska, A.; Leśnikowski, Z.J. Synthesis and in vitro antiviral activity of lipophilic pyrimidine nucleoside/carborane conjugates. *J. Organomet. Chem.* **2015**, *798*, 99–105. [CrossRef]

21. Żołnierczyk, J.D.; Olejniczak, A.B.; Mieczkowski, A.J.; Błoński, J.Z.; Kiliańska, Z.M.; Robak, T.; Leśnikowski, Z.J. In vitro antileukemic activity of novel adenosine derivatives bearing boron cluster modification. *Bioorg. Med. Chem.* **2016**, *24*, 5076–5087. [CrossRef] [PubMed]

22. Bours, M.J.L.; Swennen, E.L.R.; Di Virgilio, F.; Cronstein, B.N.; Dagnelie, P.C. Adenosine 5′-triphosphate and adenosine as endogenous signaling molecules in immunity and inflammation. *Pharmacol. Ther.* **2006**, *112*, 358–404. [CrossRef] [PubMed]

23. Cronstein, B.N.; Levin, R.I.; Belanoff, J.; Weissmann, G.; Hirschhorn, R. Adenosine: An endogenous inhibitor of neutrophil-mediated injury to endothelial cells. *J. Clin. Investig.* **1986**, *78*, 760–770. [CrossRef] [PubMed]

24. Hasko, G.; Pacher, P.J. A2A receptors in inflammation and injury: Lessons learned from transgenic animals. *J. Leukoc. Biol.* **2008**, *83*, 447–455. [CrossRef] [PubMed]

25. Areias, F.M.; Brea, J.; Gregori-Puigjané, E.; Zaki, M.E.A.; Carvalho, M.A.; Domínguez, E.; Gutiérrez-de-Terán, H.; Proença, M.F.; Loza, M.I.; Mestres, J. In silico directed chemical probing of the adenosine receptor family. *Bioorg. Med. Chem.* **2010**, *18*, 3043–3052. [CrossRef] [PubMed]

26. Knight, A.; Hemmings, J.L.; Winfield, I.; Leuenberger, M.; Frattini, E.; Frenguelli, B.G.; Dowell, S.J.; Lochner, M.; Ladds, G. Discovery of novel adenosine receptor agonists that exhibit subtype selectivity. *J. Med. Chem.* **2016**, *59*, 947–964. [CrossRef] [PubMed]

27. Sirci, F.; Goracci, L.; Rodriguez, D.; van Muijlwijk-Koezen, J.; Gutierrez-de-teran, H.; Mannhold, R. Ligand-, structure- and pharmacophore-based molecular fingerprints: A case study on adenosine A1, A 2A, A2B and A3 receptor antagonists. *J. Comput. Aided Mol. Des.* **2012**, *26*, 1247–1266. [CrossRef] [PubMed]

28. Lebon, G.; Warne, T.; Edwards, P.C.; Bennett, K.; Langmead, C.J.; Leslie, A.G.; Tate, C.G. Agonist-bound adenosine A2A receptor structures reveal common features of GPCR activation. *Nature* **2011**, *474*, 521–525. [CrossRef] [PubMed]

29. Pettersen, E.F.; Goddard, T.D.; Huang, C.C.; Couch, G.S.; Greenblatt, D.M.; Meng, E.C.; Ferrin, T.E. UCSF Chimera—A visualization system for exploratory research and analysis. *J. Comput. Chem.* **2004**, *13*, 1605–1612. [CrossRef] [PubMed]

30. Almerico, A.M.; Tutone, M.; Pantano, L.; Lauria, A. A3 adenosine receptor: Homology modeling and 3D-QSAR studies. *J. Mol. Graph. Model.* **2013**, *42*, 60–72. [CrossRef] [PubMed]

31. Vyas, V.K.; Ukawala, R.D.; Ghate, M.; Chintha, C. Homology modeling a fast tool for drug discovery: Current perspectives. *Indian J. Pharm. Sci.* **2012**, *74*, 1–17. [CrossRef] [PubMed]

32. Biasini, M.; Bienert, S.; Waterhouse, A.; Arnold, K.; Studer, G.; Schmidt, T.; Kiefer, F.; Gallo Cassarino, T.; Bertoni, M.; Bordoli, L.; et al. SWISS-MODEL: Modelling protein tertiary and quaternary structure using evolutionary information. *Nucl. Acids Res.* **2014**, *42*, W252–W258. [CrossRef] [PubMed]

33. Wu, S.; Zhang, Y. LOMETS: A local meta-threading-server for protein structure prediction. *Nucl. Acids Res.* **2007**, *35*, 3375–3382. [CrossRef] [PubMed]

34. Segala, E.; Guo, D.; Cheng, R.K.Y.; Bortolato, A.; Deflorian, F.; Doré, A.S.; Errey, J.C.; Heitman, L.H.; Jzerman, A.P.; Marshall, F.H.; et al. Controlling the dissociation of ligands from the adenosine A2A receptor through modulation of salt bridge strength. *J. Med. Chem.* **2016**, *59*, 6470–6479. [CrossRef] [PubMed]

35. Grosdidier, A.; Zoete, V.; Michielin, O. SwissDock, a protein-small molecule docking web service based on EADock DSS. *Nucl. Acids Res.* **2011**, *39*, W270–W277. [CrossRef] [PubMed]

36. Eisenberg, D.; Lüthy, R.; Bowie, J.U. VERIFY3D: Assessment of protein models with three-dimensional profiles. *Methods Enzymol.* **1997**, *277*, 396–404. [CrossRef] [PubMed]

37. Lovell, S.C.; Davis, I.W.; Arendall III, W.B.; de Bakker, P.I.W.; Word, J.M.; Prisant, M.G.; Richardson, J.S.; Richardson, D.C. Structure validation by Calpha geometry: Phi, psi and Cbeta deviation. *Proteins* **2003**, *50*, 437–450. [CrossRef] [PubMed]

38. Lee, M.W.; Sevryugina, Y.V.; Khan, A.; Ye, S.Q. Carboranes increase the potency of small molecule inhibitors of nicotinamide phosphoribosyltranferase. *J. Med. Chem.* **2012**, *55*, 7290–7294. [CrossRef] [PubMed]

39. Wilkinson, S.M.; Gunosewoyo, H.; Barron, M.L.; Boucher, A.; McDonnell, M.; Turner, P.; Morrison, D.E.; Bennett, M.R.; Mc Gregor, I.S.; Rendina, L.M.; et al. The first CNS-active carborane: A novel P2X7 receptor antagonist with antidepressant activity. *ACS Chem. Neurosci.* **2014**, *5*, 335–339. [CrossRef] [PubMed]

40. Kracke, G.R.; VanGordon, M.R.; Sevryugina, Y.V.; Kueffer, P.J.; Kabytaev, K.; Jalisatgi, S.S.; Hawthorne, M.F. Carborane-derived local anesthetics are isomer dependent. *Chem. Med. Chem.* **2015**, *10*, 62–67. [CrossRef] [PubMed]

41. Tiwari, R.; Mahasenan, K.; Pavlovicz, R.; Li, C.; Tjarks, W.J. Carborane clusters in computational drug design: A comparative docking evaluation using AutoDock, FlexX, Glide and Surflex. *J. Chem. Inf. Model.* **2009**, *49*, 1581–1589. [CrossRef] [PubMed]

42. Byun, Y.; Thirumamagal, B.T.; Yang, W.; Eriksson, S.; Barth, R.F.; Tjarks, W. Preparation and biological evaluation of 10B-enriched 3-[5-{2-(2,3-dihydroxyprop-1-yl)-o-carboran-1-yl}pentan-1-yl]thymidine (N5-2OH), a new boron delivery agent for boron neutron capture therapy of brain tumors. *J. Med. Chem.* **2006**, *18*, 5513–5523. [CrossRef] [PubMed]

43. Johnsamuel, J.; Byun, Y.; Jones, T.P.; Endo, Y.; Tjarks, W. A new strategy for molecular modeling and receptor-based design of carborane containing compounds. *J. Organomet. Chem.* **2003**, *680*, 223–231. [CrossRef]

44. Narayanasamy, S.; Thirumamagal, B.T.; Johnsamuel, J.; Byun, Y.; Al-Madhoun, A.S.; Usova, E.; Cosquer, G.Y.; Yan, J.; Bandyopadhyaya, A.K.; Tiwari, R.; et al. Hydrophilically enhanced 3-carboranyl thymidine analogues (3CTAs), for boron neutron capture therapy BNCT, of cancer. *Bioorg. Med. Chem.* **2006**, *14*, 6886–6899. [CrossRef] [PubMed]

45. Sedlák, R.; Fanfrlik, J.; Pecina, A.; Hnyk, D.; Hobza, P.; Lepšík, M. Noncovalent Interactions of Heteroboranes. In *Boron: The Fifth Element (Challenges and Advances in Computational Chemistry and Physics)*, 1st ed.; Hnyk, D., McKee, M.L., Eds.; Springer International Publishing: Basel, Switzerland, 2015; Volume 20, pp. 219–239. ISBN 978-3-319-22281-3.

46. Schneidman-Duhovny, D.; Inbar, Y.; Nussinov, R.; Wolfson, H.J. PatchDock and SymmDock: Servers for rigid and symmetric docking. *Nucl. Acids Res.* **2005**, *33*, W363–W367. [CrossRef] [PubMed]

47. Rork, T.H.; Wallace, K.L.; Kennedy, D.P.; Marshall, M.A.; Lankford, A.R.; Linden, J. Adenosine A2A receptor activation reduces infarct size in the isolated, perfused mouse heart by inhibiting resident cardiac mast cell degranulation. *Am. J. Physiol. Heart Circ. Physiol.* **2008**, *5*, H1825–H1833. [CrossRef] [PubMed]

48. Chen, J.F.; Eltzschig, H.K.; Fredholm, B.B. Adenosine receptors as drug targets—What are the challenges? *Nat. Rev. Drug Discov.* **2013**, *12*, 265–286. [CrossRef] [PubMed]

49. SCH 58261 Product Datasheet. Available online: https//www.tocris.com/products/sch-58261_2270; https//documents.tocris.com/pdfs/tocris_coa/2270_9_coa.pdf (accessed on 21 August 2017).

50. Nishikawa, S.; Kumazawa, Z.; Kashimura, N.; Niskimi, Y.; Uemura, S. Alternating Dependency of Cytokinin Activity on the Number of Methylene Units in ω-Phenylalkyl Derivatives of Some Purine Cytokinins and 4-Substituted Pyrido[3,4-*d*]pyrimidme. *Agric. Biol. Chem.* **1986**, *50*, 2243–2249. [CrossRef]

51. Wojtczak, B.; Semenyuk, A.; Olejniczak, A.B.; Kwiatkowski, M.; Lesnikowski, Z.J. General method for the synthesis of 2′-*O*-carboranyl-nucleosides. *Tetrahedron Lett.* **2005**, *46*, 3969–3972. [CrossRef]

52. Sproat, B.S. Synthesis of 2′-*O*-Alkyloligoribonucleotides. In *Protocols for Oligonucleotides and Analogs. Synthesis and Properties*, 1st ed.; Agrawal, S., Ed.; Humana Press Inc.: Totowa, NJ, USA, 1993; pp. 115–141. ISBN 978-0-89603-281-1.

53. Hovinen, J.; Azhayeva, E.; Azhayev, A.; Guzaev, A.; Lönberg, H. Synthesis of 3′-*O*-(ω-aminoalkoxymethyl, thymidine 5′-triphosphates, terminators of DNA synthesis that enable 3′-labelling. *J. Chem. Soc. Perkin Trans.* **1994**, *2*, 211–217. [CrossRef]

54. Volpini, R.; Costanzi, S.; Lambertucci, C.; Taffi, S.; Vittori, S.; Klotz, K.-N.; Cristalli, G. N(6,-alkyl-2-alkynyl derivatives of adenosine as potent and selective agonists at the human adenosine A3, receptor and a starting point for searching A2B, ligands. *J. Med. Chem.* **2002**, *45*, 3271–3279. [CrossRef] [PubMed]

55. Dal Ben, D.; Buccioni, M.; Lambertucci, C.; Marucci, G.; Volpini, R.; Cristalli, G. The importance of alkynyl chain presence for the activity of adenine nucleosides/nucleotides on purinergic receptors. *Curr. Med. Chem.* **2011**, *18*, 1444–1463. [CrossRef] [PubMed]

12

Binding Affinity via Docking: Fact and Fiction

Tatu Pantsar [1] **and Antti Poso** [1,2,*] (iD)

[1] School of Pharmacy, University of Eastern Finland, P.O. BOX 1627, 70211 Kuopio, Finland;
    tatu.pantsar@uef.fi
[2] Department of Internal Medicine VIII, University Hospital Tübingen, Otfried-Müller-Strasse 14,
    72076 Tübingen, Germany
*   Correspondence: antti.poso@uef.fi

Academic Editor: Rebecca C. Wade

Abstract: In 1982, Kuntz et al. published an article with the title "A Geometric Approach to Macromolecule-Ligand Interactions", where they described a method "to explore geometrically feasible alignment of ligands and receptors of known structure". Since then, small molecule docking has been employed as a fast way to estimate the binding pose of a given compound within a specific target protein and also to predict binding affinity. Remarkably, the first docking method suggested by Kuntz and colleagues aimed to predict binding poses but very little was specified about binding affinity. This raises the question as to whether docking is the right tool to estimate binding affinity. The short answer is no, and this has been concluded in several comprehensive analyses. However, in this opinion paper we discuss several critical aspects that need to be reconsidered before a reliable binding affinity prediction through docking is realistic. These are not the only issues that need to be considered, but they are perhaps the most critical ones. We also consider that in spite of the huge efforts to enhance scoring functions, the accuracy of binding affinity predictions is perhaps only as good as it was 10–20 years ago. There are several underlying reasons for this poor performance and these are analyzed. In particular, we focus on the role of the solvent (water), the poor description of H-bonding and the lack of the systems' true dynamics. We hope to provide readers with potential insights and tools to overcome the challenging issues related to binding affinity prediction via docking.

Keywords: docking; solvent effect; binding affinity; scoring function; molecular dynamics

1. Introduction

Docking, originally introduced by Kuntz et al. [1], is a computational method that virtually tries to predict a complex of (usually) two binding partners. Typically, these binding partners are biological macromolecules (e.g., protein, DNA/RNA, peptide) or small molecules (e.g., endogenous ligands, drugs). Although nowadays specific docking methods are available for distinct binding partners, such as HADDOCK for protein-protein docking [2], here we focus on the more traditional small-molecule molecular docking methods, such as GOLD [3–5], Surflex-Dock [6], AutoDock [7] and Glide [8–10], that are regularly utilized in structure-based drug design to predict ligand interactions with the target protein. In structure-based small-molecule docking a small ligand molecule is aligned inside the binding cavity of the target protein and the resulting docking pose is evaluated by a specific scoring function. The scoring function generates a score for each pose, and the resulting values are used to rank the different poses and ligands. In a methodological sense, there are two independent stages in the docking process: the pose generation and the scoring. The first refers to the methods which are used to create different ligand and protein conformations and aligning different ligand conformations within the binding site of the protein. The latter, the scoring, is required in the docking process for a quantitative estimation of the pose quality. As docking is typically utilized to screen extensive small-molecule (up to millions) chemical libraries, the pose generation and the pose quality evaluation must be carried out by fast methods i.e., the computational cost should be low. To fulfill this, several simplifications are needed in the overall docking process.

The first simplification in docking is related to water, as this solvent is neglected by most docking programs. Only very recently, have several docking methods been introduced where individual water molecules are included in the pose generation and evaluation phase [10,11]. The challenge in water description is related to the fact that these abundant molecules are fast moving and rotating and they participate in hydrogen bonding (H-bonding) as a donor and acceptor. This means that a change in the orientation of a single water molecule in the binding site not only has an effect on the neighboring waters, but also extends to the surrounding multiple hydration layers, thereby affecting the whole water network. In addition, the differentiation of strong and weak H-bonds in these interactions should be considered. Thus, the abundant possibilities in the water arrangement prohibit a feasible, explicit evaluation of all the potential water interactions. The current state of how the water can be treated explicitly in docking is reviewed in [12].

The lack of motion is another simplification in docking. However, the dynamic nature of the whole system in terms of entropy and enthalpy should be acknowledged. Whereas the ligand flexibility is typically included in the docking process, the same does not hold true for protein. Usually, protein is considered as rigid, with the exception of the rotating hydroxyl groups of serine, threonine and tyrosine residues. Obviously, these simplifications affect the quality of the generated poses, which may be artificial [13]. As a result, different approaches that consider the protein flexibility have been developed, such as ensemble docking [14], where docking is conducted in an ensemble of different protein conformations.

The third simplification in docking is related to the analysis of the interactions between the protein and the ligand. The different types of protein–ligand interactions (for non-covalent binders) include ionic interactions, hydrogen bonds and van der Waals interactions (including dispersion, polar and induced interactions). The most accurate way to estimate these interactions is with a quantum mechanics (QM) based approach [15]. However, in most cases QM methods are computationally too expensive for docking purposes. To speed up the interaction analysis, calculations are typically conducted with simple potential energy functions, usually related to force-fields or statistical potentials. While the current force fields and scoring functions are well parametrized, polarization effects and a detailed proton affinity estimation are still lacking.

Docking programs produce one (or several) different poses for every ligand, and further rank different compounds based on their scoring functions. A comparison of different docking programs is difficult as the data sets to estimate the docking performance are often of low quality and there is no consensus on which metrics to apply in these comparisons. For instance, binding affinities predicted by the docking might be incorrect, despite the correctly predicted binding pose. Another example is the case in which a particular docking method performs reasonably with one protein but with another protein, docking poses are constantly mispredicted. These problems are well explained in the work of Cheng et al. [16], in which the frequently used CASF-2007 data set was employed to evaluate docking performance. In the same work, evaluation problems were solved by using three different metrics, namely "docking power", "ranking power" and "scoring power". Recently Li et al. [17], described a fourth metric, called "screening power". "Docking power" is the power to identify the native docking pose among the decoy poses, while "scoring power" is the ability to predict the binding affinity. In virtual screening campaigns, employment of "ranking power" is usually more appropriate, as it is the ability to correctly rank compounds according to their binding affinity. Also, the "screening power" is highly relevant for virtual screening, as it measures how well the method is able to identify the true binders from a random pool of ligands, including non-binders.

Docking is utilized as a tool in both virtual screening and compound optimization. There are several very comprehensive reports indicating unreliable binding affinity predictions by docking;

a good summary of those studies has been published by Pagadala et al. [18]. Additionally, in order to achieve reliable binding affinity predictions, the old docking methods being updated and new methods have been published. New parametrizations and methods are based on ever increasing datasets and increased computational capacities, such as the implementation of QM in docking [19] and moving towards dynamic docking [20]. In our laboratory, we have carried out different virtual screening and docking experiments since the early 2000s. While we used docking as a stand-alone approach in our early studies, nowadays, to increase the quality of our results we are increasingly employing diverse methods in parallel to docking. In the following, we briefly discuss those theoretical aspects which have directed us to use docking in combination with other methods. First, we focus on the pose generation in docking and then, we provide a short overview of scoring function caveats. Finally, we discuss the role of water and (the lack of) dynamics in the docking process.

2. Pose Generation and Scoring Functions in Docking

2.1. Pose Generation

Ligand and protein conformational freedom is a huge challenge in docking. Widely used docking programs handle ligands as fully flexible, thus typically generating a very large number of conformations during the docking process. In addition to conformation generation methods, the quality of the docking will depend on the force field. It is evident that the conformational aspect of the process is well optimized as all the widely used docking methods are able to identify the correct bioactive conformation of a ligand (i.e., to recreate the X-ray pose) in several instances. For example, this was shown by Li et al. [17], Warren et al. [21], and the same conclusions were reached in the CASP2 competition [22]. However, these results do not necessarily imply that the pose generation produces the correct ligand conformation and binding pose in all instances. Based on these observations, it is apparent that the focus should be placed on the scoring function to increase the quality of docking.

2.2. Scoring Functions

The strength of a protein-ligand complex is related to the intermolecular interactions between these binding partners, solvent effects and dynamics. The most conservative method to estimate all of these simultaneously, is to apply all-atom molecular dynamics (MD) simulations. However, in order to avoid the significant computational costs related to these simulations, molecular docking utilizes scoring functions to provide a fast and crude estimation of the binding affinity. There are three main types of scoring functions: force-field based, knowledge-based statistical functions, and empirical scoring functions [23]. Force-field based methods utilize molecular mechanics functions for evaluating the direct interactions between a ligand and the protein, and solvent effects are typically evaluated by a generalized Born/surface area (GB/SA) type of approach [24], which is often based on the work of Wesson and Eisenberg [25]. Knowledge-based methods rely on statistical information derived from the existing ligand-receptors complex structures [26] in the form of distance-dependent atom-pair potentials. The third approach, empirical scoring functions [8] is based on the idea that all the relevant factors affecting the binding are expressed in the form of (preferably simple) equations, like those describing H-bonding, rotational/translational degrees of freedom and polar/lipophilic effects. In addition, these equations are balanced by using a regression-type approach; in the literature this approach is sometimes referred to as regression-based scoring.

2.2.1. Enthalpy and Entropy

Scoring functions attempts to estimate the binding affinity, which is directly related to the Gibbs energy of binding. There are several ways to describe the partitions of binding energy and one of these is described in Equation (1). This partition by Ajay and Murcko [27], describes the binding energy as individual components: the solvation/desolvation energy ($\Delta G_{solvent}$); the change in energy of the receptor and ligand due to complex formation (ΔG_{conf}); the change in energy due to specific interactions between the ligand and the receptor (ΔG_{int}); and the contribution due to changes in movement (rotational, translational, vibrational) (ΔG_{motion}).

$$\Delta G_{bind} = \Delta G_{solvent} + \Delta G_{conf} + \Delta G_{int} + \Delta G_{motion} \tag{1}$$

What can we conclude based on Equation (1)? First, one must note that it is inadequate to only study the protein-ligand interactions. Additionally, it is important to understand how both interact with water before the formation of the complex and how water mediates this process. Also, one must recognize the fact that binding energy includes the conformational aspects of ligand and protein, and also changes in motion (this is mainly an entropic effect). As entropy is directly related to the motion and the temperature, a single protein-ligand complex pose may not provide enough information to reliably predict binding affinity.

Finally, how reliable is the estimation of the strength of the direct interactions (ΔG_{int}) between different binding partners (protein-ligand-solvent)? This depends greatly on the description of the ionic interactions and van der Waals interactions (including H-bonding). The entropic component is thought to be related mainly to the conformational and rotational/translational aspects, but we believe this is an optimistic view. More emphasis should be placed on how much the protein flexibility contributes to the stability of the protein-ligand complex and how the water affects the binding energy.

2.2.2. Direct Interactions

Direct polar interactions between ligand, protein and water are enthalpic in nature. In scoring functions, these interactions are considered by specific terms such as H-bonding, Lennard-Jones type of functions and ionic interactions. Dispersion-type interactions (erroneously called van der Waals interaction) are usually reasonably described by classical Lennard-Jones potential. As indirect proof of how precise these equations are, even the latest parametrization of the OPLS (Optimized Potentials for Liquid Simulations) force field family, OPLS3 [28], utilizes the formerly developed Lennard-Jones parameters. Indeed, many of the scoring functions use this approach to model dispersion [5,24,29], although GOLD uses softer 8–4 potential while Dock and Glide prefer 12–6 potential [1,3–5,8–10]. Other approaches do exist, for example Surflex uses surface-based description (derived from van der Waals surface) [6] and FlexX has a scoring term based on separate attractive terms for H-bonds, ionic, aromatic and lipophilic interactions and atom-center distance-based repulsive function [30].

One could argue that a proper description of dispersion is needed for accurate ligand-binding prediction, however, a precise understanding of the H-bonding is even more important. Recently, Raschka et al. analyzed the type of interactions found in 136 non-homologous protein-ligand complexes [31], concluding that strong H-bonds are required for most high-affinity ligands. In addition, they disclosed that the protein prefers to act as a H-bonding donor. As an explanation of why the protein prefers to act as a H-bond donor for high-affinity ligands, the authors speculated that geometrically more constrained H-bonding donors were enriched during the evolution. Consequently, a proper H-bonding description in scoring functions is required.

In all scoring functions, both distance and angular parameters are included in the H-bond potentials in similar fashion, in several force fields. In addition, the type of H-bonding (e.g., charged, neutral) is considered. One of the most detailed forms of H-bonding potential is implemented within the Glide XP [10], where three different types of H-bond are used: neutral-neutral, neutral-charged and charged-charged. The functional form of the Glide XP includes several H-bond class-specific modifications and environment-based restrictions. As a result, enhanced recognition of the "false positive" H-bonds is achieved.

2.2.3. Hydrogen Bond Strength and Classification

The hydrogen-bond (H-bond) is mainly an electrostatic interaction, which is typically modelled via Coulomb-type-equations, and for this, the dielectric constant is a critical factor.

Unfortunately, a reliable and fast method to calculate the dielectric constant does not exist. One approach is to use QM/MD-methods but this approach is currently too slow for docking purposes [15]. Furthermore, the challenge in H-bond modeling is the high variability of different H-bond types and strengths. Even the environment has a huge effect as demonstrated with water–water H-bonds [32]. In a neutral (pH 7) environment a single water-water H-bond is weak but in basic or acidic medium it becomes a 6-fold stronger and 15% shorter charge assisted H-bond. In fact, H-bond strengths can usually be estimated based on the proton affinities or pK_a-values of both the donor and acceptor site. Based on this approach, a 6-class classification for H-bonding has been developed [33]. Normal, weak H-bonds (those without charge) are the most common type of H-bonds found in biological systems. These are unassisted by charge or resonance and thus are weak, asymmetric and driven by electrostatic force. On the other hand, all the H-bonds assisted by charge (either negative or positive or both) are strong and short, if the pK_a-values of donor and acceptor match. When pK_a-values mismatch (>2 pK_a units), these H-bonds are classified as regular H-bonds. In a biological system this means that the ionization state of the protein and ligand are needed for proper H-bonding evaluation, and also the pK_a-values are required. Most of the current H-bond potentials produce reliable predictions for uncharged H-bonds. The same does not hold true for the H-bonds that include ionizable donor and acceptor groups. There are several computational methods available for both protein and ligand pK_a-value calculations [34–36]. Nevertheless, proton affinities are hardly ever considered in scoring functions. Therefore, the Glide approach [10] of differently scoring H-bonds with charge, is probably correct.

3. Water, Dynamics and Docking

Water has an important role in the biological environment, especially in the protein matrix [37,38]. The crucial role of the water in the ligand binding process has long been acknowledged [39]. Water has an important role in ligand binding thermodynamics [40,41], even in the environment of a lipophilic binding cavity [42] and displacing specific water molecules from the binding site may play an important role in the ligand optimization process [43]. Moreover, water related H-bonding networks have a significant influence in the structure-activity relationship [44], and optimizing the ligand taking into account the surrounding water network may result in enhanced binding affinity and prolonged residence time [45]. The problem is that detailed information of how water is located within and around the ligand binding site is mostly unavailable. The most common tool for determining 3D structure, X-ray crystallography, can only provide partial information because the resolution and low-quality electron density limits water detection. Those water molecules which are detected by X-ray are often entropically stabilized [46]. In addition, crystallization conditions are typically far from the biologically relevant ones, and also the co-crystallized ligand molecule(s) may influence the observed hydration network (differently when compared to a docked ligand).

Easy application of water placement in docking is restricted because the water in the binding site is heterogenous. In different locations, an individual water molecule has restricted rotational freedom and H-bonding capabilities. The terminology, "happy" and "unhappy" water has been introduced to describe the individual water energies compared to bulk water [47]. Happy and unhappy water refer to low-energy and high-energy water, respectively. The unhappy water molecules within the binding site have either lost their degree of freedom (entropic penalty) or they are incapable of fulfilling all possible H-bonds (enthalpic penalty), which result in higher energies compared to the bulk water. Therefore, displacing unhappy water molecules from the binding site with the ligand results in a gain in binding affinity [48]. On the other hand, displacing a happy water molecule from the site is typically unfavorable. Furthermore, not all regions within the binding sites are hydrated and occupied by water molecules [49,50]. Areas exist that are energetically so unfavorable for water to occupy that there is no water present; instead, they appear as dry void regions (also referred to as vacuum or dewetted regions) [51,52]. Occupying these regions with a ligand molecule results in both more favorable enthalpy and entropy of binding. The reason for this gain in binding affinity is the fact that the increased protein-ligand interaction surface results in stronger van der Waals interaction. In addition, filling the dewetted region increases entropy. In accordance with this, we have noticed that these vacuum sites played a significant role in determining the compound activity in our series of Autotaxin inhibitors [53].

Even though water has been acknowledged to play an important role in binding, de novo placement of water has not been explicitly included in docking methods, with the exception of Glide XP [10]. The Glide XP includes terms for the hydrophobic enclosure, which promotes the insertion of the lipophilic parts of the ligand in the protein's lipophilic cavities; thereby, simulating the displacement of potential high-energy water. Moreover, in this method, by utilizing a grid-based methodology "virtual waters" are placed into the binding site, and penalties for are given for improperly solvated hydrophilic (polar or charged) groups and for the water that makes an unusual number of hydrophobic contacts.

As already stated, docking is usually unable to provide a good estimate of the role of the solvation penalties related to the binding. As a result, several complementary computational methods have been developed to identify and analyze water molecules around the protein-ligand complex to estimate its role in binding. Different approaches have been reported and the most popular methods are reviewed by Bodnarchuk [54]. For instance, the Schrödinger's WaterMap uses a short MD simulation and the estimation of the energies of the hydration sites are derived based on the simulation [48,50], whereas in the Molecular Operating Environment (MOE), the binding desolvation penalties can be estimated by 3D reference interaction site model (3D-RISM), which is based on the density functional theory of liquids [55,56]. The main limitation of these methods is that they are heavily dependent on the protein conformation used in the calculation. To exemplify this, a parallel calculation of the hydration site energy with the same protein may produce totally different results, even if only minor protein conformational change occurs or only one side of the chain conformation is altered. This limitation should be kept in mind when utilizing these methods, as for example, a conformational "induced-fit" effect upon ligand binding (via docking) might hamper the results [57]. Although these methods are now becoming increasingly popular and have demonstrated usefulness in explaining lead molecule structure–activity relationships [58], it is still unclear if these methods are applicable in virtual screening campaigns. One of the first attempts to include these computational approaches directly into scoring functions is WScore [11]. In WScore, a default WaterMap calculation with the apo-protein is utilized to gain insight into the hydration site positions and their corresponding energies. The occupancy of these hydration sites by a ligand are included in the scoring. Moreover, an ensemble docking is carried out that aims to take into account the protein flexibility, which as mentioned above, is the major issue with the WaterMap. The usefulness of WScore and other related methods remain to be seen. Furthermore, conventional MD simulations can be applied to evaluate the hydration networks; thus, some errors related to force field accuracy may arise [59]. In a way, we agree with the statement by Hummer [60], that the contribution of water for the ligand binding may be substantial but its evaluation is challenging.

One of the shortcomings of docking is that it produces only a snapshot of the putative binding conformation. This is a notable limitation, as in real-life the binding event is not a static event, it is dynamic. For instance, we observed a good example of this in a study of 1-/2-monoacylglycerol hydrolysis by Monoacylglycerol lipase (MAGL) [61]. Whereas the wild-type MAGL hydrolyzes both substrates at an identical rate, a C242A mutation in the active site impairs the hydrolysis of the 1-acylglycerol but not the 2-acylglycerol. This mutation had no effect on the binding conformations obtained by the docking; but, it was unable to provide an explanation for the observed difference in the hydrolysis among the substrates. However, in this case even short MD-simulations were capable of highlighting the differences in the substrate binding dynamics that arose due to the mutation.

Perhaps due to the fact that docking is currently unable to consider the impact of water and the dynamic nature of binding, applying MD simulations for the docking pose validation has attracted growing interest in the scientific community [62–67]. This is probably also due to increased accessibility to adequate computing resources (e.g., GPUs) that are required for simulations with a reasonable time-scale. Another factor that has made MD simulations more relevant is the improvement in the force fields that are now capable of handling both small molecules and proteins with reasonable accuracy. These improvements have led to more relevant observations from the simulations. Interestingly, MD simulations appear to provide the solution to the two issues that docking is incapable of handling—water and the dynamics.

4. Solution

In this opinion paper, we have exemplified the underlying issues in predicting binding affinity via docking. The main issues are related to the H-bonding and the water description, and how water and the protein-ligand complex should be considered as a dynamic system. While describing the H-bond is clearly an issue, we should also acknowledge that this has already been quite well described in modern force fields. For example, the new OPLS3 and recent AMBER (Assisted Model Building with Energy Refinement) and CHARMM (Chemistry at Harvard Macromolecular Mechanics) force fields include a better H-bonding description [28,68,69]. Additionally, MD is becoming an increasingly robust method to study individual protein-ligand complexes. Unfortunately, the computational costs of MD are still too high to allow virtual screening.

What can be done to increase the accuracy of the binding affinity prediction? With current methods, resolving this issue is extremely challenging. For H-bonding, it is feasible to include a more precise energy evaluation method that would allow recognition and differentiation of the strong and weak H-bonds. However, this requires a fast and reliable pK_a-value calculation that also considers conformational and environmental aspects of the binding cavity. Furthermore, due to the active role of the water in binding, it is obvious that water needs to be explicitly included in the docking process. All the current evidence contradicts docking in the gas phase. WaterMap and other related methods have partially resolved this issue but a more comprehensive solution is required. Finally, implementation of dynamics in scoring functions remains challenging. In future, scoring functions need to be reinvented so that they are able to describe the dynamics related to the binding. Overall, new approaches are required to address the issues discussed above.

Our current solution is based on two comprehensive approaches, one to use docking tools in more efficient ways [53,70], and the other is to use MD simulations to validate the results of the classical docking [71,72]. Prior to any docking experiment, one should explore the flexibility of the target protein, based on both the existing protein structures and MD simulations. At the same time, it is of utmost importance to determine the solvation status of the binding cavity and the energy levels of the potentially happy and unhappy water. Subsequently, this information is further applied in docking by utilizing suitable constraints. This approach can help us to identify more reliable binding poses. Finally, the most promising poses are further analyzed by short (usually 200 ns) MD simulations and followed by WaterMap analysis. However, our approach has two major shortcomings: It is slow and difficult to implement. These shortcomings are tolerable, as long as we have sufficient computing resources and an adequate amount of time to work with the target. We resolve the H-bond issue by estimating the pK_a-values with different computational methods (e.g., QM-polarized docking). Lastly, even after implementing all of these user-based interventions, we always use the most sophisticated scoring function, the eye. If you trust your docking pose, you might be right.

Author Contributions: Writing-Original Draft Preparation, T.P. and A.P.; Writing-Review & Editing, T.P. and A.P.; Funding Acquisition, A.P.

Acknowledgments: We would like to thank Thales Kronenberger for the critical reading and useful comments on the manuscript, and CSC-IT Center for Science Ltd. (Espoo, Finland) for computational resources.

References

1.   Kuntz, I.; Blaney, J.; Oatley, S.; Langridge, R.; Ferrin, T. A geometric approach to macromolecule-ligand interactions. *J. Mol. Biol.* **1982**, *161*, 269–288. [CrossRef]

2.   Van Zundert, G.C.P.; Rodrigues, J.P.G.L.M.; Trellet, M.; Schmitz, C.; Kastritis, P.L.; Karaca, E.; Melquiond, A.S.J.; van Dijk, M.; de Vries, S.J.; Bonvin, A.M.J.J.M. The HADDOCK2.2 Web Server: User-Friendly Integrative Modeling of Biomolecular Complexes. *J. Mol. Biol.* **2016**, *428*, 720–725. [CrossRef] [PubMed]

3.   Jones, G.; Willett, P.; Glen, R.C. Molecular recognition of receptor sites using a genetic algorithm with a description of desolvation. *J. Mol. Biol.* **1995**, *245*, 43–53. [CrossRef]

4.   Jones, G.; Willett, P.; Glen, R.; Leach, A.; Taylor, R. Development and validation of a genetic algorithm for flexible docking. *J. Mol. Biol.* **1997**, *267*, 727–748. [CrossRef] [PubMed]

5.   Verdonk, M.; Cole, J.; Hartshorn, M.; Murray, C.; Taylor, R. Improved protein—Ligand docking using GOLD. *Proteins* **2003**, *52*, 609–623. [CrossRef] [PubMed]

6.   Jain, A.N. Surflex: Fully Automatic Flexible Molecular Docking Using a Molecular Similarity-Based Search Engine. *J. Med. Chem.* **2003**, *46*, 499–511. [CrossRef] [PubMed]

7.   Trott, O.; Olson, A.J. AutoDock Vina: Improving the speed and accuracy of docking with a new scoring function, efficient optimization, and multithreading. *J. Comput. Chem.* **2010**, *31*, 455–461. [CrossRef] [PubMed]

8.   Friesner, R.A.; Banks, J.L.; Murphy, R.B.; Halgren, T.A.; Klicic, J.J.; Mainz, D.T.; Repasky, M.P.; Knoll, E.H.; Shaw, D.E.; Shelley, M.; et al. Glide: A New Approach for Rapid, Accurate Docking and Scoring. 1. Method and Assessment of Docking Accuracy. *J. Med. Chem.* **2004**, *47*, 1739–1749. [CrossRef] [PubMed]

9.   Halgren, T.A.; Murphy, R.B.; Friesner, R.A.; Beard, H.S.; Frye, L.L.; Pollard, W.T.; Banks, J.L. Glide: A New Approach for Rapid, Accurate Docking and Scoring. 2. Enrichment Factors in Database Screening. *J. Med. Chem.* **2004**, *47*, 1750–1759. [CrossRef] [PubMed]

10.   Friesner, R.A.; Murphy, R.B.; Repasky, M.P.; Frye, L.L.; Greenwood, J.R.; Halgren, T.A.; Sanschagrin, P.C.; Mainz, D.T. Extra Precision Glide: Docking and Scoring Incorporating a Model of Hydrophobic Enclosure for Protein-Ligand Complexes. *J. Med. Chem.* **2006**, *49*, 6177–6196. [CrossRef] [PubMed]

11.   Murphy, R.; Repasky, M.; Greenwood, J.; Tubert-Brohman, I.; Jerome, S.; Annabhimoju, R.; Boyles, N.; Schmitz, C.; Abel, R.; Farid, R.; et al. WScore: A Flexible and Accurate Treatment of Explicit Water Molecules in Ligand-Receptor Docking. *J. Med. Chem.* **2016**, *59*, 4364–4384. [CrossRef] [PubMed]

12.   Hu, X.; Maffucci, I.; Contini, A. Advances in the Treatment of Explicit Water Molecules in Docking and Binding Free Energy Calculations. *Curr. Med. Chem.* **2018**, *25*, 1–23. [CrossRef] [PubMed]

13.   Chen, Y.C. Beware of docking! *Trends Pharmacol. Sci.* **2015**, *36*, 78–95. [CrossRef] [PubMed]

14.   Amaro, R.E.; Baudry, J.; Chodera, J.; Demir, Ö.; McCammon, J.A.; Miao, Y.; Smith, J.C. Ensemble Docking in Drug Discovery. *Biophys. J.* **2018**, *114*, 2271–2278. [CrossRef] [PubMed]

15.   Raha, K.; Peters, M.B.; Wang, B.; Yu, N.; Wollacott, A.M.; Westerhoff, L.M.; Merz, K.M. The role of quantum mechanics in structure-based drug design. *Drug Discov. Today* **2007**, *12*, 725–731. [CrossRef] [PubMed]

16.   Cheng, T.; Li, X.; Li, Y.; Liu, Z.; Wang, R. Comparative assessment of scoring functions on a diverse test set. *J. Chem. Inf. Model.* **2009**, *49*, 1079–1093. [CrossRef] [PubMed]

17.   Li, Y.; Han, L.; Liu, Z.; Wang, R. Comparative Assessment of Scoring Functions on an Updated Benchmark: 2. Evaluation Methods and General Results. *J. Chem. Inf. Model.* **2014**, *54*, 1717–1736. [CrossRef] [PubMed]

18.   Pagadala, N.; Syed, K.; Tuszynski, J. Software for molecular docking: A review. *Biophys. Rev.* **2017**, *9*, 91–102. [PubMed]

19.   Adeniyi, A.A.; Soliman, M.E.S. Implementing QM in docking calculations: Is it a waste of computational time? *Drug Discov. Today* **2017**, *22*, 1216–1223. [CrossRef] [PubMed]

20.   Gioia, D.; Bertazzo, M.; Recanatini, M.; Masetti, M.; Cavalli, A. Dynamic Docking: A Paradigm Shift in Computational Drug Discovery. *Molecules* **2017**, *22*. [CrossRef] [PubMed]

21.   Warren, G.; Andrews, C.; Capelli, A.-M.; Clarke, B.; LaLonde, J.; Lambert, M.; Lindvall, M.; Nevins, N.; Semus, S.; Senger, S.; et al. A Critical Assessment of Docking Programs and Scoring Functions. *J. Med. Chem.* **2006**, *49*, 5912–5931. [CrossRef] [PubMed]

22.   Dixon, J.S. Evaluation of the CASP2 docking section. *Proteins* **1997**, *29* (Suppl. S1), 198–204. [CrossRef]

23.   Kitchen, D.; Decornez, H.; Furr, J.; Bajorath, J. Docking and scoring in virtual screening for drug discovery: Methods and applications. *Nat. Rev. Drug Discov.* **2004**, *3*, 935–949. [PubMed]

24.   Morris, G.; Goodsell, D.; Halliday, R.; Huey, R.; Hart, W.; Belew, R.; Olson, A. Automated docking using a Lamarckian genetic algorithm and an empirical binding free energy function. *J. Comput. Chem.* **1998**, *19*, 1639–1662. [CrossRef]

25. Wesson, L.; Eisenberg, D. Atomic solvation parameters applied to molecular dynamics of proteins in solution. *Protein Sci.* **1992**, *1*, 227–235. [CrossRef] [PubMed]

26. Gohlke, H.; Hendlich, M.; Klebe, G. Knowledge-based scoring function to predict protein-ligand interactions. *J. Mol. Biol.* **2000**, *295*, 337–356. [CrossRef] [PubMed]

27. Ajay; Murcko, M.A. Computational Methods to Predict Binding Free Energy in Ligand-Receptor Complexes. *J. Med. Chem.* **1995**, *38*, 4953–4967. [PubMed]

28. Harder, E.; Damm, W.; Maple, J.; Wu, C.; Reboul, M.; Xiang, J.Y.; Wang, L.; Lupyan, D.; Dahlgren, M.K.; Knight, J.L.; et al. OPLS3: A Force Field Providing Broad Coverage of Drug-like Small Molecules and Proteins. *J. Chem. Theory Comput.* **2016**, *12*, 281–296. [CrossRef] [PubMed]

29. Shoichet, B.; Kuntz, I.; Bodian, D. Molecular docking using shape descriptors. *J. Comput. Chem.* **1992**, *13*, 380–397. [CrossRef]

30. Rarey, M.; Kramer, B.; Lengauer, T.; Klebe, G. A fast flexible docking method using an incremental construction algorithm. *J. Mol. Biol.* **1996**, *261*, 470–489. [CrossRef] [PubMed]

31. Raschka, S.; Wolf, A.; Bemister-Buffington, J.; Kuhn, L. Protein—Ligand interfaces are polarized: Discovery of a strong trend for intermolecular hydrogen bonds to favor donors on the protein side with implications for predicting and designing ligand complexes. *J. Comput. Aided Mol. Des.* **2018**, *32*, 511–528. [CrossRef] [PubMed]

32. Gilli, P.; Pretto, L.; Bertolasi, V.; Gilli, G. Predicting Hydrogen-Bond Strengths from Acid−Base Molecular Properties. The pKa Slide Rule: Toward the Solution of a Long-Lasting Problem. *Acc. Chem. Res.* **2009**, *42*, 33–44. [CrossRef] [PubMed]

33. Gilli, P.; Gilli, G. Hydrogen bond models and theories: The dual hydrogen bond model and its consequences. *J. Mol. Struct.* **2010**, *972*, 2–10. [CrossRef]

34. Kilambi, K.; Gray, J. Rapid Calculation of Protein pKa Values Using Rosetta. *Biophys. J.* **2012**, *103*, 587–595. [CrossRef] [PubMed]

35. Song, Y.; Mao, J.; Gunner, M. MCCE2: Improving protein pKa calculations with extensive side chain rotamer sampling. *J. Comput. Chem.* **2009**, *30*, 2231–2247. [CrossRef] [PubMed]

36. Shelley, J.C.; Cholleti, A.; Frye, L.; Greenwood, J.R.; Timlin, M.R.; Uchimaya, M. Epik: A software program for pKa prediction and protonation state generation for drug-like molecules. *J. Comput. Aided Mol. Des.* **2007**, *21*, 681–691. [CrossRef] [PubMed]

37. Ball, P. Water is an active matrix of life for cell and molecular biology. *Proc. Natl. Acad. Sci. USA* **2017**, *114*, 13327–13335. [CrossRef] [PubMed]

38. Spyrakis, F.; Ahmed, M.H.; Bayden, A.S.; Cozzini, P.; Mozzarelli, A.; Kellogg, G.E. The Roles of Water in the Protein Matrix: A Largely Untapped Resource for Drug Discovery. *J. Med. Chem.* **2017**, *60*, 6781–6827. [CrossRef] [PubMed]

39. Ladbury, J.E. Just add water! The effect of water on the specificity of protein-ligand binding sites and its potential application to drug design. *Chem. Biol.* **1996**, *3*, 973–980. [CrossRef]

40. Snyder, P.W.; Mecinovic, J.; Moustakas, D.T.; Thomas, S.W., 3rd; Harder, M.; Mack, E.T.; Lockett, M.R.; Héroux, A.; Sherman, W.; Whitesides, G.M. Mechanism of the hydrophobic effect in the biomolecular recognition of arylsulfonamides by carbonic anhydrase. *Proc. Natl. Acad. Sci. USA* **2011**, *108*, 17889–17894. [CrossRef] [PubMed]

41. Breiten, B.; Lockett, M.R.; Sherman, W.; Fujita, S.; Al-Sayah, M.; Lange, H.; Bowers, C.M.; Heroux, A.; Krilov, G.; Whitesides, G.M. Water networks contribute to enthalpy/entropy compensation in protein—Ligand binding. *J. Am. Chem. Soc.* **2013**, *135*, 15579–15584. [CrossRef] [PubMed]

42. Baron, R.; Setny, P.; McCammon, A. Water in Cavity—Ligand Recognition. *J. Am. Chem. Soc.* **2010**, *132*, 12091–12097. [CrossRef] [PubMed]

43. Michel, J.; Tirado-Rives, J.; Jorgensen, W.L. Energetics of Displacing Water Molecules from Protein Binding Sites: Consequences for Ligand Optimization. *J. Am. Chem. Soc.* **2009**, *131*, 15403–15411. [CrossRef] [PubMed]

44. Biela, A.; Nasief, N.N.; Betz, M.; Heine, A.; Hangauer, D.; Klebe, G. Dissecting the hydrophobic effect on the molecular level: The role of water, enthalpy, and entropy in ligand binding to thermolysin. *Angew. Chem. Int. Ed. Engl.* **2013**, *52*, 1822–1828. [CrossRef] [PubMed]

45. Krimmer, S.G.; Cramer, J.; Betz, M.; Fridh, V.; Karlsson, R.; Heine, A.; Klebe, G. Rational Design of Thermodynamic and Kinetic Binding Profiles by Optimizing Surface Water Networks Coating Protein-Bound Ligands. *J. Med. Chem.* **2016**, *59*, 10530–10548. [CrossRef] [PubMed]

46. Beuming, T.; Che, Y.; Abel, R.; Kim, B.; Shanmugasundaram, V.; Sherman, W. Thermodynamic analysis of water molecules at the surface of proteins and applications to binding site prediction and characterization. *Proteins* **2012**, *80*, 871–883. [CrossRef] [PubMed]

47. Mason, J.S.; Bortolato, A.; Congreve, M.; Marshall, F.H. New insights from structural biology into the druggability of G protein-coupled receptors. *Trends Pharmacol. Sci.* **2012**, *33*, 249–260. [CrossRef] [PubMed]

48. Abel, R.; Young, T.; Farid, R.; Berne, B.J.; Friesner, R.A. The role of the active site solvent in the thermodynamics of factor Xa-ligand binding. *J. Am. Chem. Soc.* **2008**, *130*, 2817–2831. [CrossRef] [PubMed]

49. Homans, S.W. Water, water everywhere—Except where it matters? *Drug Discov. Today* **2007**, *12*, 534–539. [CrossRef] [PubMed]

50. Young, T.; Abel, R.; Kim, B.; Berne, B.J.; Friesner, R.A. Motifs for molecular recognition exploiting hydrophobic enclosure in protein-ligand binding. *Proc. Natl. Acad. Sci. USA* **2007**, *104*, 808–813. [CrossRef] [PubMed]

51. Young, T.; Hua, L.; Huang, X.; Abel, R.; Friesner, R.; Berne, B.J. Dewetting Transitions in Protein Cavities. *Proteins* **2010**, *78*, 1856–1869. [CrossRef] [PubMed]

52. Wang, L.; Berne, B.J.; Friesner, R.A. Ligand binding to protein-binding pockets with wet and dry regions. *Proc. Natl. Acad. Sci. USA* **2011**, *108*, 1326–1330. [CrossRef] [PubMed]

53. Pantsar, T.; Singha, P.; Nevalainen, T.J.; Koshevoy, I.; Leppänen, J.; Poso, A.; Niskanen, J.M.A.; Pasonen-Seppänen, S.; Savinainen, J.R.; Laitinen, T.; et al. Design, synthesis, and biological evaluation of 2,4-dihydropyrano[2,3-c]pyrazole derivatives as autotaxin inhibitors. *Eur. J. Pharm. Sci.* **2017**, *107*, 97–111. [CrossRef] [PubMed]

54. Bodnarchuk, M.S. Water, water, everywhere... It's time to stop and think. *Drug Discov. Today* **2016**, *21*, 1139–1146. [CrossRef] [PubMed]

55. Kovalenko, A.; Hirata, F. Self-consistent description of a metal—Water interface by the Kohn-Sham density functional theory and the three-dimensional reference interaction site model. *J. Chem. Phys.* **1999**, *110*, 10095–10112. [CrossRef]

56. Luchko, T.; Gusarov, S.; Roe, D.R.; Simmerling, C.; Case, D.A.; Tuszynski, J.; Kovalenko, A. Three-Dimensional Molecular Theory of Solvation Coupled with Molecular Dynamics in Amber. *J. Chem. Theory Comput.* **2010**, *6*, 607–624. [CrossRef] [PubMed]

57. Pearlstein, R.; Sherman, W.; Abel, R. Contributions of water transfer energy to protein-ligand association and dissociation barriers: Watermap analysis of a series of p38α MAP kinase inhibitors. *Proteins* **2013**, *81*, 1509–1526. [CrossRef] [PubMed]

58. Bucher, D.; Stouten, P.; Triballeau, N. Shedding Light on Important Waters for Drug Design: Simulations versus Grid-Based Methods. *J. Chem. Inf. Model.* **2018**, *58*, 692–699. [CrossRef] [PubMed]

59. Betz, M.; Wulsdorf, T.; Krimmer, S.G.; Klebe, G. Impact of Surface Water Layers on Protein—Ligand Binding: How Well Are Experimental Data Reproduced by Molecular Dynamics Simulations in a Thermolysin Test Case? *J. Chem. Inf. Model.* **2016**, *56*, 223–233. [CrossRef] [PubMed]

60. Hummer, G. Molecular binding: Under water's influence. *Nat. Chem.* **2010**, *2*, 906–907. [CrossRef] [PubMed]

61. Laitinen, T.; Navia-Paldanius, D.; Rytilahti, R.; Marjamaa, J.J.; Kařízková, J.; Parkkari, T.; Pantsar, T.; Poso, A.; Laitinen, J.T.; Savinainen, J.R. Mutation of Cys242 of human monoacylglycerol lipase disrupts balanced hydrolysis of 1- and 2-monoacylglycerols and selectively impairs inhibitor potency. *Mol. Pharmacol.* **2014**, *85*, 510–519. [CrossRef] [PubMed]

62. Alonso, H.; Bliznyuk, A.A.; Gready, J.E. Combining docking and molecular dynamic simulations in drug design. *Med. Res. Rev.* **2006**, *26*, 531–568. [CrossRef] [PubMed]

63. Bartuzi, D.; Kaczor, A.A.; Targowska-Duda, K.M.; Matosiuk, D. Recent Advances and Applications of Molecular Docking to G Protein-Coupled Receptors. *Molecules* **2017**, *22*. [CrossRef] [PubMed]

64. Cavalli, A.; Bottegoni, G.; Raco, C.; De Vivo, M.; Recanatini, M. A computational study of the binding of propidium to the peripheral anionic site of human acetylcholinesterase. *J. Med. Chem.* **2004**, *47*, 3991–3999. [CrossRef] [PubMed]

65. Colizzi, F.; Perozzo, R.; Scapozza, L.; Recanatini, M.; Cavalli, A. Single-Molecule Pulling Simulations Can Discern Active from Inactive Enzyme Inhibitors. *J. Am. Chem. Soc.* **2010**, *132*, 7361–7371. [CrossRef] [PubMed]

66. Ruiz-Carmona, S.; Schmidtke, P.; Luque, F.J.; Baker, L.; Matassova, N.; Davis, B.; Roughley, S.; Murray, J.; Hubbard, R.; Barril, X. Dynamic undocking and the quasi-bound state as tools for drug discovery. *Nat. Chem.* **2017**, *9*, 201–206. [CrossRef] [PubMed]

67. Sabbadin, D.; Ciancetta, A.; Moro, S. Bridging Molecular Docking to Membrane Molecular Dynamics to Investigate GPCR—Ligand Recognition: The Human A2A Adenosine Receptor as a Key Study. *J. Chem. Inf. Model.* **2014**, *54*, 169–183. [CrossRef] [PubMed]

68. Cerutti, D.; Rice, J.; Swope, W.; Case, D. Derivation of Fixed Partial Charges for Amino Acids Accommodating a Specific Water Model and Implicit Polarization. *J. Phys. Chem. B* **2013**, *117*, 2328–2338. [CrossRef] [PubMed]

69. Best, R.; Zhu, X.; Shim, J.; Lopes, P.; Mittal, J.; Feig, M.; MacKerell, A. Optimization of the Additive CHARMM All-Atom Protein Force Field Targeting Improved Sampling of the Backbone φ, ψ and Side-Chain $\chi 1$ and $\chi 2$ Dihedral Angles. *J. Chem. Theory Comput.* **2012**, *8*, 3257–3273. [CrossRef] [PubMed]

70. Käsnänen, H.; Myllymäki, M.; Minkkilä, A.; Kataja, A.; Saario, S.; Nevalainen, T.; Koskinen, A.; Poso, A. 3-Heterocycle-Phenyl *N*-Alkylcarbamates as FAAH Inhibitors: Design, Synthesis and 3D-QSAR Studies. *Chem. Med. Chem.* **2010**, *5*, 213–231. [CrossRef] [PubMed]

71. Jyrkkärinne, J.; Küblbeck, J.; Pulkkinen, J.; Honkakoski, P.; Laatikainen, R.; Poso, A.; Laitinen, T. Molecular dynamics simulations for human CAR inverse agonists. *J. Chem. Inf. Model.* **2012**, *52*, 457–464. [CrossRef] [PubMed]

72. Küblbeck, J.; Jyrkkärinne, J.; Molnár, F.; Kuningas, T.; Patel, J.; Windshügel, B.; Nevalainen, T.; Laitinen, T.; Sippl, W.; Poso, A.; et al. Newin VitroTools to Study Human Constitutive Androstane Receptor (CAR) Biology: Discovery and Comparison of Human CAR Inverse Agonists. *Mol. Pharm.* **2011**, *8*, 2424–2433. [CrossRef] [PubMed]

13

Computational Insight into the Effect of Natural Compounds on the Destabilization of Preformed Amyloid-β(1–40) Fibrils

Francesco Tavanti * [ID]**, Alfonso Pedone and Maria Cristina Menziani** * [ID]

Department of Chemical and Geological Sciences, University of Modena and Reggio Emilia, Via G. Campi 103, 41125 Modena, Italy; alfonso.pedone@unimore.it
* Correspondence: francesco.tavanti@unimore.it (F.T.); mariacristina.menziani@unimore.it (M.C.M.);
Academic Editors: Rebecca C. Wade and Outi Salo-Ahen

Abstract: One of the principal hallmarks of Alzheimer's disease (AD) is related to the aggregation of amyloid-β fibrils in an insoluble form in the brain, also known as amyloidosis. Therefore, a prominent therapeutic strategy against AD consists of either blocking the amyloid aggregation and/or destroying the already formed aggregates. Natural products have shown significant therapeutic potential as amyloid inhibitors from in vitro studies as well as in vivo animal tests. In this study, the interaction of five natural biophenols (curcumin, dopamine, (-)-epigallocatechin-3-gallate, quercetin, and rosmarinic acid) with amyloid-β(1–40) fibrils has been studied through computational simulations. The results allowed the identification and characterization of the different binding modalities of each compounds and their consequences on fibril dynamics and aggregation. It emerges that the lateral aggregation of the fibrils is strongly influenced by the intercalation of the ligands, which modulates the double-layered structure stability.

Keywords: molecular dynamics simulation; biophenols; natural compounds; amyloid fibrils; Alzheimer's disease; ligand–protofiber interactions

1. Introduction

The pathological hallmark of Alzheimer's disease (AD) is the extracellular accumulation of insoluble proteinaceous deposits called amyloid fibrils [1] that induce cytotoxicity. The formation of mature amyloid fibrils (Aβ) proceeds through a nucleation-dependent process, where monomers and oligomers aggregate together, forming β-sheet-rich protein structures. The most common fibrils are Aβ(1–40) and Aβ(1–42), which are composed of 40 and 42 amino acids, respectively, and are characterized by β-strand units aligned perpendicularly to the main fibril axis [2]. Destabilization and clearance of amyloid aggregates by small molecules is one of the promising approaches towards the development of AD therapies [3].

In recent years, epidemiological studies on the effects of the diet against AD and dementia suggested that the high intake of flavonoids and polyphenols found in fruits and vegetables reduces the risk of AD and cognitive impairments, and several natural molecules have been identified as promoting cognitive health and interfering with the amyloidogenic activity in AD [4].

A detailed knowledge of how these molecules interact with Aβ fibrils is a prerequisite for the design of new efficient drugs. Unfortunately, despite intensive research, the experimental characterization of full-length Aβ oligomers/inhibitor complexes at a high level of resolution remains a great challenge.

Atomistic computer simulations are well-suited to provide molecular-level details of amyloid oligomer and fibril interactions with ligands, helping in the future development and characterization of druggable modalities [5]. Basically, four aspects of the flavonoid–amyloid interactions have been studied by computational methods: (1) the effect of ligands on the conformational transitions of Aβ monomers from an initial random coil or α-helix into β-sheet structures [6,7] and ligand-mediated conformational changes of the Aβ dimer [8] by means of replica exchange molecular dynamics (REMD) simulations; (2) the effect of ligands on the aggregation of Aβ(17–36) using coarse-grained simulations [9]; (3) the effect of ligands on the conformation and stability of amyloid-beta mutants [10] by molecular dynamics (MD) simulations; (4) the preferential binding sites of ligands and their effect on amyloid structure dynamics [11], Aβ fragments, and full-length single Aβ protofilaments [12–18] by means of docking experiments, MD simulations, and free energy calculations.

Although recently, a few studies devoted their attention to the interaction of ligands (mainly markers for amyloid detection [19–21]) with multiple Aβ protofilaments, to the best of our knowledge, this aspect has not been investigated thoroughly for natural polyphenol ligands, except for curcumin [12].

In this study, the binding modalities of five natural biophenols (curcumin, dopamine, (-)-epigallocatechin-3-gallate, quercetin, and rosmarinic acid) with single Aβ(1–40) protofilaments and double-layer oligomer aggregates will be studied through atomistic computational simulations, in order to explore structural changes in aggregate pathways upon binding.

First, putative binding sites on the Aβ(1–40) protofibril will be explored by replica exchange molecular dynamics (REMD) simulations. Then, binding free energies (ΔG_{bind}) will be computed on the complexes to determine the thermodynamically favored binding modalities. Finally, the structural effects caused by the binding of polyphenols to two double-layer protofilament polymorphs will be assessed. To this goal, the determination of the stability of the sheet-to-sheet associations of the double-layered organizations with and without the polyphenols will be computed by means of the potential of mean force (PMF) methodology.

2. Methods

2.1. Molecular Dynamics Simulations

Molecular dynamics simulations were performed with GROMOS 54a7 force field [22]. The structural model of amyloid fibrils was retrieved from the Protein Data Bank [23] (PDB ID: 2LMN [24]). From this structure, an Aβ monomer was isolated and the missing N-terminal peptide region of the Aβ(1–40) monomer ([1]DAEFRHDS[8]) was built using the Molefacture plugin in the VMD package [25] as random coils as predicted by both the Jpred web server [26] and by the Modeller package [27] for protein secondary structure assignments. Standard protonation states corresponding to pH 7 were assigned to ionizable residues. The Aβ(1–40) protofibril was composed by repeating 10 monomeric units along its principal axis, obtaining a continuous structure 5 nm long.

The force field assigned to each ligand in their standard protonation states at pH 7 was built in the GROMACS format [28] by using the Automated Topology Builder [29,30] web server.

The simulation box (7.5 × 9.7 × 8.0 nm) contains one Aβ(1–40) protofibril composed by repeating 10 monomeric units, with one ligand placed in a random position with respect to the fibril, and about 30,000 simple point charge water molecules [31]. Counter ions (Na^+ and Cl^-) were added at random locations to neutralize the systems, with an ion concentration of 150 mM, close to the physiological value.

All the simulations were carried out at physiological temperature (310 K) and pressure of 1 bar. The systems were first equilibrated for 2 ns in the NVT ensemble, then 10 n runs were carried out in the NPT ensemble. The temperature was controlled using a velocity-rescaling thermostat with a coupling time of 0.1 ps. During equilibration, the Berendsen barostat was used to control the pressure, while during the production run, the Parrinello–Rhaman barostat was used with coupling time of 2 ps and an isothermal compressibility of 4.5×10^{-5} bar^{-1}, and the timestep used was 2.0 fs. The particle-mesh Ewald algorithm was used to calculate long-range electrostatics [32], with a fourth-order cubic interpolation, a grid spacing of 0.16 nm, and a real-space cutoff of 1 nm [33]. Both Van der Waals and neighbor list cutoffs describing short-range interactions were set to 1.0 nm.

A production run of 50 ns was used to identify the ligand binding sites (Section 2.2), whereas trajectories of 100 ns were necessary for the computation of the stability of the different protofibril polymorphs (Section 2.4). Data analysis was performed using the GROMACS-5.0.4 package [34].

2.2. Ligand Binding Sites

Temperature replica exchange MD (REMD) simulations were used to define the most probable interacting sites of each compound with the Aβ(1–40) protofibrils. The temperatures used for replicas were obtained by the work of Patriksson and van der Spoel [35] and are reported below: 300.00, 301.16, 302.32, 303.49, 304.66, 305.83, 307.01, 308.19, 309.38, 310.57, 311.76, 312.96, 314.16, 315.37, 316.57, 317.78, 319.00, 320.22, 321.44, 322.66, 323.89, 325.12, 326.36, 327.60, 328.85, 330.09, 331.34, 332.60, 333.86, 335.12, 336.39, 337.66, 338.93, 340.21.

An acceptance ratio of 20% was chosen, as previously suggested by Ngo et al. [36]. Each REMD simulation replica was equilibrated with an NVT and an NPT ensemble with the same parameters as for MD simulations. Then, a 50 ns run (i.e., the production run) was performed for each replica, and exchanges between neighboring replicas were checked every 500 steps corresponding to 1 ps [36]. The 50 ns simulations were used for data analysis.

2.3. Ligand Binding Energy

The Molecular Mechanics Poisson–Boltzmann surface area (MM-PBSA) method [37] was used to calculate the binding energy of each ligand to the protofibril. This method is based on the single-trajectory approach. Thus, 100 snapshots collected consecutively over the course of the 50 ns simulations, once the ligand reached a stable binding (i.e., Root Mean Square Displacement of its center of mass <5 Å; Figure S1), were used. The binding free energy ($\Delta G_{binding}$) is described as the free energy difference between the complex, $G_{complex}$, and the summation of the free energy of the protein, $G_{protein}$, and ligand, G_{ligand}:

$$\Delta G_{binding} = G_{complex} - (G_{protein} - G_{ligand}) \tag{1}$$

The free energy of each molecule is given by

$$G = E_{MM} + G_{solvation} - T\Delta S \tag{2}$$

where T and S represent the temperature and entropy, respectively; and the mechanical energy, E_{MM}, of the solute in the gas phase is given by the summation of bond, angles, dihedrals, Van der Waals, and electrostatic terms:

$$E_{MM} = E_{bond} + E_{angle} + E_{dihedral} + E_{electr} + E_{VdW} \tag{3}$$

The solvation energy, $G_{solvation}$, is calculated as follows:

$$G_{solvation} = G_{surf} + G_{PB} \tag{4}$$

where the nonpolar solvation term, G_{surf}, is approximated on the solvent-accessible-surface area (SASA) derived from the Shrake–Rupley numerical method [38]:

$$G_{surf} = \gamma SASA + \beta \tag{5}$$

with $\gamma = 0.0072$ kcal/mol $Å^2$ and $\beta = 0$ [39].

The term comprising the electrostatic potential between the solute and the solvent, G_{PB}, is calculated using the continuum solvent approximation [40] by the APBS package [41].

The entropy term, $T\Delta S$, is computed using the quasi-harmonic formula [42].

2.4. Aβ(1–40) Oligomer Double-Layered Structures

Two possible double-layered structures were built by stacking the β-sheets of each monomer onto each other in an antiparallel fashion [43,44], as shown in Figure 1. The C-terminal–C-terminal and N-terminal–N-terminal interfaces were thus obtained. The intersheet distance was computed as the distance between the centers of mass of the two β-sheets that are in contact. The amino acids that were considered for the calculations of the center of mass are H13, H14, Q15, K16, L17, V18, F19, F20, A21, and E22 for the N-terminal–N-terminal interface (β-1 β-sheets) (Figure 1a) and A30, I31, I32, G33, L34, M35, V36, G37, G38, and V39 for the C-terminal–C-terminal interface (β-2 β-sheets) (Figure 1b).

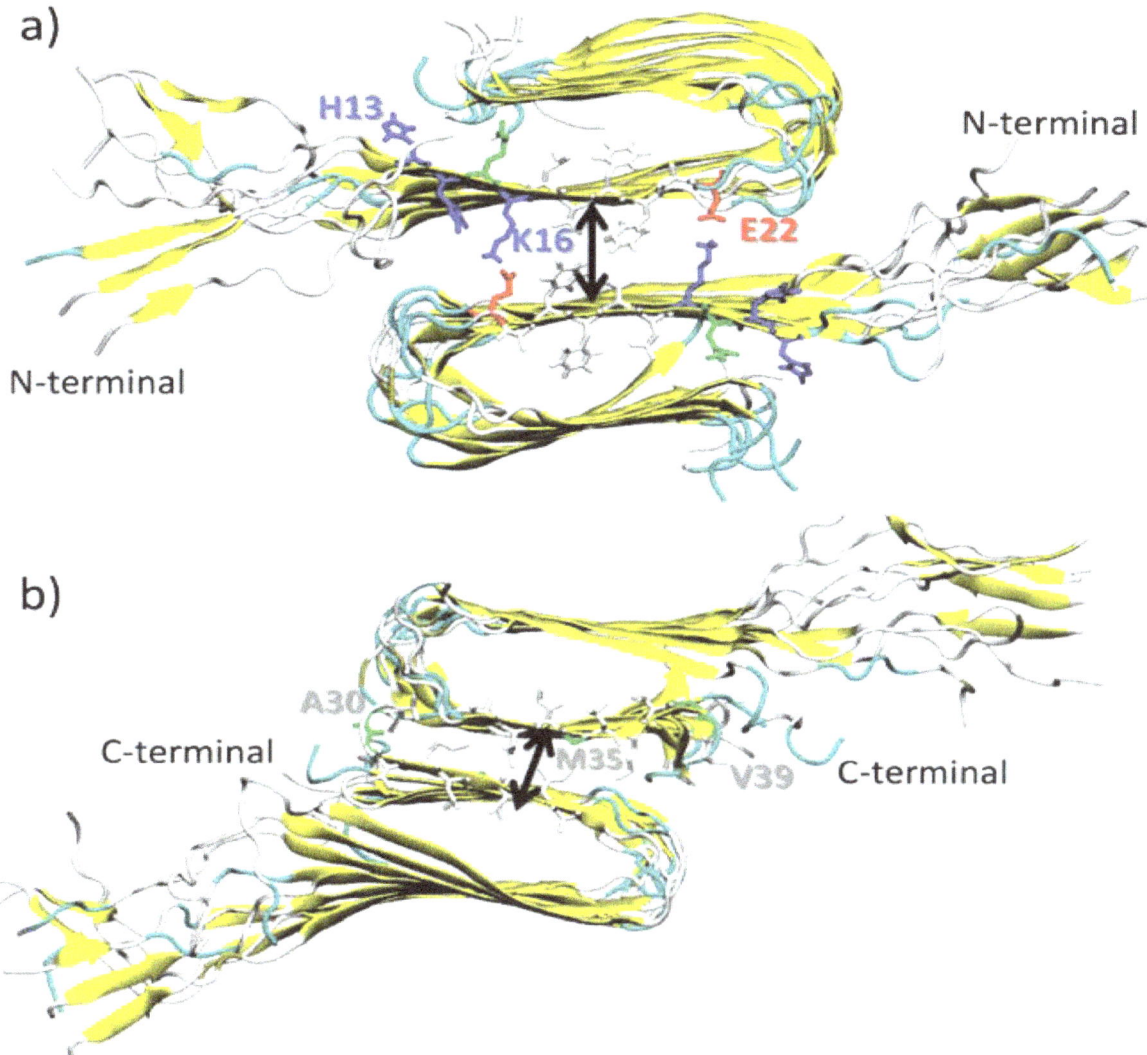

Figure 1. Cartoon representation of double-layered structures of Aβ(1–40) oligomers facing through their β-1, in (**a**), and β-2 β-sheets, in (**b**). Fibrils are colored according to their secondary structures. Amino acids at the interface are explicitly represented (color code: blue for positively charged, red for negatively charged, and white for hydrophobic amino acid residues). Black arrows roughly represent the intersheet distance.

In order to evaluate the influence of the ligands on the stability of the different protofibrils polymorphs, the potential of mean force (PMF) method implemented in the GROMACS program was used [45,46].

The backbone of protofibril (1) was restrained in its starting position, while a force increasing with time was assigned to the center of mass of protofibril (2). Three directions were taken into account, as shown in Figure 2: the x-axis (i.e., outward), the y-axis (i.e., lateral), and the z-axis (i.e., vertical). For each ligand and for both protofibril contact modes (β-1 and β-2 β-sheets), three runs were performed, using as the starting configurations the ones at 90, 95, and 100 ns, ensuring good sampling. The starting force used at the beginning of the simulation was 1000 kJ/mol nm^2, and the rate at which the application point of the force moves was 0.01 nm/ps.

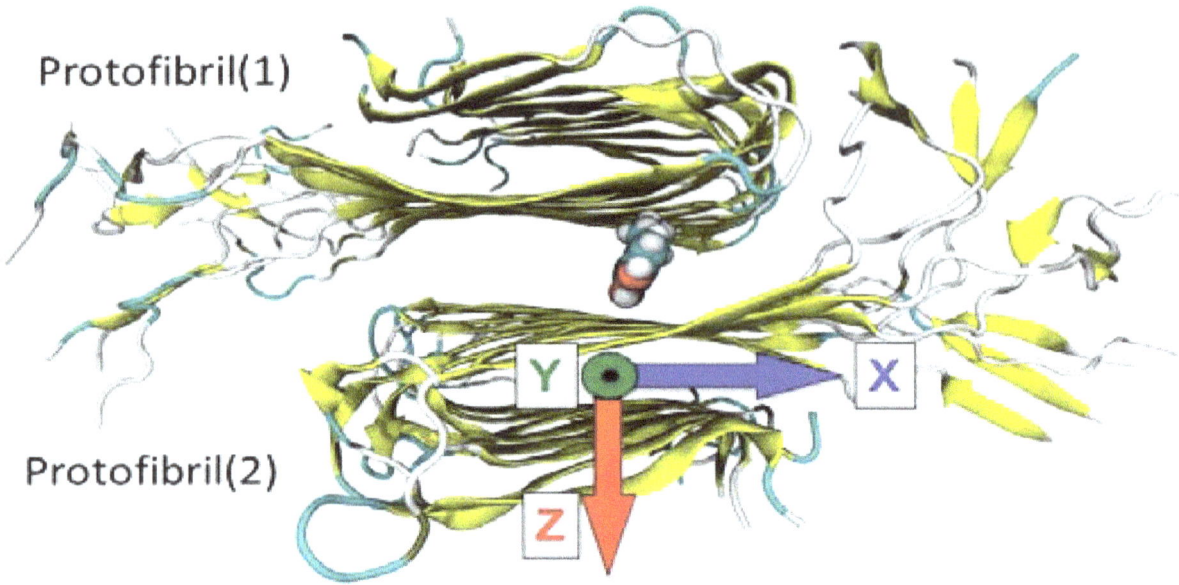

Figure 2. Pulling directions applied to protofibril (2) during the calculation of the forces needed for double-layered destruction: along the x-axis (i.e., outward shift of the protofibril (2) along its secondary axes), the y-axis (i.e., lateral shift of the protofibril (2) along its primary axes), and the z-axis (i.e., vertical shift, progressive removal of protofibril (2)).

3. Results and Discussion

The five natural compounds studied are listed in Table 1, together with their effective concentrations (EC$_{50}$) for the formation, extension, and destabilization of preformed Aβ(1–40) (fAβ(1–40)).

The overall in vitro activities of curcumin (CUR) and rosmarinic acid (ROSM) are similar [47]. Moreover, in vivo observations suggest that curcumin may be beneficial even after the disease has developed, reducing the amyloid levels and plaque burden of aged mice with advanced amyloid accumulation [48]. Quercetin (QUER) shows moderate in vitro preformed fAβ(1–40) destabilization effects with respect to CUR [49]. (-)-Epigallocatechin-3-gallate (EGCG) is undergoing phase II–III clinical trials as an inhibitor of Aβ fibrillogenesis. It decreases plaque burdens in the brain and reduces soluble and insoluble preformed fAβ(1–40)s [50]. Finally, dopamine (DOPA) proved to be a potent anti-amyloidogenic agent at all the different levels of formation, extension of amyloid fibrils, and destabilization of preformed fAβ(1–40)s [51].

Heterogeneity in the experimental conditions (i.e., peptide concentrations, incubation condition, and procedure of fAβ preparation) used in different laboratories or different experiments in the same laboratory gives rise to discrepancies in effective EC$_{50}$ concentrations, thus preventing a quantitative rationalization of the observed experimental trend by means of the results of the computational simulations. However, some interesting qualitative structure–activity relationships could be considered, as shown in the following.

Table 1. The effective concentrations (EC_{50}) of the ligands studied for the formation, extension, and destabilization of fAβ(1–40).

Compound	Acronym	Structure	Aβ(1–40) Formation (EC_{50}) μM	Aβ(1–40) Extension (EC_{50}) μM	Aβ(1–40) Destabilization (EC_{50}) μM
Curcumin diketo form	CUR-di		0.19 [47]	0.19 [47]	0.42 [47]
Curcumin ketoenol form	CUR-ke		0.81 [48]	0.19 [47]	1.00 [48]
Dopamine	DOPA		0.01 [51]	0.03 [51]	0.21 [51]
(-)-Epigallocatechin-3-gallate	EGCG		0.18 [4]	–	15 * [50]
Quercetin	QUER		0.24 [49]	0.25 [49]	2.1 [49]
Rosmarinic acid	ROSM		0.29 [47]	0.26 [47]	0.83 [47]

* Referred to Aβ(1–42) fibrils.

3.1. Putative Binding Sites and Binding Free Energies

Six main binding sites have been highlighted by means of the REMD method applied to the ligands considered. They are located at the surface of the protofibril:

1. β-1 β-sheet corresponding to the amino-acid sequence: [16]KLVFFAEDV[24],
2. β-2 β-sheet corresponding to the amino-acid sequence: [31]IIGLMVG[37],
3. Elbow connecting the two β-sheets with the corresponding amino-acid sequence: [22]EDVGSN[27],
4. top of the protofibril, over the two β-sheets of the terminal Aβ(1–40) monomer ("Over"),
5. disordered tails located at the N-terminal,
6. end of the β-2 β-sheet, on the C-terminal (entry of the cleft).

For each binding site, amino acids that make persistent interactions (in this work, an interaction is considered as persistent if the amino acid residue remains in contact with the ligand for at least 60% of the total simulation time) with the ligands and that contribute more than 1 kcal/mol to the binding energy are highlighted in Figure 3. The probability of the occupancy of each site is shown in Figure 4a.

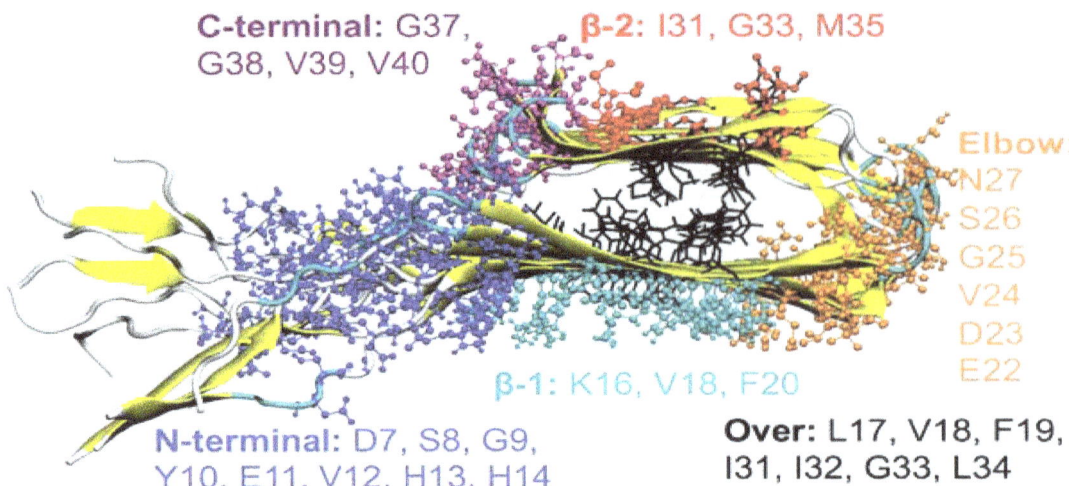

Figure 3. Ball-and-stick representation of the ligand binding sites obtained by REMD. Amino acids (single-letter code) involved in the interactions are reported for each binding site with different colors: Amino acids belonging to the N-terminal site are in blue, to the the β-1 site in cyan, to the Elbow site in orange, to the β-2 site in red, to the C-terminal site in purple, and to the Over site in black).

It is interesting to note the different occupancy preferences of the two forms of curcumin. The CUR-di form predominantly interacts with the N-terminal, whereas CUR-ke is mainly found at the β-2 site.

Multiple binding sites have been previously described in the literature for curcumin derivatives and other related compounds. In particular, the β-2 site has been very recently targeted in a combined computational and experimental study by Battisti et al. [15], aimed at the design of curcumin-like amyloid beta peptide inhibitors. Binding to the N-terminal and Over positions have been observed for curcumin and other ligands by means of site map analysis by Kundaikar et al. [52]. Moreover, the β-1 binding site has previously been suggested as a possible binding site for curcumin on the basis of solid-state NMR experiments [53] and computational studies on the Aβ hexapeptide [16]KLVFFA[21] and full-length Aβ fibrils [12,15].

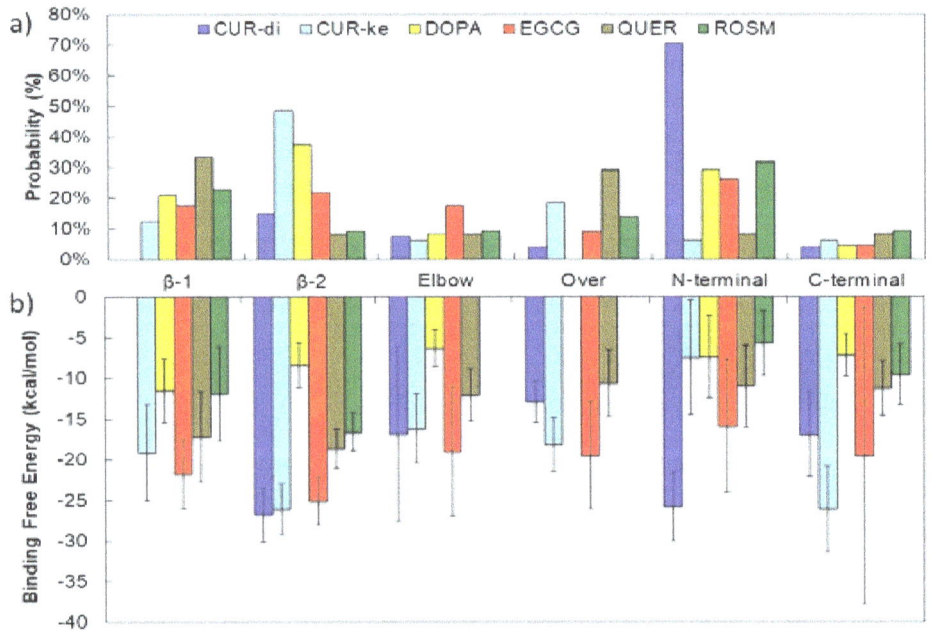

Figure 4. Probability of occupancy of each binding site (**a**) and binding free energy (**b**) for each ligand considered.

Although a few studies in the literature proposed the cavity formed by the two β-sheets and the turn as a possible binding site for curcumin [17,18] and other compounds such as Orange-G [19], this site is never occupied by the ligands considered in the present study. However, small portions of the CUR-ke, EGCG, QUER, and ROSM ligands can occasionally penetrate this cavity during the dynamic simulations runs, when they are interacting with the Aβ(1–40) protofibril in the Over position.

By considering the probability of the occupancy of each binding sites (Figure 4a) together with the corresponding binding free energies (Figure 4b), it emerges that:

-   CUR-ke, the predominant form in aqueous solution on the basis of the recent results obtained by Manolova et al. [54], shows a strong propensity to dock at the β-2 site and realizes at this site strong interactions ($\Delta G_{bind} > -20$ kcal/mol) with the fAβ(1–40) fibril. However, moderate to strong ($-10 < \Delta G_{bind} > -20$ kcal/mol) free energies of binding are found for all the binding sites, with the exception of the N-terminal (N-ter) one.

-   DOPA shows a preference for docking at the β-2 and N-ter sites. However, by considering the free energy of binding, it does not show selectivity among the six sites studied, realizing moderate to weak interactions ($G_{bind} < -10$ kcal/mol) with all of them.

-   EGCG preferentially targets the N-ter and β-2 sites and secondarily, the Elbow and β-1 sites. However, this ligand is able to realize strong binding with all six possible sites. The most stable complexes ($\Delta G_{bind} > -20$ kcal/mol) are obtained at the β-2 and β-1 sites. The ability of EGCG to bind to the N-terminal amino acids (residues 1–16) is confirmed by results obtained by isothermal titration calorimetry experiments [55]. Moreover, recent findings by solution NMR indicate that EGCG preferentially binds to Aβ oligomers and shields them at the β-1 and β-2 sites [56], where it remodels the oligomer surface, altering the interactions with the monomers.

-   QUER is found almost equally distributed between the β-1 and Over sites, with significantly lower probability for the other sites. However, it realizes moderate binding free energies ($\Delta G_{bind} \sim -10$ kcal/mol) in all sites, with the most stable complexes ($\Delta G_{bind} \sim -20$ kcal/mol) involving the β-2 and β-1 sites. These results are in agreement with the finding of a computational study recently reported by Ren et al. [13] for a structurally homologous compound, genistein. They showed that genistein prefers to bind the β-sheet grooves to interfere with their self-aggregation.

-   ROSM has higher probability for docking at the N-ter and β-1 sites, but realizes the most stable interactions with moderate binding free energies at the β-2 and β-1 sites. Indeed, NMR investigations suggest that a ROSM hairpin-like structure would allow the intercalation into the Aβ oligomers structure at the interprotofilament (β−β zippers) interface [57].

Thus, taking the error in the computation of the ΔG_{bind} into account, it can be stated that the β-2 groove is a common structural target for all the ligands studied; at this site, the ligands realize their most stable interactions with residues M35, G33, and I31. The β-1 site is also targeted for energetically favored complexes, realized mainly by the interaction with the K16, V18, and F20 residues.

These regions are particularly interesting since they constitute the junction between protofilaments in common Aβ(1–40) polymorphs [24,58]. Several recent computational studies employing different multiple protofilament structures and a variety of ligands, used as markers for amyloid detection, indicate the interfacial pockets at the junction between protofilaments as preferential binding sites [19–21]. Binding of ligands at these sites can interfere with the formation or induce the disruption of the aggregates, as discussed in the next section.

In agreement with the previous studies on related compounds [9,47], the binding free energies obtained for these complexes are driven by more favorable nonpolar interactions rather than by electrostatic ones (Figure S2).

Visual inspection of all MD trajectories shows that the random-coil N-terminal [1]DAEFRHDS[8] sequence does not appreciably alter the conformation and the usual behavior of the rest of the fibril, despite its high flexibility, promoting the nomadism of the ligands that bind preferentially to D7 and S8. Moreover, overall, the binding of the ligands does not disturb the structural integrity of the Aβ protofibrils, their overall U-shaped conformations being retained with or without interacting ligands. The secondary structure of the Aβ monomers forming the core of the protofibrils remains unperturbed upon ligand binding in the time length of the simulations, whereas β-sheet unfolding is observed for the first two monomers at the top and bottom of the protofibril. This is shown in Figure 5, where the time evolution of the Aβ(1–40) secondary structure upon EGCG binding on the β-2 β-sheet groove is reported: chain 3 is representative of the core monomers from 3 to 8 in the simulated protofibril, whereas chains 1 and 2 are representative of the top two (bottom two) monomers.

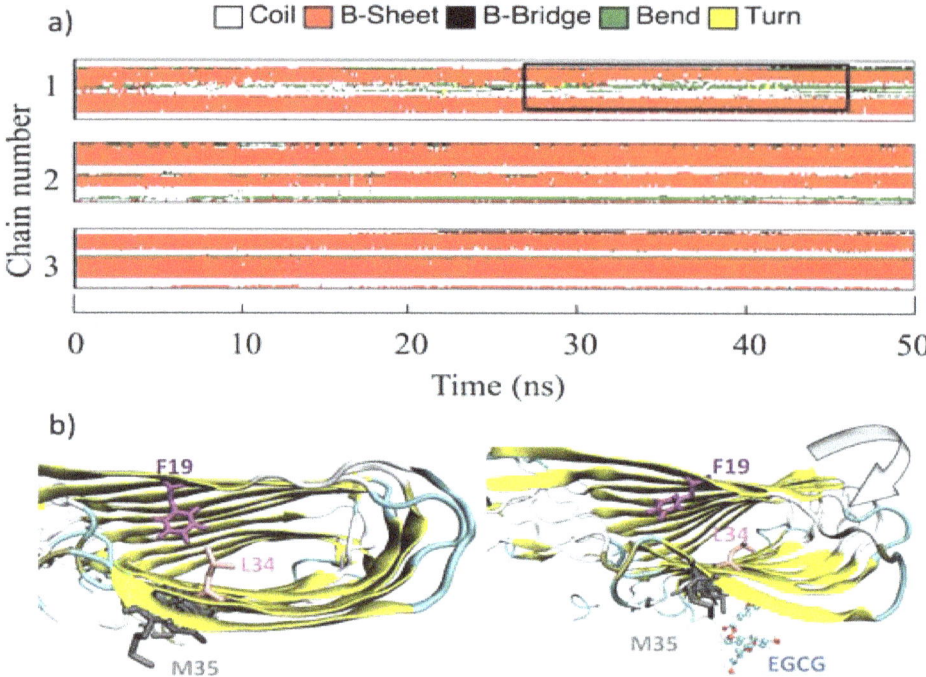

Figure 5. (**a**) Time evolution of the Aβ(1–40) secondary structure (computed with the GROMACS DSSP tool) upon EGCG binding on the β-2 β-sheet groove. The perturbation induced at the monomers lying at the head of the protofibril is highlighted by a black box. For clarity's sake, only the top three Aβ(1–40) monomers are shown. (**b**) Conformation of F19 and L34 before (left) and after (right) the interaction of EGCG with M35.

However, in a few cases (EGCG, CUR-ke, and QUER) when the ligands, during the dynamic run, migrate from the Over site to the β-2 β-sheet in proximity to M35, a perturbation of the fibril secondary structure in the terminal monomers lying at the head of the protofibril is observed. This perturbation, observed in the time of the simulations, especially in the elbow region, induces a bend in the long fibril axis that can impair the process of fibril elongation.

Figure 5 explains the phenomenon for the complex formed by EGCG and the Aβ protofibrils. The side chain of M35, interacting with the ligand, chaperones it in the search for the best interactions in the β-2 groove, causing a bending of the protofibril and altering the Aβ protofibril secondary structure in the Elbow region. Moreover, the dynamics of the M35 side chain, induced by the interacting ligand, disrupts the hydrophobic interaction between L34 and F19, which is found to influence a broad range of different processes including the initiation of fibrillation, oligomer stability, fibril elongation, and cellular toxicity [59]. In addition, it is worth underlining that M35 itself is also known to be responsible for the hierarchical assembly of amyloid fibrils.

3.2. Influence of the Ligands on the Stability of the Aβ(1–40) Oligomer Double-Layered Structures

The effect of the ligands on the stability of the protofibril double-layered structures has been quantified by the calculation of the forces (PMF) for protofibril(1)/ligand–protofibril(2) unbinding.

On the basis of the binding site preferences discussed in the previous section, the intercalation of the ligands into the C-terminal–C-terminal and N-terminal–N-terminal interfaces of the protofibrils have been considered. Moreover, three possible ways for complex disruption have been examined by applying the forces along the x-axis (i.e., outward shift of the protofibril(2) along its secondary axes), the y-axis (i.e., lateral shift of the protofibril(2) along its primary axes), and the z-axis (i.e., vertical shift, progressive removal of protofibril(2)), as shown in Figure 2.

The results are reported in Table 2, together with the force needed to separate the pristine protofibril–protofibril aggregation, taken as the control.

It is worth noting that the C-terminal–C-terminal interface of the double-layered Aβ-sheets consists of highly hydrophobic patches of I31, I41, and M35, with an average intermolecular distance between the two β-sheets of ~9.1 Å (see Table 3), whereas the N-terminal–N-terminal interface consists of both hydrophobic patches of V18 and F20 and K16–E22 salt bridges, with an average intermolecular distance of ~14.3 Å, in agreement with previous computational studies on Aβ17–42 [60] and on different segmental polymorphs (Aβ 35–42, Aβ 16–21, Aβ 27–32) modelled by Berhanu et al. [61]. These characteristics determine the stability of the β-sheet–β-sheet interfaces, which is significantly higher for the N-terminal–N-terminal arrangement with respect to the C-terminal–C-terminal one, as indicated by results from the PMF for protofibril(1)–protofibril(2) unbinding, at least for the vertical and outward directions (Table 2).

Table 2. Computed force (expressed in kJ/mol) needed for protofibril(1)–protofibril(2) (control) and protofibril(1)/ligand–protofibril(2) unbinding along the x, y, and z-axes.

Force Direction	Lateral (x-axis)		Vertical (y-axis)		Outward (z-axis)	
Ligand/binding site	β-1	β-2	β-1	β-2	β-1	β-2
Control	2743 ± 115	2772 ± 140	3520 ± 200	2013 ± 30	3413 ± 250	2387 ± 330
CUR-di	2573 ± 40	1913 ± 110	1570 ± 70	1843 ± 35	2810 ± 10	2107 ± 140
CUR-ke	2600 ± 100	2167 ± 280	1653 ± 60	1733 ± 150	2760 ± 70	2633 ± 250
DOPA	2356 ± 95	2180 ± 190	1663 ± 55	1927 ± 420	2150 ± 95	2540 ± 90
EGCG	2968 ± 93	2407 ± 75	1967 ± 25	1610 ± 115	2570 ± 30	2533 ± 60
QUER	2493 ± 90	2043 ± 155	1726 ± 75	1720 ± 30	2553 ± 120	2650 ± 100
ROSM	2888 ± 173	1677 ± 55	2053 ± 40	1367 ± 15	2767 ± 70	2310 ± 105

Overall, the binding of the ligands to β-sheet–β-sheet interfacial pockets located between two protofilaments produces a reduction of the stability of the protofibril dimeric structures. However, this cannot be directly correlated to the increasing in the intermolecular distances between the two interacting protofibrils. In fact, for the N-terminal–N-terminal interface, the distance increase upon ligand binding is in the order of 2 Å, while for the C-terminal–C-terminal one, initially characterized by a tight binding due to hydrophobic interactions, it is ~4–5 Å (Table 3).

Table 3. Intersheet distance in the Aβ(1–40) oligomer double-layered structures.

	β-1	β-2
Control	14.3 ± 0.3	9.1 ± 0.3
CUR-di	15.7 ± 0.4	13.4 ± 0.3
CUR-ke	15.4 ± 0.3	13.6 ± 0.4
DOPA	16.5 ± 0.5	14.1 ± 0.4
EGCG	16.6 ± 0.5	13.0 ± 0.4
QUER	16.3 ± 0.4	12.7 ± 0.4
ROSM	15.8 ± 0.3	14.1 ± 0.3

On the other hand, the maximum destabilization of the double-layered Aβ-sheet aggregates is observed for the β1-arrangements, when the forces are applied along the vertical (y), outward (z), and lateral (x) axes, in that (descending) order.

The binding of ligands at the C-terminal–C-terminal interface results in a moderate destabilization of the double-layered Aβ-sheet aggregates with respect to the lateral and vertical modalities, whereas for the outward disruption, it appears that the ligands have no effect or confer a small stabilization of the complexes; the large errors obtained do not allow further lucubration.

It is worth noting that the intersheet separation produced by DOPA, the smallest ligand, is larger or comparable to the one observed for more cumbersome ligands, and its effect on the destabilization of the protofibril dimeric aggregates is also overall stronger than the other ligands.

4. Concluding Remarks

The results of the systematic computational study carried out on the interaction of five natural biophenols with single Aβ(1–40) protofilaments by means of REMD simulations allowed the individuation of multiple binding sites for each ligand, located at the surface of the protofibril near to the β-1 β-sheet, β-2 β-sheet, elbow connecting the two β-sheets, top of the protofibril, disordered N-terminal, and the C-terminal.

The REMD methodology used does not allow the biophenols to enter into the hydrophobic core of the preformed protofibril, probably because the energy penalty associated with the penetration process cannot be overcome using conventional MD. The absence of binding sites in the cavity of the preformed protofibril prevents the study of destabilizing effects of the ligands by promotion of disruption of the native backbone hydrogen bonds in the protofibril interior.

The MM-PBSA energetic analysis of the binding shows that the β-1 and β-2 binding sites at the exposed surface of the Aβ(1–40) protofibrils, shared by all the five ligands studied, are thermodynamically favored. At these sites, the anti-amyloid activity of biophenols consists in the inhibition of fibril thickening and elongation.

In fact, although no significant perturbation of the overall protofibril secondary structure is observed in the periods of time studied, interesting conformational changes of the terminal peptides with subsequent bending of the principal axis of the protofibril are induced by ligands that migrate during the dynamic run from the Over binding site to the β-2 binding site. This effect is more marked for EGCG, but is observed also for CUR-ke and QUER and may preclude the association of an incoming Aβ peptide inhibiting the fibril elongation.

Moreover, ligand binding at the β-2 binding site may inhibit the amyloidogenic process by shielding the M35, which is responsible for the hierarchical assembly of amyloid fibrils, and disrupting the hydrophobic interaction between L34 and F19, which is found to influence a broad range of different processes including the initiation of fibrillation, oligomer stability, fibril elongation, and cellular toxicity.

Finally, the stability of the β-sheet–β-sheet interfaces of the Aβ(1–40) oligomer double-layered structures is significantly affected by the intercalation of the biophenols. The force needed for disruption of the aggregates is halved by all the ligands binding the N-terminal–N-terminal interface when the forces are applied along the principal axis of the protofibril. The most remarkable effect is observed for DOPA on the double-layered structure in the N-terminal–N-terminal arrangement, whatever the force direction; whereas ROSM and EGCG exert a stronger destabilization at the double-layered structure in the C-terminal–C-terminal arrangement.

These structural insights may serve as a molecular guide for setting up further rational drug design in close collaboration with experimentalists in order to obtain effective inhibitors targeting fibril formation in Alzheimer's disease.

Author Contributions: Conceptualization, M.C.M.; formal analysis, F.T.; funding acquisition, M.C.M.; investigation, F.T.; methodology, F.T.; project administration, M.C.M.; supervision, M.C.M. and A.P.; visualization, F.T.; writing of the original draft, F.T.; writing review and editing, A.P.

Acknowledgments: The authors gratefully acknowledge the CINECA supercomputing center (Italy) and HPC projects ISCRA-C HP10CATCHE for computational resources. APC was sponsored by MDPI.

References

1. Knowles, T.P.J.; Vendruscolo, M.; Dobson, C.M. The amyloid state and its association with protein misfolding diseases. *Nat. Rev. Mol. Cell Biol.* **2014**, *15*, 384–396. [CrossRef] [PubMed]

2. Chiti, F.; Dobson, C.M. Protein Misfolding, Functional Amyloid, and Human Disease. *Annu. Rev. Biochem.* **2006**, *75*, 333–366. [CrossRef] [PubMed]

3. Han, X.; He, G. Toward a Rational Design to Regulate β-Amyloid Fibrillation for Alzheimer's Disease Treatment. *ACS Chem. Neurosci.* **2018**, *9*, 198–210. [CrossRef] [PubMed]

4. Yair, P.; Adel, A.; Ehud, G. Inhibition of Amyloid Fibril Formation by Polyphenols: Structural Similarity and Aromatic Interactions as a Common Inhibition Mechanism. *Chem. Biol. Drug Des.* **2005**, *67*, 27–37. [CrossRef]

5. Lemkul, J.A.; Bevan, D.R. The Role of Molecular Simulations in the Development of Inhibitors of Amyloid β-Peptide Aggregation for the Treatment of Alzheimer's Disease. *ACS Chem. Neurosci.* **2012**, *3*, 845–856. [CrossRef] [PubMed]

6. Liu, F.-F.; Dong, X.-Y.; He, L.; Middelberg, A.P.J.; Sun, Y. Molecular Insight into Conformational Transition of Amyloid β-Peptide 42 Inhibited by (−)-Epigallocatechin-3-gallate Probed by Molecular Simulations. *J. Phys. Chem. B* **2011**, *115*, 11879–11887. [CrossRef] [PubMed]

7. Zhao, L.N.; Chiu, S.-W.; Benoit, J.; Chew, L.Y.; Mu, Y. The Effect of Curcumin on the Stability of Aβ Dimers. *J. Phys. Chem. B* **2012**, *116*, 7428–7435. [CrossRef] [PubMed]

8. Zhang, T.; Zhang, J.; Derreumaux, P.; Mu, Y. Molecular Mechanism of the Inhibition of EGCG on the Alzheimer Aβ1–42 Dimer. *J. Phys. Chem. B* **2013**, *117*, 3993–4002. [CrossRef] [PubMed]

9. Wang, Y.; Latshaw, D.C.; Hall, C.K. Aggregation of Aβ(17–36) in the Presence of Naturally Occurring Phenolic Inhibitors Using Coarse-Grained Simulations. *J. Mol. Biol.* **2017**, *429*, 3893–3908. [CrossRef] [PubMed]

10. Awasthi, M.; Singh, S.; Pandey, V.P.; Dwivedi, U.N. Modulation in the conformational and stability attributes of the Alzheimer's disease associated amyloid-beta mutants and their favorable stabilization by curcumin: Molecular dynamics simulation analysis. *J. Biomol. Struct. Dyn.* **2017**, 1–16. [CrossRef] [PubMed]

11. Chebaro, Y.; Jiang, P.; Zang, T.; Mu, Y.; Nguyen, P.H.; Mousseau, N.; Derreumaux, P. Structures of Aβ17–42 Trimers in Isolation and with Five Small-Molecule Drugs Using a Hierarchical Computational Procedure. *J. Phys. Chem. B* **2012**, *116*, 8412–8422. [CrossRef] [PubMed]

12. Rao, P.P.N.; Mohamed, T.; Teckwani, K.; Tin, G. Curcumin Binding to Beta Amyloid: A Computational Study. *Chem. Biol. Drug Des.* **2015**, *86*, 813–820. [CrossRef] [PubMed]

13. Ren, B.; Liu, Y.; Zhang, Y.; Cai, Y.; Gong, X.; Chang, Y.; Xu, L.; Zheng, J. Genistein: A Dual Inhibitor of Both Amyloid β and Human Islet Amylin Peptides. *ACS Chem. Neurosci.* **2018**. [CrossRef] [PubMed]

14. Taguchi, R.; Hatayama, K.; Takahashi, T.; Hayashi, T.; Sato, Y.; Sato, D.; Ohta, K.; Nakano, H.; Seki, C.; Endo, Y.; Tokuraku, K.; Uwai, K. Structure–activity relations of rosmarinic acid derivatives for the amyloid β aggregation inhibition and antioxidant properties. *Eur. J. Med. Chem.* **2017**, *138*, 1066–1075. [CrossRef] [PubMed]

15. Battisti, A.; Piccionello, A.P.; Sgarbossa, A.; Vilasi, S.; Ricci, C.; Ghetti, F.; Spinozzi, F.; Gammazza, A.M.; Giacalone, V.; Martorana, A.; Lauria, A.; Ferrero, C.; Bulone, D.; Rosalia Mangione, M.; Biagio, P.L.S.; Grazia Ortore, M. Curcumin-like compounds designed to modify amyloid beta peptide aggregation patterns. *RSC Adv.* **2017**, *7*, 31714–31724. [CrossRef]

16. Espargaró, A.; Ginex, T.; Vadell, M.D.; Busquets, M.A.; Estelrich, J.; Muñoz-Torrero, D.; Luque, F.J.; Sabate, R. Combined in Vitro Cell-Based/in Silico Screening of Naturally Occurring Flavonoids and Phenolic Compounds as Potential Anti-Alzheimer Drugs. *J. Nat. Prod.* **2017**, *80*, 278–289. [CrossRef] [PubMed]

17. Ngo, S.T.; Li, M.S. Curcumin Binds to Aβ1–40 Peptides and Fibrils Stronger Than Ibuprofen and Naproxen. *J. Phys. Chem. B* **2012**, *116*, 10165–10175. [CrossRef] [PubMed]

18. Ngo, S.T.; Fang, S.-T.; Huang, S.-H.; Chou, C.-L.; Huy, P.D.Q.; Li, M.S.; Chen, Y.-C. Anti-arrhythmic Medication Propafenone a Potential Drug for Alzheimer's Disease Inhibiting Aggregation of Aβ: In Silico and in Vitro Studies. *J. Chem. Inf. Model.* **2016**, *56*, 1344–1356. [CrossRef] [PubMed]

19. Kawai, R.; Araki, M.; Yoshimura, M.; Kamiya, N.; Ono, M.; Saji, H.; Okuno, Y. Core Binding Site of a Thioflavin-T-Derived Imaging Probe on Amyloid β Fibrils Predicted by Computational Methods. *ACS Chem. Neurosci.* **2018**. [CrossRef] [PubMed]

20. Murugan, N.A.; Halldin, C.; Nordberg, A.; Långström, B.; Ågren, H. The Culprit Is in the Cave: The Core Sites Explain the Binding Profiles of Amyloid-Specific Tracers. *J. Phys. Chem. Lett.* **2016**, *7*, 3313–3321. [CrossRef] [PubMed]

21. Peccati, F.; Pantaleone, S.; Riffet, V.; Solans-Monfort, X.; Contreras-García, J.; Guallar, V.; Sodupe, M. Binding of Thioflavin T and Related Probes to Polymorphic Models of Amyloid-β Fibrils. *J. Phys. Chem. B* **2017**, *121*, 8926–8934. [CrossRef] [PubMed]

22. Schmid, N.; Eichenberger, A.P.; Choutko, A.; Riniker, S.; Winger, M.; Mark, A.E.; van Gunsteren, W.F. Definition and testing of the GROMOS force-field versions 54A7 and 54B7. *Eur. Biophys. J. EBJ* **2011**, *40*, 843–856. [CrossRef] [PubMed]

23. Berman, H.M.; Westbrook, J.; Feng, Z.; Gilliland, G.; Bhat, T.N.; Weissig, H.; Shindyalov, I.N.; Bourne, P.E. The Protein Data Bank. *Nucleic Acids Res.* **2000**, *28*, 235–242. [CrossRef] [PubMed]

24. Petkova, A.T.; Yau, W.-M.; Tycko, R. Experimental Constraints on Quaternary Structure in Alzheimer's β-Amyloid Fibrils†. *Biochemistry (Mosc.)* **2006**, *45*, 498–512. [CrossRef] [PubMed]

25. Humphrey, W.; Dalke, A.; Schulten, K. VMD: Visual molecular dynamics. *J. Mol. Graph.* **1996**, *14*, 33–38. [CrossRef]

26. Drozdetskiy, A.; Cole, C.; Procter, J.; Barton, G.J. JPred4: A protein secondary structure prediction server. *Nucl. Acids Res.* **2015**, *43*, W389–W394. [CrossRef] [PubMed]

27. Eswar, N.; Webb, B.; Marti-Renom, M.A.; Madhusudhan, M.S.; Eramian, D.; Shen, M.-Y.; Pieper, U.; Sali, A. Comparative protein structure modeling using Modeller. *Curr. Protoc. Bioinform.* **2006**, *15*, 5–6. [CrossRef] [PubMed]

28. Pronk, S.; Páll, S.; Schulz, R.; Larsson, P.; Bjelkmar, P.; Apostolov, R.; Shirts, M.R.; Smith, J.C.; Kasson, P.M.; van der Spoel, D.; Hess, B.; Lindahl, E. GROMACS 4.5: A high-throughput and highly parallel open source molecular simulation toolkit. *Bioinforma. Oxf. Engl.* **2013**, *29*, 845–854. [CrossRef] [PubMed]

29. Koziara, K.B.; Stroet, M.; Malde, A.K.; Mark, A.E. Testing and validation of the Automated Topology Builder (ATB) version 2.0: Prediction of hydration free enthalpies. *J. Comput. Aided Mol. Des.* **2014**, *28*, 221–233. [CrossRef] [PubMed]

30. Malde, A.K.; Zuo, L.; Breeze, M.; Stroet, M.; Poger, D.; Nair, P.C.; Oostenbrink, C.; Mark, A.E. An Automated Force Field Topology Builder (ATB) and Repository: Version 1.0. *J. Chem. Theory Comput.* **2011**, *7*, 4026–4037. [CrossRef] [PubMed]

31. Berendsen, H.J.C.; Postma, J.P.M.; van Gunsteren, W.F.; Hermans, J. Interaction Models for Water in Relation to Protein Hydration. In *Intermolecular Forces: Proceedings of the Fourteenth Jerusalem Symposium on Quantum Chemistry and Biochemistry Held in Jerusalem, Israel, April 13–16, 1981*; Pullman, B., Ed.; Springer: Dordrecht, The Netherlands, 1981; pp. 331–342. ISBN 978-94-015-7658-1.

32. Darden, T.; York, D.; Pedersen, L. Particle mesh Ewald: An N·log(N) method for Ewald sums in large systems. *J. Chem. Phys.* **1993**, *98*, 10089–10092. [CrossRef]

33. Essmann, U.; Perera, L.; Berkowitz, M.L.; Darden, T.; Lee, H.; Pedersen, L.G. A smooth particle mesh Ewald method. *J. Chem. Phys.* **1995**, *103*, 8577–8593. [CrossRef]

34. Abraham, M.J.; Murtola, T.; Schulz, R.; Páll, S.; Smith, J.C.; Hess, B.; Lindahl, E. GROMACS: High performance molecular simulations through multi-level parallelism from laptops to supercomputers. *SoftwareX* **2015**, *1*, 19–25. [CrossRef]

35. Patriksson, A.; Spoel, D. van der A temperature predictor for parallel tempering simulations. *Phys. Chem. Chem. Phys.* **2008**, *10*, 2073–2077. [CrossRef] [PubMed]

36. Ngo, S.T.; Hung, H.M.; Tran, K.N.; Nguyen, M.T. Replica exchange molecular dynamics study of the amyloid beta (11–40) trimer penetrating a membrane. *RSC Adv.* **2017**, *7*, 7346–7357. [CrossRef]

37. Kumari, R.; Kumar, R.; Lynn, A. g_mmpbsa—A GROMACS Tool for High-Throughput MM-PBSA Calculations. *J. Chem. Inf. Model.* **2014**, *54*, 1951–1962. [CrossRef] [PubMed]

38. Shrake, A.; Rupley, J.A. Environment and exposure to solvent of protein atoms. Lysozyme and insulin. *J. Mol. Biol.* **1973**, *79*, 351–371. [CrossRef]

39. Sitkoff, D.; Sharp, K.A.; Honig, B. Accurate Calculation of Hydration Free Energies Using Macroscopic Solvent Models. *J. Phys. Chem.* **1994**, *98*, 1978–1988. [CrossRef]

40. Sharp, K.A.; Honig, B. Electrostatic interactions in macromolecules: Theory and applications. *Annu. Rev. Biophys. Biophys. Chem.* **1990**, *19*, 301–332. [CrossRef] [PubMed]

41. Baker, N.A.; Sept, D.; Joseph, S.; Holst, M.J.; McCammon, J.A. Electrostatics of nanosystems: Application to microtubules and the ribosome. *Proc. Natl. Acad. Sci.* **2001**, *98*, 10037–10041. [CrossRef] [PubMed]

42. Baron, R.; Hünenberger, P.H.; McCammon, J.A. Absolute Single-Molecule Entropies from Quasi-Harmonic Analysis of Microsecond Molecular Dynamics: Correction Terms and Convergence Properties. *J. Chem. Theory Comput.* **2009**, *5*, 3150–3160. [CrossRef] [PubMed]

43. Tycko, R. Molecular structure of amyloid fibrils: Insights from solid-state NMR. *Q. Rev. Biophys.* **2006**, *39*, 1–55. [CrossRef] [PubMed]

44. Tycko, R. Amyloid Polymorphism: Structural Basis and Neurobiological Relevance. *Neuron* **2015**, *86*, 632–645. [CrossRef] [PubMed]

45. Trzesniak, D.; Kunz, A.P.; van Gunsteren, W.F. A Comparison of Methods to Compute the Potential of Mean Force. *ChemPhysChem* **2006**, *8*, 162–169. [CrossRef] [PubMed]

46. Roux, B. The calculation of the potential of mean force using computer simulations. *Comput. Phys. Commun.* **1995**, *91*, 275–282. [CrossRef]

47. Ono, K.; Hasegawa, K.; Naiki, H.; Yamada, M. Curcumin has potent anti-amyloidogenic effects for Alzheimer's β-amyloid fibrils in vitro. *J. Neurosci. Res.* **2004**, *75*, 742–750. [CrossRef] [PubMed]

48. Yang, F.; Lim, G.P.; Begum, A.N.; Ubeda, O.J.; Simmons, M.R.; Ambegaokar, S.S.; Chen, P.P.; Kayed, R.; Glabe, C.G.; Frautschy, S.A.; et al. Curcumin inhibits formation of amyloid beta oligomers and fibrils, binds plaques, and reduces amyloid in vivo. *J. Biol. Chem.* **2005**, *280*, 5892–5901. [CrossRef] [PubMed]

49. Ono, K.; Yoshiike, Y.; Takashima, A.; Hasegawa, K.; Naiki, H.; Yamada, M. Potent anti-amyloidogenic and fibril-destabilizing effects of polyphenols in vitro: Implications for the prevention and therapeutics of Alzheimer's disease. *J. Neurochem.* **2003**, *87*, 172–181. [CrossRef] [PubMed]

50. Bieschke, J.; Russ, J.; Friedrich, R.P.; Ehrnhoefer, D.E.; Wobst, H.; Neugebauer, K.; Wanker, E.E. EGCG remodels mature α-synuclein and amyloid-β fibrils and reduces cellular toxicity. *Proc. Natl. Acad. Sci.* **2010**, *107*, 7710–7715. [CrossRef] [PubMed]

51. Ono, K.; Hasegawa, K.; Naiki, H.; Yamada, M. Anti-Parkinsonian agents have anti-amyloidogenic activity for Alzheimer's beta-amyloid fibrils in vitro. *Neurochem. Int.* **2006**, *48*, 275–285. [CrossRef] [PubMed]

52. Kundaikar, H.S.; Degani, M.S. Insights into the Interaction Mechanism of Ligands with Aβ42 Based on Molecular Dynamics Simulations and Mechanics: Implications of Role of Common Binding Site in Drug Design for Alzheimer's Disease. *Chem. Biol. Drug Des.* **2015**, *86*, 805–812. [CrossRef] [PubMed]

53. Masuda, Y.; Fukuchi, M.; Yatagawa, T.; Tada, M.; Takeda, K.; Irie, K.; Akagi, K.; Monobe, Y.; Imazawa, T.; Takegoshi, K. Solid-state NMR analysis of interaction sites of curcumin and 42-residue amyloid β-protein fibrils. *Bioorg. Med. Chem.* **2011**, *19*, 5967–5974. [CrossRef] [PubMed]

54. Manolova, Y.; Deneva, V.; Antonov, L.; Drakalska, E.; Momekova, D.; Lambov, N. The effect of the water on the curcumin tautomerism: A quantitative approach. *Spectrochim. Acta. A. Mol. Biomol. Spectrosc.* **2014**, *132*, 815–820. [CrossRef] [PubMed]

55. Wang, S.-H.; Dong, X.-Y.; Sun, Y. Thermodynamic Analysis of the Molecular Interactions between Amyloid β-Protein Fragments and (−)-Epigallocatechin-3-gallate. *J. Phys. Chem. B* **2012**, *116*, 5803–5809. [CrossRef] [PubMed]

56. Ahmed, R.; Melacini, G. A solution NMR toolset to probe the molecular mechanisms of amyloid inhibitors. *Chem. Commun.* **2018**. [CrossRef] [PubMed]

57. Airoldi, C.; Sironi, E.; Dias, C.; Marcelo, F.; Martins, A.; Rauter, A.P.; Nicotra, F.; Jimenez-Barbero, J. Natural Compounds against Alzheimer's Disease: Molecular Recognition of Aβ1–42 Peptide by Salvia sclareoides Extract and its Major Component, Rosmarinic Acid, as Investigated by NMR. *Chem. Asian J.* **2013**, *8*, 596–602. [CrossRef] [PubMed]

58. Bertini, I.; Gonnelli, L.; Luchinat, C.; Mao, J.; Nesi, A. A New Structural Model of Aβ40 Fibrils. *J. Am. Chem. Soc.* **2011**, *133*, 16013–16022. [CrossRef] [PubMed]

59. Korn, A.; McLennan, S.; Adler, J.; Krueger, M.; Surendran, D.; Maiti, S.; Huster, D. Amyloid β (1-40) Toxicity Depends on the Molecular Contact between Phenylalanine 19 and Leucine 34. *ACS Chem. Neurosci.* **2017**. [CrossRef] [PubMed]

60. Zheng, J.; Jang, H.; Ma, B.; Tsai, C.-J.; Nussinov, R. Modeling the Alzheimer Aβ17-42 Fibril Architecture: Tight Intermolecular Sheet-Sheet Association and Intramolecular Hydrated Cavities. *Biophys. J.* **2007**, *93*, 3046–3057. [CrossRef] [PubMed]

61. Berhanu, W.M.; Hansmann, U.H.E. Structure and Dynamics of Amyloid-β Segmental Polymorphisms. *PLoS ONE* **2012**, *7*, e41479. [CrossRef] [PubMed]

14

Targeting Difficult Protein-Protein Interactions with Plain and General Computational Approaches

Mariarosaria Ferraro [1] and Giorgio Colombo [1,2,*]

[1] Istituto di Chimica del Riconoscimento Molecolare, CNR, Via Mario Bianco 9, 20131 Milano, Italy; mariar.ferraro@gmail.com

[2] Dipartimento di Chimica, Università di Pavia, V.le Taramelli 10, 27100 Pavia, Italy

[*] Correspondence: giorgio.colombo@icrm.cnr.it or g.colombo@unipv.it

Abstract: Investigating protein-protein interactions (PPIs) holds great potential for therapeutic applications, since they mediate intricate cell signaling networks in physiological and disease states. However, their complex and multifaceted nature poses a major challenge for biochemistry and medicinal chemistry, thereby limiting the druggability of biological partners participating in PPIs. Molecular Dynamics (MD) provides a solid framework to study the reciprocal shaping of proteins' interacting surfaces. Here, we review successful applications of MD-based methods developed in our group to predict interfacial areas involved in PPIs of pharmaceutical interest. We report two interesting examples of how structural, dynamic and energetic information can be combined into efficient strategies which, complemented by experiments, can lead to the design of new small molecules with promising activities against cancer and infections. Our advances in targeting key PPIs in angiogenic pathways and antigen-antibody recognition events will be discussed for their role in drug discovery and chemical biology.

Keywords: molecular dynamics; proteins; molecular recognition; protein protein interactions

1. Introduction

The existence of complex wirings in protein-protein interaction (PPI) networks finely modulates the inner working of the circuits at the basis of cell life. Their correct or incorrect regulation is naturally linked to the evolution of cells towards normal or diseases states. Being so important in disparate aspects of cellular functions, it comes as no surprise that PPIs have been the subject of intense studies over the last few years [1–6]. Understanding protein-protein recognition and binding entails shedding light on the regulatory mechanisms, as well as deepening our knowledge of the relationships between protein sequences, structure and their interactions [7]. From the practical point of view, our ability to master PPIs could play a key role in the fields of medicinal chemistry, chemical and synthetic biology. Indeed, not only could there be room for new strategies aimed at rewiring signaling pathways for synthetic biology, but also to develop new molecules against complex or yet undrugged targets, for diagnostic and therapeutic purposes [2].

In general, PPIs represent a class of interactions of high complexity. Structural and biophysical studies have shown that the features of the regions involved in interactions with other partners are diverse and multifaceted: contact surfaces may be large compared to the ones involved in protein-small molecule interactions; they are often flat and lack the grooves and crevices which are engaged by small molecules, and finally, they can be highly dynamic to favor adaptation to alternative binding partners [1,8]. Nonetheless, several methods and strategies to discover orthosteric, adaptive and allosteric inhibitors, as well as those pointing at PPIs promoters and stabilizers have been developed and excellently reviewed by Cesa and coworkers [9,10].

From the experimental point of view, mutational studies have shown that limited subsets of interface residues actually contribute to the affinity between the binding partners. In the context of targeting interface plasticity, flexible peptides selected by high-throughput screening (HTS) methods (such as phage display or large library screenings) have shown the ability to outcompete the natural partner by adapting to the interaction surface [2,11]. Similarly, HTS of small molecules against biochemically-reconstituted complexes have led to the identification of useful compounds with phenotypic effects when tested in cells. However, in this case, instead of directly monitoring physical interaction, researchers set out to characterize the functional consequences of the inhibition of a particular class of PPIs as a surrogate for binding measurements [9,12]. This is an interesting example of application of HTS methods to find modulators of PPI networks that highlights the importance of considering with care approaches to target challenging PPIs, like those intrinsically characterized by weak or transient interactions and for which classical HTS-based detection is not suitable [9,10].

These facts vividly portray a situation in which many aspects of protein-protein interactions have been investigated with success. Despite this sophistication and advancement, there is still no experimental technique that can predict at atomic level the determinants of what makes a protein surface an interacting one, or defines rules for the design of new molecular entities with applications in chemical biology or drug development. To tackle these problems, we have little choice but to turn to theoretical and computational approaches.

Theoretical methods to predict interacting surfaces of a protein of known structure fall into three main classes: (a) statistical approaches, (b) structural techniques, and (c) molecular dynamics (MD)-based methods. Statistical approaches relate an amino acid sequence to known 3-D structures and known tendencies for specific sequence motifs to be localized within interaction areas. Nowadays, these methods are widely used also in combination with coevolution concepts [13]. However, they provide no information regarding possible alternative conformations. Structural techniques use information on the geometric patterns of backbones and side chains involved in PPIs to recognize whether they are present in previously uncharacterized instances [14]. However, these methods cannot be used to describe the dynamics underlying the recognition process.

MD simulations represent a prime tool to characterize both the networks of interactions and the range of alternative states that can determine whether a protein surface may actually be an interacting one, and/or the dynamics of the processes of molecular recognition with binding partners [15–19]. In some cases, MD simulations can be integrated with quantum calculations to describe complex reactive processes at the basis of downstream recognition events [20]. In this focused perspective, we will discuss cases from our own experience where MD-based approaches have been used to derive compact physico-chemical descriptors of peptide-protein interactions that could be efficaciously translated into the discovery of new active small molecules, and to predict specific types of protein-protein interaction interfaces (namely those involved in antigen-antibody binding). In general, our framework entails the use of computational results for the design and experimental tests of active chemical tools to probe a certain PPI. Such chemical probes, indeed, represent the direct products of our ability to understand and suitably mimic the determinants of an interaction: in this view, they are designed to target and perturb a specific area and to report on the effects of such perturbation in cells. At the end of this paper, we will discuss possible perspectives in the development of novel therapeutics, such as drugs with novel mechanisms and synthetic antigens for vaccination.

2. MD-Based Methods for Studying PPIs: Studying Peptides to Develop Novel Small-Molecule Anticancer Drug Candidates

The availability of a general framework to design molecules that meet the specific structural/dynamical requirements to perturb a certain function is both a necessity and an opportunity towards innovative discovery of therapeutics and chemical tools. A full understanding of the roles of different sub-states of a molecular interacting system will allow a more rational design of the chemical probes we need to target a specific PPI; this can potentially translate into our ability to control the responses obtained by any system in which the interaction is involved.

Building on these considerations, we built a pipeline for the design of small molecules mimics of peptides known to interrupt relevant PPIs in the control of angiogenesis, the process of vascular growth widely exploited by tumors to support their own development and diffusion. To proceed along these lines, we started from the experimentally characterized interactions between the protein Fibroblast Growth Factor-2 (FGF2) and peptides derived from two large extracellular multi-domain proteins known to interact with it, namely Thrombospondin-1 (TSP1) and Pentraxin-3 (PTX-3) [21–26]. TSP1 and PTX3 are two distinct endogenous inhibitors of FGF2, which engage the target with different mechanisms at different interfaces [25,26] (Figure 1).

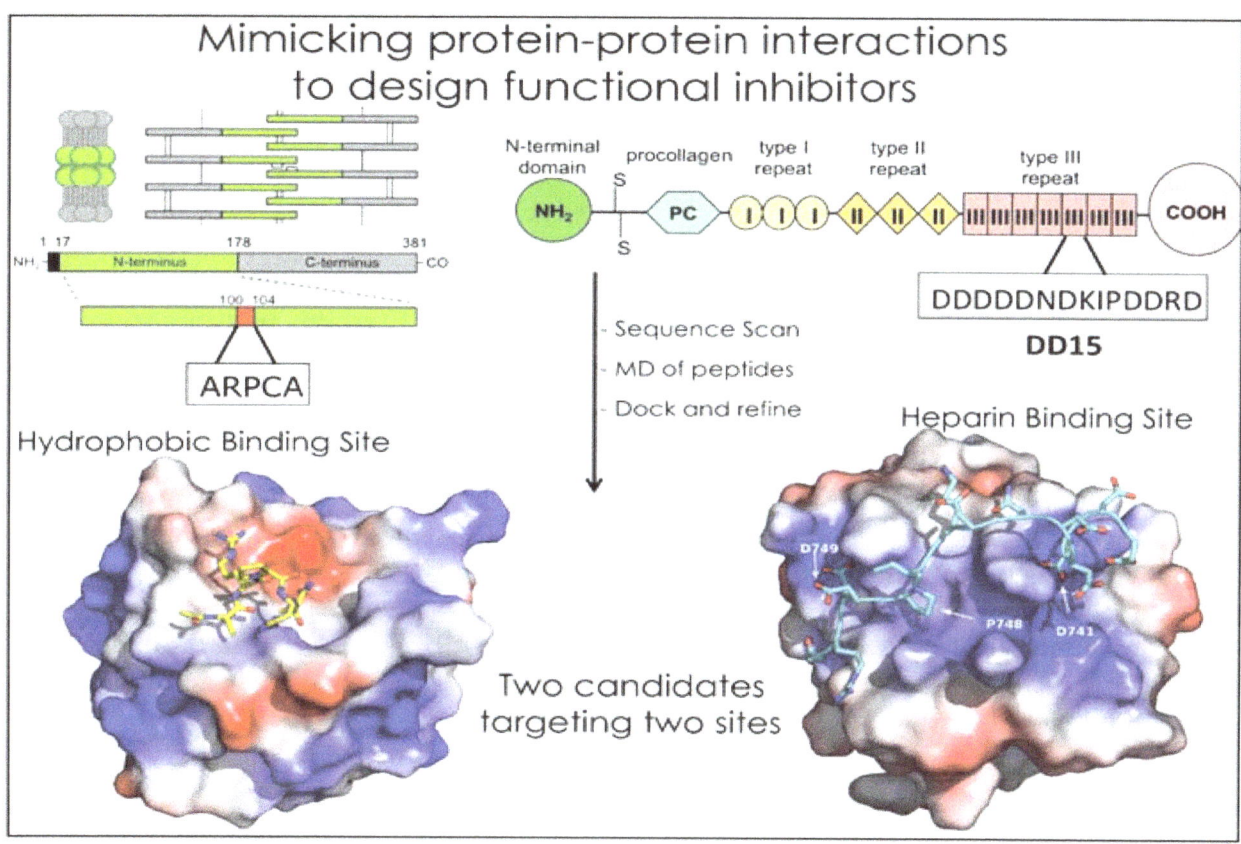

Figure 1. Simplified scheme depicting the identification of specific binding sequences in large multidomain proteins. Here, the cases of TSP1 and PTX3 binding to FGF2 are shown.

Although both proteins inhibit FGF-dependent angiogenic responses, in mechanisms related to tumor onset and development such inhibitory activity is not present and FGF2 is free to engage tyrosine kinase (TK) FGF Receptors (FGFR1-4). In presence of heparan sulphate proteoglycans (HSPGs), FGF2 binds the TKR subtypes to form HSPG/FGF/FGFR ternary complexes [27]. Activation of the FGF/FGFR system is implicated in key steps of tumor growth and progression [27]. Furthermore, compensatory up-regulation of the FGF/FGFR system may facilitate the escape from endothelial growth factor (VEGF) blockade [27]. Thus, the development of anti-FGF/FGFR targeting agents represents an urgent medical need in cancer therapy.

In this context, we started by examining the possibility of exploiting the dynamic cross-talk between FGF2 and a binding peptide in drug-candidate selection [24,28]. Our reasoning was based on the idea that molecular recognition entails a two-way influence between the interacting partners, whereby FGF2 flexibility determines the peptide conformation while the peptide poses dictate the stereochemical organization of the binding site. This dynamic adaptation is used to define the principal pharmacophoric determinants responsible for forming a stable complex. To dissect the sequence determinants of the interaction between TSP1 and FGF2, we first analyzed the binding profile of an array of peptides from a library of TSP1-derived synthetic compounds. The peptide array was designed based on the sequence of the type III repeats: 237 20-mer peptides with partially overlapping sequences (19-amino acid overlaps) were synthesized and covalently linked to polypropylene cards. The binding of biotinylated FGF2 (10 μg/mL) to the peptides was then tested. Bound FGF-2 was detected with peroxidase-conjugated streptavidin and the peroxidase substrate 2,2′-azino-di-3-ethylbenzthiazoline sulfonate (ABTS). Color development was quantified with a CCD camera, which reported on the affinities of different sequences for the target FGF2 [24].

Upon focusing on the best binding sequences, SPR identified peptide DDDDDNDKIPDDRDN, labeled DD15, as the one with the highest affinity. Sensorgrams indicated a dose-dependent binding of DD15 to FGF-2, with an association rate K_{on} of 19.7 ± 2.0 $M^{-1} \cdot s^{-1}$ and a dissociation rate K_{off} of $(5.5 \pm 0.8) \times 10^{-4}$ s^{-1}, with a resulting K_d of 28.0 μM. The peptide was located in the type III repeats of TSP1 [24,28] (Figure 1).

MD simulations were extensively performed on DD15 to obtain a pool of conformations, which were grouped into clusters. Simulations for DD15 were started from a fully extended conformation of the peptide to eliminate possible conformational biases. An initial representative conformation for the peptide was obtained by conformational search using the Systematic Unbounded Multiple Minimum (SUMM) method with the AMBER force field and the Polak-Ribiere Conjugate Gradient (PRCG) minimization method [29]. The minimum conformation obtained from this preliminary calculation was then subjected to MD refinement in explicit water solvent. The resulting trajectories were analyzed by the structural clustering method described by Daura et al. [30]. The most representative structures of DD15 obtained after cluster analysis of the trajectory were subjected to multiple docking runs on the surface of FGF-2 (PDB code 1fq9) using the program AUTODOCK, as described in [31]. The representative structure of the most populated cluster obtained from the docking runs, corresponding also to the free energy minimum, was used for successive MD refinement, which was carried out at 300K in explicit SPC water using the GROMACS software. This step was aimed essentially to characterize ligand-receptor reciprocal adaptation at atomic level.

Statistical analyses of the trajectories were next used to identify the stereochemical requirements the peptide must satisfy to ensure a stable binding to FGF. This information was translated into a pharmacophore model used to screen the NCI2003 small molecule databases. Briefly, the model was created using the central structure of the most populated cluster for the DD15·FGF-2 complex as a template on which to cast the design. The relative distances, orientations (dihedral angles) among the different groups of DD15, and the contacts (hydrophilic/hydrophobic) associated to the most persistent interactions with FGF-2 were retained as pharmacophoric determinants. The details of the procedure can be found in [24]. The screening of the NCI repository eventually led to the identification of three FGF-2-binding small molecules (Figure 2).

The lead compounds inhibited the angiogenic activity of FGF-2 in vitro, and in the Chick Chorioallantoic Membrane (CAM) assay, in vivo. Importantly, the discovered leads showed inhibiting properties comparable to the ones of the full length TSP-1 protein domain, which they were discovered from, at the same time featuring drug-like properties.

These results demonstrate the feasibility of integrating structure and dynamics to develop small molecule mimics of endogenous proteins as therapeutic agents [24,28]. It is important to underline here that MD revealed that both the small molecule and the peptide were able to engage the FGF2 interface involved in binding FGFR and heparin. Competition experiments further supported this finding.

This work was one of the first instances in which simulations and experiments were combined to target a difficult PPI. The surface on FGF2 is indeed large, flat and highly charged, all factors that together conspire against the possibility to define a druggable surface. In subsequent developments, the most potent compound, sm27, was used as a template for a similarity-based screening of small molecule libraries, followed by docking calculations and experimental studies. This allowed selecting seven binaphthalenic compounds that bound FGF2, inhibiting its binding to both heparin sulfate proteoglycans and FGFR. The compounds suppressed FGF2 activity in ex vivo and in vitro models of angiogenesis, with improved potency over sm27. Comparative analysis of the selected hits, complemented by NMR and biochemical analysis of four newly synthesized phenylamino-substituted naphthalene derivatives, allowed identifying the minimal stereochemical requirements to improve the design of naphthalene sulfonates as FGF2 inhibitors [32–35].

Next, we studied the interaction of a peptidic lead derived from the soluble pattern recognition receptor long-pentraxin 3 (PTX3) (Figures 1 and 2). Human PTX3 overexpression inhibits tumor growth, angiogenesis and metastasis in heterotopic, orthotopic and autochthonous FGF-dependent tumor models by trapping FGF2 [36]. The acetylated pentapeptide Ac-ARPCA-NH$_2$ (in single letter code, hereafter referred to as ARPCA), corresponding to the N-terminal amino acid sequence of PTX3 (100–104), was shown to act as a minimal anti-angiogenic FGF-binding peptide able to interfere with the formation of FGF/FGFR complexes [37]. We started from these observations to characterize ARPCA in solution and dock its principal conformations to FGF2. ARPCA was predicted to bind to a different region than DD15. Indeed, experimentally, it was unable to antagonize HSPGs.

Pharmacophore modeling of the interaction of ARPCA with FGF2 was next used for the identification of the first small molecule chemical (NSC12), which was shown to act as an orally active extracellular FGF trap with significant implications in cancer therapy. Indeed, in FGF-dependent murine and human tumor models, parenteral and oral delivery of NSC12 inhibits FGFR activation, tumor growth, angiogenesis and metastasis [36] (Figure 2).

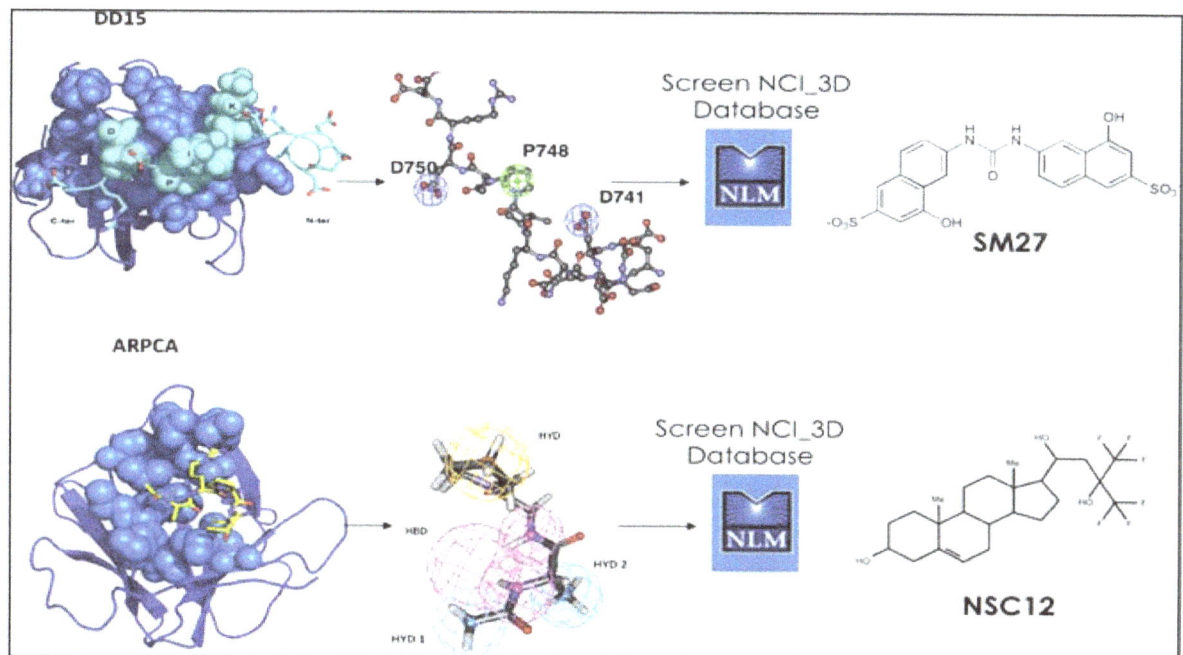

Figure 2. Definition of the most relevant contacts underlying the interaction between TSP1 and PTX3 derived peptides and FGF2, and their translation into pharmacophores for drug screening. Active small molecules sm27 and NSC12 are depicted.

Importantly, the characterization of a PPI by means of a minimal peptide led to the rational design of NSC12, which represents the first orally active small molecule ligand that can selectively prevent FGF2 from binding to FGFR and has interesting potential for anticancer drug development.

Most interestingly, the two small FGF2-targeted molecules were predicted by computational approaches to bind different regions of FGF2. This fact was verified experimentally by competition experiments and NMR analyses [32–35].

These results strongly support the validity of computational approaches to investigate hard-to-drug PPIs, showing the ability to recapitulate the determinants of the binding process involving large multi-domain proteins (TSP1 and PTX3) and their endogenous target, for drug design applications. Furthermore, the diversity of the generated chemotypes and their ability to target different interaction surfaces open up attracting perspectives for drug development and drug-combination strategies.

3. MD-Based Methods for Studying PPIs: The Case of Antibody-Antigen Interactions

As hard as they are to drug, protein-protein interaction regions offer, nonetheless, fresh opportunities for the discovery of molecules with therapeutic perspectives. This consideration may be particularly valid in the context of the development of strategies to tackle emerging pathogens or drug-resistant ones. Indeed, the spread of drug-resistance in pathogenic bacteria or the appearance of new viruses (Ebola, novel forms of aggressive influenza, Zika and dengue . . .) have severely limited the therapeutic efficacy of routinely used antibiotics, posing one of the most serious threats in modern medicine. In most of these cases, rapid diagnosis and vaccination represent the best option for the treatment of emerging infectious diseases. In fact, rapid and effective diagnosis can help preventing the spread of these threats in an increasing part of the population, while directing patients towards the best therapeutic options. In the last few years, it has become increasingly clear that, in order to develop biomolecules with both diagnostic and vaccine application potential, it is crucial to identify antigens on the surface of bacteria that are capable of eliciting a strong immune response, which is usually achieved through the production of (bactericidal) antibodies (Abs). In terms of diagnostic applications, the ability of antigens to proficiently interact with Abs can be exploited to develop probes that can reveal circulating antibodies produced in response to a specific infection in patient serum, blood or plasma samples. In terms of vaccine development, reactive protein antigens can be exploited in formulations aimed to elicit protective responses against successive pathogenic challenges. Even if vaccines have traditionally suffered from slow routine studies, sometimes providing viable products well after the peak in the epidemics, the advent of 'Reverse Vaccinology' (RV) has revolutionized the field, introducing a whole new strategy of antigen selection [38–41]. Starting from the full genome analysis of a pathogen or from the analysis of multiple pathogens of a certain family, RV antigen candidates that show key properties required for vaccine development (e.g., cell-surface exposure, ability to interact with/elicit Abs, protein stability, possibility to produce the protein antigens in recombinant form) are selected. To achieve such selection, RV makes use of complementary and synergistic methods, such as functional genomics, protein microarrays, and bioinformatics/computational biology. The reach of RV can be dramatically extended by the exploitation of atomic-level 3D information to engineer new biomolecules with improved immunoreactivity and/or biochemical properties [42–45].

In chemical and physico-chemical terms, this comes down to identifying which regions in a protein antigen are the ones most likely to be immunoreactive with Abs. Such regions are called epitopes. In other words, one should detect the parts of the antigen that have the highest tendency to bind Abs. In this context, the problem is a particular case-study of protein-protein interactions (Figure 3). To meet this challenge, we have developed a simple computational strategy that aims at predicting Abs-binding epitopes starting only from the consideration of the structure, interactions and conformational dynamics of the antigen [46,47] (Figure 4).

Our approach starts from the idea that recognition sites may correspond to localized regions on the surface with low-intensity energetic couplings with the folding core of the protein which antigen belongs to: such minimal coupling to the rest of the structure can in principle allow the regions to sustain the conformational changes necessary to adapt to a binding partner. Indeed, in many cases, PPI regions have been shown to be endowed with flexibility features.

We thus set out to identify non-optimized, low-intensity energetic interaction networks in the protein structure isolated in solution and then to benchmark the results against antibody complexes. Interestingly, it was found that the method could successfully identify binding sites located on the protein surface that are accessible to putative binding partners.

To identify localized surface regions with non-optimized interactions, we combined the analysis of internal protein energetics with the topological structural information obtainable from the contact matrix of either the crystal structure of the protein or the representative structure extracted from the MD trajectory (Figure 4).

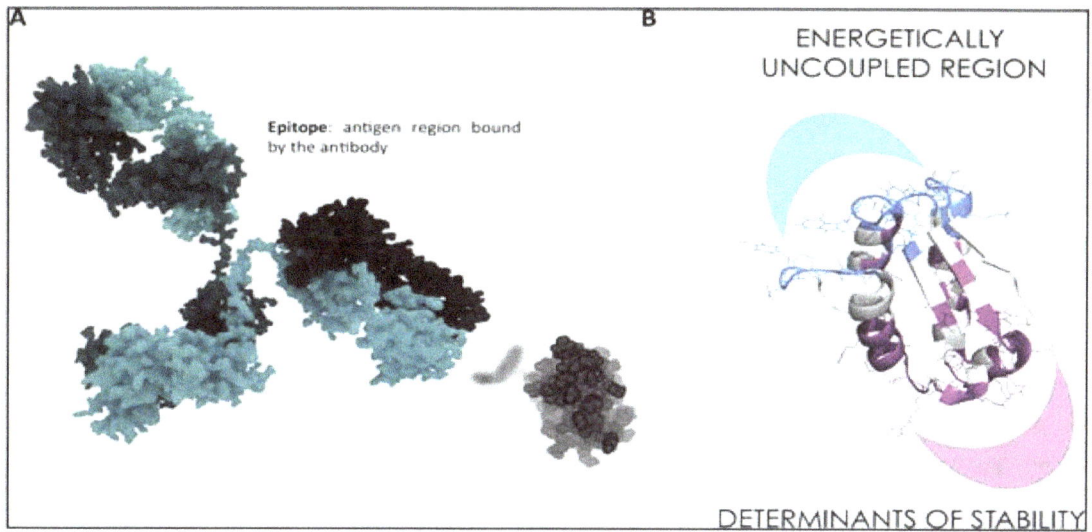

Figure 3. (**A**) Identification of the epitope region of an antigen binding to an antibody. (**B**) Simplified representation of the separation between stability and recognition regions in one protein antigen.

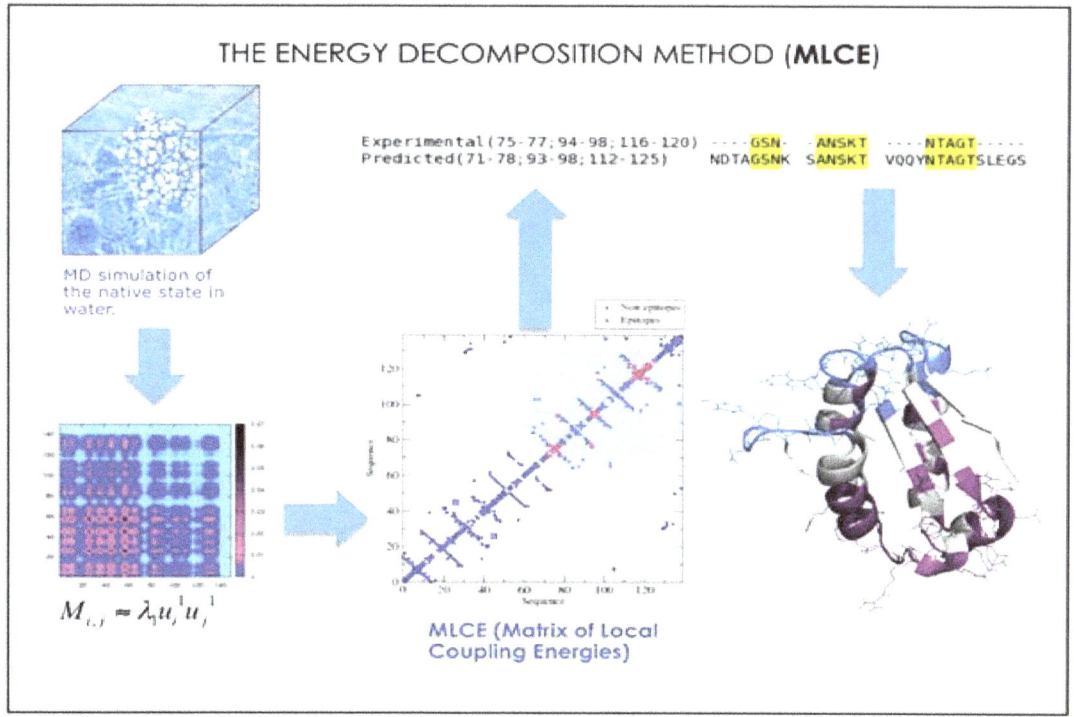

Figure 4. Schematics of how the MLCE algorithm works.

The analysis of energetics derives from the energy decomposition method (EDM) [48–55]: specifically, the method provides a simplified view of residue-residue pair interactions, extracting the strongest and weakest residue pair-interactions and their contributions to energetic stability of a certain 3D structure.

In the case of a protein of length N, the $N \times N$ matrix (M_{ij}) of average nonbonded interactions between pairs of residues is built first. This energy matrix is then simplified through eigenvalue decomposition.

Analysis of the N components of the eigenvector associated with the lowest eigenvalue was shown to identify strong interaction centers. This map of pair interactions is subsequently analyzed in light of the topological information summarized by the contact matrix associated to a certain structure. The resulting filtered matrix can be used to identify local couplings characterized by energetic interactions of minimal intensities. In fact, while local low-energy couplings identify those sites in which interaction-networks are not energetically optimized, low-intensity couplings between distant residues in the structure are only a trivial consequence of the distance-dependence of energy functions.

Local low-energy coupled regions can thus be considered as the "soft spots" needed to interact with potential binding partners (in contrast with the "hot spots" characterized by high coupling intensities). Given the low intensity constraints to the rest of the structure, these sub-structures would be characterized by dynamic properties that allow them to visit multiple conformations, a subset of which can be recognized by the antibody to form a complex [46,47].

After validating the predictions against the crystal structures of known Ab-Ag complexes, we set out to apply the matrix of local coupling energies (MLCE) approach in a predictive and design-oriented fashion. The first instance in which the method was applied focused on the discovery and design of reactive epitopes from the antigens of the bacterium *Burkholderia pseudomallei* (Bp), the etiological agent of melioidosis. The latter is a severe respiratory infection against which no rapid and efficient diagnostic method or vaccination strategy exists. Several immunoreactive proteins were identified through an RV strategy. The crystal structure of one of these antigens, OppA (Bp), was solved at 2.1 Å resolution and was the basis for MLCE analysis that returned three potential epitopes (Figure 5). Once identified, mimics of the potential epitopes were synthesized in peptidic form and successively tested for their immunoreactivity against sera from healthy seronegative, healthy seropositive, and recovered melioidosis patients. The synthetic peptides allowed the different patient groups to be distinguished, underlining the potential of this approach. These results were a first remarkable illustration of the feasibility of a structure-based epitope discovery process, whose application could effectively expand the understanding of the physico-chemical determinants of protein-protein interactions to the development of designed diagnostic molecules [56].

Starting from the resolution of the structure of a second *Burkholderia* antigen, namely BPSL2765, the approach was extended to the production of bactericidal antibodies. Based on the structure, MLCE, coupled to in vitro mass-spectrometry mapping, identified a sequence within the antigen that, when engineered as a synthetic peptide, was selectively immunorecognized to the same extent as the recombinant protein in sera from melioidosis patients. Next, the peptide was employed to elicit Abs that were subsequently tested in bacterial killing experiments and antibody-dependent agglutination tests. Importantly, the Abs produced against the designed synthetic peptide turned out to induce the killing of *B. pseudomallei* at levels higher than the Abs raised against the full length protein [57] (Figures 5 and 6). In this case, our strategy represented not only a step in the development of immunodiagnostics, but also a first step in the engineering of antigens and production of specific antibodies for vaccine development.

MLCE was further applied to proteins constituent of the flagella of the bacterium. Flagella are used by the bacterium to move in the environment and are conceivably the first parts of the pathogen that come into contact with the host. MLCE epitope prediction was applied to *B. pseudomallei* flagellar hook-associated protein (FlgK(Bp)) [58,59], allowing us to predict three antigenic regions that locate to discrete protein domains and may work as vaccine components. Another component of the flagella is the large protein flagellin (FliC(Bp)). Interestingly, in this case, three predicted epitopes, when synthesized and tested as free peptides, turned out to be both B and T cell FliC(Bp) epitopes: they were immunoreactive against human IgG antibodies and elicited cytokine production from human peripheral blood mononuclear cells. Furthermore, two of the peptides (F51-69 and F270-288) were found to be immunodominant, with their antibodies enhancing the bactericidal activities of purified human neutrophils [60]. Together with the previously reported ones, these epitopes may represent potential melioidosis vaccine components.

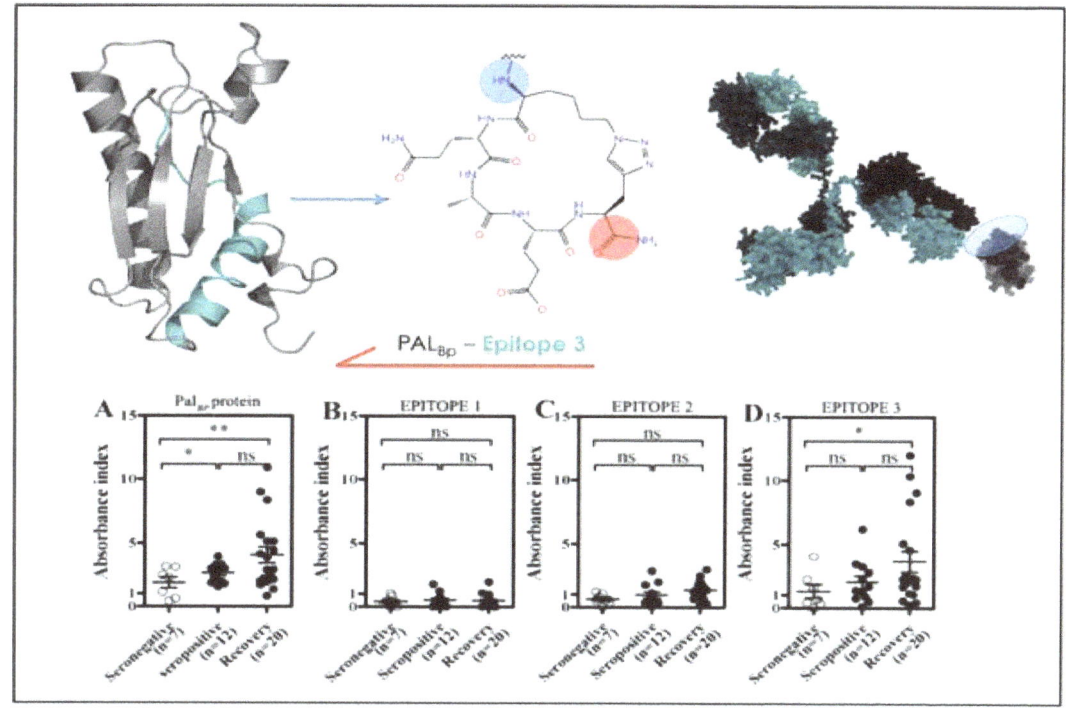

Figure 5. Identification, chemical modification and immunodiagnostic test of the epitope sequence derived from BPSL2765 (PAL$_{Bp}$).

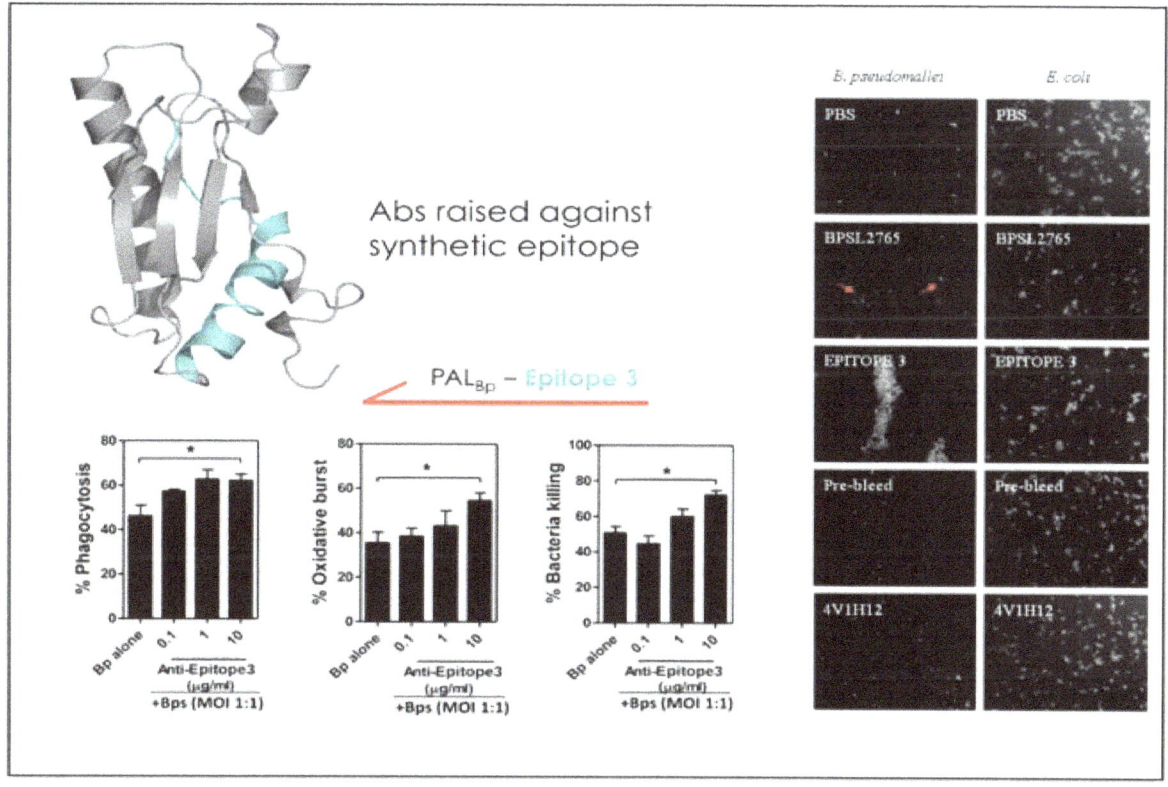

Figure 6. The epitope sequence derived from BPSL2765 (PAL$_{Bp}$) is able to elicit bactericidal antibodies.

In general, it is tempting to suggest that the possibility to predict the parts of a protein (antigens) endowed with antibody recognition/binding properties and the demonstration of their reactivity in the form of isolated peptides can open up new venues for diagnosis and treatment. In the case of diagnostics, for instance, multiple predicted binding sequences can be displayed on microarrays for medium-high throughput analysis of their interaction profiles: in a notable instance, predicted peptides were optimized for oriented display on microarray plates and proved to be efficient in the rapid diagnosis of *Burkholderia* infections in cystic fibrosis (CF) patients [61]. To mimic conformational epitopes, oriented and spatially controlled co-immobilization of predicted epitope sequences that are spatially proximal in the Zika virus NS1 protein, showed the ability to cooperatively interact to provide enhanced immunoreactivity with respect to single linear epitopes [62].

4. Conclusions and Perspectives

The data described above indicate that it is becoming possible to apply rational methods to target difficult protein-protein interactions, both through small molecules and through the harnessing of the reactivity towards large biological molecules as antibodies. We suggest that these methods of drug and peptide design could be conceivably coupled to the design of polyvalent systems that allow the simultaneous binding of multiple ligands to a certain target, mimicking the types of interactions that are widespread in biology [63]. The availability of chemical synthesis methods for the access to complex mimics of natural products or chemical-biology probes [64–68], and the explosion of chemical methods for the display of multiple ligands (through nanoparticles, bio-inspired polymers etc...) can indeed help the development of multivalent systems that we see as potentially suitable for vaccination and patient diagnostics: in these cases, the simultaneous presentation of multiple determinants of Ab-recognition from the antigens of a certain pathogen may help trigger protective response against it [69–72]. In the case of small molecule drugs, multi-presentation approaches may become particularly useful when targeting large multi-component complexes. In our view, computational chemistry approaches are set to become in the next few years more and more instrumental and integrated with chemical biology and drug design approaches, increasing our understanding of how biological systems work and translating this knowledge into new molecules with interesting therapeutic potential.

References

1. Arkin, M.R.; Randal, M.; DeLano, W.L.; Hyde, J.; Luong, T.N.; Oslob, J.D.; Raphael, D.R.; Taylor, L.; Wang, J.; McDowell, R.S.; et al. Binding of small molecules to an adaptive protein–protein interface. *Proc. Natl. Acad. Sci. USA* **2003**, *100*, 1603–1608. [CrossRef] [PubMed]

2. Arkin, M.R.; Tang, Y.; Wells, J.A. Small-Molecule Inhibitors of Protein-Protein Interactions: Progressing toward the Reality. *Chem. Biol.* **2014**, *21*, 1002–1114. [CrossRef] [PubMed]

3. Clackson, T.; Wells, J.A. A hot spot of binding energy in a hormone- receptor interface. *Science* **1995**, *267*, 383–386. [CrossRef] [PubMed]

4. Wells, J.A.; McCLendon, C.L. Reaching for high-hanging fruit in drug discovery at protein-protein interfaces. *Nature* **2007**, *450*, 1001–1009. [CrossRef] [PubMed]

5. Tuncbag, N.; Gursoy, A.; Guney, E.; Nussinov, R.; Keskin, O. Architectures and functional coverage of protein-protein interfaces. *J. Mol. Biol.* **2008**, *381*, 785–802. [CrossRef] [PubMed]

6. Wei, G.; Xi, W.; Nussinov, R.; Ma, B. Protein Ensembles: How Does Nature Harness Thermodynamic Fluctuations for Life? The Diverse Functional Roles of Conformational Ensembles in the Cell. *Chem. Rev.* **2016**, *116*, 6516–6551. [CrossRef] [PubMed]

7. Aloy, P.; Bottcher, B.; Ceulemans, H.; Leutwein, C.; Mellwig, C.; Fischer, S.; Gavin, A.C.; Bork, P.; Superti-Furga, G.; Serrano, L.; et al. Structure-based assembly of protein complexes in yeast. *Science* **2004**, *303*, 2026–2029. [CrossRef] [PubMed]

8. Lo Conte, L.; Chothia, C.; Janin, J. The atomic structure of protein–protein recognition sites. *J. Mol. Biol.* **1999**, *285*, 2177–2198. [CrossRef] [PubMed]

9. Cesa, L.C.; Patury, S.; Komiyama, T.; Ahmad, A.; Zuiderweg, E.R.P.; Gestwicki, J.E. Inhibitors of Difficult Protein-Protein Interactions Identified by High-Throughput Screening of Multiprotein Complexes. *ACS Chem. Biol.* **2013**, *8*, 1988–1997. [CrossRef] [PubMed]

10. Cesa, L.C.; Mapp, A.K.; Gestwicki, J.E. Direct and propagated effects of small molecules on protein–protein interaction networks. *Front. Bioeng. Biotechnol.* **2015**, *3*, 119. [CrossRef] [PubMed]

11. Arkin, M.R.; Wells, J.A. Small-Molecule inhibitors of protein-protein interactions: Progressing towards the dream. *Nat. Rev. Drug Discov.* **2004**, *3*, 301–317. [CrossRef] [PubMed]

12. Thompson, A.D.; Dugan, A.; Gestwicki, J.E.; Mapp, A.K. Fine-Tuning Multiprotein Complexes Using Small Molecules. *ACS Chem. Biol.* **2012**, *7*, 1311–1320. [CrossRef] [PubMed]

13. Weigt, M.; White, R.A.; Szurmant, H.; Hoch, J.A.; Hwa, T. Identification of direct residue contacts in protein–protein interaction by message passing. *Proc. Natl. Acad. Sci. USA* **2009**, *106*, 67–72. [CrossRef] [PubMed]

14. Ma, B.; Elkayam, T.; Wolfson, H.; Nussinov, R. Protein–protein interactions: Structurally conserved residues distinguish between binding sites and exposed protein surfaces. *Proc. Natl. Acad. Sci. USA* **2003**, *100*, 5772–5777. [CrossRef] [PubMed]

15. Van Gunsteren, W.F.; Dolenc, J.; Mark, A. Molecular simulation as an aid to experimentalists. *Curr. Opin. Struct. Biol.* **2008**, *18*, 149–153. [CrossRef] [PubMed]

16. Van Gunsteren, W.F.; Bakowies, D.; Baron, R.; Chandrasekhar, I.; Christen, M.; Daura, X.; Gee, P.; Geerke, D.P.; Glättli, A.; Hünenberger, P.H.; et al. Biomolecular Modeling: Goals, Problems, Perspectives. *Angew. Chem. Int. Ed.* **2006**, *45*, 4064–4092. [CrossRef] [PubMed]

17. Meli, M.; Morra, G.; Colombo, G. Investigating the mechanism of peptide aggregation: Insights from mixed Monte Carlo-molecular dynamics simulations. *Biophys. J.* **2008**, *94*, 4414–4426. [CrossRef] [PubMed]

18. Monticelli, L.; Tieleman, D.P.; Colombo, G. Mechanism of helix nucleation and propagation: Microscopic view from microsecond time scale MD simulations. *J. Phys. Chem. B* **2005**, *109*, 20064–20067. [CrossRef] [PubMed]

19. Ferraro, M.; D'Annessa, I.; Moroni, E.; Morra, G.; Paladino, A.; Rinaldi, S.; Compostella, F.; Colombo, G. Allosteric Modulators of HSP90 and HSP70: Dynamics Meets Function through Structure-Based Drug Design. *J. Med. Chem.* **2018**. [CrossRef] [PubMed]

20. Melaccio, F.; del Carmen Marín, M.; Valentini, A.; Montisci, F.; Rinaldi, S.; Cherubini, M.; Yang, X.; Kato, Y.; Stenrup, M.; Orozco-Gonzalez, Y.; et al. Toward Automatic Rhodopsin Modeling as a Tool for High-Throughput Computational Photobiology. *J. Chem. Theory Comput.* **2016**, *12*, 6020–6034. [CrossRef] [PubMed]

21. Taraboletti, G.; Belotti, D.; Borsotti, P.; Vergani, V.; Rusnati, M.; Presta, M.; Giavazzi, R. The 140-kilodalton antiangiogenic fragment of thrombospondin-1 binds to basic fibroblast growth factor. *Cell Growth Differ.* **1997**, *8*, 471–479. [PubMed]

22. Taraboletti, G.; Morbidelli, L.; Donnini, S.; Parenti, A.; Granger, H.J.; Giavazzi, R.; Ziche, M. The heparin binding 25 kDa fragment of thrombospondin-1 promotes angiogenesis and modulates gelatinase and TIMP-2 production in endothelial cells. *FASEB J.* **2000**, *14*, 1674–1676. [CrossRef] [PubMed]

23. Margosio, B.; Marchetti, D.; Vergani, V.; Giavazzi, R.; Rusnati, M.; Presta, M.; Taraboletti, G. Thrombospondin 1 as a scavenger for matrix-associated fibroblast growth factor 2. *Blood* **2003**, *102*, 4399–4406. [CrossRef] [PubMed]

24. Taraboletti, G.; Rusnati, M.; Ragona, L.; Colombo, G. Targeting tumor angiogenesis with TSP-1-based compounds: Rational design of antiangiogenic mimetics of endogenous inhibitors. *Oncotarget* **2010**, *1*, 662–673. [PubMed]

25. Presta, M.; Dell'Era, P.; Mitola, S.; Moroni, E.; Ronca, R.; Rusnati, M. Fibroblast growth factor/fibroblast growth factor receptor system in angiogenesis. *Cytokine Growth Factor Rev.* **2005**, *16*, 159–178. [CrossRef] [PubMed]

26. Rusnati, M.; Presta, M. Extracellular angiogenic growth factor interactions: An angiogenesis interactome survey. *Endothelium* **2006**, *13*, 93–111. [CrossRef] [PubMed]

27. Beenken, A.; Mohammadi, M. The FGF family: Biology, pathophysiology and therapy. *Nat. Rev. Drug Discov.* **2009**, *8*, 235–253. [CrossRef] [PubMed]

28. Colombo, G.; Margosio, B.; Ragona, L.; Neves, M.; Bonifacio, S.; Annis, D.S.; Stravalaci, M.; Tomaselli, S.; Giavazzi, R.; Rusnati, M.; et al. Non-peptidic thrombospondin-1 mimics as fibroblast growth factor-2 inhibitors: An integrated strategy for the development of new antiangiogenic compounds. *J. Biol. Chem.* **2010**, *285*, 8733–8742. [CrossRef] [PubMed]

29. Senderowitz, H.; Guarnieri, F.; Still, W.C. A Smart Monte Carlo Technique for Free Energy Simulations of Multiconformational Molecules. Direct Calculations of the Conformational Populations of Organic Molecules. *J. Am. Chem. Soc.* **1995**, *117*, 8211–8219. [CrossRef]

30. Daura, X.; Gademann, K.; Jaun, B.; Seebach, D.; van Gunsteren, W.F.; Mark, A.E. Peptide folding: When simulation meets experiment. *Angew. Chem. Int. Ed.* **1999**, *38*, 236–240. [CrossRef]

31. Meli, M.; Pennati, M.; Curto, M.; Daidone, M.G.; Plescia, J.; Toba, S.; Altieri, D.C.; Zaffaroni, N.; Colombo, G. Small-Molecule Targeting of Heat Shock Protein 90 Chaperone Function: Rational Identification of a New Anticancer Lead. *J. Med. Chem.* **2006**, *49*, 7721–7730. [CrossRef] [PubMed]

32. Foglieni, C.; Torella, R.; Bugatti, A.; Pagano, K.; Ragona, L.; Ribatti, D.; Rusnati, M.; Presta, M.; Giavazzi, R.; Colombo, G.; et al. Inhibition of FGF-2 angiogenic activity by novel small molecules mimetic of thrombospondin-1 (TSP-1). *Thromb. Res.* **2012**, *129*, S193. [CrossRef]

33. Pagano, K.; Torella, R.; Foglieni, C.; Bugatti, A.; Tomaselli, S.; Zetta, L.; Presta, M.; Rusnati, M.; Taraboletti, G.; Colombo, G.; et al. Direct and Allosteric Inhibition of the FGF2/HSPGs/FGFR1 Ternary Complex Formation by an Antiangiogenic, Thrombospondin-1-Mimic Small Molecule. *PLoS ONE* **2012**, *7*, e36990. [CrossRef] [PubMed]

34. Foglieni, C.; Pagano, K.; Lessi, M.; Bugatti, A.; Moroni, E.; Pinessi, D.; Resovi, A.; Ribatti, D.; Bertini, S.; Ragona, L.; et al. Integrating computational and chemical biology tools in the discovery of antiangiogenic small molecule ligands of FGF2 derived from endogenous inhibitors. *Sci. Rep.* **2016**, *6*, 23432. [CrossRef] [PubMed]

35. Pinessi, D.; Foglieni, C.; Bugatti, A.; Moroni, E.; Resovi, A.; Ribatti, D.; Rusnati, M.; Giavazzi, R.; Colombo, G.; Taraboletti, G. Antiangiogenic small molecule ligands of FGF2 derived from the endogenous inhibitor thrombospondin-1. *Thromb. Res.* **2016**, *140*, S182. [CrossRef]

36. Ronca, R.; Giacomini, A.; Di Salle, E.; Coltrini, D.; Pagano, K.; Ragona, L.; Matarazzo, S.; Rezzola, S.; Maiolo, D.; Torrella, R.; et al. Long-Pentraxin 3 Derivative as a Small-Molecule FGF Trap for Cancer Therapy. *Cancer Cell* **2015**, *28*, 225–239. [CrossRef] [PubMed]

37. Leali, D.; Bianchi, R.; Bugatti, A.; Nicoli, S.; Mitola, S.; Ragona, L.; Tomaselli, S.; Gallo, G.; Catello, S.; Rivieccio, V.; et al. Fibroblast growth factor 2-antagonist activity of a long-pentraxin 3-derived anti-angiogenic pentapeptide. *J. Cell. Mol. Med.* **2010**, *14*, 2109–2121. [CrossRef] [PubMed]

38. Rappuoli, R. From Pasteur to genomics: Progress and challenges in infectious diseases. *Nat. Med.* **2004**, *10*, 1177–1185. [CrossRef] [PubMed]

39. Rappuoli, R.; Bottomley, M.J.; D'Oro, U.; Finco, O.; De Gregorio, E. Reverse vaccinology 2.0: Human immunology instructs vaccine antigen design. *J. Exp. Med.* **2016**, *213*, 469–481. [CrossRef] [PubMed]

40. Bloom, D.E.; Black, S.; Rappuoli, R. Emerging infectious diseases: A proactive approach. *Proc. Natl. Acad. Sci. USA* **2017**, *114*, 4055–4059. [CrossRef] [PubMed]

41. Thomas, S.; Dilbarova, R.; Rappuoli, R. Future Challenges for Vaccinologists. *Methods Mol. Biol.* **2016**, *1403*, 41–55. [PubMed]

42. Dormitzer, P.R.; Ulmer, J.B.; Rappuoli, R. Structure-based antigen design: A strategy for next generation vaccines. *Trends Biotechnol.* **2008**, *26*, 659–667. [CrossRef] [PubMed]

43. Nuccitelli, A.; Cozzi, R.; Gourlay, L.J.; Donnarumma, D.; Necchi, F.; Norais, N.; Telford, J.L.; Rappuoli, R.; Bolognesi, M.; Maione, D.; et al. Structure-based approach to rationally design a chimeric protein for an effective vaccine against Group B Streptococcus infections. *Proc. Natl. Acad. Sci. USA* **2011**, *108*, 10278–10283. [CrossRef] [PubMed]

44. Scarselli, M.; Arico, B.; Brunelli, B.; Savino, S.; Di Marcello, F.; Palumbo, E.; Veggi, D.; Ciucchi, L.; Cartocci, E.; Bottomley, M.J.; et al. Rational design of a meningococcal antigen inducing broad protective immunity. *Sci. Transl. Med.* **2011**, *3*, 91ra62. [CrossRef] [PubMed]

45. Dormitzer, P.R.; Grandi, G.; Rappuoli, R. Structural vaccinology starts to deliver. *Nat. Rev. Microbiol.* **2012**, *10*, 807–813. [CrossRef] [PubMed]

46. Scarabelli, G.; Morra, G.; Colombo, G. Predicting interaction sited from the energetics of isolated proteins: A new approach to epitope mapping. *Biophys. J.* **2010**, *98*, 1966–1975. [CrossRef] [PubMed]

47. Soriani, M.; Petit, P.; Grifantini, R.; Petracca, R.; Gancitano, G.; Frigimelica, E.; Nardelli, F.; Garcia, C.; Spinelli, S.; Scarabelli, G.; et al. Exploiting antigenic diversity for vaccine design: The chlamydia ArtJ paradigm. *J. Biol. Chem.* **2010**, *285*, 30126–30138. [CrossRef] [PubMed]

48. Colacino, S.; Tiana, G.; Colombo, G. Similar folds with different stabilization mechanisms: The cases of Prion and Doppel proteins. *BMC Struct. Biol.* **2006**, *6*, 17. [CrossRef] [PubMed]

49. Colacino, S.; Tiana, G.; Broglia, R.A.; Colombo, G. The determinants of stability in the human prion protein: Insights into the folding and misfolding from the analysis of the change in the stabilization energy distribution in different condition. *Proteins Struct. Funct. Bioinform.* **2006**, *62*, 698–707. [CrossRef] [PubMed]

50. Tiana, G.; Simona, F.; De Mori, G.M.S.; Broglia, R.A.; Colombo, G. Understanding the determinants of stability and folding of small globular proteins from their energetics. *Protein Sci.* **2004**, *13*, 113–124. [CrossRef] [PubMed]

51. Morra, G.; Genoni, A.; Colombo, G. Mechanisms of Differential Allosteric Modulation in Homologous Proteins: Insights from the Analysis of Internal Dynamics and Energetics of PDZ Domains. *J. Chem. Theory Comput.* **2014**, *10*, 5677–5689. [CrossRef] [PubMed]

52. Genoni, A.; Morra, G.; Colombo, G. Identification of Domains in Protein Structures from the Analysis of Intramolecular Interactions. *J. Phys. Chem. B* **2012**, *116*, 3331–3343. [CrossRef] [PubMed]

53. Torella, R.; Moroni, E.; Caselle, M.; Morra, G.; Colombo, G. Investigating dynamic and energetic determinants of protein nucleic acid recognition: Analysis of the zinc finger zif268-DNA complexes. *BMC Struct. Biol.* **2010**, *10*, 42. [CrossRef] [PubMed]

54. Genoni, A.; Morra, G.; Merz, K.M., Jr.; Colombo, G. Computational Study of the Resistance Shown by the Subtype B/HIV-1 Protease to Currently Known Inhibitors. *Biochemistry* **2010**, *49*, 4283–4295. [CrossRef] [PubMed]

55. Morra, G.; Colombo, G. Relationship between energy distribution and fold stability: Insights from molecular dynamics simulations of native and mutant proteins. *Proteins Struct. Funct. Bioinform.* **2008**, *72*, 660–672. [CrossRef] [PubMed]

56. Lassaux, P.; Peri, C.; Ferrer-Navarro, M.; Gourlay, L.; Gori, A.; Conchillo-Solé, O.; Rinchai, D.; Lertmemongkolchai, G.; Longhi, R.; Daura, X.; et al. A structure-based strategy for epitope discovery in Burkholderia pseudomallei OppA antigen. *Structure* **2013**, *21*, 167–175. [CrossRef] [PubMed]

57. Gourlay, L.J.; Peri, C.; Ferrer-Navarro, M.; Conchillo-Sole, O.; Gori, A.; Rinchai, D.; Thomas, R.J.; Champion, O.L.; Michell, S.L.; Kewcharoenwong, C.; et al. Exploiting the Burkholderia pseudomallei Acute Phase Antigen BPSL2765 for Structure-Based Epitope Discovery/Design in Structural Vaccinology. *Chem. Biol.* **2013**, *20*, 1147–1156. [CrossRef] [PubMed]

58. Gourlay, L.J.; Thomas, R.J.; Peri, C.; Conchillo-Sole, O.; Ferrer-Navarro, M.; Nithichanon, A.; Vila, J.; Daura, X.; Lertmemongkolchai, G.; Titball, R.; et al. From crystal structure to in silico epitope discovery in the Burkholderia pseudomallei flagellar hook-associated protein FlgK. *FEBS J.* **2015**, *282*, 1319–1333. [CrossRef] [PubMed]

59. Gourlay, L.J.; Lassaux, P.; Thomas, R.J.; Peri, C.; Conchillo-Sole, O.; Nithichanon, A.; Ferrer-Navarro, M.; Vila, J.; Daura, X.; Lertmemongkolchai, G.; et al. Flagellar subunits as targets for structure-based epitope discovery approaches and melioidosis vaccine development. *FEBS J.* **2015**, *282*, 338.

60. Nithichanon, A.; Rinchai, D.; Gori, A.; Lassaux, P.; Peri, C.; Conchillio-Sole, O.; Ferrer-Navarro, M.; Gourlay, L.J.; Nardini, M.; Vila, J.; et al. Sequence- and Structure-Based Immunoreactive Epitope Discovery for Burkholderia pseudomallei Flagellin. *PLoS Negl. Trop. Dis.* **2015**, *9*. [CrossRef] [PubMed]

61. Peri, C.; Gori, A.; Gagni, P.; Sola, L.; Girelli, D.; Sottotetti, S.; Cariani, L.; Chiari, M.; Cretich, M.; Colombo, G. Evolving serodiagnostics by rationally designed peptide arrays: The Burkholderia paradigm in Cystic Fibrosis. *Sci. Rep.* **2016**, *6*, 32873. [CrossRef] [PubMed]

62. Sola, L.; Gagni, P.; D'Annessa, I.; Capelli, R.; Bertino, C.; Romanato, A.; Damin, F.; Bergamaschi, G.; Marchisio, E.; Cuzzocrea, A.; et al. Enhancing Antibody Serodiagnosis Using a Controlled Peptide Coimmobilization Strategy. *ACS Infect. Dis.* **2018**, *4*, 998–1006. [CrossRef] [PubMed]

63. Mammen, M.; Choi, S.-K.; Whitesides, G.M. Polyvalent Interactions in Biological Systems: Implications for Design and Use of Multivalent Ligands and Inhibitors. *Angew. Chem. Int. Ed.* **1998**, *37*, 2754–2794. [CrossRef]

64. Brasile, G.; Mauri, L.; Sonnino, S.; Compostella, F.; Ronchetti, F. A practical route to long-chain non-natural alpha,omega-diamino acids. *Amino Acids* **2013**, *44*, 435–441. [CrossRef] [PubMed]

65. Chiricozzi, E.; Ciampa, M.G.; Brasile, G.; Compostella, F.; Prinetti, A.; Nakayama, H.; Ekyalongo, R.C.; Iwabuchi, K.; Sonnino, S.; Mauri, L. Direct interaction, instrumental for signaling processes, between LacCer and Lyn in the lipid rafts of neutrophil-like cells. *J. Lipid Res.* **2015**, *56*, 129–141. [CrossRef] [PubMed]

66. Franchini, L.; Compostella, F.; Colombo, D.; Panza, L.; Ronchetti, F. Synthesis of the Sulfonate Analogue of Seminolipid via Horner-Wadsworth-Emmons Olefination. *J. Org. Chem.* **2010**, *75*, 5363–5366. [CrossRef] [PubMed]

67. Vetro, M.; Costa, B.; Donvito, G.; Arrighetti, N.; Cipolla, L.; Perego, P.; Compostella, F.; Ronchetti, F.; Colombo, D. Anionic glycolipids related to glucuronosyldiacylglycerol inhibit protein kinase Akt. *Org. Biomol. Chem.* **2015**, *13*, 1091–1099. [CrossRef] [PubMed]

68. Di Brisco, R.; Ronchetti, F.; Mangoni, A.; Costantino, V.; Compostella, F. Development of a fluorescent probe for the study of the sponge-derived simplexide immunological properties. *Carbohydr. Res.* **2012**, *348*, 27–32. [CrossRef] [PubMed]

69. Compostella, F.; Pitirollo, O.; Silvestri, A.; Polito, L. Glyco gold nanoparticles: Synthesis and applications. *Beilstein J. Org. Chem.* **2017**, *13*, 1008–1021.
70. Armentano, I.; Fortunati, E.; Latterini, L.; Rinaldi, S.; Saino, E.; Visai, L.; Elisei, F.; Kenny, J.M. Biodegradable PLGA matrix nanocomposite with silver nanoparticles: Material properties and bacteria activity. *J. Nanostruct. Polym. Nanocompos.* **2010**, *6*, 110–118.
71. Legnani, L.; Compostella, F.; Sansone, F.; Toma, L. Cone Calix 4 arenes with Orientable Glycosylthioureido Groups at the Upper Rim: An In-Depth Analysis of Their Symmetry Properties. *J. Org. Chem.* **2015**, *80*, 7412–7418. [CrossRef] [PubMed]
72. Toma, L.; Legnani, L.; Compostella, F.; Giuliani, M.; Faroldi, F.; Casnati, A.; Sansone, F. Molecular Architecture and Symmetry Properties of 1,3-Alternate Calix 4 arenes with Orientable Groups at the Para Position of the Phenolic Rings. *J. Org. Chem.* **2016**, *81*, 9718–9727. [CrossRef] [PubMed]

Permissions

The contributors of this book come from diverse backgrounds, making this book a truly international effort. This book will bring forth new frontiers with its revolutionizing research information and detailed analysis of the nascent developments around the world.

We would like to thank all the contributing authors for lending their expertise to make the book truly unique. They have played a crucial role in the development of this book. Without their invaluable contributions this book wouldn't have been possible. They have made vital efforts to compile up to date information on the varied aspects of this subject to make this book a valuable addition to the collection of many professionals and students.

This book was conceptualized with the vision of imparting up-to-date information and advanced data in this field. To ensure the same, a matchless editorial board was set up. Every individual on the board went through rigorous rounds of assessment to prove their worth. After which they invested a large part of their time researching and compiling the most relevant data for our readers.

The editorial board has been involved in producing this book since its inception. They have spent rigorous hours researching and exploring the diverse topics which have resulted in the successful publishing of this book. They have passed on their knowledge of decades through this book. To expedite this challenging task, the publisher supported the team at every step. A small team of assistant editors was also appointed to further simplify the editing procedure and attain best results for the readers.

Apart from the editorial board, the designing team has also invested a significant amount of their time in understanding the subject and creating the most relevant covers. They scrutinized every image to scout for the most suitable representation of the subject and create an appropriate cover for the book.

The publishing team has been an ardent support to the editorial, designing and production team. Their endless efforts to recruit the best for this project, has resulted in the accomplishment of this book. They are a veteran in the field of academics and their pool of knowledge is as vast as their experience in printing. Their expertise and guidance has proved useful at every step. Their uncompromising quality standards have made this book an exceptional effort. Their encouragement from time to time has been an inspiration for everyone.

The publisher and the editorial board hope that this book will prove to be a valuable piece of knowledge for researchers, students, practitioners and scholars across the globe.

List of Contributors

Stephani Joy Y. Macalino
Chemistry Department, De La Salle University, 2401 Taft Avenue, Manila 0992, Philippines
OVPAA-EIDR Program, "Computer-Aided Discovery of Compounds for the Treatment of Tuberculosis In the Philippines", Department of Physical Sciences and Mathematics, College of Arts and Sciences, University of the Philippines Manila, Manila 1000, Philippines

Junie B. Billones, Voltaire G. Organo and Maria Constancia O. Carrillo
OVPAA-EIDR Program, "Computer-Aided Discovery of Compounds for the Treatment of Tuberculosis In the Philippines", Department of Physical Sciences and Mathematics, College of Arts and Sciences, University of the Philippines Manila, Manila 1000, Philippines

Maksim Kouza
Faculty of Chemistry, University of Warsaw, Pasteura 1, 02-093 Warsaw, Poland
Battelle Center for Mathematical Medicine, Nationwide Children's Hospital, Columbus, OH 43215, USA

Anirban Banerji
Battelle Center for Mathematical Medicine, Nationwide Children's Hospital, Columbus, OH 43215, USA

Andrzej Kolinski
Faculty of Chemistry, University of Warsaw, Pasteura 1, 02-093 Warsaw, Poland

Irina Buhimschi
Center for Perinatal Research, Research Institute at Nationwide Children's Hospital, Columbus, OH 43215, USA
Department of Pediatrics, The Ohio State University College of Medicine, Columbus, OH 43215, USA

Andrzej Kloczkowski
Battelle Center for Mathematical Medicine, Nationwide Children's Hospital, Columbus, OH 43215, USA
Department of Pediatrics, The Ohio State University College of Medicine, Columbus, OH 43215, USA

Gerhard Hessler
R&D, Integrated Drug Discovery, Industriepark Hoechst, 65926 Frankfurt am Main, Germany

Karl-Heinz Baringhaus
R&D, Industriepark Hoechst, 65926 Frankfurt am Main, Germany

Lucas A. Defelipe, Juan Pablo Arcon, Carlos P. Modenutti, Marcelo A. Marti and Adrián G. Turjanski
Departamento de Química Biológica, Facultad de Ciencias Exactas y Naturales, Universidad de Buenos Aires, Buenos Aires 1428, Argentina
IQUIBICEN/UBA-CONICET, Facultad de Ciencias Exactas y Naturales, Universidad de Buenos Aires, Buenos Aires 1428, Argentina

Xavier Barril
Catalan Institution for Research and Advanced Studies (ICREA), Passeig Lluís Companys 23, 08010 Barcelona, Spain
Faculty of Pharmacy and Institute of Biomedicine (IBUB), University of Barcelona, Avgda. Diagonal 643, 08028 Barcelona, Spain

Eva-Maria Krammer, Jerome de Ruyck, Goedele Roos, Julie Bouckaert and Marc F. Lensink
Unite de Glycobiologie Structurale et Fonctionnelle, UMR 8576 of the Centre National de la Recherche Scientifique and the University of Lille, 50 Avenue de Halley, 59658 Villeneuve d'Ascq, France

Wiktoria Jedwabny and Edyta Dyguda-Kazimierowicz
Department of Chemistry, Wrocław University of Science and Technology, 50370 Wrocław, Poland

Alessio Lodola
Department of Food and Drug, University of Parma, 43100 Parma, Italy

Ruyin Cao
Institute of Neuroscience and Medicine (INM-9) and Institute for Advanced Simulation (IAS-5), Forschungszentrum Jülich, Wilhelm-Johnen-Strasse, 52425 Jülich, Germany

Alejandro Giorgetti
Institute of Neuroscience and Medicine (INM-9) and Institute for Advanced Simulation (IAS-5), Forschungszentrum Jülich, Wilhelm-Johnen-Strasse, 52425 Jülich, Germany
Department of Biotechnology, University of Verona, Strada Le Grazie 15, 37134 Verona, Italy

Andreas Bauer
Institute for Neuroscience and Medicine (INM)-2, Forschungszentrum Jülich, 52428 Jülich, Germany

Bernd Neumaier
Institute for Neuroscience and Medicine (INM)-5, Forschungszentrum Jülich, 52428 Jülich, Germany

Giulia Rossetti
Institute of Neuroscience and Medicine (INM-9) and Institute for Advanced Simulation (IAS-5), Forschungszentrum Jülich, Wilhelm-Johnen-Strasse, 52425 Jülich, Germany
Jülich Supercomputing Center (JSC), Forschungszentrum Jülich, 52428 Jülich, Germany
Department of Oncology, Hematology and Stem Cell Transplantation, University Hospital Aachen, 52078 Aachen, Germany

Paolo Carloni
Institute of Neuroscience and Medicine (INM-9) and Institute for Advanced Simulation (IAS-5), Forschungszentrum Jülich, Wilhelm-Johnen-Strasse, 52425 Jülich, Germany
Department of Physics, RWTH Aachen University, 52078 Aachen, Germany
Institute for Neuroscience and Medicine (INM)-11, Forschungszentrum Jülich, 52428 Jülich, Germany
Department of Neurology, University Hospital Aachen, 52078 Aachen, Germany

Jérémie Mortier, Pratik Dhakal and Andrea Volkamer
In-Silico Toxicology Group, Institute of Physiology, Charité—Universitätsmedizin Berlin, Virchowweg 6, 10117 Berlin, Germany

Yankun Chen, Xi Chen, Ganggang Luo, Xu Zhang, Fang Lu, Liansheng Qiao, Gongyu Li and Yanling Zhang
School of Chinese Material Medica, Beijing University of Chinese Medicine, Beijing 100102, China

Wenjing He
College of Traditional Chinese Medicine Xinjiang Medical University, Urumqi 830054, China

Lucas G. Viviani and Antonia T.-do Amaral
Departamento de Química Fundamental, Instituto de Química, Universidade de São Paulo, Av. Prof. Lineu Prestes, 748, São Paulo 05508-000, Brazil

Erika Piccirillo and Leandro de Rezende
Departamento de Química Fundamental, Instituto de Química, Universidade de São Paulo, Av. Prof. Lineu Prestes, 748, São Paulo 05508-000, Brazil

Departamento de Bioquímica, Instituto de Química, Universidade de São Paulo, Av. Prof. Lineu Prestes, 748, São Paulo 05508-000, Brazil

Arquimedes Cheffer, Henning Ulrich and Ana Maria Carmona-Ribeiro
Departamento de Bioquímica, Instituto de Química, Universidade de São Paulo, Av. Prof. Lineu Prestes, 748, São Paulo 05508-000, Brazil

Marian Vincenzi and Zbigniew J. Leśnikowski
Laboratory of Molecular Virology and Biological Chemistry, Institute of Medical Biology of the Polish Academy of Sciences, 106 Lodowa St., 93-232 Lodz, Poland

Katarzyna Bednarska
Laboratory of Experimental Immunology, Institute of Medical Biology, Polish Academy of Sciences, 106 Lodowa St., 93-232 Lodz, Poland

Tatu Pantsar
School of Pharmacy, University of Eastern Finland, 70211 Kuopio, Finland

Antti Poso
School of Pharmacy, University of Eastern Finland, 70211 Kuopio, Finland
Department of Internal Medicine VIII, University Hospital Tübingen, Otfried-Müller-Strasse 14, 72076 Tübingen, Germany

Francesco Tavanti, Alfonso Pedone and Maria Cristina Menziani
Department of Chemical and Geological Sciences, University of Modena and Reggio Emilia, Via G. Campi 103, 41125 Modena, Italy

Mariarosaria Ferraro
Istituto di Chimica del Riconoscimento Molecolare, CNR, Via Mario Bianco 9, 20131 Milano, Italy

Giorgio Colombo
Istituto di Chimica del Riconoscimento Molecolare, CNR, Via Mario Bianco 9, 20131 Milano, Italy
Dipartimento di Chimica, Università di Pavia, V.le Taramelli 10, 27100 Pavia, Italy

Index

www.ingramcontent.com/pod-product-compliance
Lightning Source LLC
Chambersburg PA
CBHW080411190526

45161CB00003B/201